Communications in Computer and Information Science 2650

Series Editors

Gang Li◉, *School of Information Technology, Deakin University, Burwood, VIC, Australia*

Joaquim Filipe◉, *Polytechnic Institute of Setúbal, Setúbal, Portugal*

Zhiwei Xu, *Chinese Academy of Sciences, Beijing, China*

Rationale

The CCIS series is devoted to the publication of proceedings of computer science conferences. Its aim is to efficiently disseminate original research results in informatics in printed and electronic form. While the focus is on publication of peer-reviewed full papers presenting mature work, inclusion of reviewed short papers reporting on work in progress is welcome, too. Besides globally relevant meetings with internationally representative program committees guaranteeing a strict peer-reviewing and paper selection process, conferences run by societies or of high regional or national relevance are also considered for publication.

Topics

The topical scope of CCIS spans the entire spectrum of informatics ranging from foundational topics in the theory of computing to information and communications science and technology and a broad variety of interdisciplinary application fields.

Information for Volume Editors and Authors

Publication in CCIS is free of charge. No royalties are paid, however, we offer registered conference participants temporary free access to the online version of the conference proceedings on SpringerLink (http://link.springer.com) by means of an http referrer from the conference website and/or a number of complimentary printed copies, as specified in the official acceptance email of the event.

CCIS proceedings can be published in time for distribution at conferences or as post-proceedings, and delivered in the form of printed books and/or electronically as USBs and/or e-content licenses for accessing proceedings at SpringerLink. Furthermore, CCIS proceedings are included in the CCIS electronic book series hosted in the SpringerLink digital library at http://link.springer.com/bookseries/7899. Conferences publishing in CCIS are allowed to use our online conference service (Meteor) for managing the whole proceedings lifecycle (from submission and reviewing to preparing for publication) free of charge.

Publication process

The language of publication is exclusively English. Authors publishing in CCIS have to sign the Springer CCIS copyright transfer form, however, they are free to use their material published in CCIS for substantially changed, more elaborate subsequent publications elsewhere. For the preparation of the camera-ready papers/files, authors have to strictly adhere to the Springer CCIS Authors' Instructions and are strongly encouraged to use the CCIS LaTeX style files or templates.

Abstracting/Indexing

CCIS is abstracted/indexed in DBLP, Google Scholar, EI-Compendex, Mathematical Reviews, SCImago, Scopus. CCIS volumes are also submitted for the inclusion in ISI Proceedings.

How to start

To start the evaluation of your proposal for inclusion in the CCIS series, please send an e-mail to ccis@springer.com

Sicong Liu · Xiaolong Zheng · Dong Ma ·
Yuezhong Wu
Editors

Artificial Intelligence of Things and Systems

Third International Conference, AIoTSys 2025
Lanzhou, China, August 15–17, 2025
Proceedings

Editors
Sicong Liu
Northwestern Polytechnical University
Xi'an, China

Dong Ma
Singapore Management University
Singapore, Singapore

Xiaolong Zheng
Beijing University of Posts
and Telecommunications
Beijing, China

Yuezhong Wu
Fuzhou University
Fuzhou, China

ISSN 1865-0929　　　　　　ISSN 1865-0937 (electronic)
Communications in Computer and Information Science
ISBN 978-981-95-2580-5　　ISBN 978-981-95-2581-2 (eBook)
https://doi.org/10.1007/978-981-95-2581-2

© The Editor(s) (if applicable) and The Author(s), under exclusive license
to Springer Nature Singapore Pte Ltd. 2026

This work is subject to copyright. All rights are solely and exclusively licensed by the Publisher, whether the whole or part of the material is concerned, specifically the rights of translation, reprinting, reuse of illustrations, recitation, broadcasting, reproduction on microfilms or in any other physical way, and transmission or information storage and retrieval, electronic adaptation, computer software, or by similar or dissimilar methodology now known or hereafter developed.
The use of general descriptive names, registered names, trademarks, service marks, etc. in this publication does not imply, even in the absence of a specific statement, that such names are exempt from the relevant protective laws and regulations and therefore free for general use.
The publisher, the authors and the editors are safe to assume that the advice and information in this book are believed to be true and accurate at the date of publication. Neither the publisher nor the authors or the editors give a warranty, expressed or implied, with respect to the material contained herein or for any errors or omissions that may have been made. The publisher remains neutral with regard to jurisdictional claims in published maps and institutional affiliations.

This Springer imprint is published by the registered company Springer Nature Singapore Pte Ltd.
The registered company address is: 152 Beach Road, #21-01/04 Gateway East, Singapore 189721, Singapore

If disposing of this product, please recycle the paper.

Preface

Welcome to the 3rd International Conference on Artificial Intelligence of Things and Systems, AIoTSys 2025!

With the rapid advancement and convergence of Internet of Things (IoT), big data, and artificial intelligence (AI) technologies, the integration of AI into IoT (AIoT) has emerged as a promising frontier with substantial developmental potential. This dynamic field is increasingly reshaping and deepening its impact across a wide range of economic and social domains, driving the transformation of traditional industries into a new era of intelligence that goes beyond mere digitalization and connectivity. AIoTSys 2025 was dedicated to exploring the intersection of AI and IoT, and brought together leading researchers, industry experts, and practitioners from around the world to share cutting-edge research, insights, and innovations.

Our technical program committee, comprising 12 Track chairs (covering six tracks of different topics) and 75 members from 7 countries, played a critical role in shaping this year's program. We are truly grateful for their diligent efforts in providing thorough reviews in a timely fashion.

This year, the conference received 82 submissions. After an initial screening by the TPC chairs and Track chair to identify papers that did not comply with formatting requirements, or lacked the necessary maturity or relevance to AIoTSys, 78 papers underwent a thorough review. Each paper was evaluated in a single-round review process by three reviewers. The type of peer review was single-blind. After constructive discussions among the TPC chairs and Track chairs, 39 full papers were accepted, resulting in an overall acceptance rate of 47.6%. Subsequently, two accepted papers were withdrawn after the notifications were sent, bringing the final number of accepted papers to 37 and the acceptance rate to 45.1%. We extend our heartfelt appreciation to all TPC members for their active engagement and insightful contributions.

We also express our gratitude to several members of the Organizing Committee, especially (a) the Local chairs Zhongyu Ma, Yuankun Tang, and Di Zhang, (b) the Publications chairs Yuezhong Wu and Wen Sun, and (c) the Web chairs Xiaochen Li and Han Wang, all of whom helped immensely through the entire process of reviewing and program creation. Without their dedication and support, this program would not have come together. We are deeply grateful for their efforts. Finally, we sincerely hope you find the AIoTSys 2025 technical program both insightful and inspiring.

August 2025

Chenren Xu
Zhanjun Hao
Sicong Liu
Xiaolong Zheng
Dong Ma

Organization

Honorary Chairs

Xingshe Zhou — Northwestern Polytechnical University, China
Jie Liu — Harbin Institute of Technology, China

General Chairs

Chenren Xu — Peking University, China
Zhanjun Hao — Northwest Normal University, China

Program Committee Chairs

Sicong Liu — Northwestern Polytechnical University, China
Xiaolong Zheng — Beijing University of Posts and Telecommunications, China
Dong Ma — Singapore Management University, Singapore

Local Chairs

Zhongyu Ma — Northwest Normal University, China
Yuankun Tang — Northwest Normal University, China
Di Zhang — Northwest Normal University, China

Industry Chair

Longbiao Chen — Xiamen University, China

Publicity Chairs

JeongGil Ko — Yonsei University, South Korea
Yin Chen — Reitaku University, Japan

Publication Chairs

Yuezhong Wu	Fuzhou University, China
Wen Sun	Northwestern Polytechnical University, China

Rising Star Chairs

Takuya Maekawa	Osaka University, Japan
Dongyao Chen	Shanghai Jiao Tong University, China
Zhenyu Yan	Chinese University of Hong Kong, China

Web Chairs

Xiaochen Li	Northwestern Polytechnical University, China
Han Wang	Northwestern Polytechnical University, China

Steering Committee

Zhiwen Yu	Harbin Engineering University/Northwestern Polytechnical University, China
Bin Guo	Northwestern Polytechnical University, China

Track Chairs

Track 1: Grand IoT Applications

Lei Xie	Nanjing University, China
Chuyu Wang	Nanjing University, China

Track 2: IoT Networking

Fusang Zhang	Institute of Software, Chinese Academy of Sciences, China
Chi Lin	Dalian University of Technology, China

Track 3: Edge Computing and IoT

Dongbo Li	Harbin Institute of Technology, China
Pengfei Wang	Dalian University of Technology, China

Track 4: LLMs in AIoT

Zhenzhe Zheng	Shanghai Jiao Tong University, China
Xinyi Li	Tsinghua University, China

Track 5: AIoT Data Analysis

Zhiyong Yu	Fuzhou University, China
Yuezhong Wu	Fuzhou University, China

Track 6: IoT and Industrial Security

Yanru Chen	Sichuan University, China
Zhiwen Pan	Chinese Academy of Sciences, China

Program Committee

Zhenlin An	Hong Kong Polytechnic University, China
Yanling Bu	Nanjing University of Aeronautics and Astronautics, China
Zhaoxin Chang	Institut Polytechnique de Paris, France
Dongyao Chen	Shanghai Jiao Tong University, China
Pengpeng Chen	China University of Mining and Technology, China
Shuyi Chen	Harbin Institute of Technology, China
Yanru Chen	Sichuan University, China
Nan Cheng	Xidian University, China
Cheng Dai	Sichuan University, China
Shengxin Dai	Sichuan University, China
Yongheng Deng	Tsinghua University, China
Xinxin Fan	Chinese Academy of Sciences, China
Chao Feng	Northwest University, China
Jiayao Gao	Tongji University, China
Changzhan Gu	Shanghai Jiao Tong University, China
Xiaonan Guo	George Mason University, USA
Yue Guo	ByteDance, China

Jiawei Hu	University of New South Wales, Australia
Hong Jia	University of Melbourne, Australia
Dongbo Li	Harbin Institute of Technology, China
Xinyi Li	Northwest University, China
Youqi Li	Beijing Institute of Technology, China
Yuanchun Li	Tsinghua University, China
Chi Lin	Dalian University of Technology, China
Qi Lin	Shandong University, China
Jia Liu	Nanjing University, China
Sicong Liu	Northwestern Polytechnical University, China
Bingxian Lu	Dalian University of Technology, China
Li Lu	Zhejiang University, China
Jun Luo	Nanyang Technological University, Singapore
Dong Ma	Singapore Management University, Singapore
Junqi Ma	Chinese Academy of Sciences, China
Jingyi Ning	Nanjing University, China
Srinivas Nomula	Indian Institute of Science, India
Zhiwen Pan	Chinese Academy of Sciences, China
Xiaoyi Pang	Hong Kong University of Science and Technology, China
Cheng Qiao	Guangzhou University, China
Chunming Rong	University of Stavanger, Norway
Minyu Shen	Southwestern University of Finance and Economics, China
Yiran Shen	Shandong University, China
Biyun Sheng	Nanjing University of Posts and Telecommunications, China
Shuyao Shi	Chinese University of Hong Kong, China
Geng Sun	Jilin University, China
Ke Sun	University of Michigan, USA
Weifeng Sun	Dalian University of Technology, China
Yimao Sun	Sichuan University, China
Yu Sun	North China Electric Power University, China
Yuyan Sun	Chinese Academy of Sciences, China
Chuyu Wang	Nanjing University, China
Ge Wang	Xi'an Jiaotong University, China
Haozhe Wang	University of Exeter, UK
Lei Wang	Soochow University, China
Pengfei Wang	Dalian University of Technology, China
Wei Wang	Wuhan University, China
Xiong Wang	University of Science and Technology of China, China

Xuan Wang	Northwest University, China
Yanyan Wang	Nanjing University, China
Zhu Wang	Northwestern Polytechnical University, China
Zi Wang	Augusta University, USA
Yuezhong Wu	Fuzhou University, China
Lei Xie	Nanjing University, China
Jie Xiong	Nanyang Technological University, Singapore
Chenren Xu	Peking University, China
Mengwei Xu	Beijing University of Posts and Telecommunications, China
Xiangyu Xu	Southeast University, China
Biao Xue	Nanjing University of Science and Technology, China
Wanli Xue	University of New South Wales, Australia
Xinyu Xue	Télécom SudParis, France
Tiantian Yan	Dalian University, China
Zhenyu Yan	Chinese University of Hong Kong, China
Fan Yang	Qinghai Normal University, China
Kang Yang	Shenyang Aerospace University, China
Leyou Yang	Nanjing University of Information Science & Technology, China
Liang Yang	Dalian University of Technology, China
Yanbing Yang	Sichuan University, China
Yanni Yang	Shandong University, China
Junchen Ye	Beihang University, China
Yimeng Feng	Macquarie University, Australia
Zhimeng Yin	City University of Hong Kong, China
Zhisheng Yin	Xidian University, China
Zhiyong Yu	Fuzhou University, China
Sheng Yue	Sun Yat-sen University, China
Fusang Zhang	Chinese Academy of Sciences, China
Guanglin Zhang	Donghua University, China
Jing Zhang	Changchun University of Science and Technology, China
Yang Zhang	Macquarie University, Australia
Yin Zhang	University of Electronic Science and Technology of China, China
Yusi Zhang	National University of Defense Technology, China
Tianyue Zheng	Southern University of Science and Technology, China
Xiaolong Zheng	Beijing University of Posts and Telecommunications, China

Zhenzhe Zheng Shanghai Jiao Tong University, China
Ming Zhou Nanjing University of Science and Technology, China
Yongpan Zou Shenzhen University, China

Additional Reviewers

Yupu Yang
Haoxuan Zhang
Yangfan Zhang
Mengda Zhao

Contents

IMU-Accelerated Image Stitching: Fast and Artifact-Free Panorama Construction 1
   Xinyuan He, Yanling Bu, and Saibing Han

Intrusion Detection for Unmanned Aerial Vehicle Systems with Deep Reinforcement Learning 17
   Lixin Liu, Xiaozhen Lu, Yanling Bu, and Qihui Wu

Seeing from Within: Ego-Centric 3D Orientation Prediction for Autonomous Driving 31
   Xinhai Li, Chenxu Meng, Qizhong Chan, Zhicheng Wang, Shihan Zhao, and Lei Yang

Tag in Bloom: Fast RFID Membership via Software-Defined Readers 48
   Chenxu Meng, Yao Yuan, Wenping Liu, Haoran Yan, Donghui Dai, and Lei Yang

FedDEK: Federated Domain-Incremental Learning via Expert Knowledge Construction 65
   Lu Liu, Juan Li, and Tianzi Zang

An Accurate Indoor Depth Estimation Method Based on Iterative Pose Refinement 82
   Yi Le, Xiang Gao, and Hao Sun

A Dynamic Stress Assessment Framework via Multi-scale Feature Fusion 99
   Kang Yu, Wenjing Hu, Meng Tian, Peng Tian, Jun Zhang, and Yunfeng Wang

Regularized Offline Reinforcement Learning for Energy Efficient Urban Rail Transit System Control 114
   Han Chen, Changkai Zhang, Jinfeng Ma, and Bolei Zhang

P2MFDS: A Privacy-Preserving Multimodal Fall Detection System for Elderly People in Bathroom Environments 129
   Haitian Wang, Yiren Wang, Xinyu Wang, Yumeng Miao, Yuliang Zhang, Yu Zhang, and Atif Mansoor

Enhancing Indoor Trajectory Tracking with XGBoost-Based Classification
on mmWave Radar Point Clouds 147
   Yuru Lu, Zhanjun Hao, Yuejiao Wang, Guowei Wang, and Xiangyu Wang

Towards Large-Scale Wireless Sensing in Smart Buildings Using LoRa
Signals 160
   *Xinyu Xue, Zhaoxin Chang, Xujun Ma, Pei Wang, Fusang Zhang,
Badii Jouaber, and Daqing Zhang*

Wi-CLIP: Toward Zero-Shot Air Gesture Recognition Based on RF-Text
Foundation Model 174
   Haoyu Zhang, Yifan Guo, Zhu Wang, Zhuo Sun, Bin Guo, and Zhiwen Yu

Object Size Classification in Garbage Disposal Sensing System Using
Monocular Depth Estimation 190
   Takashi Ito, Wenhao Huang, Yin Chen, and Jin Nakazawa

Memory-Aware Structured Pruning for DL with Joint Optimization
of L1-norm and Peak Memory 209
   Yu Gong and Ling Wang

SAULC: Semantic-Driven Adaptive Method for UAV-LEO Satellite
Communication 224
   *Liangwei Qin, Dongbo Li, Chongrong Li, Yibo Hou, Jiahe Gao,
and Bo Yin*

A Low Migration and Low Energy Consumption Fog Computing
Workflow Scheduling Framework for Multiple Constraints 238
   Tianqi Zhao, Wei Duan, Li He, and Ruihan Hu

A Method for Recognition and Analysis of Industrial Sewing Machine
Operating States Based on Edge Computing 255
   Huojin Xie and Yangbo Wu

Trustworthy Distributed Decision-Making for Multi-view Sensing Data ... 267
   *Zishuo Song, Yuzhu Pan, Mingshu Zhao, Muhammad Ameen,
Zhenwei Wang, and Pengfei Wang*

From Sparse Labels to Accurate Models: Active Semi-Supervised
Learning for mmWave Radar Target Classification 282
   Liyang Zheng, Miao Zhang, Siyuan Fang, Lishen Guan, and Xing Chen

Runtime Heterogeneous Sensor Selection with Data Reuse in Multi-device
Environments 298
   Chun Li, Yu Zhang, Hira Khyzer, Yu Yan, and Xingshe Zhou

Can Time-Series Foundation Models Enhance Wireless Sensing Data
Analytics? An Empirical Study 318
    Shuangping Li, Ruifeng Wang, Ke Xu, and Jiangtao Wang

DGA-Based Power Transformer Fault Diagnosis via Knowledge
Distillation of Large Language Model 333
    *Xinhai Li, Lingcheng Zeng, Qingzhu Zeng, Yunan Lu, Yi Guo,
and Lei Yang*

GraphSAGE-Enhanced Reinforcement Learning for Optimizing
Load-Aware Microservice Deployment 348
    *Keli Liu, Jing Yang, Xiaoli Ruan, Qing Hou, Xianghong Tang,
and Jianhong Cheng*

STCL-Dynamic Sparsity-Driven Transition Feature Replay Continual
Learning for Edge Devices 364
    *Peng Zhang, Jing Yang, Xiaoli Ruan, Qing Hou, Xianghong Tang,
and Jianhong Cheng*

Towards On-Device NPU-Friendly Neural Network Operator Optimization 380
    Wei Ye, Jinrui Zhang, Deyu Zhang, Huan Yang, and Yin Tang

AIGC in Mobile Edge Networks: A Survey 397
    Tingting Long and Deyu Zhang

Large Language Models-Driven Personalized Adaptation Framework
for Intelligent Agents 411
    *Zhelin Xu, Congle Fu, Nan Sun, Honglan Huang, Bing He,
and Xianyang Zhang*

Learning-Based Taxi Selection for Opportunistic Street Parking Sensing 424
    Yongbin Huang, Yuezhong Wu, Zhiyong Yu, and Fangwan Huang

Eye-PPG: Remote PPG Signal Generation for Heart-Rate Estimation
and User Identification in Virtual Reality 442
    Rao Fu, Guangrong Zhao, and Yiran Shen

Temporal Decision-Making Optimization for Intelligent Agents
with Gradually Clarified Objectives 458
    *Chen Gu, Guo Chen, Fangwan Huang, Xuanyun Liu, Zhiyong Yu,
and Yuezhong Wu*

A LoRa Positioning Algorithm Based on the Integration of Kalman
Filtering and Neural Networks and Its Implementation 477
 *Yiwei Li, Zhanjun Hao, Yuejiao Wang, Guowei Wang, Jiang Zhang,
 and Fenfang Li*

MSGR-DCM:Multi-Scale Graph Relational Learning and DC-Mamba
for Multivariate Time Series Anomaly Detection in Industrial Control
Systems 493
 *Xinjie Wang, Kaixiang Liu, Shijie Li, Zhiwen Pan, Shichao Lv,
 and Limin Sun*

A Review on Temporal Knowledge Graph Completion in the Context
of Internet of Things and Industrial Security 505
 Runze Li, Sha Xiang, Shuo Zhu, and Banglie Yang

Federated Learning for AIoT Security: Current Advances and Future
Challenges 520
 Hao Yin and Peng Wang

Lightweight Chaos-Based Image Encryption Algorithm for the Internet
of Things 532
 *Wangcan Liu, Dawei Ding, Yuanyuan Wang, Zongli Yang,
 Chaoma Qian, and Jingwen Zhao*

Image Watermarking Encryption Algorithm for IoT Utilizing Chaotic
Neural Network 545
 *Yuan Zhu, Dawei Ding, Yuanyuan Wang, Zongli Yang, Chaoma Qian,
 and Jingwen Zhao*

BluePLP: Dynamic Vulnerability Patching for Heterogeneous BLE Devices . . . 557
 *Xupu Hu, Zhongfeng Jin, Tongjie Wei, Peng Zhang, Chonghua Wang,
 and Ming Zhou*

Author Index 573

IMU-Accelerated Image Stitching: Fast and Artifact-Free Panorama Construction

Xinyuan He[1], Yanling Bu[1,2(✉)], and Saibing Han[1]

[1] College of Computer Science and Technology,
Nanjing University of Aeronautics and Astronautics, Nanjing, China
{byling,hsbing}@nuaa.edu.cn
[2] State Key Laboratory for Novel Software Technology, Nanjing University,
Nanjing, China

Abstract. Image stitching is widely used in panoramic image generation and mobile visual processing. However, during image capture with handheld devices, variations in camera pose and orientation often lead to inconsistent perspectives among images. This frequently results in distortions and misalignments in the stitched result. Most existing methods rely solely on visual information for image registration and stitching. They are not only computationally expensive, but also prone to geometric distortions when significant changes in camera pose occur. To address this, we propose an *I*MU-*A*ccelerated *I*mage *S*titching method, named *IAIS*. *IAIS* leverages the built-in IMU of the camera to compute rotational and translational information between images, which is then used as a prior for image registration and stitching. This effectively alleviates image distortions caused by variations in camera pose. Based on the rotation and translation information, we estimate the overlapping region between the images to be stitched. Feature point extraction and matching are performed only within this region, thus significantly accelerating the registration process. Furthermore, we select a global transformation that closely aligns with the rotation angle and combine it with grid-based local homographies to enhance stitching quality and continuity. We have implemented a prototype system of *IAIS* and performed sufficient evaluations. The results demonstrate that our method improves stitching speed while maintaining high accuracy, and achieves superior visual quality compared to existing approaches.

Keywords: Multi-Modal Fusion · Image Stitching · IMU Assisted

1 Introduction

Image stitching is a fundamental task in the field of computer vision and has been widely applied in scenarios such as panoramic image generation [2] and mobile visual processing [14]. With the rapid proliferation of mobile devices, users increasingly capture scenes from multiple viewpoints using smartphones or tablets. As a result, panoramic stitching has become a common feature in image editing software and social media applications. However, image stitching still faces many challenges. In particular, under handheld photography conditions, cameras are often in unstable states, which can easily lead to pose variations such as translation and rotation [19]. These changes in pose result in significant viewpoint variations between images, causing stitching artifacts and distortions. For example, when capturing urban architecture or street scenes, large pose differences between frames can lead to visible artifacts and bending at edges. Traditional stitching methods not only suffer from slow processing speed under such conditions, but are also more prone to producing visual distortions, which severely affect the quality of the final stitched result.

Researchers have conducted extensive studies on image stitching, which can be broadly categorized into three main approaches. Image region-based image stitching methods [1,6–8,13,17] rely solely on visual information to stitch images. While effective in many cases, they struggle with multi-angle camera movements, often resulting in irregular image deformation and distortion. IMU-based image stitching methods [5,9,11] combine visual and IMU data to effectively compensate for motion, but they do not yet fully exploit the potential of IMU measurements. Deep learning-based image stitching methods [10,12,15,16,18] leverage deep learning to significantly improve stitching accuracy, but these methods are typically computationally intensive and time consuming. Overall, despite these advances, current research still faces challenges like abnormal image distortion and long processing times.

Fig. 1. *IAIS*: An IMU-accelerated image stitching method for fast and artifact-free panorama construction.

To address this, this paper proposes an *I*MU-*A*ccelerated *I*mage *S*titching method, named *IAIS*. It leverages orientation data captured by the built-in Inertial Measurement Unit (IMU) during image acquisition to guide image registration and fusion. Specifically, *IAIS* first utilizes gyroscope-based rotational data to perform pre-rotation on the images, aligning their viewpoints and estimating the overlapping region. This region is then used as the area for feature point extraction and matching. This design significantly reduces computational overhead. In the fusion stage, we introduce a global-local hybrid transformation method. First, we use the rotation angle estimated from the IMU to identify the global transformation that most closely matches this prior. Then, this global transformation is combined with locally computed homographies derived from grid-based subdivisions to generate a natural and coherent stitched result.

There are two main challenges in the design of this method: Firstly, how to accurately align images using IMU data? IMU measurements are typically affected by drift and noise, which can degrade the accuracy of image alignment if used directly. Moreover, IMU data contains three-dimensional measurements along the X, Y, and Z axes, which cannot be applied directly to image alignment. To mitigate these issues, we maintain a brief stationary period during image capture to extract a stable segment of IMU data. Additionally, for computational efficiency and robustness, we compute only the two-dimensional rotation around the Z-axis and use accelerometer data to estimate the two-dimensional translation component. Secondly, how to fully utilize the estimated rotation angles to accelerate stitching and enhance quality? Without analyzing image content, it is difficult to determine overlapping regions between frames. As shown in Fig. 1, we align the images using rotational corrections and then estimate the overlapping region based on the aligned images. The overlap area is set to cover 60% of a single image's area, ensuring sufficient shared content for registration and fusion. Furthermore, the rotation angle is used to guide the selection of the global transformation model, enhancing the accuracy of the transformation process. This strategy improves both the geometric alignment and the visual consistency of the final stitched result.

In summary, we make three key contributions in this paper. Firstly, we propose an IMU-based pre-rotation compensation method for image stitching, which improves perspective consistency and enhances stitching efficiency. Secondly, we design an IMU-guided global transformation selection strategy that enables robust image registration under significant pose variations. Thirdly, we implement a prototype system and evaluate it on multiple real-world datasets. Experimental results demonstrate that *IAIS* achieves superior performance in both processing speed and visual quality compared to existing state-of-the-art techniques.

2 Related Work

Current image stitching can be categorized into image region-based image stitching, IMU-based image stitching, and deep learning-based image stitching.

2.1 Image Region-Based Image Stitching

Image region-based image stitching estimates local homographies across image regions to better handle depth variations and parallax [1,6–8,13,17]. APAP [17] adopts grid-based local transformations to handle parallax and improves alignment accuracy by varying homographies in different regions. AANAP [8] further incorporates global constraints to enhance the naturalness of the result. SPW [7]

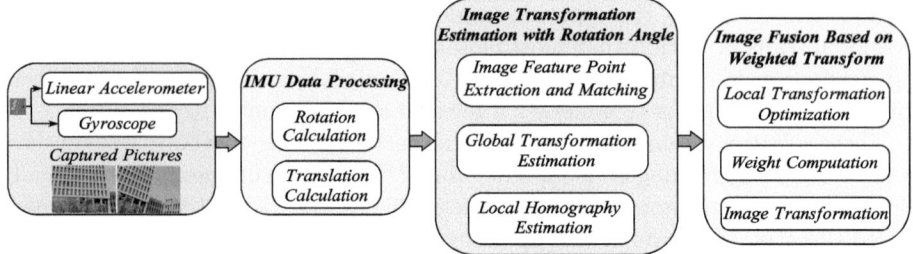

Fig. 2. System architecture

introduces an optimization framework to balance alignment and deformation, although it is highly computationally costly. LPC [6] leverages line constraints to reduce distortion, but pays limited attention to global structural consistency. Despite these advances, edge distortions and structural inconsistencies remain challenges in many scenarios, motivating further improvement.

2.2 IMU-Based Image Stitching Methods

Some researchers have enhanced pose estimation and motion compensation by introducing IMU [5,9,11]. For example, ORB-SLAM2 [11] integrates IMU data to improve real-time performance and pose accuracy in mobile scenarios. However, it does not apply IMU to the field of image stitching. Other studies have used IMU data to assist in image stitching, compensating for the shortcomings of pure visual information [5,9], but IMU has not been fully exploited. In contrast, our approach makes more effective use of IMU measurements to directly assist image alignment and blending, enabling faster and more natural results.

2.3 Deep Learning-Based Image Stitching

In addition to IMU-based methods, deep learning has been introduced to improve image stitching [10,12,15,16,18]. These approaches typically achieve more accurate image alignment, but often involve large-scale model training and feature extraction, resulting in high computational costs, long training times, and significant hardware resource requirements. This can pose challenges in certain real-world applications.

To address image deformation and time efficiency issues, we propose *IAIS*, which leverages IMU to guide pose estimation and enhance stitching robustness.

3 System Overview

As shown in Fig. 2, it illustrates the architecture of *IAIS*, which consists of three modules: IMU data processing, image transformation estimation based on rotation angle, and image fusion based on weighted transform.

IMU Data Processing. Module is to acquire rotation information of the images to be stitched using IMU data. Specifically, images are captured using a mobile camera equipped with a built-in IMU. The corresponding IMU data segments are extracted using the *Physics Toolbox Sensor Suite* mobile application, and the gyroscope data is used to calculate the three-dimensional rotation matrix via integration. From this matrix, the rotation component around the Z-axis is extracted to calculate the rotation angle θ and its associated rotation matrix N. Meanwhile, the translation vector T on the X-Y plane is estimated based on the acceleration data.

Image Transformation Estimation Based on Rotation Angle. Module aims to accelerate the extraction and matching of feature points, and subsequently estimate the global transformation and local transformations after grid partitioning. First, the rotation angle θ is used to estimate the overlapping region between images, and feature point extraction and matching are performed only within this estimated region. The RANSAC algorithm is applied to filter the matched feature points, selecting the subset whose estimated global transformation yields a rotation angle closest to the rotation angle θ. This transformation is then adopted as the final global transformation. Finally, the images are divided into grids and compute the local transformations for each grid.

Image Fusion Based on Weighted Transform. Module integrates both global and local transformations to achieve seamless image stitching. Anchor points and reference points are first defined, and weights are computed based on the reference points. The image is then divided into distinct regions, each undergoing a different transformation strategy. The non-overlapping region of image I_1 does not require any transformation. In the overlapping region between the two images, a weighted combination of the global and local transformations, determined by the weights, is applied. For the non-overlapping region of image I_2, the transformation incorporates weighted combination and further integrates anchor point constraints to suppress boundary distortions.

4 IMU Data Processing

We capture image frames using a moving camera while simultaneously recording the camera's motion via the IMU. In this module, both gyroscope and accelerometer data are extracted and used to compute the rotational and translational information of the images to be stitched.

Traditional image stitching methods perform image registration on the entire image, which is time consuming. Using the IMU, it is possible to roughly estimate the overlapping region between images, and then feature point extraction and matching are performed only within this region. This significantly reduces the computational cost. Moreover, the estimated rotation angle can serve as a prior to further mitigate edge distortions and improve the quality of stitching. Therefore, this section presents how to compute both the rotational and translational information of the images using IMU.

Specifically, we first track the motion of the camera and record the IMU data during image capture. Since IMU data are represented in its own coordinate system, it must be transformed into a global coordinate system. We assume that the IMU coordinate system at the initial moment serves as the global coordinate system. Let the camera pose at time t be denoted as $C(t)$. Then, the angular velocity measured by the IMU over the time interval δ is given by $\boldsymbol{\omega} = \begin{bmatrix} \omega_x & \omega_y & \omega_z \end{bmatrix}^T$, which is used to update the pose as:

$$C(t+\delta t) = C(t)\left(I + \frac{\sin\varphi}{\varphi}R + \frac{1-\cos\varphi}{\varphi^2}R^2\right), \tag{1}$$

where $R = \begin{bmatrix} 0 & -\omega_z\delta t & \omega_y\delta t \\ \omega_z\delta t & 0 & -\omega_x\delta t \\ -\omega_y\delta t & \omega_x\delta t & 0 \end{bmatrix}$, $\varphi = |\boldsymbol{\omega}\delta t|$, I is the identity matrix, and $C(0) = I$.

The camera pose between the images to be stitched is represented by the matrix C. C denotes a three-dimensional transformation in the global coordinate system. However, in the captured scene, the variation in viewpoint is primarily caused by rotation around the Z-axis. For computational convenience, the camera pose is given by $C = \begin{bmatrix} c_{11} & c_{12} & c_{13} \\ c_{21} & c_{22} & c_{23} \\ c_{31} & c_{32} & c_{33} \end{bmatrix}$, then $N_\theta = \begin{bmatrix} c_{11} & c_{12} \\ c_{21} & c_{22} \end{bmatrix}$ represents a rotation around the Z-axis by an angle θ [4], where $c_{11} = \cos\theta$, and $c_{21} = \sin\theta$. The specific rotation angle θ is then computed.

Similarly, as shown in Fig. 4, using the accelerometer data of IMU, and after coordinate transformation and integration, we obtain the three-dimensional displacement matrix. We then extract the components in the X-Y plane, resulting in the two-dimensional translation vector between the images, denoted T (Fig. 3).

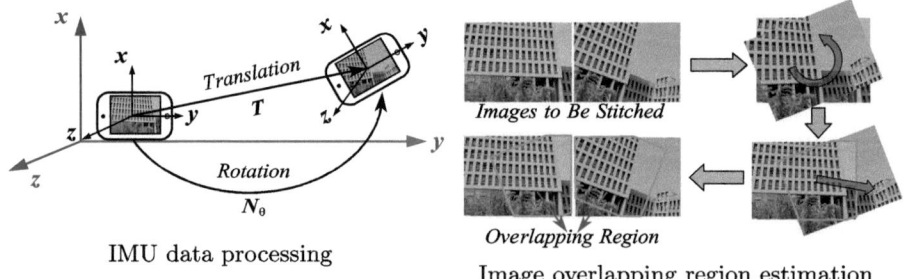

Fig. 3. Image overlapping region estimation based on IMU.

5 Image Transformation Estimation with Rotation Angle

The rotational information of the images to be stitched has been obtained in the previous module. In this module, the rotation information is utilized for image transformation estimation, which includes image feature points extraction and matching based on the rotation angle, global transformation estimation using the rotation angle, and local homography estimation based on grid transformation.

5.1 Image Feature Points Extraction and Matching Based on Rotation

Using the precomputed rotation matrix N_θ, we apply a transformation to image I_2. Let the two images to be stitched be denoted as I_{o_1} and I_{o_2}. The transformation of image I_2 can be expressed as:

$$I_{a_2} = \begin{bmatrix} N_\theta & \mathbf{0} \\ \mathbf{0} & 1 \end{bmatrix} I_{o_2}. \tag{2}$$

As shown in Fig. 5, the two images are roughly aligned. At this point, a simple translation applied to image I_2 can achieve coarse alignment between the images to be stitched. Therefore, in order to estimate the approximate overlapping region between the images to be stitched, image I_2 is first rotated and then translated along the direction indicated by the computed displacement matrix. The translation is performed in such a way that the overlapping area accounts for approximately 60% of each image. This region is then defined as the estimated overlapping region between these two images. Next, the Scale-Invariant Feature Transform (SIFT) algorithm is used to extract and match feature points within the estimated overlapping region. This results in two sets of feature points and their corresponding descriptors for both images, where the descriptors capture the local feature information around each feature point in both image I_1 and image I_2. Feature matching is performed based on the similarity of the descriptors, yielding a set of highly reliable feature point correspondences. This step leverages the precomputed rotation angle to estimate the overlapping region, significantly improving the efficiency of image registration.

5.2 Global Transformation Estimation Based on Rotation

Although high-confidence feature point correspondences have been obtained, these feature points often lie at different depths, and directly using them may affect the accuracy of the global transformation. Therefore, we use the Random Sample Consensus (RANSAC) algorithm [3] to compute the global transformation. Unlike traditional methods, we do not select the global transformation with the smallest rotation angle, but instead use the rotation angle θ computed from the IMU data as a prior, selecting the global transformation whose rotation angle is closest to this angle. This approach not only reduces the randomness in traditional RANSAC selection but also makes the final global transformation more accurate, further reducing edge distortion and improving the stitching accuracy.

Specifically, we first use all feature points to compute the transformation matrix while discarding those points whose differences after transformation exceed ε_a. Next, a subset of the remaining matched points is randomly selected to recompute the matrix. Points with pixel differences less than ε_s are considered inliers, where $\varepsilon_s < \varepsilon_a$. When the number of inliers exceeds n_s, we record the transformation matrix S_i. This process is repeated until the number of obtained transformation matrices in $\{S_i\}$ exceeds n_r. Finally, we compare the rotation angles of the global transformations computed from these matrices and select the one whose rotation angle is closest to the one computed from the IMU data θ as the preliminary global transformation matrix S.

5.3 Local Homography Estimation Based on Grid Transform

We first divide the image frames into multiple grids and compute the local transformation for each grid based on the selected feature point correspondences, in order to ensure the alignment of internal details within the image.

Specifically, we calculate the distance between the center of each grid, denoted as m_i, and each of the selected feature point correspondences, and compute the weights based on these distances. Points closer to the center m have a greater influence. Let the local transformation corresponding to each grid center be denoted as H, where $H \in \mathbb{R}^{3\times 3}$. Then, the weight of the jth feature point p_j relative to the grid center m is computed, denoted as:

$$\omega_{m_i, p_j} = e^{\alpha \|m_i - p_j\|^2}, \tag{3}$$

where $\|\cdot\|^2$ represents the squared Euclidean distance, α is a negative scale factor. For computational convenience, H is flattened column-wise to a vector $\mathbf{h} \in \mathbb{R}^{9\times 1}$. The solution to $\mathbf{h}$ is formulated as an optimization problem:

$$\mathbf{h} = \arg\min_{\mathbf{h}} \|WA\mathbf{h}\|^2, \tag{4}$$

where W is a diagonal weight matrix based on ω_{m, p_i}, A encodes the geometric constraints of the matched feature points. This optimization is solved via Singular Value Decomposition (SVD), and the eigenvector corresponding to the smallest singular value is taken as the solution. The vector $\mathbf{h}$ is then normalized to remove scale ambiguity and reshaped back into 3×3 matrix form to obtain the local matrix H_i.

6 Image Fusion Based on Weighted Transform

After obtaining the global and local transformations for each grid, this section aims to further improve the quality of image stitching. To achieve this, we introduce anchor points and divide the image into different regions, with different transformations applied to each region. This section is divided into three parts: local transformation optimization based on anchor points, weight computation based on reference points, and image transformation based on regions.

6.1 Local Transformation Optimization Based on Anchor Points

In non-overlapping regions, there are no matched feature point pairs. Even with the inclusion of global, relying solely on feature points from overlapping regions to infer transformations can still result in unnatural distortions at the image boundaries. Therefore, we introduce boundary anchor point constraints to refine local transformations in these areas [8].

We evenly place anchor points along the edges of the image and compute a transformation matrix for each anchor point. For each anchor point a_z, based on the local transformation $\mathbf{h}_{m_i}$ at its corresponding grid center m_i, we use a Taylor series expansion to compute both the transformation matrix $\mathbf{h}_{m_i}(a_z)$ and the Jacobian matrix $J_{\mathbf{h}_{m_i}}(a_z)$. The refined local transformation for grid center m is then obtained:

$$\mathbf{h}_{m_i} = \sum_{a_z} w_z \left(\mathbf{h}_{m_i}(a_z) + J_{\mathbf{h}_{m_i}}(a_z)(m_i - a_z) \right), \tag{5}$$

where the weight w_z is defined as: $w_z = \left(1 + \frac{\|m_i - a_z\|^2}{\beta}\right)^{-\frac{\beta+1}{2}}$, representing the influence of anchor point a_z on the grid center m_i, and β is a constant. The final transformation obtained is denoted as H_{l_i}.

6.2 Weight Computation Based on Reference Points

To perform weighted transformations on different regions, we first need to compute four reference points as weight baselines. Following the AANAP algorithm, we first apply the global transformation to stitch these two images together and compute the centers of the two stitched images, denoted $O_1(x_{O_1}, y_{O_1})$, $O_2(x_{O_2}, y_{O_2})$. Based on this, we establish a linear relationship between the centers of the two images, obtaining the slope k and intercept b:

$$k = \frac{y_{O_2} - y_{O_1}}{x_{O_2} - x_{O_1}}, b = y_{O_1} - k \cdot x_{O_1}. \tag{6}$$

Next, we select the leftmost and rightmost horizontal coordinates of image I_1, as well as the center and rightmost horizontal coordinates of image I_2, and use the line equation $y = kx + b$ to compute the corresponding vertical coordinates, thereby obtaining four reference points. For each center point m_i of the grid, we compute the distances to these four reference points, denoted as $\gamma_1, \gamma_2, \gamma_3, \gamma_4$, and then calculate the weight relationships, which are expressed as follows: $a_i = \frac{\gamma_1}{\gamma_1+\gamma_2}$, $b_i = \frac{\gamma_2}{\gamma_1+\gamma_2}$, $c_i = \frac{\gamma_3}{\gamma_3+\gamma_4}$, $d_i = \frac{\gamma_4}{\gamma_3+\gamma_4}$. These weights will later be used for weighted stitching.

6.3 Image Transformation Based on Regions

We divide the images to be stitched into four regions: the non-overlapping region of image I_1, the overlapping region of image I_1, the overlapping region of image

I_2, and the non-overlapping region of image I_2. Different transformation strategies are applied to each region. For the non-overlapping region of image I_1, no transformation is applied, and the image remains unchanged. For the overlapping region of image I_2, a weighted combination of the global transformation S_i and the local transformation H is used:

$$\hat{H}_{2i} = b_i S_i + a_i H. \tag{7}$$

For the overlapping region of image I_1, the inverse of the above transformation is applied:

$$\hat{H}_{1i} = \hat{H}_{2i}^{-1}. \tag{8}$$

In Sect. 6.1, the transformation H_{l_i} involving anchor points is obtained by setting anchor points. We first update the local transformation by weighting H_{l_i} with the global transformation H:

$$H_{2i} = c_i H_{l_i} + d_i H. \tag{9}$$

Furthermore, we incorporate the local transformations, and the final effective transformation applied to the non-overlapping region of image I_2 is:

$$\hat{H}_{2i} = b_i S_i + a_i H_{2i} \tag{10}$$

Finally, we apply the corresponding transformations to these regions, obtaining the final stitched result.

7 Performance Evaluation

7.1 Experiment Settings

Dataset, Hardware and Software. We use a HUAWEI nova 7 SE smartphone to capture image frames and simultaneously record IMU data. To facilitate evaluation, we create a dataset of 30 frame sets, each containing two images, the corresponding IMU data, and a panoramic image as the ground truth. Scenes include buildings, parks, and neighborhood areas. All frames are resized to approximately 1,000,000 pixels to speed up processing. Experiments are conducted on a laptop with an Intel Core i5-11320H CPU (3.20GHz), 16 GB RAM, and an NVIDIA GeForce MX450 GPU.

Metrics. To comprehensively evaluate the performance of the image stitching algorithm, we adopt a combination of subjective visual inspection and objective metric analysis. For subjective evaluation, we visually examine the stitched images with a focus on the transition at the seams, geometric consistency, and overall visual naturalness, to identify any noticeable distortions, ghosting, or misalignments. For objective evaluation, we mainly rely on feature-based geometric consistency metrics. Specifically, we compute the inlier ratio by first extracting matching feature points between the stitched image and the reference panorama using the SIFT, and then identifying inliers through RANSAC. The inlier ratio is calculated as the number of inliers divided by the total number of matched feature points. This ratio provides a strong quantitative indicator of stitching accuracy. A higher inlier ratio typically reflects better structural preservation and fewer distortions or misalignments in the stitched result.

7.2 Experiment Analysis

Overall Accuracy Comparison. To quantitatively evaluate the geometric alignment accuracy of different image stitching algorithms, we select five representative image stitching methods, including APAP [17], SPHP [1], AANAP [8], SPW [7], LPC [6] for comparison with *IAIS*. We analyze a total of 30 pairs of collected images and apply each image stitching algorithm to the same data set to generate stitching results. To evaluate stitching accuracy, we use the SIFT algorithm to extract matching feature points between each stitched image and its corresponding reference panorama, and employ the RANSAC algorithm to filter out inliers from the initial matches. These inliers generally exhibit high matching precision and a strong fit to the homography of the stitching process. Therefore, we compute the inlier ratio, defined as the number of inliers divided by the total number of matched feature points, to quantify the geometric consistency between the stitched image and the real scene. A higher inlier ratio indicates better alignment accuracy with fewer geometric distortions and ghosting artifacts. Across all test cases, we calculate the average inlier ratio for each method, as shown in Table 1. Notably, our proposed *IAIS* method achieves the highest inlier ratio in all test cases, reaching 88.60%, and significantly outperforms the other methods. This result shows that *IAIS* produces stitching outputs that are more consistent with the actual scene, which verifies the effectiveness and robustness of our IMU-assisted alignment strategy when dealing with various scene structures.

Table 1. Average Number of Matched Features with the Ground Truth Panorama

Method	IAIS	SPHP	APAP	AANAP	SPW	LPC
Rate(%)	88.60	70.22	87.91	88.15	88.01	88.39

Furthermore, to visually compare the performance of these methods, we present a set of representative stitching results generated by each algorithm, as illustrated in Fig. 6. The results of AANAP show obvious stitching artifacts, SPHP introduces substantial image distortion, and APAP suffers from severe geometric deformation along the image boundaries. Although SPW and LPC produce relatively minor edge distortions, both methods require more computational time. We compare their efficiency in the next section. In summary, compared to other methods, *IAIS* exhibits noticeably less geometric distortion along the image boundaries and effectively eliminates ghosting artifacts, significantly improving stitching quality through the integration of IMU information.

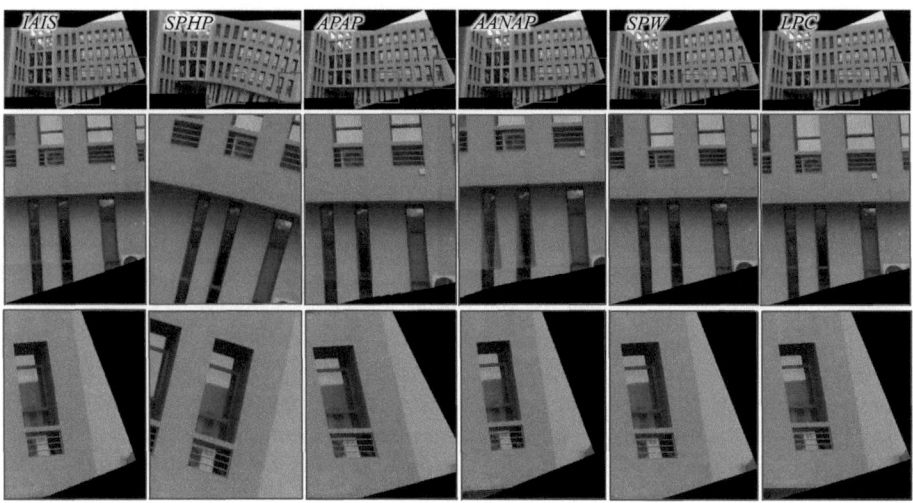

Fig. 4. Overall Comparison of an example

Efficiency Comparison. To enhance the efficiency of stitching, *IAIS* leverages IMU data to estimate the overlapping region between image pairs and confines the processes of feature extraction and matching to this region, thus reducing computational overhead. We compare the improved SIFT with the conventional SIFT-based feature extraction and matching strategy, and the results are presented in Fig. 7a. Benefiting from the estimation and spatial restriction of the overlapping area, *IAIS* achieves a significant speedup in the feature matching stage, with an average runtime of only 1.25 s. In contrast, traditional SIFT processes the entire image, resulting in a slower runtime of 2.21 s.

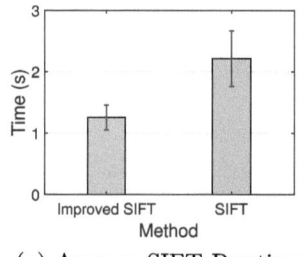
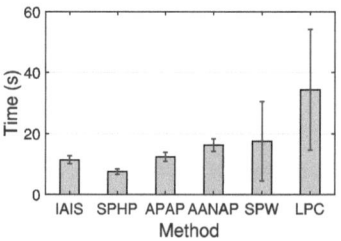

(a) Average SIFT Runtime   (b) Average Overall Runtime

Fig. 5. Efficiency Comparison

To further accelerate stitching, *IAIS* incorporates IMU guidance during the estimation of global transformation based on RANSAC. Specifically, during RANSAC process, *IAIS* prioritizes the transformation that is most consistent

Fig. 6. Stitching results of different coverage rate

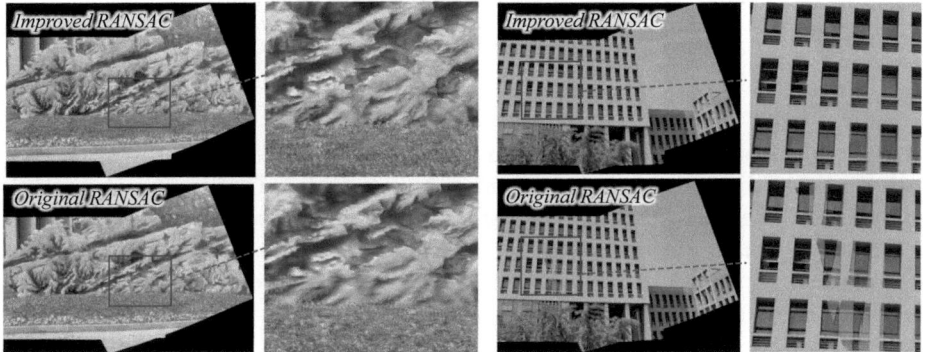

Fig. 7. Impact of Global Transformation Computation

with the camera pose estimated from IMU data. An angular threshold of 3° is defined and when the angular difference between the estimated transformation and the IMU orientation falls below this threshold, the RANSAC process is terminated early. This strategy significantly improves overall runtime efficiency. We perform a comparative analysis of the overall runtime between *IAIS* and other methods, as shown in Fig. 7b. *IAIS* outperforms APAP, AANAP, LPC, and SPW in terms of average runtime. It should be noted that although SPHP is slightly faster than *IAIS*, it does not incorporate grid partitioning or other advanced geometric modeling techniques, resulting in significantly lower stitching quality compared to *IAIS*.

Impact of Different Coverage Rate. *IAIS* estimates the overlapping region between images using IMU data. Specifically, this estimation is based on the rotation and translation directions provided by the IMU, and the overlapping region is defined along this direction according to a specified proportion of the image. We evaluated three overlap ratios: 50%, 60%, and 70%, and conducted a comparative analysis of stitching quality and runtime performance under different levels of overlap. As shown in Fig. 6, stitching performance improves progressively with increasing overlap, but tends to stabilize once the overlap exceeds 60%. Meanwhile, as shown in Table 2, the runtime increases as the estimated overlapping region grows. Therefore, to achieve a balance between stitching accuracy and computational efficiency, we selected an overlap ratio of 60%.

Table 2. Runtime of Different Coverage

Coverage	50%	60%	70%
Time(s)	11.21	11.83	12.61

Table 3. Average Rotation Error Comparison

Method	IAIS-RANSAC	RANSAC
Average Rotation Error(°)	1.82	23.63

Impact of Global Transformation Computation. *IAIS* improves the original RANSAC-based global transformation estimation by incorporating IMU guidance. While the original RANSAC method always selects the transformation with the smallest rotation angle, *IAIS* chooses the one closest to the rotation angle estimated by the IMU. As shown in Fig. 7, our method achieves better performance in reducing artifacts. Furthermore, we calculate the absolute angular differences between the global transformations obtained by both methods and the IMU-estimated rotation angle. As presented in Table 3, the transformations selected by our method are closer to the actual motion of the camera.

8 Conclusion

Image stitching is widely applied in mobile visual processing, but variations in camera pose often cause distortions and misalignments. Most methods rely solely on visual information, leading to high computational cost and susceptibility to geometric distortions under significant camera pose variations. To address this, we propose *IAIS* which leverages the built-in IMU to perform pre-rotation, effectively mitigating distortions. By limiting feature matching to the roughly aligned 60% overlapping region, *IAIS* greatly speeds up registration. Furthermore, *IAIS* selects a global transformation and combine it with grid-based local homographies to improve stitching quality and continuity. The experiment shows that *IAIS* achieves faster and more accurate stitching with superior visual results.

Acknowledgement. This work was supported in part by National Natural Science Foundation of China (Grant No. 62402217); Jiangsu Natural Scicnce Foundation (Grant No. BK20241377); Collaborative Innovation Center of Novel Software Technology and Industrialization.

References

1. Chang, C.H., Sato, Y., Chuang, Y.Y.: Shape-preserving half-projective warps for image stitching. In: 2014 IEEE Conference on Computer Vision and Pattern Recognition, pp. 3254–3261 (2014). https://doi.org/10.1109/CVPR.2014.422

2. Duan, Y., Han, C., Tao, X., Geng, B., Du, Y., Lu, J.: Panoramic image generation: from 2-D sketch to spherical image. IEEE J. Sel. Top. Sig. Process. **14**(1), 194–208 (2020)
3. Fischler, M.A., Bolles, R.C.: Random sample consensus: a paradigm for model fitting with applications to image analysis and automated cartography. Commun. ACM **24**(6), 381–395 (1981). https://doi.org/10.1145/358669.358692
4. Higham, N.J.: Computing real square roots of a real matrix. Linear Algebra Appl. **88-89**, 405–430 (1987). https://doi.org/10.1016/0024-3795(87)90118-2
5. Iz, S.A., Unel, M.: Aerial image stitching using IMU data from a UAV. In: 2023 8th International Conference on Image, Vision and Computing (ICIVC), pp. 513–518 (2023). https://doi.org/10.1109/ICIVC58118.2023.10269879
6. Jia, Q., et al.: Leveraging line-point consistence to preserve structures for wide parallax image stitching. In: 2021 IEEE/CVF Conference on Computer Vision and Pattern Recognition (CVPR), pp. 12181–12190 (2021). https://doi.org/10.1109/CVPR46437.2021.01201
7. Liao, T., Li, N.: Single-perspective warps in natural image stitching. IEEE Trans. Image Process. **29**, 724–735 (2020). https://doi.org/10.1109/TIP.2019.2934344
8. Lin, C.C., Pankanti, S.U., Ramamurthy, K.N., Aravkin, A.Y.: Adaptive as-natural-as-possible image stitching. In: 2015 IEEE Conference on Computer Vision and Pattern Recognition (CVPR), pp. 1155–1163 (2015). https://doi.org/10.1109/CVPR.2015.7298719
9. Liu, T., Chen, B., Wang, B., Liu, D.: An improved image stitching method based on vision-inertial fusion. In: 2024 IEEE International Instrumentation and Measurement Technology Conference (I2MTC), pp. 1–6. IEEE (2024)
10. Ma, J., Liu, W., Chen, T.: DQN-based stitching algorithm for unmanned aerial vehicle images. In: Yang, X., et al. (eds.) Advanced Data Mining and Applications, pp. 125–138. Springer Nature Switzerland, Cham (2023)
11. Mur-Artal, R., Tardós, J.D.: ORB-SLAM2: an open-source slam system for monocular, stereo, and RGB-D cameras. IEEE Trans. Rob. **33**(5), 1255–1262 (2017)
12. Nie, L., Lin, C., Liao, K., Liu, S., Zhao, Y.: Unsupervised deep image stitching: reconstructing stitched features to images. IEEE Trans. Image Process. **30**, 6184–6197 (2021)
13. Shi, Z., Wang, P., Cao, Q., Ding, C., Luo, T.: Misalignment-eliminated warping image stitching method with grid-based motion statistics matching. Multimedia Tools Appl. **81**(8), 10723–10742 (2022). https://doi.org/10.1007/s11042-022-12064-2
14. Thabet, R., Mahmoudi, R., Bedoui, M.H.: Image processing on mobile devices: an overview. In: International Image Processing, Applications and Systems Conference, pp. 1–8. IEEE (2014)
15. Xie, Z., Lai, X., Zhao, W., Jiang, S., Liu, X., Hou, W.: Modification takes courage: seamless image stitching via reference-driven inpainting (2025). https://arxiv.org/abs/2411.10309
16. Xie, Z., Zhao, W., Liu, X., Zhao, J., Jia, N.: Reconstructing the image stitching pipeline: integrating fusion and rectangling into a unified inpainting model (2024). https://arxiv.org/abs/2404.14951
17. Zaragoza, J., Chin, T.J., Brown, M.S., Suter, D.: As-projective-as-possible image stitching with moving DLT. In: 2013 IEEE Conference on Computer Vision and Pattern Recognition, pp. 2339–2346 (2013). https://doi.org/10.1109/CVPR.2013.303

18. Zhang, J., et al.: Content-aware unsupervised deep homography estimation. In: Vedaldi, A., Bischof, H., Brox, T., Frahm, J.-M. (eds.) ECCV 2020. LNCS, vol. 12346, pp. 653–669. Springer, Cham (2020). https://doi.org/10.1007/978-3-030-58452-8_38
19. Zhuang, B., Tran, Q.H.: Image stitching and rectification for hand-held cameras. In: European Conference on Computer Vision, pp. 243–260. Springer (2020)

Intrusion Detection for Unmanned Aerial Vehicle Systems with Deep Reinforcement Learning

Lixin Liu[1], Xiaozhen Lu[1(✉)], Yanling Bu[1], and Qihui Wu[2]

[1] College of Computer Science and Technology, Nanjing University of Aeronautics and Astronautics, Nanjing, China
{liulixin,luxiaozhen,byling}@nuaa.edu.cn
[2] College of Electronic and Information Engineering, Nanjing University of Aeronautics and Astronautics, Nanjing, China
wuqihui@nuaa.edu.cn

Abstract. With the widespread deployment and application of unmanned aerial vehicles (UAVs), a large amount of traffic is generated in highly dynamic UAV systems, along with numerous malicious attacks such as identity impersonation, denial of service, and GPS spoofing, which can cause significant security threats. Existing intrusion detection schemes suffer from slow detection speed or low detection accuracy in highly dynamic UAV systems. In this paper, we propose an intrusion detection scheme for UAV systems with deep reinforcement learning to accurately detect malicious attacks and optimize the detection policy (i.e., the selection of classifiers and their parameters) based on the feature information from UAV traffic. By applying a pre-learning method, this scheme uses a pre-trained model to obtain parameters of the evaluated network to avoid random unnecessary exploration and accelerate learning. This scheme formulates a punishment function using the F1 score, precision, accuracy, recall, latency, miss detection rate, and false alarm rate as metrics to evaluate the immediate risk of each detection policy and thus avoids risky detection policies that result in degradation of intrusion detection performance or even failure. Performance evaluation based on the CICIDS-2017 dataset shows that our proposed scheme outperforms baseline and classical classifiers with higher detection performance.

Keywords: Intrusion detection · UAV systems · malicious attacks · deep reinforcement learning

1 Introduction

Due to the rapid growth and wide application of unmanned aerial vehicles (UAVs) in diverse fields such as disaster rescue, road monitoring, logistics delivery, and even battlefield communications [11], UAV systems generate a significant amount of traffic. This traffic data may potentially carry numerous malicious attacks that pose security threats, such as identity impersonation attacks that steal data, denial of service (DoS) attacks that block communication channels, and GPS spoofing attacks that affect task efficiency [3].

Existing intrusion detection methods for UAV systems mainly rely on classical methods such as Bayesian, decision tree (DT), random forest (RF), and support vector machine (SVM). For example, a Bayesian game-based intrusion detection system (IDS) in [19] adjusts the monitoring, eviction, and malicious behavior policies to reduce the system overhead and detection time. An intelligent deployment mechanism based on SVM and deep learning in [4] uses SVM to deploy UAV-BSs depending on the characteristics of DoS and jamming attacks to reduce the probability of their exposure in dangerous areas. Further, analyzing UAV traffic using statistical analysis techniques can detect abnormal traffic. For instance, a lightweight data integrity scheme in UAV networks in [2] calculates the authentication parameters based on the XOR operation of the received data and compares them with those provided by the UAV to achieve intrusion detection and reduce communication overhead. However, due to the high flexibility and volatility of the topology of UAV networks [17], traditional methods often have significant limitations in the balance between real-time data processing and high precision detection, making it difficult to cope with more complex and variable attack scenarios.

Reinforcement learning (RL) has been applied to develop intelligent intrusion detection schemes to adaptively identify malicious attacks in UAV systems in real time [8,9,14]. For example, a novel adaptive IDS designed in [14] uses Bayesian game theory and deep Q-network algorithms to optimize detection thresholds and achieve the trade-off between detection accuracy and efficiency. An intrusion detection scheme in UAV networks proposed in [9] applies a deep deterministic policy gradient algorithm and a digital twin network to optimize the selection of UAVs, data size, and transmission power to improve the security and privacy of intrusion detection. On the other hand, for UAV information monitoring in farmland, a deep reinforcement learning (DRL) based IDS in [8] uses a double deep Q-network algorithm to optimize UAV deployment positions to ensure that the data collection process is free from malicious attacks or intrusions. This method hardly meets the requirements of fast and high precision detection, and neglects the avoidance of risky policies in the learning process, and thus still faces the problem of low efficiency or even failure of detection in UAV systems with large traffic.

To solve the above problems, this paper proposes a DRL-based intrusion detection scheme named RLMUD for detecting malicious attacks in UAV systems. This scheme obtains the location and identities (IDs) information of the UAVs based on the periodically broadcast Automatic Dependent Surveillance-Broadcast (ADS-B) system [12] and analyzes the transmitted packets to get packet inter-arrival times and packet counts through the transceiver of the BS [5] to formulate traffic information. An evaluated network with an input layer, a hidden layer, and an output layer is designed to evaluate the long-term expected reward based on intrusion detection metrics. This network takes the state constructed from UAV traffic as input to optimize the detection policy, including the selection and parameter optimization of the classifiers.

The designed intrusion detection algorithm uses a pre-learning method to improve detection efficiency [15]. Specifically, the evaluated network is pretrained on a source intrusion detection task, and the trained weight parameters are transferred to the UAV systems' target intrusion detection task. By training only the higher-layer parameters of the evaluated network for the target task, the training overhead of the detection policy can be reduced. To satisfy the performance requirements of intrusion detection in UAV systems and timely avoid risks in exploration, RLMUD designs a punishment function using the F1 score, precision, accuracy, recall, latency, miss detection rate, and false alarm rate as criteria to estimate the immediate risk of selecting each detection policy.

Experiments are performed on the CICIDS-2017 dataset [18], and the results show that our proposed RLMUD outperforms the benchmark and classical classifiers in intrusion detection performance. Besides, we conducted ablation experiments to verify the effectiveness of the pre-learning method and the punishment function. Our main contributions are summarized as follows:

- We propose a DRL-based intrusion detection scheme for UAV systems to detect potential malicious attacks, which dynamically selects the classifiers and optimizes their parameters, such as the number of trees and the regularization parameter, to adapt to the highly dynamic UAV systems.
- We apply a pre-learning method by dividing the learning process into two tasks, where the evaluated network is first trained in a source task and then transferred to the target detection task to improve training efficiency.
- We design a punishment function to evaluate the short-term risks of each detection policy, which is based on the F1 score, precision, accuracy, recall, latency, miss detection rate, and false alarm rate. This function guides UAV systems to avoid risky policies and improve detection performance.

The rest of this paper is organized as follows. We provide the related work in Sect. 2. A DRL-based intrusion detection scheme is designed in Sect. 3. We provide the experimental results in Sect. 4, followed by the conclusion in Sect. 5.

2 Related Work

Recently, fixed-rule and authentication methods have been used in intrusion detection in UAV networks to identify and defend against attackers [1,21,22]. IDS can analyze the routing behavior of UAV networks based on given rules to detect malicious attacks [21]. For example, a lightweight data integrity detection technique in [1] periodically generates authentication parameters based on data packets and verifies the legality of UAVs with the test of generalized likelihood ratio to reduce transmission bandwidth and latency. A time delay attack detection framework for UAV networks in [22] collects features from four dimensions and evaluates node reputation values using single-class classification training to detect malicious nodes by clustering.

Fixed models and rules combined with optimization methods can improve the performance of detecting attackers [6,23]. For instance, a malicious node identification scheme in [23] uses rules and data consistency to detect abnormal behavior and dynamically adjusts path selection and data transmission policies to reduce communication overhead and processing latency. To effectively detect positional spoofing attacks of UAVs, a semi-positive definite relaxation method is designed in [6] to detect coordinated attacks between malicious UAVs using a cooperative detection and identification algorithm.

RL has also been widely applied to the intrusion detection of UAV systems [10,16,20]. For example, an intrusion detection scheme for UAV aerial computing networks in [20] uses a deep deterministic policy gradient algorithm to optimize the detection policy and improve the accuracy and efficiency of malicious attacker detection. An autonomous UAV system that can detect intrusions in [16] uses the state-action-reward-state-action algorithm to evaluate the optimal detection policy based on the UAVs' location to facilitate autonomous UAV detection. To enhance the detection performance of the Internet of Drones, an RL and blockchain-based IDS in [10] utilizes Q-learning to optimize the detection category using radial basis function neural networks as the agent to improve storage capacity. A lightweight UAV intrusion detection and prevention system called DRL_IDS in [7] selects the type of UAV activity and designs reward functions to give higher rewards for minority class attacks, encouraging detection of minority classes and improving the accuracy of suspicious activity detection. However, this scheme fails to take into account the potential hazardous policies in the learning process and ignores the performance requirements of the miss detection rate and false alarm rate in the RL algorithms for UAV systems.

3 Deep Reinforcement Learning-Based Intrusion Detection Scheme for UAV Systems

We propose an intrusion detection scheme for UAV systems named RLMUD based on DRL to optimize the detection policy against malicious attacks. This scheme formulates a Markov decision process (MDP) for each BS, using a prelearning method to avoid unnecessary exploration and improve learning efficiency. Further, this scheme uses the accuracy, precision, F1 score, recall, miss detection rate, false alarm rate, and latency as metrics to design a punishment function that evaluates the risk of each state-action pair to avoid dangerous detection policies in the learning process.

3.1 Intrusion Detection for UAV Systems

A UAV intrusion detection system consists of M UAVs and N BSs. UAV i generates traffic information, such as packet data and location information, based

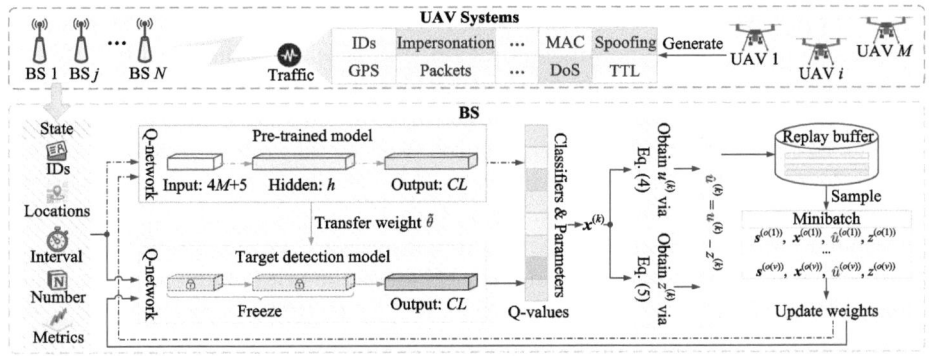

Fig. 1. Illustration of DRL-based intrusion detection scheme for UAV systems.

on service requirements and completes the service through other UAVs or BS j, with $1 \leq i \leq M$, $1 \leq j \leq N$.

The M UAVs may generate traffic that includes both normal behaviors and multiple attacks, causing several types of security problems. First, the attacker could forge a legitimate identity and send malicious or false commands to the BS, causing it to be incorrectly recognized as a trusted device. This could interfere with the normal communication process and lead to the leakage of sensitive information. Second, the attacker may also send a large volume of meaningless packets or make high-frequency requests to the BS, occupying the communication channel and reducing the efficiency of the BS. Besides, the attacker can spoof GPS signals so that BS acquires incorrect location information and may fail to maintain effective communications.

At time slot k, BS j obtains UAVs' IDs and location information from the ADS-B messages broadcast by M UAVs. The BS's transceiver timestamps the arrival time of each packet received during T time slots, computes the packet inter-arrival times according to the time difference between two consecutive packets of the UAV to generate the time differences sequence, and obtains the packet count based on the number of packets received within T time slots. BS can identify potential malicious attacks based on the above UAV traffic information.

Therefore, UAV intrusion detection can be formulated as an optimization problem, which improves the adaptability to highly dynamic UAV systems based on the classical scheme to ensure communication security.

3.2 DRL-Based Intrusion Detection Scheme for UAV Systems

This scheme employs the BS as an agent to formulate an MDP model via

$$\mathcal{M} = \left\{ \mathbf{S}, \mathbf{A}, u^{(k)}, z^{(k)} \right\}, \tag{1}$$

where $\mathbf{S}$ is the set of states, $\mathbf{A}$ is the set of actions of the BS, $u^{(k)}$ is the reward based on the selected detection policy and $z^{(k)}$ is the punishment function used to evaluate the risk of each state-action pair.

Pre-learning Method: To further improve the performance of intrusion detection, avoid unnecessary random exploration, and accelerate the learning process, this scheme uses a pre-learning method within the DRL-based intrusion detection scheme, in which the source and target tasks are intrusion detection for UAV systems. Specifically, the DRL algorithm is used in the source task to pre-train the BS, learning the relationship between UAV traffic and attack behaviors to obtain an optimal Q-value function estimator with generalization capability. The pre-trained model with the trained evaluated network parameters $\tilde{\boldsymbol{\theta}}$ is saved.

In the target intrusion detection task, BS loads the pre-trained model and initializes the neural network parameters, i.e., $\boldsymbol{\theta}^{(0)} \leftarrow \tilde{\boldsymbol{\theta}}$. By freezing the parameters of the lower layer neural network and engaging the higher layers to participate in training, the intrusion detection model optimizes the detection policy to improve performance and reduce learning exploration costs.

State: At time slot k, the BS evaluates the previous F1 score $f^{(k-1)}$, miss detection rate $m^{(k-1)}$, false alarm rate $l^{(k-1)}$, precision $p^{(k-1)}$, recall $r^{(k-1)}$ based on previous traffic of the M UAVs. The detection performance information $\boldsymbol{c}^{(k-1)}$ of the previous time slot can be denoted as

$$\boldsymbol{c}^{(k-1)} = \left[f^{(k-1)}, m^{(k-1)}, l^{(k-1)}, p^{(k-1)}, r^{(k-1)} \right]. \tag{2}$$

The identity $\boldsymbol{G}$ and coordinates $\boldsymbol{d}^{(k)}$ of UAVs are obtained from the messages sent by the ADS-B out device. BS parses the packets to measure the packet inter-arrival time $\boldsymbol{\tau}^{(k)}$ and packet counts $\boldsymbol{n}^{(k)}$ of packets transmitted by each of the M UAVs. The current state $\boldsymbol{s}^{(k)}$ could be formulated as

$$\boldsymbol{s}^{(k)} = \left[\boldsymbol{c}^{(k-1)}, \boldsymbol{G}, \boldsymbol{d}^{(k)}, \boldsymbol{\tau}^{(k)}, \boldsymbol{n}^{(k)} \right]. \tag{3}$$

Action: The state of dimension $4M+5$ is input into the evaluated network with weights $\boldsymbol{\theta}^{(k)}$, which is used to evaluate the long-term expected reward (i.e. Q-values) $\mathbf{Q}(\boldsymbol{s}^{(k)}, \cdot; \boldsymbol{\theta}^{(k)})$ to guide detection policy learning. As shown in Fig. 1, the evaluated network consists of an input layer with size $4M+5$, a hidden layer with h neurons, and an output layer with size CL. Where the dimension of the output layer is equal to the number of detection policies. Specifically, the detection policy includes C classifiers and their L optimization parameters (such as number of trees and regularization parameter), in which $a_1^{(k)} \in \{1, 2, ..., C\}$ denotes the selected classifier and $a_2^{(k)} \in \left[\hat{L}_{a_1^{(k)}}, \check{L}_{a_1^{(k)}} \right]$ represents the optimization parameters of the classifier $a_1^{(k)}$, i.e., $\boldsymbol{x}^{(k)} = [a_1^{(k)}, a_2^{(k)}] \in \mathbf{A}$. BS uses a ϵ-greedy to explore the detection policy, and the exploration randomness decreases over time slots.

Reward: After choosing the detection policy, BS analyzes the behavior based on traffic information of M UAVs to calculate the F1 score $f^{(k)}$, the miss detection rate $m^{(k)}$, the false alarm rate $l^{(k)}$, the precision $p^{(k)}$, the accuracy $q^{(k)}$, and the recall $r^{(k)}$, and measures the latency $t^{(k)}$. The reward $u^{(k)}$ is evaluated via

$$u^{(k)} = f^{(k)} + w_1 p^{(k)} + w_2 q^{(k)} + w_3 r^{(k)} - w_4 l^{(k)} - w_5 m^{(k)} - w_6 t^{(k)} \tag{4}$$

where $w_1 \sim w_6$ are coefficients to balance the precision, accuracy, recall, false alarm rate, miss detection rate, and latency in the reward formulation.

Punishment Function: To avoid dangerous explorations that lead to the degradation of intrusion detection performance or even failure in the learning process, we design a punishment function to measure the risk level of the selected detection policy $x^{(k)}$. This scheme lets $\mathbb{I}(\cdot)$ represent the indicator function and uses the F1 score, miss detection rate, false alarm rate, precision, recall, and latency to evaluate whether it satisfies the performance requirements of intrusion detection for UAV systems. Specifically, the minimum F1 score, precision, accuracy, and recall are limited to $\hat{f}$, $\hat{p}$, $\hat{q}$, $\hat{r}$, respectively, and the maximum tolerable false alarm rate, miss detection rate and latency of the BS are $\hat{l}$, $\hat{m}$, and $\hat{t}$, respectively. The punishment function $z^{(k)}$ could be computed as

$$z^{(k)} = \sum_{i=1}^{4} \mathbb{I}\left(\mu_i^{(k)} \leq \hat{\mu}_i\right) + \sum_{j=1}^{3} \mathbb{I}\left(\rho_i^{(k)} \geq \hat{\rho}_j\right) \quad (5)$$

where $\boldsymbol{\mu}^{(k)} = [f^{(k)}, p^{(k)}, q^{(k)}, r^{(k)}]$ and its corresponding threshold $\hat{\boldsymbol{\mu}} = [\hat{f}, \hat{p}, \hat{q}, \hat{r}]$, $\boldsymbol{\rho}^{(k)} = [l^{(k)}, m^{(k)}, t^{(k)}]$ and its maximum limit $\hat{\boldsymbol{\rho}} = [\hat{l}, \hat{m}, \hat{t}]$.

This scheme uses a punishment function $z^{(k)}$ to modify the reward $u^{(k)}$ to avoid immediate risk exploration in the intrusion detection learning process, i.e., $\hat{u}^{(k)} = u^{(k)} - z^{(k)}$.

Network Update: In the network update, the experience replay technique is applied to the backpropagation process, and Adam is used as a gradient descent algorithm to minimize the loss between the evaluated Q-values and the target Q-values [13]. BS formulates an experience sequence $\boldsymbol{\varphi}^{(k)} = \{\boldsymbol{s}^{(k)}, \boldsymbol{x}^{(k)}, \hat{u}^{(k)}, z^{(k)}\}$ based on the current state $\boldsymbol{s}^{(k)}$, the selected detection policy $\boldsymbol{x}^{(k)}$, the short-term reward $\hat{u}^{(k)}$, and the punishment function value $z^{(k)}$, and store it in the experience pool $\mathcal{D}$, i.e., $\mathcal{D} \leftarrow \mathcal{D} \cup \{\boldsymbol{\varphi}^{(k)}\}$. To adequately extract the learning experience using the experience replay technique and ensure the safety and stability in the learning process, the scheme randomly sampling V detection experiences from the replay pool $\mathcal{D}$ to build a minibatch $\mathcal{B} = \{\boldsymbol{s}^{(o(v))}, \boldsymbol{x}^{(o(v))}, \hat{u}^{(o(v))}, z^{(o(v))}\}_{1 \leq v \leq V}$, where $o(v)$ follows a uniform distribution $U(1, k)$.

$$L(\boldsymbol{\theta}^{(k)}) = \min_{\boldsymbol{\theta}^{(k)}} \frac{1}{V} \sum_{v=1}^{V} \left[\left(\hat{u}^{(o(v))} - \mathbf{Q}\left(\boldsymbol{s}^{(o(v))}, \boldsymbol{x}^{(o(v))}; \boldsymbol{\theta}^{(k)}\right) \right. \right.$$
$$\left. \left. + \gamma \max_{\hat{\boldsymbol{x}} \in \mathbf{A}} \mathbf{Q}\left(\boldsymbol{s}^{(o(v)+1)}, \hat{\boldsymbol{x}}; \hat{\boldsymbol{\theta}}^{(k)}\right) \right)^2 \right]. \quad (6)$$

The update of weights $\boldsymbol{\theta}^{(k)}$ for the evaluated network by minimizing the loss function is given by Eq. (6), where γ is a discount factor and $\hat{\boldsymbol{\theta}}^{(k)}$ is the weights of the target network. Our proposed scheme is summarized in Algorithm 1.

Algorithm 1. DRL-based intrusion detection

1: Input $\tilde{\boldsymbol{\theta}}$ from pre-trained model
2: Initialize target intrusion detection model $\boldsymbol{\theta}^{(0)} \leftarrow \tilde{\boldsymbol{\theta}}$
3: Freeze parameters of the input layer and hidden layer in $\boldsymbol{\theta}^{(0)}$
4: Initialize $f^{(0)}$, $p^{(0)}$, $r^{(0)}$, $l^{(0)}$, $m^{(0)}$, γ, ϵ.
5: **for** $k = 1, 2, \cdots$ **do**
6:     Formulate $c^{(k-1)}$ via (2)
7:     Obtain IDs $\boldsymbol{G}$ and coordinates $\boldsymbol{d}^{(k)}$ of M UAVs from ADS-B messages
8:     Measure $\boldsymbol{\tau}^{(k)}$ and $\boldsymbol{n}^{(k)}$ of packets transmitted by UAVs
      Form the state $\boldsymbol{s}^{(k)}$ via (3)
9:     Input $\boldsymbol{s}^{(k)}$ to the evaluated network with weight $\boldsymbol{\theta}^{(k-1)}$
10:    Get $\mathbf{Q}(\boldsymbol{s}^{(k)}, \cdot; \boldsymbol{\theta}^{(k)})$
11:    Choose $\boldsymbol{x}^{(k)} = \left[a_1^{(k)}, a_2^{(k)}\right] \in \mathbf{A}$ with ϵ-greedy algorithm
12:    Compute accuracy $q^{(k)}$
13:    Evaluate precision $p^{(k)}$ and recall $r^{(k)}$
14:    Calculate F1 score $f^{(k)}$
15:    Measure the miss detection rate $m^{(k)}$ and false alarm rate $l^{(k)}$
16:    Estimate the latency $t^{(k)}$
17:    Calculate $u^{(k)}$ via (4)
18:    Compute $z^{(k)}$ via (5)
19:    $\hat{u}^{(k)} = u^{(k)} - z^{(k)}$
20:    Store $\boldsymbol{\varphi}^{(k)} = \left\{\boldsymbol{s}^{(k)}, \boldsymbol{x}^{(k)}, \hat{u}^{(k)}, z^{(k)}\right\}$ into the replay buffer $\mathcal{D}$
21:    Sample $\mathcal{B}$ from replay buffer
22:    Update $\boldsymbol{\theta}^{(k)}$ via (6)
23: **end for**

4 Experimental Results

4.1 Experimental Settings

Experiments were performed based on Pytorch 2.3.0 and an Intel Core i5-13400 workstation. In the experiments, the CICIDS-2017 dataset is used to simulate the traffic of UAVs, which contains approximately 3×10^6 samples divided into normal types and 14 types of attacks. BS selects a classifier from four options: Bayesian, DT, RF, and SVM, and optimizes its smoothing parameter, maximum tree depth, number of trees, and regularization parameter respectively to classify the traffic data of UAVs accurately. F1 score, recall, precision, accuracy, miss detection rate, false alarm rate, and latency are used as metrics to evaluate the performance of this scheme.

To ensure full exploration of detection policies and to find the optimal policy, the ϵ-greedy decays exponentially from 0.8 to 1×10^{-3} within 400 time slots. In the Algorithm, the input layer of the evaluated network with input size of 517 and an output size of 64, the hidden layer has an input size of 64 and an output size of 128, and the output layer with input size of 128 and an output size of 12. The discount factor is set to 0.8, and the replay buffer stores up to

1×10^4 detection experiences, with 64 experiences randomly selected to form a minibatch for network updating.

Since the imbalanced distribution of the 15 behavior types in CICIDS-2017, where normal behavior accounts for about 80% and attack behavior accounts for 20%, the class distribution of UAV behavior after data cleaning and normalization is shown in Fig. 2. To solve this problem, this scheme uses a class-balanced stratified batch sampling method when sampling from the dataset. Specifically, each batch allocates a similar number of samples to each class to enhance the ability to learn from minority classes, avoid class imbalance, and improve the stability of the detection process.

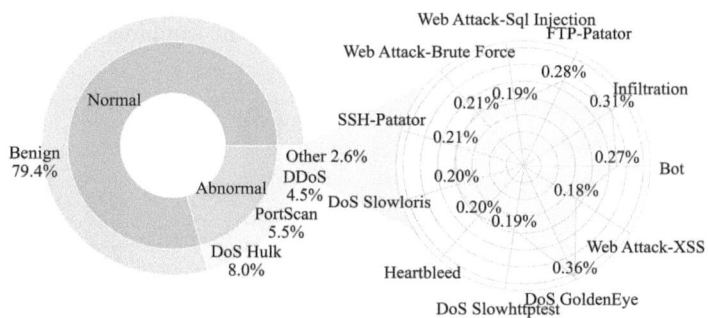

Fig. 2. Class distribution of CICIDS-2017 dataset after data preprocessing.

The experiments evaluate the performance of our scheme by measuring true positives, true negatives, false positives, and false negatives. Based on these metrics, we calculate accuracy, precision, recall, F1 score, miss detection rate, and false alarm rate to assess the accuracy and reliability of the proposed intrusion detection system.

4.2 Intrusion Detection Performance

The accuracy of the UAV intrusion detection scheme with different sampling batches of datasets with varying sizes is shown in Fig. 3. The results show that the accuracy increases steadily with the increase of the sampled data volume. For instance, in Fig. 3, the accuracy has increased by 0.41% from 128 data extracted from 20% dataset to 512 data extracted from the entire dataset, from 99.34% to 99.77%. This is because the scheme is insensitive to the dataset size and batch size, but increasing the sampling size mitigates the class imbalance and makes model training more stable.

As shown in Fig. 4, our scheme outperforms the benchmark in terms of F1 score, recall, precision, and accuracy. For example, RLMUD improves the F1 score by 27.22%, recall by 17.37%, precision by 36.67%, and accuracy by 17.01% compared to DRL_IDS [7]. The performance improvement comes from the optimization of the classifiers and their parameters, as well as the incorporation

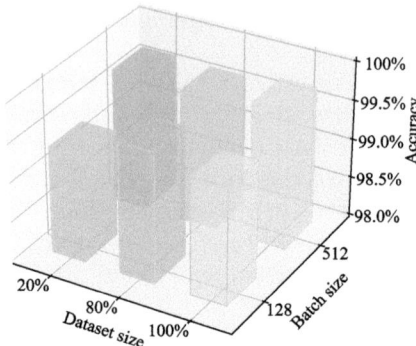

Fig. 3. Accuracy of the proposed scheme with different sampling sizes based on the CICIDS-2017 dataset.

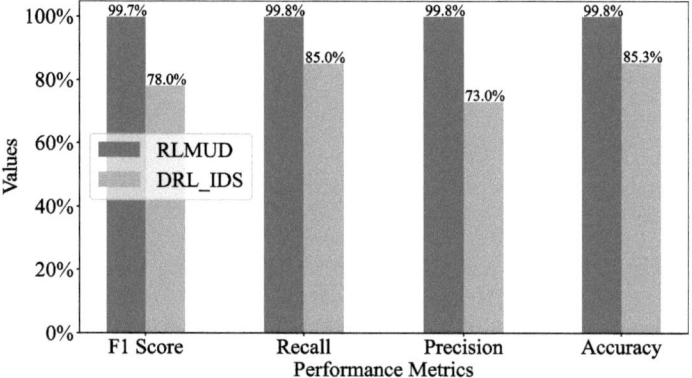

Fig. 4. Performance of the proposed scheme and the benchmark based on the CICIDS-2017 dataset.

of the pre-learning method in the learning process. Moreover, the punishment function designed by RLMUD reduces the exploration of dangerous detection policies and ensures the safety of the scheme.

4.3 Performance Comparison with Classical Classifier

4.4 Ablation

We also conduct a series of ablation experiments to evaluate the performance of the punishment function and pre-learning method in our scheme. As shown in Fig. 6, where RLMUD^{-S} does not use the punishment function to modify the reward, it directly uses the reward calculated by Eq. (4) in the DRL. RLMUD^{-P}

Table 1. Performance Comparison with Four Classifiers Based on CICIDS-2017 Dataset

Metrics	RLMUD	Bayesian	DT	RF	SVM
Accuracy (%)	**99.77**	97.11	98.24	99.35	95.47
Precision (%)	**99.83**	83.78	93.75	94.93	63.29
Recall (%)	**99.76**	97.10	98.23	99.34	90.37
F1-Score (%)	**99.73**	89.83	95.85	97.04	72.74
Miss Detection Rate (%)	**0.23**	2.89	1.76	0.65	4.53
False Alarm Rate (%)	**0.10**	21.68	8.71	6.99	15.26

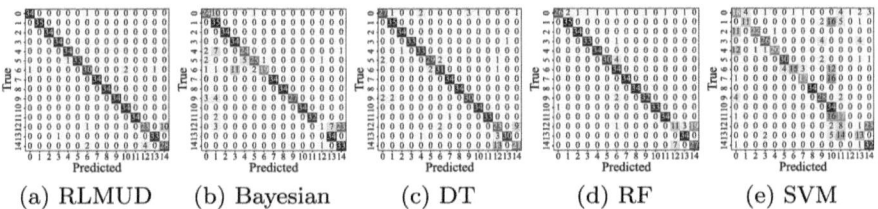

(a) RLMUD (b) Bayesian (c) DT (d) RF (e) SVM

Fig. 5. Confusion matrices of the proposed scheme and four classifiers.

does not adopt the pre-learning method and learns the intrusion detection task completely from scratch. RLMUD^{-PS} uses neither the punishment function nor pre-learning, and selects detection policies only through the ϵ-greedy algorithm. The experimental results show that the punishment function designed in the scheme enables the BS to avoid detection policies with immediate risks. For example, RLMUD^{-P} improves F1 score and precision by 1.89% and 3.44%, and reduces miss detection rate and false alarm rate by 68.75% and 84.75%, respectively, compared to RLMUD^{-PS}. In addition, the pre-learning method could utilize prior knowledge and experience to facilitate learning the intrusion detection task, improving training efficiency and performance metrics. For instance, RLMUD^{-S} improves F1 score and precision by 2.14% and 3.85%, and reduces miss detection rate and false alarm rate by 87.50% and 81.36%, respectively, compared to RLMUD^{-PS} (Table 1 and Fig. 5).

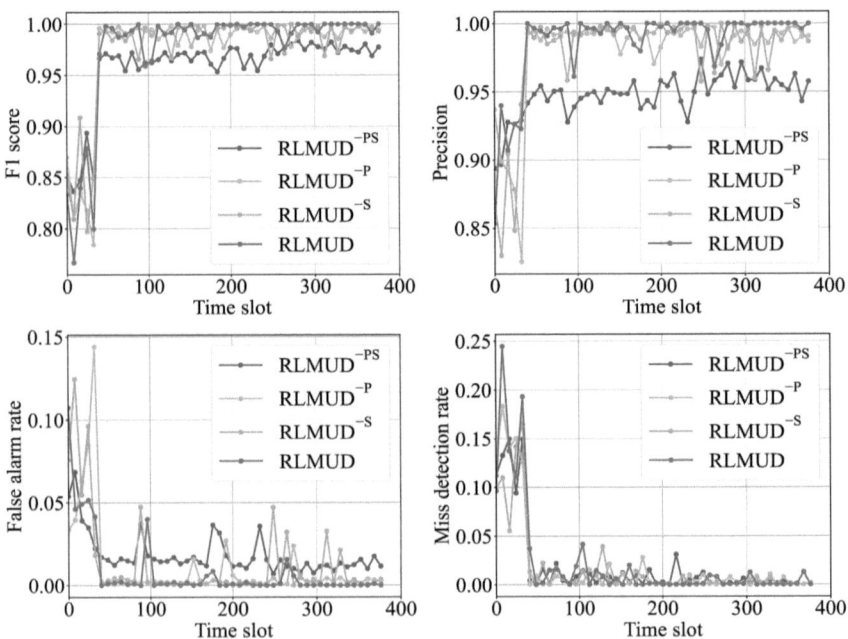

Fig. 6. Performance of ablation study for RLMUD.

5 Conclusion

In this paper, we have proposed a DRL-based intrusion detection scheme for UAV systems to detect malicious attacks in UAV traffic. This scheme uses an evaluated network to evaluate the long-term expected reward to optimize the classifier and its parameters. By training some parameters of the evaluated network in advance through a pre-learning method, this scheme reduces training overhead and achieves higher detection performance. A punishment function has been used to evaluate the short-term risks of each state-action pair based on detection performance and latency, thus guiding the BS to avoid exploring risky detection policies. Experimental results based on the CICIDS-2017 dataset show that the proposed scheme outperforms the benchmark and classical classifier methods. For example, our scheme improves the F1 score, recall, precision, and accuracy by 27.22%, 17.37%, 36.67%, and 17.01% at time slot 400, compared to the benchmark DRL_IDS in [7].

Acknowledgment. This work was supported in part by the Jiangsu Provincial Key Research and Development Program under Grant BE2022068-1, in part by the Natural Science Foundation of China under Grant U22B2062 and Grant 62202222, in part by the Natural Science Foundation of Jiangsu Province under Grant BK20220880, and in part by the Natural Science Foundation on Frontier Leading Technology Basic Research Project of Jiangsu under Grant BK20222001. Xiaozhen Lu is the corresponding author.

References

1. Abhishek, N.V., Aman, M.N., Lim, T.J., Sikdar, B.: MaDe: malicious aerial vehicle detection using generalized likelihood ratio test. In: Proc. IEEE Int. Conf. on Commun. (ICC), Seoul, Korea, pp. 1–6 (2022)
2. Abhishek, N.V., Aman, M.N., Lim, T.J., Sikdar, B.: PIC: preserving data integrity in UAV assisted communication. In: Proc. IEEE Conf. Comput. Commun. Workshops (INFOCOM WKSHPS), New York, USA, pp. 1–6 (2022)
3. Adil, M., Song, H., Mastorakis, S., Abulkasim, H., Farouk, A., Jin, Z.: UAV-assisted IoT applications, cybersecurity threats, AI-enabled solutions, open challenges with future research directions. IEEE Trans. Intell. Veh. **9**(4), 4583–4605 (2023)
4. Aftab, A., Ashraf, N., Qureshi, H.K., Hassan, S.A., Jangsher, S.: BLOCK-ML: blockchain and machine learning for UAV-BSs deployment. In: Proc. IEEE Veh. Technol. Conf. (VTC2020-Fall), Victoria, BC, Canada, pp. 1–5 (2020)
5. Baltaci, A., Klügel, M., Geyer, F., Duhovnikov, S., Bajpai, V., Ott, J., Schupke, D.: Experimental UAV data traffic modeling and network performance analysis. In: Proc. IEEE Conf. on Comput. Commun. (INFOCOM), Vancouver, BC, Canada, pp. 1–10 (2021)
6. Bi, S., Li, K., Hu, S., Ni, W., Wang, C., Wang, X.: Detection and mitigation of position spoofing attacks on cooperative UAV swarm formations. IEEE Trans. Inf. Forensics Security **19**, 1883–1895 (2023)
7. Bouhamed, O., Bouachir, O., Aloqaily, M., Al Ridhawi, I.: Lightweight ids for UAV networks: a periodic deep reinforcement learning-based approach. In: Proc. IFIP/IEEE Int. Symp. Integr. Netw. Manag. (IM), Bordeaux, France, pp. 1032–1037 (2021)
8. Fu, R., Ren, X., Li, Y., Wu, Y., Sun, H., Al-Absi, M.A.: Machine-learning-based UAV-assisted agricultural information security architecture and intrusion detection. IEEE Internet Things J. **10**(21), 18589–18598 (2023)
9. He, X., et al.: Stacked broad learning system empowered FCL assisted by DTN for intrusion detection in UAV networks. In: Proc. IEEE Glob. Commun. Conf. (GLOBECOM), Kuala Lumpur, Malaysia, pp. 5372–5377 (2023)
10. Heidari, A., Navimipour, N.J., Unal, M.: A secure intrusion detection platform using blockchain and radial basis function neural networks for Internet of Drones. IEEE Internet Things J. **10**(10), 8445–8454 (2023)
11. Javaid, S., et al.: Communication and control in collaborative UAVs: recent advances and future trends. IEEE Trans. Intell. Transp. Syst. **24**(6), 5719–5739 (2023)
12. Khan, H.A., Khan, H., Ghafoor, S., Khan, M.A.: A survey on security of automatic dependent surveillance-broadcast (ADS-B) protocol: challenges, potential solutions and future directions. IEEE Commun. Surveys Tuts., 1–1 (2024)
13. Kingma, D.P., Ba, J.: Adam: a method for stochastic optimization. In: Proc. Int. Conf. Learn. Repr. (ICLR), San Diego, CA, USA, pp. 1–15 (2015)
14. Liang, J., Ma, M., Tan, X.: GaDQN-IDS: a novel self-adaptive IDS for VANETs based on Bayesian game theory and deep reinforcement learning. IEEE Trans. Intell. Transp. Syst. **23**(8), 12724–12737 (2021)
15. Lu, X., Xiao, L., Dai, C., Dai, H.: UAV-aided cellular communications with deep reinforcement learning against jamming. IEEE Wireless Commun. **27**(4), 48–53 (2020)
16. Masadeh, A., Alhafnawi, M., Salameh, H.A.B., Musa, A., Jararweh, Y.: Reinforcement learning-based security/safety UAV system for intrusion detection under

dynamic and uncertain target movement. IEEE Trans. Eng. Manag. **71**, 12498–12508 (2022)
17. Mu, J., Zhang, R., Cui, Y., Gao, N., Jing, X.: UAV meets integrated sensing and communication: challenges and future directions. IEEE Commun. Mag. **61**(5), 62–67 (2023)
18. Panigrahi, R., Borah, S.: A detailed analysis of CICIDS2017 dataset for designing intrusion detection systems. Int. J. Eng. Technol. **7**(3), 479–482 (2018)
19. Sedjelmaci, H., Senouci, S.M., Ansari, N.: Intrusion detection and ejection framework against lethal attacks in UAV-aided networks: a Bayesian game-theoretic methodology. IEEE Trans. Intell. Transp. Syst. **18**(5), 1143–1153 (2017)
20. Tao, J., Han, T., Li, R.: Deep-reinforcement-learning-based intrusion detection in aerial computing networks. IEEE Netw. **35**(4), 66–72 (2021)
21. Wang, L., Chen, Y., Wang, P., Yan, Z.: Security threats and countermeasures of unmanned aerial vehicle communications. IEEE Commun. Stand. Mag. **5**(4), 41–47 (2022)
22. Zhai, W., Liu, L., Ding, Y., Sun, S., Gu, Y.: ETD: an efficient time delay attack detection framework for UAV networks. IEEE Trans. Inf. Forensics Secur. **18**, 2913–2928 (2023)
23. Zilberman, A., Stulman, A., Dvir, A.: Identifying a malicious node in a UAV network. IEEE Trans. Netw. Serv. Manag. **21**(1), 1226–1240 (2023)

Seeing from Within: Ego-Centric 3D Orientation Prediction for Autonomous Driving

Xinhai Li[1], Chenxu Meng[1], Qizhong Chan[1], Zhicheng Wang[2], Shihan Zhao[2], and Lei Yang[2(✉)]

[1] Zhongshan Power Supply Bureau, China Southern Power Grid Co., Ltd., Zhongshan, China
[2] Department of Computing and The Shenzhen Research Institute, The Hong Kong Polytechnic University, Hong Kong, China
young@tagsys.org

Abstract. Despite recent advances in autonomous driving, most self-driving systems primarily emphasize short-term trajectory prediction for motion planning, often neglecting the critical role of future 3D vehicle orientation (posture) in achieving safe and stable navigation—especially in dynamic and cluttered environments. Accurate posture forecasting is vital for anticipating vehicle behavior, enabling smoother control, and avoiding unsafe maneuvers. In this work, we present PosFormer, a novel transformer-based architecture that predicts future vehicle orientation by jointly leveraging visual perception and motion intent. PosFormer extracts rich spatial features from six in-vehicle camera views using a convolutional backbone, and aligns them with planned trajectory information through a cross-attention mechanism within a Transformer decoder. This fusion enables precise, ego-centric 3D orientation forecasting in quaternion form at future horizons (e.g., 3 and 6 steps ahead). Unlike traditional approaches that decouple perception from planning or rely on heuristic filters, PosFormer offers a unified, data-driven framework that captures context-aware motion dynamics. Experiments on real-world autonomous driving datasets demonstrate that PosFormer achieves average quaternion angular errors of 0.87°, 0.79°, and 2.26° across future frames.

Keywords: 3D Orientation Prediction · Full self-driving model · Transformer · Multi-View Perception

1 Introduction

Autonomous driving has emerged as a key application in robotics, aiming to enable vehicles to perceive, reason, and act safely in dynamic environments without human intervention. A core capability of self-driving systems lies in their ability to perform trajectory planning and collision avoidance. Trajectory planning generates smooth, feasible paths that guide the vehicle toward its destination, while collision avoidance ensures that these paths remain safe in the presence of dynamic agents and environmental uncertainties. These functions are tightly coupled and form the decision-making backbone of autonomous navigation, enabling robust operation in real-world, unstructured driving scenarios.

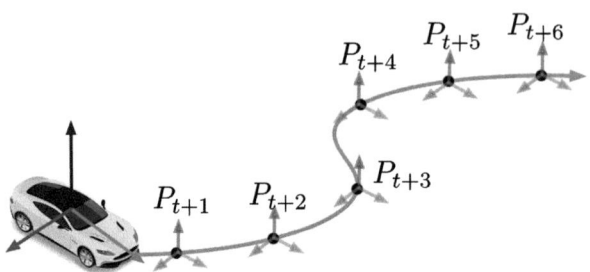

Fig. 1. Illustration of Ego-centric 3D Orientation Prediction. While existing self-driving models primarily focus on position prediction, they often overlook the critical role of orientation in trajectory planning. Our work bridges this gap by fusing multi-view vision data with historical posture measurements to improve orientation awareness.

Trajectory planning in autonomous driving relies heavily on accurate short-term predictions of the ego vehicle's future positions to generate feasible and safe motion paths. Modern self-driving models are typically designed to forecast these positions over a short time horizon, serving downstream planning and control modules. However, they often overlook the prediction of vehicle posture. This omission can lead to suboptimal planning, particularly in scenarios that require precise maneuvering, such as lane changes, sharp turns, or navigation in tight spaces. Predicting the future posture of the vehicle is essential not only for ensuring the physical feasibility of planned trajectories but also for enabling smoother motion control, better alignment with dynamic context, and safer interaction with surrounding agents. Addressing this gap is critical for enhancing the robustness and behavioral fidelity of autonomous driving systems.

To bridge this gap, we propose PosFormer, a novel end-to-end Transformer-based framework for short-term 3D orientation forecasting in autonomous vehicles. PosFormer takes as input multi-view images from in-vehicle cameras, real-time IMU readings, and the planned trajectory to predict the vehicle's future posture along its intended path. By integrating visual perception, inertial sensing, and motion intent, PosFormer delivers accurate and context-aware posture predictions that improve the fidelity of downstream motion planning.

Yet, engineering PosFormer in practice is non-trivial due to two key challenges:

- *What sensory data should be utilized for posture prediction?* Traditional posture estimation methods often rely solely on historical IMU measure-

ments, processed through heuristic filters such as the Kalman filter. While this approach is effective in smooth, predictable environments—owing to the inherent temporal continuity of vehicle motion—it often fails in complex or dynamic driving scenarios, such as sharp turns, uneven road surfaces, or interactions with unpredictable agents. Modern autonomous vehicles are typically equipped with a rich suite of sensors, including cameras, LiDAR, and mmWave radar, which offer complementary environmental context. These sensing modalities, particularly vision, can provide cues about upcoming maneuvers or road geometry that may influence future posture. Motivated by this, PosFormer integrates three key modalities: historical IMU-based postures, visual features from multi-view cameras, and planned future trajectories.
- *How can heterogeneous modality data be effectively fused?* The three input modalities—IMU-derived postures, camera-based visual embeddings, and planned trajectory coordinates—differ significantly in structure, semantics, and temporal characteristics. A central challenge lies in how to align and fuse these heterogeneous inputs in a unified framework. To address this, we propose a Transformer decoder architecture with modality-aware cross-attention mechanisms. Our design projects all modalities into a shared latent space and uses planned trajectory tokens as predictive queries. By attending to temporally aligned posture histories and contextual visual features, the decoder dynamically captures cross-modal dependencies and learns to infer physically consistent orientation forecasts.

Contributions. This work makes three key contributions to the field of autonomous driving and motion prediction.

- First, we propose PosFormer, a novel framework that leverages intermediate representations from existing self-driving models to predict vehicle posture. By reusing features already extracted for planning, our design avoids redundant sensor data processing, thereby addressing both computational efficiency and potential security vulnerabilities and enhancing system stability.
- Second, we introduce a transformer-based architecture equipped with cross-modal attention mechanisms to fuse heterogeneous data sources, including trajectories, ego-motion histories, and visual road geometry. Our temporal decoder effectively models the interaction between vehicle kinematics and terrain-induced posture changes, enabling accurate, context-aware orientation prediction.
- Finally, we validate the proposed framework through extensive experiments on the public nuScenes dataset. PosFormer demonstrates strong predictive performance.

Orgnization. In the following sections, we first present the background in Sect. 2, followed by the system design in Sect. 3. Section 4 details the implementation and evaluation. Related work is discussed in Sect. 5.

2 Background

2.1 Motivation

Accurate prediction of a vehicle's dynamic posture is crucial for maintaining stability and control, especially during aggressive maneuvers (e.g., sharp turns or evasive actions) and when traversing uneven terrain. For example, roll angle directly affects tire traction and braking efficiency, while errors in yaw prediction can cause significant drift during trajectory tracking. Particularly, modern intelligent chassis systems [1], commonly integrated into vehicles equipped with advanced driver assistance systems (ADAS), depend on real-time sensor inputs (e.g., IMU) [2] to actively regulate vehicle behavior through digital control signals. These systems enables precise coordination among the chassis, cockpit, and powertrain subsystems. Predicting 3D orientation in advance empowers the intelligent chassis to anticipate posture changes, allowing for proactive optimization of suspension settings and power output. This not only mitigates control latency but also improves ride smoothness, handling performance, and overall safety.

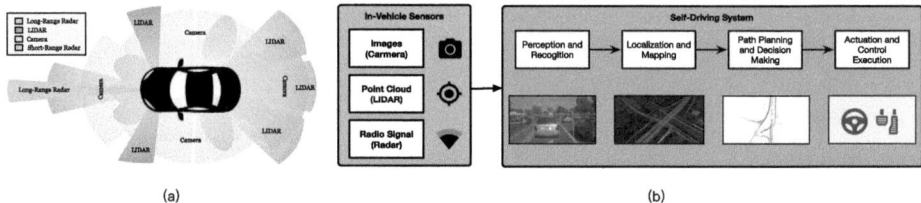

Fig. 2. The architecture of a typical self-driving model, which is a neural network that processes multi-sensor data to generate planned trajectories for autonomous navigation.

Beyond motion control, orientation forecasting also plays a vital role in maintaining high-quality wireless communication. In Vehicle-to-Infrastructure (V2I) systems using millimeter-wave (mmWave) [3] technology, and Vehicle-to-Satellite (V2S) systems with direct satellite links, timely orientation prediction facilitates optimal beam alignment, ensuring robust and high-throughput connectivity under dynamic vehicular conditions.

Thus, 3D posture prediction is essential not only for ensuring physical stability but also for enabling reliable communication and coordinated autonomy [4] in next-generation intelligent vehicles. These practical demands motivate us to explore a new paradigm for posture prediction in self-driving vehicles.

2.2 Posture Representation

The posture of a vehicle refers to its full orientation in 3D space, describing how the vehicle is rotated relative to a fixed reference frame (e.g., Earth or a map-based coordinate system). Accurate posture representation is critical for motion planning, control, and localization in autonomous systems. Conventionally, posture is characterized by three angular components: roll (ϕ, rotation about the x-axis), pitch (θ, about the y-axis), and yaw (ψ, about the z-axis). These can be expressed using several mathematical representations, including Euler angles, rotation matrices, and quaternions [5].

To efficiently and robustly represent 3D orientation, we adopt quaternions. Quaternions offer several advantages over Euler angles, such as avoiding gimbal lock and enabling smooth interpolation, making them particularly well-suited for real-time applications. A quaternion q is a four-dimensional extension of complex numbers, defined as:

$$q = w + x\mathbf{i} + y\mathbf{j} + z\mathbf{k} \tag{1}$$

where w is the scalar part, and (x, y, z) form the vector part. The imaginary units $\mathbf{i}, \mathbf{j}, \mathbf{k}$ follow the Hamilton product rules:

$$\mathbf{i}^2 = \mathbf{j}^2 = \mathbf{k}^2 = \mathbf{ijk} = -1 \tag{2}$$

A unit quaternion $\hat{q}$ (satisfying $\|\hat{q}\| = 1$) is commonly used to represent a rotation. It can be parameterized by a rotation angle θ_q about a unit axis vector $\mathbf{u} = (x, y, z)$ as:

$$\hat{q} = \cos\left(\frac{\theta_q}{2}\right) + \sin\left(\frac{\theta_q}{2}\right)(x\mathbf{i} + y\mathbf{j} + z\mathbf{k}) \tag{3}$$

or equivalently in tuple form:

$$\begin{aligned}\hat{q} &= (w, x, y, z) \\ &= \left(\cos\left(\frac{\theta_q}{2}\right), x\sin\left(\frac{\theta_q}{2}\right), y\sin\left(\frac{\theta_q}{2}\right), z\sin\left(\frac{\theta_q}{2}\right)\right)\end{aligned} \tag{4}$$

Given a unit quaternion $\hat{q} = (w, x, y, z)$, the equivalent Euler angles—roll ϕ, pitch θ, and yaw ψ—can be computed as:

$$\begin{cases} \phi &= \arctan 2\left(2(wx + yz), 1 - 2(x^2 + y^2)\right) \\ \theta &= \arcsin\left(2(wy - zx)\right) \\ \psi &= \arctan 2\left(2(wz + xy), 1 - 2(y^2 + z^2)\right) \end{cases} \tag{5}$$

The above equation allows for seamless conversion between quaternion-based internal representations and Euler-angle-based interfaces used by many planning and control modules.

In practice, vehicle posture is measured by an in-vehicle IMU, which combines data from gyroscopes, accelerometers, and, in some cases, magnetometers. Gyroscopes capture angular velocity around the vehicle's principal axes (roll, pitch, and yaw), while accelerometers measure linear acceleration, including the gravitational component, which provides information about tilt. Magnetometers, when present, sense the Earth's magnetic field and assist in determining the heading direction. By integrating these sensor readings over time, the IMU produces a continuous estimate of the vehicle's orientation, typically represented as a quaternion for its numerical stability and robustness in 3D rotation modeling.

2.3 Self-Driving Model

Modern autonomous vehicles are equipped with a comprehensive suite of sensors to enable robust perception, localization, and decision-making, as illustrated in Fig. 2(a). This sensor array typically includes six surround-view cameras to provide 360° environmental coverage, three to six LiDAR units for high-resolution 3D mapping, and mmWave radars for both short- and long-range object detection. In addition, GPS and IMU modules are integrated to deliver accurate global positioning and real-time inertial measurements, supporting precise vehicle localization and motion estimation.

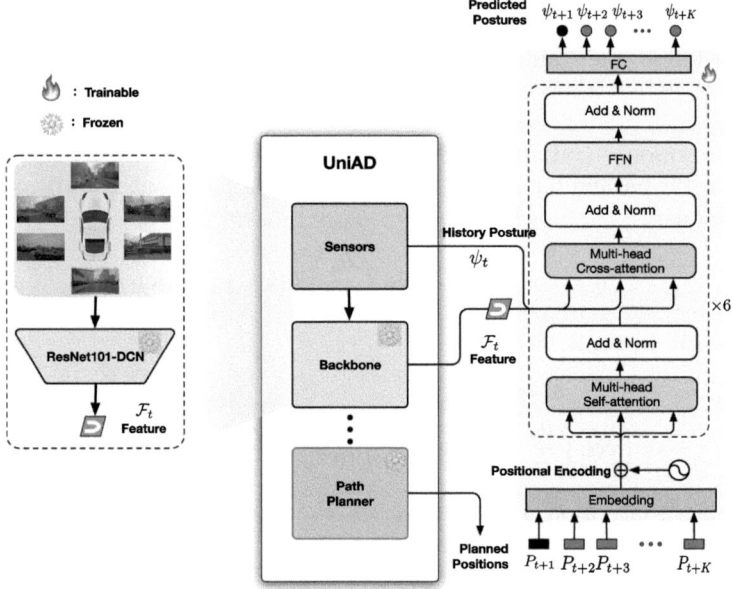

Fig. 3. The Architecture of PosFormer.

The self-driving model is a neural network that fuses multi-sensor inputs to produce a planned trajectory for autonomous navigation. Taking UniAD [6] as a representative example, an sefl-driving model typically consists of the following core modules (illustrated in Fig. 2(b)):

- **Perception and Fusion:** This module processes raw sensor data—primarily camera inputs in vision-centric systems—through a convolutional backbone (e.g., ResNet) to extract high-level perspective features. When additional modalities such as LiDAR are available, a sensor fusion component integrates these streams to enhance spatial awareness.
- **Localization and Mapping:** Perspective features are transformed into a unified bird's-eye view (BEV) representation, which serves as a spatially consistent map of the environment. This module simultaneously performs object

detection, tracking, and semantic segmentation of roadway elements, facilitating situational understanding.
- **Path Planning:** Leveraging the BEV representation and detected dynamic agents, this module generates feasible and safe trajectories. It accounts for vehicle dynamics, environmental constraints, and predicted interactions with surrounding traffic participants, optimizing for safety, comfort, and efficiency.
- **Actuation and Control Execution:** The final stage translates planned trajectories into actionable control commands for steering, acceleration, and braking. It continuously adapts to real-time feedback from the vehicle's state estimation system to ensure stability and accurate execution.

Our module integrates seamlessly with the existing self-driving models by utilizing intermediate features already produced during the standard processing pipeline.

3 PosFormer Design

3.1 Overview

PosFormer is developed on top of the open-source Unified Autonomous Driving (UniAD) framework [6], an end-to-end self-driving system that processes surround-view camera inputs to produce planned trajectories. UniAD adopts a vision-centric design [7], using only cameras as the sole sensor modality for perception and planning. In line with this principle, our system is also designed as a camera-only solution.

The self-driving pipeline operates in a frame-by-frame manner: at each timestep, a set of six temporally synchronized images—captured from the surrounding cameras—is processed by the self-driving model to generate a new trajectory plan and execute the corresponding control actions. This discrete update scheme ensures that trajectory planning and actuation remain tightly coupled with the camera input rate.

Figure 3 illustrates the overall architecture of PosFormer. It adopts a Transformer decoder-based design that operates on three complementary input modalities: historical vehicle postures, current visual features extracted from multi-view camera images, and the planned future positions along the intended trajectory. These modalities are projected into a shared embedding space and processed jointly through cross-attention mechanisms, allowing the model to learn spatial-temporal correlations and motion intent. The output of the Transformer is then used to predict the vehicle's 3D posture—represented as unit quaternions—for the next K future frames. In the following, we delve into the details.

3.2 Input: Three Modalities

At each time step t, the PosFormer model processes three types of input modalities:

- **Historical Postures:** $\Psi_t \in \mathbb{R}^{K \times d_o}$ denotes a sequence of K past vehicle orientations (e.g., roll, pitch, yaw encoded as unit quaternions). Each posture at time $t - i$ is represented as a d_o-dimensional vector $\psi_{t-i} \in \mathbb{R}^{d_o}$ for $i = 0, \ldots, K - 1$. The full sequence is written as $\Psi_t = [\psi_{t-K+1}, \psi_{t-K+2}, \ldots, \psi_{t-1}, \psi_t]^\top$. These historical postures capture the vehicle's recent motion and rotation dynamics. Since vehicle movement in physical space is inherently continuous and exhibits temporal smoothness, historical posture information provides strong inductive priors for predicting future orientation. Leveraging this temporal coherence enables the model to anticipate future motion more accurately, especially in complex driving scenarios involving curves, merges, or gradual heading changes.
- **Visual Context:** $\mathcal{F}_t \in \mathbb{R}^{d_v}$ is a single vector encoding the environmental context at time t. UniAD utilizes a CNN-based backbone to extract features from the six surrounding-view camera images, enabling downstream perception tasks such as dynamic object detection and tracking (e.g., vehicles, pedestrians), panoptic segmentation for mapping, and motion forecasting. Rather than reprocessing raw images, PosFormer directly reuses the intermediate fused feature vector $\mathcal{F}_t$ produced by UniAD's visual perception module. This approach reduces computational overhead and maintains alignment with the planning context. These vision features offer rich spatial cues from the vehicle's surroundings, which are essential for predicting motion-consistent orientation.
- **Planned Trajectory:** $\mathcal{P}_t \in \mathbb{R}^{K \times d_p}$ contains K future vehicle positions or waypoints, predicted by the self-driving system's path planner. Each position $P_{t+i} \in \mathbb{R}^{d_p}$ corresponds to the target location at time $t + i$, and the full sequence is written as $\mathcal{P}_t = [P_{t+1}, P_{t+2}, \ldots, P_{t+K}]^\top$. These waypoints are generated based on environmental context, perceived dynamic agents, and map constraints, and serve as targets for the vehicle's control system. The execution module uses these planned positions to issue low-level control commands (e.g., steering, acceleration, braking) that guide the vehicle along the intended path. Incorporating the planned trajectory into our orientation prediction task ensures that posture forecasts are not only temporally coherent but also aligned with the vehicle's navigational intent.

To enable cross-modal fusion, each modality is first projected into a shared d-dimensional latent space using learned linear transformations. Let $W_o \in \mathbb{R}^{d_o \times d}$, $W_v \in \mathbb{R}^{d_v \times d}$, and $W_p \in \mathbb{R}^{d_p \times d}$ be the projection matrices for the historical posture, visual feature, and planned trajectory, respectively.

- **Embedded Postures:** $H_o = \Psi_t W_o \in \mathbb{R}^{K \times d}$, where each posture vector ψ_{t-i} is transformed into a d-dimensional posture token.
- **Embedded Vision:** $H_v = \mathcal{F}_t W_v \in \mathbb{R}^{1 \times d}$, yielding a single d-dimensional visual token.
- **Embeded Trajectory**: $H_p = \mathcal{P}_t W_p \in \mathbb{R}^{K \times d}$, where each planned position is projected into the d-simension token.

We concatenate the posture tokens and the visual token to form the memory matrix used as key-value pairs:

$$M_t = [H_o; H_v] \in \mathbb{R}^{(K+1) \times d}, \quad (6)$$

where M_t contains K posture embeddings and one visual embedding. To preserve temporal structure, we add positional encoding to the projected trajectory. Let $E_{\text{pos}}(i) \in \mathbb{R}^d$ be the positional encoding for the i-th timestep. The final input to the Transformer decoder is:

$$Q_t^{(0)} = H_p + E_{\text{pos}} \in \mathbb{R}^{K \times d}, \quad (7)$$

where $E_{\text{pos}} \in \mathbb{R}^{K \times d}$ encodes temporal indices across the K future steps. This query matrix $Q_t^{(0)}$ serves as the initial representation for the decoder to predict the corresponding future orientations.

3.3 Transformer Decoder Architecture

We employ a standard Transformer decoder architecture composed of L stacked layers to fuse the input modalities. Each decoder layer $\ell \in \{1, \ldots, L\}$ sequentially applies the following sublayers:

(1) Self-Attention (Query Sequence): The decoder initiates with multi-head self-attention applied to the query sequence (planned trajectory embeddings). This mechanism captures temporal dependencies and correlations within the planned motion trajectory. Let $Q_t^{(\ell-1)} \in \mathbb{R}^{K \times d}$ denote the input to layer ℓ, with $Q_t^{(0)}$ representing the initial query embeddings. The multi-head self-attention updates these embeddings as:

$$\tilde{Q}_t^{(\ell)} = \text{MHA}_{\text{self}}(Q_t^{(\ell-1)}) \in \mathbb{R}^{K \times d}. \quad (8)$$

In multi-head attention (MHA), the dimension d is split across h heads, each with dimension $d_h = d/h$. Within each head, scaled dot-product self-attention is computed among query tokens. For each head with query Q, key K, and value V (each $\in \mathbb{R}^{K \times d_h}$), the attention output is:

$$\text{Attention}(Q, K, V) = \text{softmax}\left(\frac{QK^\top}{\sqrt{d_h}}\right) V, \quad (9)$$

with $Q = K = V$ for self-attention. Outputs from all heads are concatenated and linearly transformed by an output matrix W^O. A residual connection and layer normalization are subsequently applied to yield the final self-attention output:

$$Q_t^{(\ell)} = \text{LayerNorm}(Q_t^{(\ell-1)} + \tilde{Q}_t^{(\ell)}). \quad (10)$$

(2) Cross-Attention (Fusion of Modalities): Next, the decoder employs multi-head cross-attention, where the queries originate from the self-attention outputs ($Q_t^{(\ell)}$), and the key–value pairs derive from memory M_t, representing the historical postures and visual features. This step critically fuses modalities: each query token (representing a future trajectory step) attends across the memory tokens (past K postures plus the visual feature token).

Each head projects queries $Q_t^{(\ell)}$ into $Q^h \in \mathbb{R}^{K \times d_h}$, and memory M_t into keys $K^h \in \mathbb{R}^{(K+1) \times d_h}$ and values $V^h \in \mathbb{R}^{(K+1) \times d_h}$. Scaled dot-product attention computes:

$$A^h = \text{softmax}\left(\frac{Q^h (K^h)^\top}{\sqrt{d_h}}\right) V^h \in \mathbb{R}^{K \times d_h}, \tag{11}$$

where attention scores aggregate information from memory tokens into query positions. The concatenated outputs from all heads $A^1, \ldots, A^h$ are projected back to dimension d, resulting in the cross-attention output:

$$Z_t^{(\ell)} = \text{MHA}_{\text{cross}}(Q_t^{(\ell)}, M_t) \in \mathbb{R}^{K \times d}. \tag{12}$$

This result undergoes residual connection and layer normalization:

$$Z_t^{(\ell)} = \text{LayerNorm}(Q_t^{(\ell)} + Z_t^{(\ell)}). \tag{13}$$

(3) Position-Wise Feed-Forward Network: Finally, a position-wise feed-forward network (FFN) independently processes each sequence position. The FFN typically comprises two linear layers with a nonlinear activation (e.g., ReLU or GELU) between them. For each token $\mathbf{x} \in \mathbb{R}^d$, the transformation is defined as:

$$\text{FFN}(\mathbf{x}) = \max(0, \mathbf{x}W_1 + b_1)W_2 + b_2, \tag{14}$$

with learned parameters W_1, W_2, and biases b_1, b_2. Applying the FFN to each token results in:

$$U_t^{(\ell)} = \text{FFN}(Z_t^{(\ell)}) \in \mathbb{R}^{K \times d}. \tag{15}$$

Another residual connection and layer normalization finalize the layer output:

$$Q_t^{(\ell)} = \text{LayerNorm}(Z_t^{(\ell)} + U_t^{(\ell)}). \tag{16}$$

After passing through all L decoder layers, the Transformer decoder outputs a sequence of K embeddings representing the future vehicle orientations:

$$X_{t+1:t+K} = Q_t^{(L)} = [x_{t+1}, x_{t+2}, \ldots, x_{t+K}]^\top \in \mathbb{R}^{K \times d}, \tag{17}$$

where each embedding x_{t+i} integrates information from the planned trajectory (P_{t+i}), historical postures (Ψ_t), and the visual context ($\mathcal{F}_t$), effectively modulating future orientation predictions based on multimodal context.

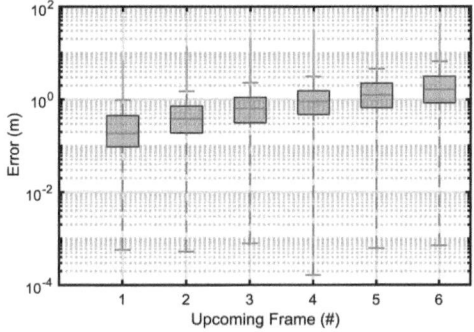

Fig. 4. Predicted Trajectory

3.4 Output: Unit Quaternion Sequences

The final decoder output $X_{t+1:t+K}$ is then mapped to the desired output format—a sequence of unit quaternions representing the vehicle's posture for the next K frames. We apply a linear projection from the model dimension d to $\mathbb{R}^4$ for each time step, followed by a normalization to unit length. Formally, for each $i = 1, \ldots, K$:

$$z_{t+i} = W_{\text{out}} x_{t+i} + b_{\text{out}} \in \mathbb{R}^4, \tag{18}$$

$$\hat{q}_{t+i} = \frac{z_{t+i}}{\| z_{t+i} \|} \in \mathbb{R}^4, \tag{19}$$

where $W_{\text{out}} \in \mathbb{R}^{d \times 4}$ and $b_{\text{out}} \in \mathbb{R}^4$ are learned output projection parameters. The vector z_{t+i} is the unnormalized quaternion prediction, and $\hat{q}_{t+i}$ is the normalized unit quaternion (enforcing $|\hat{q}_{t+i}| = 1$) which represents the predicted orientation at time $t + i$. The result of the decoder is thus the sequence of quaternions:

$$\hat{\Psi}_{t+1:t+K} = \left[\hat{q}_{t+1}, \hat{q}_{t+2}, \ldots, \hat{q}_{t+K} \right]^\top,$$

which are the fused predictions of the vehicle's future orientation over the next K frames. Each $\hat{q}_{t+i}$ encapsulates information from the planned trajectory as well as the historical and visual context, as learned by the Transformer decoder.

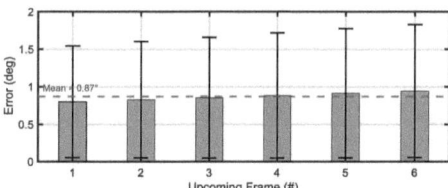

Fig. 5. Pitch Angle Error

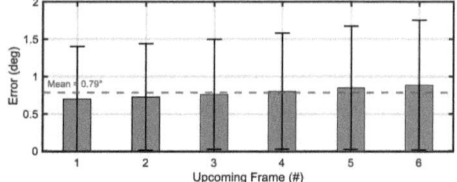

Fig. 6. Roll Angle Error

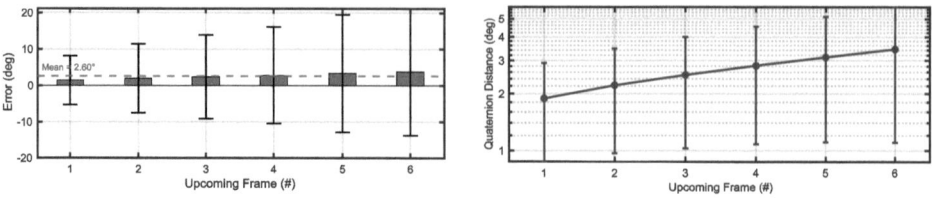

Fig. 7. Yaw Angle Error **Fig. 8.** Quaternion Distance

3.5 Loss Functions

At each future timestep, the predicted orientation is normalized to a unit quaternion $\hat{q}$. During training, we treat each predicted quaternion as a 4D vector on the unit hypersphere. To supervise the model, we use a loss that directly reflects the angular distance between the predicted quaternion $\hat{q}_{t+i}$ and the ground truth quaternion q_{t+i}. The training loss is defined as:

$$\mathcal{L} = \frac{1}{K} \sum_{i=1}^{K} \left(1 - |\hat{q}_{t+i} \cdot q_{t+i}|\right) \tag{20}$$

where $|\hat{q}_{t+i} \cdot q_{t+i}|$ computes the absolute value of the dot product between the predicted and ground truth unit quaternions, which corresponds to the cosine of the angle between the two orientations [8]. This loss formulation ensures rotational symmetry (i.e., q and $-q$ represent the same rotation) and penalizes deviations in orientation regardless of sign. By averaging across all K future frames, the model is encouraged to produce temporally consistent and accurate posture forecasts.

4 Implementation and Evaluation

4.1 Implementation

We implement PosFormer on top of the pre-trained self-driving model UniAD, which was originally trained on the nuScenes dataset [9]. The dataset comprises 1,000 driving sequences, each approximately 20 s long and annotated at 2 Hz with vehicle orientations represented as unit quaternions. These ground-truth quaternions are used as supervision targets in our training.

All experiments are conducted on a single NVIDIA GeForce RTX 4090 GPU (24 GB) hosted on an AMD Ryzen Threadripper PRO 16-core workstation. Our training setup requires only 6 GB of GPU memory, offering a significantly more efficient alternative to the original UniAD training pipeline, which consumes up to 50 GB of GPU memory during the first-stage training and 17 GB per GPU during second-stage end-to-end training on 8 NVIDIA A100 GPUs.

For optimization, we use the AdamW optimizer with a learning rate of 1×10^{-4} and a batch size of 4. The model is trained for up to 50 epochs with early stopping based on validation loss. This lightweight configuration highlights the efficiency and practicality of our framework, while still achieving strong performance in posture prediction.

4.2 Evaluation Metrics

We employ three complementary metrics to evaluate posture prediction performance: quaternion distance, mean angular error (MAE), and root mean squared error (RMSE). The quaternion distance directly measures the angular difference between the predicted and ground truth orientations in quaternion space:

$$Q_d = 2 \cdot \arccos(|q \cdot q_{\text{gt}}|) \tag{21}$$

The MAE assesses the average per-axis error in Euler angle space:

$$\text{MAE}(\theta) = \text{Avg}\left(\sum |\varphi - \varphi_{\text{gt}}|\right) \tag{22}$$

where $\varphi \in \{\phi, \theta, \psi\}$ represents the roll, pitch, and yaw angles, obtained by converting quaternions to Euler angles using Eq. 5. The MAE captures orientation discrepancies along each rotational axis. The RMSE quantifies the combined Euclidean deviation in Euler angle space:

$$\text{RMSE} = \sqrt{(\phi - \phi_{\text{gt}})^2 + (\theta - \theta_{\text{gt}})^2 + (\psi - \psi_{\text{gt}})^2} \tag{23}$$

Due to its quadratic formulation, RMSE is more sensitive to large errors, making it suitable for capturing outlier behavior in posture prediction.

4.3 Results

(1) Trajectory Prediction: Figure 4 demonstrates the trajectory prediction accuracy achieved by UniAD, evaluated against GNSS-collected groundtruth data. Our analysis on the nuScenes validation dataset examines prediction performance across six future frames. The results reveal a characteristic temporal error progression, where prediction inaccuracies accumulate over time due to error propagation from preceding frames [10]. Despite this expected behavior, the model maintains practical accuracy with mean errors of 0.1872 m and 0.3807 m for the first two frames respectively, well within acceptable operational thresholds for autonomous driving applications.

(2) Posture Prediction: Our model demonstrates strong performance in predicting future vehicle posture. Specifically, Figs. 5, 6, and 7 illustrate the mean angular errors (MAEs) for pitch, roll, and yaw predictions, respectively. PosFormer achieves average MAEs of 0.87°, 0.79°, and 2.26° for pitch, roll, and yaw. Notably, the yaw angle exhibits a higher error—approximately 2° greater than the other axes. This discrepancy arises from yaw's heightened sensitivity to rapid changes in vehicle heading, such as during turns or lane changes, which

Table 1. Compared with Baseline Models (RMSE)

Model	Upcoming Frames (angle degrees)		
	1st	2nd	3rd
MLP	2.4643±(0.0823)	2.8141±(0.0901)	2.9598±(0.1002)
LSTM	4.0624±(0.0994)	4.1051±(0.1065)	4.0260±(0.1146)
TCN	2.9140±(0.0794)	2.5967±(0.0870)	3.1193±(0.1003)
PosFormer	**1.7729±(0.5580)**	**2.2107±(0.8473)**	**2.6184±(1.0724)**
Model	Upcoming Frames (angle degrees)		
	4th	5th	6th
MLP	3.2697±(0.1154)	4.1488±(0.1472)	4.2546±(0.1471)
LSTM	4.6340±(0.1228)	5.2073±(0.1470)	5.1996±(0.1525)
TCN	3.9559±(0.1333)	7.1604±(0.2176)	4.4306±(0.1527)
PosFormer	**3.0290 ± (1.2590)**	**3.5592 ± (1.5680)**	**3.9910 ± (1.7074)**

introduce greater modeling complexity compared to the relatively smooth and stable dynamics of pitch and roll. The accuracy of orientation prediction is further validated using quaternion distance, as shown in Fig. 8. Compared to Euler angles, the quaternion representation offers a more compact and numerically stable measure of 3D rotation, avoiding singularities (e.g., gimbal lock) and providing smoother interpolation over time. From the figure, we observe that the quaternion angular error gradually increases across the prediction horizon, from $1.88°$ at $t+1$ to $3.32°$ at $t+6$. This trend is attributed to two primary factors: first, the accuracy of future position predictions by the self-driving model decreases over time, introducing uncertainty into downstream posture estimation; second, temporal correlations between past and future postures diminish as the prediction horizon extends, making it more challenging to forecast precise orientations at longer time scales.

(3) Compared against Baselines: We compare PosFormer against three state-of-the-art baseline models—MLP, LSTM [11], and TCN [12]—in terms of RMSE. The MLP baseline consists of a three-layer fully connected neural network, each layer containing 128 neurons. The LSTM model is a standard recurrent neural network architecture that captures temporal dependencies in sequential data. The TCN baseline leverages 1D dilated convolutions to model long-range temporal patterns, offering an alternative to recurrent architectures with improved parallelism and stability. All three baselines are trained using only historical posture sequences as input. The results are shown in Table 1, PosFormer consistently outperforms all baseline models in posture prediction accuracy.

This performance gain is primarily attributed to two factors. First, unlike the baselines, PosFormer leverages three complementary modalities, enabling

Table 2. Comparison of Euler Angle Errors (in degrees)

Frame	Method	Orientation Roll	Pitch	Yaw	MAE
1	w/o img	0.6°	0.6°	3.2°	1.5°
	w/o traj	0.8°	0.7°	2.4°	1.3°
	PosFormer	0.7°	0.8°	1.9°	1.2°
2	w/o img	0.7°	0.6°	4.4°	1.9°
	w/o traj	0.8°	0.7°	2.6°	1.4°
	PosFormer	0.8°	0.9°	2.4°	1.3°
3	w/o img	0.7°	0.6°	5.1°	2.1°
	w/o traj	0.7°	0.7°	3.0°	1.5°
	PosFormer	0.8°	0.9°	2.8°	1.5°
4	w/o img	0.7°	0.7°	5.9°	2.4°
	w/o traj	0.8°	0.7°	3.3°	1.6°
	PosFormer	0.8°	0.9°	3.3°	1.6°
5	w/o img	0.7°	0.7°	6.1°	2.5°
	w/o traj	0.8°	0.7°	4.2°	1.9°
	PosFormer	0.9°	0.9°	3.8°	1.9°

richer spatial-temporal context for prediction. Second, PosFormer employs a Transformer-based architecture with modality-aware attention mechanisms, allowing the model to dynamically attend to the most informative temporal and contextual cues across modalities. This results in more accurate and robust orientation forecasts, especially in complex driving scenarios.

(4) Ablation Study. Finally, we conduct an ablation study by selectively removing the visual context and planned trajectory components from PosFormer to assess the contribution of each input modality. This allows us to quantify the individual impact of each modality on posture prediction accuracy.

The results are presented in Table 2. Removing either modality leads to a noticeable degradation in performance, with MAE increasing by approximately 0.2° to 1.3° across the Euler angles. These findings highlight the complementary roles of the visual context and planned trajectory [13]. The visual context provides rich environmental cues—such as road geometry, lane curvature, and nearby objects—that help anticipate posture changes, particularly in scenes involving turns or elevation changes. Meanwhile, the planned trajectory offers a high-level motion intent that anchors the posture prediction to the expected vehicle path, improving temporal consistency and forward-looking precision.

5 Related Work

Recent research has advanced perception and motion estimation in autonomous systems. For object detection under foggy conditions, LiDAR-augmented 3D vision models have shown effectiveness [14]. Hybrid learning models accelerate flapping flight analysis by reducing CFD dependence [15]. Sparse representations improve self-driving efficiency, as demonstrated by DiFSD [16]. Egocentric pose estimation has progressed via head-body decomposition [17,18] and environmental-aware motion inference using sparse observations [19,20]. Multimodal fusion enhances trajectory prediction for vehicles [21], while deep learning frameworks generalize real-time 6DoF orientation estimation [22,23]. Integrated perception systems combine detection, tracking, and motion forecasting [24]. Transformer-enhanced PINNs provide accurate quaternion-based orientation [25], and remote teleoperation benefits from predictive ego-vehicle modeling [26].

6 Conclusion

This paper presents PosFormer, a transformer-based architecture that predicts future 3D vehicle posture. The framework enhances autonomous systems with efficient and accurate orientation prediction, supporting proactive control and improved planning, and outperforms baseline methods with minimal computational overhead

Acknowledgment. The research presented in this paper was supported by Science and Technology Project of China Southern Power Grid Limited Liability Company (Grant No. 032000KC23120050/GDKJXM20231537).

References

1. ResearchInChina, Passenger car intelligent chassis and chassis domain controller research report, p. 410 (2023)
2. Leiyang, X., Xiaolong, Z., Liang, L.: Vortex EM wave-based rotation speed monitoring on commodity WiFi. Chin. J. Electr. (2024)
3. Zheng, X., Yang, K., Xiong, J., Liu, L., Ma, H.: Pushing the limits of WiFi sensing with low transmission rates. IEEE TMC (2024)
4. Xu, L., Zheng, X., Du, X., Liu, L., Ma, H.: WiCamera: vortex electromagnetic wave-based WiFi imaging. IEEE TMC (2024)
5. Li, R., Zheng, X., Liu, L., Ma, H.: Plug-and-play indoor GPS positioning system with the assistance of optically transparent metasurfaces. In: Proc. of ACM MobiCom, pp. 875–889 (2024)
6. Hu, Y., et al.: Planning-oriented autonomous driving. In: Proc. of IEEE/CVF CVPR, pp. 853–862 (2023)
7. Xu, C., Zheng, X., Ren, Z., Liu, L., Ma, H.: UHead: driver attention monitoring system using UWB radar. Proc. ACM Interact., Mob., Wearable Ubiquit. Technol. **8**(1), 1–28 (2024)

8. Xia, D., Zheng, X., Liu, L., Huang, S., Ma, H.: WiCast: parallel cross-technology transmission for connecting heterogeneous IoT devices. IEEE TMC (2025)
9. Caesar, H., et al.: nuScenes: a multimodal dataset for autonomous driving, pp. 621–631. In: Proc. of IEEE/CVF CVPR (2020)
10. Zheng, X., Xia, D., Yu, F., Liu, L., Ma, H.: Enabling cross-technology communication from WiFi to LoRa with IEEE 802.11 ax. IEEE/ACM ToN **32**(3), 1936–1950 (2023)
11. Yu, Y., Si, X., Hu, C., Zhang, J.: A review of recurrent neural networks: LSTM cells and network architectures. Neural Comput. **31**(7), 1235–1270 (2019)
12. Lea, C., Flynn, M.D., Vidal, R., Reiter, A., Hager, G.D.: Temporal convolutional networks for action segmentation and detection. In: Proc. of IEEE/CVF CVPR, pp. 156–165 (2017)
13. Xia, D., Zheng, X., Liu, L., Wang, C., Ma, H.: c-Chirp: towards symmetric cross-technology communication over asymmetric channels. IEEE/ACM ToN **29**(3), 1169–1182 (2021)
14. Tahir, A., Mumtaz, R., Irshad, M.: 3D vision object detection for autonomous driving in fog using lidar. Simul. Model. Pract. Theory (2025)
15. Lan, B., Lai, Y.: Accelerating flapping flight analysis: reducing CFD dependency with a hybrid decision tree approach for swift velocity predictions. Physica D: Nonlinear Phenomena (2025)
16. Su, H., Wu, W., Yan, J.: DiFSD: ego-centric fully sparse paradigm with uncertainty denoising and iterative refinement for efficient end-to-end autonomous driving. arXiv preprint: arXiv:2409.09777 (2024)
17. Li, J., Liu, K., Wu, J.: Ego-body pose estimation via ego-head pose estimation. In: Proc. of IEEE/CVF CVPR (2023)
18. Yang, L., Chen, Y., Li, X.-Y., Xiao, C., Li, M., Liu, Y.: Tagoram: real-time tracking of mobile RFID tags to high precision using cots devices. In: Proc. of ACM MobiCom, pp. 237–248 (2014)
19. Xia, S., Zhang, Y., Su, Z., Zheng, X., Lv, Z., Wang, G.: EnvPoser: environment-aware realistic human motion estimation from sparse observations with uncertainty modeling. arXiv preprint: arXiv:2412.10235 (2024)
20. Zhao, X., An, Z., Pan, Q., Yang, L.: NeRF2: neural radio-frequency radiance fields. In: Proc. of ACM MobiCom, pp. 1–15 (2023)
21. Ge, L., Wang, S., Wang, G.: Fast multimodal trajectory prediction for vehicles based on multimodal information fusion. Actuators **14**(3), 136 (2025)
22. Golroudbari, A., Sabour, M.: Generalizable end-to-end deep learning frameworks for real-time attitude estimation using 6dof inertial measurement units. Measurement (2023)
23. Dai, D., An, Z., Gong, Z., Pan, Q., Yang, L.: {RFID+}: spatially controllable identification of {UHF}{RFIDs} via controlled magnetic fields. In: 21st USENIX Symposium on Networked Systems Design and Implementation (NSDI 24), pp. 1351–1367 (2024)
24. Günther, G., et al.: Perception system for autonomous vehicles: object detection, motion tracking, and future position prediction. In: IEEE ICRA (2025)
25. Golroudbari, A.: TE-PINN: quaternion-based orientation estimation using transformer-enhanced physics-informed neural networks. arXiv preprint: arXiv:2409.16214 (2024)
26. Sharma, G., Rajamani, R.: Teleoperated steering using estimated position and orientation of remote ego vehicle. In: 2024 American Control Conference (ACC). IEEE (2024)

Tag in Bloom: Fast RFID Membership via Software-Defined Readers

Chenxu Meng[1], Yao Yuan[1], Wenping Liu[1], Haoran Yan[2], Donghui Dai[2], and Lei Yang[2(✉)]

[1] Zhongshan Power Supply Bureau, Electric Power Research Institute, China Southern Power Grid Co., Ltd, Guangzhou, China
[2] Department of Computing, Shenzhen Research Institute, The Hong Kong Polytechnic University, Shenzhen, China
Michaelyen2024@yeah.net, donghdai@polyu.edu.hk, young@tagsys.org

Abstract. RFID is widely adopted for tracking and identifying objects, yet conventional tag inventory procedures remain inefficient due to frequent signal collisions and resulting latency. This paper introduces TagMap+, a practical and cost-effective framework leveraging Bloom Filters to significantly accelerate RFID tag detection. TagMap+ eliminates the need for per-tag EPC retrieval by shifting computational hashing to the reader side, using precomputed memory values within commercial off-the-shelf tags. By selectively activating tags through carefully controlled RN16 slot responses, TagMap+ avoids traditional anti-collision mechanisms, achieving rapid and collision-resilient presence detection. The system is implemented on a fully programmable, software-defined RFID reader compatible with the EPCglobal Gen2 protocol. Extensive real-world experiments demonstrate that TagMap+ reduces acquisition overhead by up to 79.53% compared to standard inventory methods, maintains robust compatibility with existing tags, and exhibits reliable performance in dense tag environments. This work represents the first hardware implementation of Bloom Filter-based acquisition in RFID, offering a scalable, efficient, and backward-compatible solution for high-performance RFID deployments.

Keywords: Radio Frequency Identification · Bloom Filter · Physical-Layer Acquisition · COTS RFID Tags · Software-Defined Radio

1 Introduction

Radio-Frequency Identification (RFID) has emerged as a core enabler within the broader Internet of Things (IoT) ecosystem [1]. By facilitating contactless, wireless communication between fixed readers and tags affixed to physical objects, RFID serves as a critical bridge between the physical and digital worlds. Its scalability and economic viability are largely attributed to the widespread deployment of passive, battery-free tags, which offer a low-cost and maintenance-free solution for tracking vast numbers of items in real time [2].

At the heart of RFID applications lies the inventory process, in which a reader interrogates and identifies all tags within its coverage area. However, this process is fundamentally constrained by the lack of inter-tag coordination: since passive tags cannot communicate with each other, simultaneous replies often result in signal collisions. To manage this, RFID systems commonly adopt anti-collision protocols-most notably, Framed-Slotted ALOHA (FSA). While simple and hardware-efficient, FSA suffers from low channel utilization. Its theoretical maximum throughput is limited to $1/e \approx 36.8\%$ [3], implying that more than 60% of time slots are wasted due to collisions or idle transmissions. This inefficiency directly translates to increased inventory latency, particularly in dense-tag environments.

Although more sophisticated multiple access techniques (such as CDMA or FDMA) could, in theory, improve performance, their complexity and energy requirements exceed the capabilities of conventional passive RFID tags. As a result, FSA and its variants remain the de facto standard, despite their limited efficiency. As a result, this persistent inventory bottleneck has significant implications in time-sensitive applications, including high-throughput warehouse logistics, automated baggage handling at airports, real-time manufacturing asset tracking, and large-scale ticketing or crowd management systems. In such scenarios, the need for faster, collision-resilient inventory mechanisms is increasingly critical to unlocking the full potential of RFID technology in next-generation IoT deployments.

In this work, we introduce TagMap+—a practical and cost-effective hardware implementation that fully leverages the capabilities of Bloom Filters within UHF RFID systems to eliminate the need for time-consuming per-tag inventory operations. A Bloom Filter is a space-efficient probabilistic data structure used to represent a set and determine whether a given element is a member of that set. Unlike conventional inventory procedures that require the sequential identification of each tag, Bloom Filters allow for fast, batch-oriented membership testing with constant-time query complexity, making them ideally suited for dense tag environments.

Figure 1 shows the key idea: the reader constructs a temporal Bloom Filter by selectively activating tags and observing whether any respond in specific time slots. Each slot corresponds to a bit in the filter, and the presence of any tag mapped to that bit is recorded by setting the bit to one. Once constructed, the resulting filter can be queried to verify the presence or absence of any known tag identifier without re-engaging in the full inventory protocol. By avoiding the exhaustive anti-collision and EPC retrieval processes typical of EPC Gen2, Bloom FilteråÄŞbased presence detection significantly reduces communication overhead, lowers latency, and improves scalability. As a result, TagMap+ enables fast, efficient, and scalable RFID querying while remaining fully compatible with commercial off-the-shelf tags and the standard Gen2 protocol.

TagMap+ builds upon our prior work, TagMap [4], but introduces significant new efforts to realize a fully functional and standards-compliant implementation. In particular, we developed a fully programmable, customized software-defined

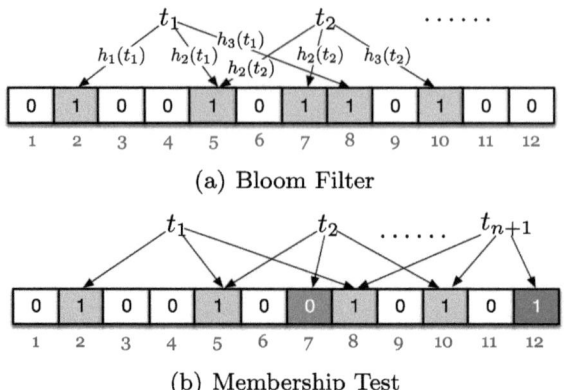

Fig. 1. Bloom Filter. (a) A Bloom Filter is represented as an M-bit array initialized to zeros. Each element in the target set T is mapped to K positions in the array using K independent hash functions $\{h_1, \ldots, h_K\}$. The bits at these positions are set to 1, constructing the filter. (b) To check membership of a query element, it is hashed using the same K functions. If any of the corresponding bits is 0, the element is definitely not in the set. If all bits are 1, the element is considered probably present, with a tolerable false positive.

RFID reader specifically designed to support the acquisition of Bloom Filters in real-world deployments. As shown in Fig. 2, the prototype integrates fine-grained control over Gen2 protocol operations, including precise timing, command sequencing, and slot-level signal monitoring—all essential for constructing Bloom Filters from selectively activated tag responses. To support the hash functionality, the reader firmware was extended to enable iterative `Select` commands with configurable bitmasking, allowing for targeted activation of tag subsets based on emulated hash values. On the hardware side, the reader—built atop a low-cost FPGA and SDR platform—was modified to detect slot-level `RN16` responses without proceeding to full EPC interrogation. This behavior aligns with the lightweight interaction model required for efficient Bloom Filter construction. These enhancements preserve full compatibility with the EPCglobal Gen2 standard while enabling new capabilities for fast, scalable, and collision-resilient tag presence detection, thereby advancing the practicality of Bloom FilterâĂŞbased RFID applications.

Summary of Results. In summary, TagMap+ achieves a compelling balance between practical deployability and operational performance. Experimental evaluations using a cost-effective, software-defined reader prototype yield the following key insights:

- **Backward Compatibility:** TagMap+ preserves full compliance with the EPCglobal Gen2 standard and demonstrates compatibility with 97.5% of commercial RFID tags tested—including legacy models deployed over decades.

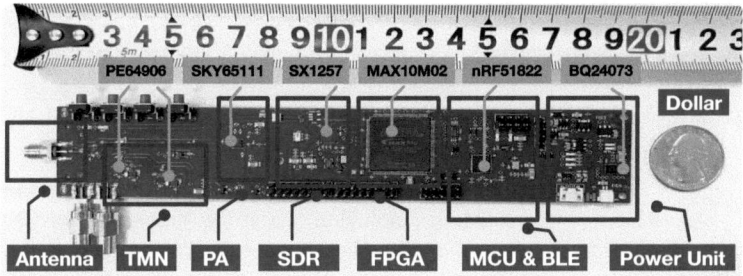

Fig. 2. UHF Reader Prototype for TagMap+.

- **Efficiency Optimization:** The system reduces acquisition overhead by up to 79.53% compared to conventional inventory procedures, while maintaining strong communication reliability with an average bit error rate below 0.4%.
- **Real-Time Detection Capability:** In missing-tag scenarios where 20% of tags are removed, TagMap+ achieves a detection success rate of 96.5% ± 0.6%, significantly outperforming traditional Gen2-based inventory in both accuracy and latency.

Contribution. To the best of our knowledge, this work is the first to implement Bloom Filter acquisition in RFID systems using customized hardware. By offloading the hashing process to the reader, introducing selective RN16-based slot responses, and deploying the design on a fully programmable software-defined Gen2-compliant reader, TagMap+ eliminates the need for per-tag EPC retrieval. Collectively, these contributions provide a lightweight, backward compatible alternative to traditional RFID inventory protocols, enabling efficient and scalable tag presence detection.

2 Background

2.1 Bloom Filter

A Bloom Filter is a compact, probabilistic data structure used to test whether an element belongs to a given set, typically denoted as $T = \{t_1, t_2, \ldots, t_n\}$, where each t_i represents a tag. It achieves high space efficiency by avoiding the need to store the actual elements. As shown in Fig. 1, the structure consists of a bit array of length M, initialized with all bits set to 0, and utilizes K independent hash functions $\{h_1, h_2, \ldots, h_K\}$, each mapping an input element to an index in the range $[1, M]$. To insert a tag $t \in T$ into the filter, each hash function is applied to t, and the corresponding K bit positions in the array are set to 1. Importantly, setting a bit to 1 multiple times has no additional effect. To check whether a tag is a member of the set, the same hash functions are applied, and the bits at the resulting positions are inspected. If any of these bits are 0, the tag is definitively not in the set, thereby ensuring that Bloom Filters never produce

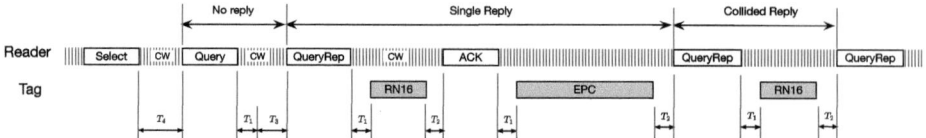

Fig. 3. Inventory Procedure in Gen2 Protocol. The EPC Gen2 standard employs an adaptive Q-algorithm for tag collision resolution. Each tag transmits its EPC packet only during an assigned time slot when no other tags are responding, ensuring efficient identification.

false negatives. If all bits are 1, the tag is assumed to be in the set, although this inference may be incorrect due to bit collisions, leading to a false positive.

The false positive rate depends on several parameters, including the bit array size M, the number of hash functions K, and the number of inserted elements n. As more hash functions are used, the likelihood of a non-member element mapping to at least one zero bit increases, thereby reducing false positives. However, using more hash functions also accelerates the saturation of the bit array, potentially increasing the overall false positive rate as more elements are added. On the other hand, using fewer hash functions preserves more unset bits but increases the risk that a non-member's hash indices coincide entirely with previously set bits. The trade-offs among these parameters have been studied extensively. For instance, the false positive rate is minimized when the number of hash functions K is approximately proportional to the ratio M/n, specifically around $K = \ln 2 \cdot (M/n)$. Additionally, to constrain the false positive probability to within a target threshold ε, the bit array size should be chosen proportional to $n \cdot \log_2(1/\varepsilon)$. These design principles guide practical deployments of Bloom Filters in systems where approximate set membership testing with bounded error ε is acceptable.

2.2 UHF RFID

The EPC Gen2 RFID protocol adopts a "reader-initiated" communication paradigm, in which the reader orchestrates the entire exchange process, and passive tags respond strictly in accordance with the reader's instructions. This approach reflects the fundamental asymmetry in power and control inherent to the system's physical and MAC layers. To manage responses from multiple tags and mitigate collisions, Gen2 implements a slotted ALOHA variant known as the Q-algorithm.

As illustrated in Fig. 3, the inventory phase typically begins with a `Select` command, allowing the reader to filter and target a subset of tags for participation. The subsequent issuance of a `Query` or `QueryAdjust` command signals the beginning of a new inventory frame. Each tag selected for this round independently and randomly chooses a slot within the frame to transmit its initial response. Slot transitions are driven by periodic `QueryRep` commands from the reader.

During its assigned slot, a tag attempts to transmit a short 16-bit message known as the RN16, which enables the reader to detect and resolve collisions. If exactly one tag responds and the reader successfully decodes the RN16, the slot is classified as a singleton. The reader then replies with an ACK, prompting the tag to send its complete identifier, including the PC, EPC, and CRC fields. If, however, the reader cannot decode the RN16, it infers either a collision—caused by multiple simultaneous responses—or an empty slot. In both cases, the reader does not proceed with further communication in that slot and simply advances to the next. As a result, singleton slots require more time to complete than either collision or idle slots. At the conclusion of a frame, if unresolved responses remain due to collisions or other errors, the reader may issue a QueryAdjust command to reattempt identification. This command can dynamically alter the frame size to better adapt to observed contention. The process repeats iteratively until all targeted tags are successfully inventoried.

3 Design of TagMap+

3.1 Motivation

In warehouse management, a fundamental challenge is the timely detection of missing items, particularly RFID-tagged goods that have been removed or stolen. To address this issue, inventory systems typically require RFID readers to continuously scan and re-inventory the tags to ensure the integrity of stock. This task aligns closely with the classic membership testing problem, which determines whether a known tag remains present in a monitored area. Crucially, full identification of every individual tag is unnecessary. Instead, the objective can be efficiently achieved using a probabilistic approach by using a Bloom Filter.

To this end, we propose constructing a temporal Bloom Filter to perform fast presence detection without exhaustive identification. We treat each reader frame as a temporary Bloom Filter, where tags randomly select time slots to transmit responses. Each time slot is conceptually mapped to a bit in the filter. If a response or a collision is detected in a given slot, the corresponding bit is set to 1; otherwise, it remains 0. This process allows the construction of a one-shot Bloom Filter over a single reader frame, enabling rapid, coarse-grained membership testing with minimal communication overhead. This strategy significantly reduces inventory latency and resource usage, while still allowing the system to flag potentially missing tags with high probability. Follow-up verification procedures can then be selectively triggered for those tags identified as likely absent.

3.2 Basic Idea

The construction of a temporal Bloom Filter relies on enabling each element to respond in a time slot that appears random yet remains deterministically computable by both parties. This is achieved by applying a hash function that maps each tag to specific slots in a predictable manner. In particular, the mapping

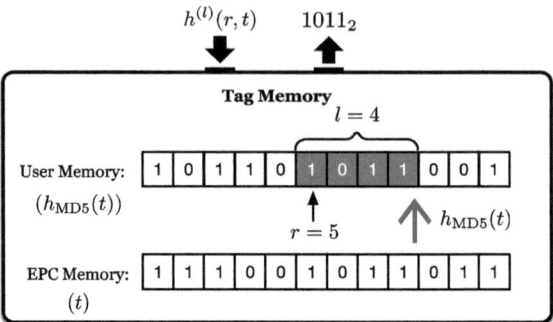

Fig. 4. Emulation of Hash Function. EachThe figure illustrates a tag's memory structure, highlighting `EPC memory` and `User memory`, which holds a pre-computed hash value. An example is provided: given input parameters $r = 6$ and $l = 4$, the function $h^{(4)}(t,6)$ yields the output 1010_2. This output is precisely the sub-bitstring of $H(t)$ stored in `MemBank3`, specifically covering the bit range from index 6 up to 9.

must be randomized per session while remaining reproducible by any party that knows the tag identifier and the random seed.

To this end, we define a family of hash functions $h^{(l)}(t,r)$ that produce an l-bit output based on two inputs: the tag's identifier t (e.g., its EPC, of length l bits) and a session-specific random seed r (provided by the system, of length l_r bits). Formally, each function is defined as:

$$h^{(l)}(t,r) : \{0,1\}^{l_t} \times \{0,1\}^{l_r} \rightarrow \{0,1\}^l \qquad (1)$$

The output is a binary string of length l, interpreted as an integer in the range $[0, 2^l - 1)$ and assumed to be uniformly distributed. To construct the Bloom Filter with K hash functions, we instantiate K independent hash functions by using K distinct seeds $r_1, r_2, \ldots, r_K$. Each tag computes K hash values $h_k^{(l)}(t, r_k)$, which determine the positions in the bit array where it contributes. These slot selections are fully deterministic given (t, r_k), allowing the system to infer which slots correspond to which tags. For clarity, the superscript l is sometimes omitted in the following.

3.3 Hash Emulation

Unfortunately, commercial RFID tags lack the on-chip circuitry necessary to perform cryptographic or custom hash computations. To overcome this limitation, we adopt the hash emulation technique introduced in our prior work [5] and further refined in TagMap [4]. As illustrated in Fig. 4, the core idea is to pre-compute a full-length hash of the tag's EPC identifier t using a standard hashing algorithm such as MD5. For example, given a tag with EPC $t = 1110010110011_2$, we compute $h_{\text{MD5}}(t) = 101101011001_2$. This hash value is then stored in the tag's user memory, which is typically reserved for application-specific data.

At runtime, when a hash output of length l is required, seeded by a value r, the emulated hash $h^{(l)}(t,r)$ is defined as the l-bit substring of $h_{\text{MD5}}(t)$ starting from bit position $r-1$. For instance, if $r = 5$ and $l = 4$, the resulting hash is $h^{(4)}(t,5) = 1011_2$. This method allows each tag to simulate multiple hash functions through simple memory access, avoiding the need for real-time computation.

This emulation approach satisfies the key properties required of hash functions in the context of Bloom Filter construction. First, the resulting hash values are intrinsically tied to the tag's EPC since they are derived from a cryptographic digest of t. Second, different seed values r extract distinct substrings from the high-entropy MD5 digest, effectively simulating independent hash functions with sufficient randomness. Thus, the emulated hashes maintain both determinism and sufficient entropy to ensure reliable and efficient Bloom Filter behavior.

3.4 Acquiring Temporal Bloom Filter

To ensure that tags respond in the correct slots based on their hash values—as required by the Bloom Filter construction—the system must provide a mechanism to selectively activate a subset of tags within a given inventory frame. This coordination is achieved through the Select command, a core feature of the EPC Gen2 protocol.

The Select command enables the reader to control tag participation by matching specific bit mask in tag memory. It can activate, deactivate, or ignore tags based on matching conditions. The command is formally defined as:

$$\boxed{\texttt{Select(Target, Action, MemBank, Pointer, Length, Mask)}}$$

Each field in the command serves a distinct purpose:

- **Target** (3 bits): Specifies which internal flag (e.g., SL, Inventoried A, or Inventoried B) will be modified based on the match result. This field is unused in our case.
- **Action** (3 bits): Defines how the tag updates its state depending on whether the mask matches. We set this field to 100, which activates the tag if the match is successful.
- **MemBank** (2 bits): Indicates the memory bank to match against. In our case, we set it to 11_2 to access the user memory, where hash values are stored.
- **Pointer** (32 bits): Specifies the starting bit address within the selected memory bank. We set this to the hash seed r, which defines the bit offset for matching.
- **Length** (8 bits): Specifies the number of bits to compare, starting from the pointer. We set this to l, which defines the length of the emulated hash value.
- **Mask** (variable length): Contains the actual bit pattern to match, which must be l bits long. This value corresponds to the emulated hash output $h^{(l)}(t,r)$ derived from the tag's stored hash.

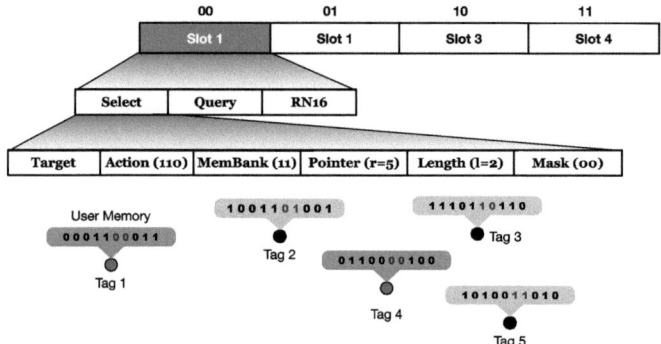

Fig. 5. Illustration of Acquiring a Temporal Bloom Filter Using Emulated Hash Functions. To guide tags to respond in specific time slots based on their hash values, the reader broadcasts a Select command that activates only those tags whose emulated hash values match the specified mask. In this example, tags 1 and 4 respond in the current slot because their hash values satisfy $h^{(2)}(t,5) = 00_2$.

Using the Select command, the reader applies a specific hash value as a conditional mask to the tag's memory, effectively filtering in only those tags that contain a matching l-bit substring (i.e., $h^{(l)}(t,r)$) in user memory. As a result, only the matching tags respond with an RN16 in the current slot. The reader then checks whether a response is detected and sets the corresponding Bloom Filter bit to 1 (if a response is heard) or leaves it at 0 (if the slot is silent). By iteratively issuing Select commands for all possible 2^l hash values, the reader can construct a complete 2^l-bit Bloom Filter that encodes the presence of all tags in the current set.

We illustrate the process with an example in Fig. 5. In this scenario, the reader issues a Select command with parameters Select($-$, 110_2, 11_2, 5, 2, 00_2), which targets tags whose emulated hash values satisfy $h^{(2)}(t,5) = 00_2$. This corresponds to a 2-bit substring starting at bit position 5 of the stored hash. As a result, only tags 1 and 4 match this condition and are therefore selected to participate in this time slot. These selected tags respond with an RN16, and the reader detects activity in the first slot. Consequently, the corresponding bit in the Bloom Filter is set to 1, indicating that one or more tags were mapped (hashed) into this slot.

4 Implementation

Inspired by prior work [6], we develop a prototype of the TagMap+ reader using a low-cost, software-defined mobile RFID platform. As shown in Fig. 2, the reader is built on a custom-designed two-layer PCB that supports modular, flexible signal processing and control.

4.1 Hardware Architecture

The major components of the software defined TagMap+ reader are described below:

(1) Power Unit: The reader is powered from a standard 5 V USB Micro-B connector, feeding an on-board PMU that generates the required supply rails for all subsystems. Low-cost regulators are used to provide stable 3.3 V and other voltages for the RF, digital, and analog components. This simple USB-powered design supports portable use (e.g. via battery pack) while keeping cost low. The PMU's regulated outputs ensure the power amplifier and transceiver receive adequate current (up to hundreds of mA for $\sim$0.5 W RF output) and a low-noise supply, which is critical for maintaining receiver sensitivity. By using common off-the-shelf regulators and avoiding expensive power management ICs, the PMU contributes to the overall cost-effectiveness of the system without sacrificing performance.

(2) Microcontroller: At the heart of the control logic is a Nordic nRF51822 ARM Cortex-M0 microcontroller (MCU) with integrated Bluetooth Low Energy (BLE). This MCU executes the high-level EPC Gen2 protocol and associated system software, and wirelessly interfaces with a smartphone via BLE for configuration, inventory control, and real-time monitoring. This BLE-enabled design supports intuitive user interaction and aligns well with personal or home-scale RFID applications. The MCU communicates directly with the FPGA over an SPI bus and can also indirectly access the software-defined radio (SDR) front-end via an SPI-to-SDR bridge implemented within the FPGA. This bridge provides register-level control over the transceiver, allowing the MCU to adjust parameters such as gain dynamically—particularly useful for tuning the transmit leakage cancellation network in real time. Together, the MCU and FPGA implement all mandatory EPC Gen2 reader commands. During operation, the MCU orchestrates inventory rounds and tag queries while leveraging the BLE link for fast reporting. Thanks to its programmable nature, the MCU enables the reader to support advanced inventory techniques, such as hash-based Bloom Filter queries or other accelerated tag-identification schemes.

(3) FPGA Digital Baseband: An ultra-low-cost Intel/Altera MAX10 FPGA handles the RFID digital baseband processing and timing-critical tasks. The design uses a 10M02 family device with about 2K logic elements. The FPGA implements the Gen2 protocol's lower-level functions: encoding the reader's DSB-ASK modulation for command transmissions, generating the precise interrogation timing, and demodulating/decoding the tag responses (FM0 or Miller-coded backscatter). It provides baseband filtering and clock/data recovery for tag signals, achieving correct decoding down to the specified sensitivity. Additionally, the FPGA includes the SPI bridge that interfaces the MCU to the SX1257 transceiver's registers, allowing coordinated control of RF parameters (e.g. gain settings during cancellation tuning). The use of an FPGA makes the reader software-defined and flexible âĂŞ new modulation schemes or protocol tweaks can be accommodated by updating the FPGA logic, a key advantage over fixed ASIC readers. Despite its small size, this FPGA design successfully implements

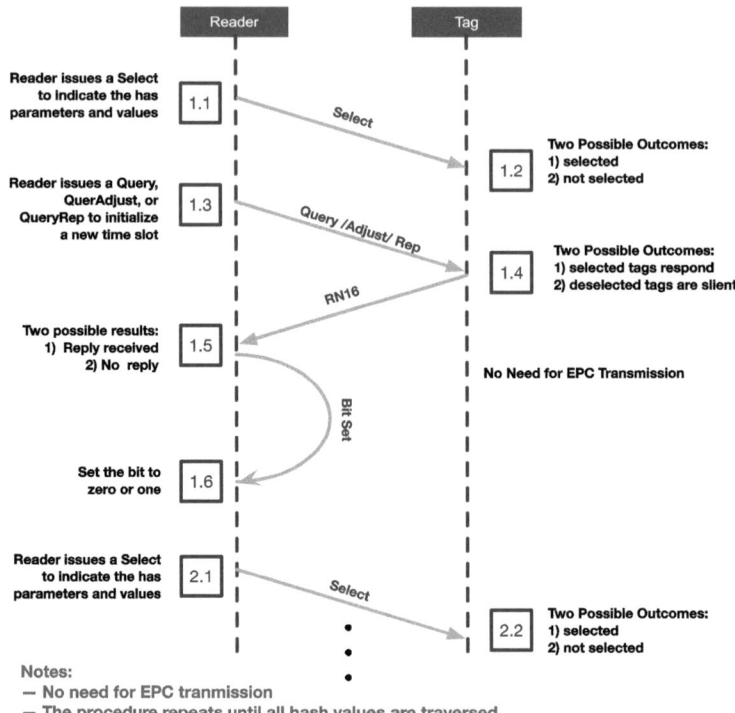

Fig. 6. Implementation of TagMap+ in terms of the Protocol Stack.

all essential Gen2 operations (inventory, read, write, etc.) and meets timing requirements. The programmable digital baseband is crucial for supporting fast inventory rounds and potential future enhancements like Bloom-Filter-based tag identification, all while keeping hardware costs to a minimum.

(4) **Radio Front End**: The RF transmit/receive front-end is centered around the Semtech SX1257, a low-cost, sub-GHz software-defined radio (SDR) transceiver ASIC. This chip supports the full UHF RFID band (902ăĂŞ928 MHz in the current implementation) and operates with programmable 500 kHz channels. Crucially, it enables full-duplex operation: the SX1257 can transmit a continuous wave (CW) to power passive tags while simultaneously receiving backscattered responses. Under FPGA control, the SX1257 synthesizes the RF carrier and applies amplitude shift keying (ASK) modulation for downlink transmissions. On the uplink, it performs I/Q downconversion and digitization of tag responses via integrated dual 13-bit ADCs, providing approximately 72 dB dynamic range. This high-resolution digitization allows the system to capture weak tag echoes even in the presence of substantial transmit leakage. Real-time configuration of the SX1257's transmit and receive signal chains, including mixer settings and gain control, is performed through an SPI interface managed jointly by the FPGA and MCU. The flexibility of the SDR front-end allows the system

to generate EPC Gen2-compliant waveforms and enables future extension to custom modulation formats or frequency-hopping protocols via software updates. In conjunction with the FPGA, the SX1257 forms a programmable and cost-efficient RF subsystem that meets the sensitivity and modulation depth specifications of the Gen2 standard, making it a suitable core for scalable and adaptable RFID reader implementations.

(5) Power Amplifier: To meet the power requirements for UHF RFID and extend read range, the reader incorporates a discrete power amplifier following the SDR transceiver. The SX1257's output is amplified to approximately 26 dBm (0.4 W) at the antenna port. While the hardware can reach up to 29 dBm, the output was adjusted and limited to 26 dBm in the prototype to avoid harmonic distortion introduced by the RF switch, which could interfere with leakage cancellation.

(6) Tunable Microwave Network : To address the transmit leakage challenge in a monostatic architecture, the reader integrates a Reflected Power Canceller (RPC) based on a digitally tunable microwave network. Sharing a single antenna for both transmission and reception results in significant CW leakage into the receiver path. The TMN connects to the isolation port of a directional coupler and acts as a tunable reflective load. By adjusting multiple digitally tunable capacitors (DTCs), it generates a cancellation signal with precisely controlled magnitude and phase. The FPGA or MCU dynamically tunes the TMN during operation, minimizing residual leakage through closed-loop adjustments based on RSSI or ADC feedback. This active cancellation achieves over 50 dB suppression, and when combined with antenna return loss, reduces total leakage by more than 60 dB. As a result, the system achieves a receiver sensitivity of -73 dBm, sufficient to detect weak tag backscatter even during continuous transmission.

The prototype is designed with modularity and reconfigurability in mind, allowing rapid development and evaluation of TagMap+ while maintaining full compatibility with the EPC Gen2 standard. By leveraging low-cost components and software-defined radio, the platform enables flexible experimentation with advanced reader-side techniques such as Bloom Filter-based tag detection, leakage cancellation, and fine-grained signal control, all without requiring hardware modifications to commercial RFID tags.

4.2 Protocol Stack

TagMap+ extends the EPC Gen2 air interface to rapidly construct a Bloom filter of the tag population without requiring any EPC transmissions. Figure 6 illustrates this process step-by-step. In step 1.1, the reader first issues a Gen2 Select command configured with a mask that specifies a target hash value. This Select command activates all tags whose stored hash equals the chosen value, while all other tags remain in the non-selected state and will ignore the upcoming inventory query (step 1.2). Next, in step 1.3, the reader begins an inventory slot by transmitting a Query command. This initiates a single-slot inventory round under the Gen2 protocol. Then, in step 1.4, any tag that was selected by the earlier mask will backscatter a 16-bit `RN16` random number in response, indicating its presence; tags that were not selected simply remain silent.

The reader monitors the slot for a reply. In step 1.5, two outcomes are possible: if an RN16 reply is received, it means one or more tags match the tested hash value; if no reply is heard (the slot is empty), it means no tag in the vicinity hashes to that value. Unlike a standard Gen2 inventory cycle, TagMap+ does not proceed to acknowledge or request any EPC from the tag after receiving the RN16. Instead, the RN16 itself is treated as a sufficient presence indication and the reader immediately ends that slot. Finally, in step 1.6, the reader sets the corresponding Bloom Filter bit to bit 1 if a reply was detected (signaling at least one tag for that hash) or bit zero if the slot was empty. At this point, one bit of the Bloom Filter has been determined via a single quick interaction, with no full tag identification required. TagMap+ repeats this process iteratively for each possible hash value to fill in the Bloom Filter.

Importantly, TagMap+'s protocol-layer design is fully compatible with the EPC Gen2 standard stack. It uses only standard Gen2 commands (`Select`, `Query`, `QueryAdjust`, `QueryRep`, etc.) and adheres to normal tag behavior for inventory rounds. This means that existing off-the-shelf Gen2 tags and readers can support TagMap+ without firmware modifications. By avoiding the conventional anti-collision process and skipping the EPC exchange, TagMap+ dramatically speeds up inventory: it captures a compact Bloom Filter representation of the tag set in just M short slots (for an M-bit filter), instead of interrogating every tag individually. The result is a fast, scalable membership testing mechanism—upper-layer applications can check for the presence of specific tags or items by examining the Bloom Filter, all achieved with minimal overhead and in a fraction of the time of a full inventory.

5 Evaluation

5.1 Experimental Settings

The experimental study was conducted in a $3 \times 3 \times 2.5\,\mathrm{m}^3$ office lounge, where 300 commercial UHF RFID tags were densely packed inside a carton. The TagMap+ reader was wirelessly connected to an iPhone 15 Pro via Bluetooth Low Energy (BLE) for control and data reporting. All tag `EPC`s were pre-registered in a backend database accessible to upper-layer applications—a prerequisite for Bloom Filter-based presence detection, which assumes prior knowledge of tag identifiers. By default, the Bloom filter's false positive tolerance was set to 0.25, providing a balanced trade-off between memory efficiency and filtering accuracy.

For comparison, TagMap+ was evaluated against the conventional Gen2 inventory protocol implemented in commercial systems, using the same reader hardware to ensure consistency. The complete set of tags detected by the reader within its effective interrogation range was treated as ground truth when evaluating detection performance. Notably, the study deliberately avoids simulation-based evaluation, focusing instead on real-world measurements to reflect practical system behavior under realistic deployment conditions.

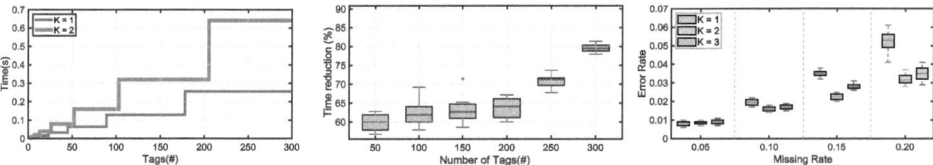

Fig. 7. Time Overhead.   **Fig. 8.** Time Gain.   **Fig. 9.** Error Rate.

5.2 Acquisition Overhead

We begin by evaluating the acquisition overhead of TagMap+ under configurations using one and two hash functions. Overhead is defined as the total number of time slots required to acquire tag presence information or to construct a complete temporal Bloom Filter. The results are summarized in Fig. 7. From the figure, we derive two key insights:

First, for a single hash function ($K = 1$), the minimal Bloom Filter length is given by the theoretical bound $M^* = \lceil n/\ln 2 \rceil$. However, for protocol timing alignment, the actual bitmap size is implemented as $M = 2^{\lceil \log_2(n/\ln 2) \rceil}$, rounding up to the nearest power of two. This leads to a stepwise growth in acquisition time as the tag population n crosses certain thresholds— specifically at $n = 89, 178, 355, 719, \ldots$— corresponding to bitmap size expansions from $M = 128 \to 256 \to 512 \to 1024 \to \ldots$

Second, increasing the number of hash functions to $K = 2$ requires scaling the Bloom Filter size to $M = 2^{\lceil \log_2(2.5n) \rceil}$. This results in new step transitions at $n = 103, 205, 410, 820, \ldots$, where the bitmap length again doubles. The acquisition overhead grows more rapidly in this case, as each Bloom filter bit now requires two separate Select commands and inventory operations—one per hash function.

These findings underscore a fundamental trade-off between acquisition latency and filtering accuracy. While higher values of K reduce the Bloom Filter's false positive rate, they proportionally increase acquisition overhead due to duplicated reader-tag interactions. Nevertheless, TagMap+ provides tunable control over this trade-off and achieves significant efficiency gains compared to traditional full EPC-based inventory, particularly in dense tag environments. In addition, while not explicitly measured, the computational overhead from hash emulation and Bloom Filter construction is minimal, involving only lightweight memory access and substring operations. For mobile or battery-powered readers, this is expected to be negligible compared to the energy saved by reducing redundant inventory rounds.

5.3 Compared with Standard Inventory

Next, we compare the acquisition time of TagMap+ against the standard EPC Gen2 inventory procedure, as shown in Fig. 8. The metric reflects total time consumption (in seconds) required to complete tag acquisition. On average,

TagMap+ achieves a 68.5% reduction in overhead, with time savings reaching up to 79.53% when the number of tags reaches $n = 300$. This considerable performance gain stems from two key innovations. First, TagMap+ relies on physical-layer presence detection using lightweight RN16 responses, eliminating the need for full EPC transmissions. Second, the design removes the need for anti-collision resolution by guiding tag responses deterministically through selective activation, avoiding contention and retransmissions.

In addition to reduced acquisition time, TagMap+ exhibits consistent and predictable behavior across varying tag densities. Even under high-collision conditions that typically degrade standard Gen2 performance, TagMap+ maintains deterministic overhead by operating independently of collision dynamics. This property is especially advantageous for scalable inventory in dense tag environments, enabling fast and reliable membership testing.

5.4 Missing Tag Identification

We conclude our evaluation by assessing the effectiveness of TagMap+ in detecting missing tags. Three test cases are considered, corresponding to different numbers of hash functions ($K = 1$, 2, and 3), with a fixed Bloom Filter tolerance of 0.25. The error rate is defined as the percentage of missing tags that are not successfully detected, while detection accuracy is evaluated across varying missing-tag rates. In each experiment, 300 tags are initially deployed within the monitored area, and subsets of 5%, 10%, 15%, and 20% are randomly removed. For statistical robustness, 50 independent trials are conducted for each condition. The results are summarized in Fig. 9. **Case 1 ($K = 1$):** Using a single hash function, the error rate gradually increases from $0.8\% \pm 0.2\%$ at a 5% missing rate to $5.1\% \pm 1.0\%$ at 20%, remaining well below the configured 25% tolerance threshold. As expected, no false negatives are observed, since Bloom Filters inherently ensure zero false negatives. **Case 2 ($K = 2$):** Introducing a second hash function increases the density of '1' bits in the Bloom Filter, thereby reducing the likelihood of false positives. This configuration reduces the error rate to $3.3\% \pm 0.5\%$ at a 20% missing rate, representing a 35.3% improvement over Case 1. **Case 3 ($K = 3$ with Bitmap Compression):** In this scenario, we use three hash functions and compress the bitmap length from 4096 to 2396 bits via modulo folding [4]. Despite the reduced bitmap size, detection performance remains comparable to Case 2, with an error rate of $3.0\% \pm 0.1\%$ at a 20% missing rate. The slight ∼0.3% increase in error is attributed to reduced randomness caused by hash collisions during compression, which biases hash outputs toward the beginning of the bitmap.

6 Related Work

Hash Functions and Bloom Filters. Prior work has shown that standard hash functions such as MD5 and SHA-1 are unsuitable for RFID tags due to their computational demands [7,8]. Although lightweight alternatives have been proposed [4,5], none have been deployed on COTS RFID tags or supported by low-cost reader platforms. TagMap+ overcomes this barrier by emulating hash functions through selective reading, enabling BF operations directly on standard Gen2 hardware. Traditional inventory protocols read all tags in bulk, resulting in high latency, while continuous-reading schemes [9,10] apply BFs to incrementally collect tag IDs but still suffer from the overhead of filter construction and querying. TagMap+ mitigates these issues by providing a lightweight service layer that streamlines Bloom Filter acquisition and significantly reduces overall latency.

RFID Network. Extensive efforts have been made to improve RFID performance in reading rate [4,5], coverage [11,12], and reliability [13–15]. Protocols like Buzz [16] and BiGroup [17] enhance inventory efficiency but rely on ideal conditions or tag modifications. In contrast, TagMap+ enhances performance with standard Gen2 tags and a low-cost reader, while maintaining reliable operation across diverse environments.

7 Conclusion

This work introduces a foundational service for commercial Gen2 RFID tags, enabling Bloom FilterâĂŞbased applications that enhance and accelerate the tag inventory process. This approach opens new directions for advancing RFID technologies and their real-world applications.

Acknowledgment. The research presented in this paper was supported by Science and Technology Project of China Southern Power Grid Limited Liability Company (Grant No. 032000KC23120050/GDKJXM20231537).

References

1. Das, R.: RFID forecasts, players and opportunities 2018–2028: the complete analysis of the global RFID industry. IDTechEx, Cambridge, MA. Tech, Rep (2018)
2. Dobkin, D.: The RF in RFID (2012)
3. Bueno-Delgado, M., Vales-Alonso, J., Gonzalez-Castano, F.: Analysis of DFSA anti-collision protocols in passive RFID environments. In: 35th Annual Conference of IEEE Industrial Electronics, pp. 2610–2617. IEEE (2009)
4. An, Z., Lin, Q., Yang, L., Lou, W.: Embracing tag collisions: acquiring bloom filters across RFIDs in physical layer. In: IEEE INFOCOM 2019-IEEE Conference on Computer Communications, pp. 1531–1539. IEEE (2019)
5. Yang, L., Lin, Q., Duan, C., An, Z.: Analog on-tag hashing: towards selective reading as hash primitives in Gen2 RFID systems. In: Proc. of ACM MobiCom, pp. 301–314. IEEE (2017)
6. Edward, A.K.: A low-cost software-defined UHF RFID reader with active transmit leakage cancellation. In: Proc. of IEEE RFID (2018)
7. Feldhofer, M., Rechberger, C.: A case against currently used hash functions in RFID protocols. In: OTM Confederated International Conferences "On the Move to Meaningful Internet Systems", pp. 372–381. Springer (2006)

8. Yoshida, H., et al.: MAME: a compression function with reduced hardware requirements. In: Paillier, P., Verbauwhede, I. (eds.) CHES 2007. LNCS, vol. 4727, pp. 148–165. Springer, Heidelberg (2007). https://doi.org/10.1007/978-3-540-74735-2_11
9. Sheng, B., Li, Q., Mao, W.: Efficient continuous scanning in RFID systems. In: Proc. of IEEE INFOCOM, pp. 1–9. IEEE (2010)
10. Xie, L., Li, Q., Chen, X., Lu, S., Chen, D.: Continuous scanning with mobile reader in RFID systems: an experimental study. In: Proc. of ACM MobiHoc, pp. 11–20 (2013)
11. Wang, J., Zhang, J., Saha, R., Jin, H., Kumar, S.: Pushing the range limits of commercial passive {RFIDs}. In: 16th USENIX Symposium on Networked Systems Design and Implementation (NSDI 19), pp. 301–316 (2019)
12. Boaventura, A.J.S., Carvalho, N.: Extending reading range of commercial RFID readers. IEEE Trans. Microw. Theory Tech. **61**(1), 633–640 (2012)
13. Yang, L., Chen, Y., Li, X.-Y., Xiao, C., Li, M., Liu, Y.: Tagoram: real-time tracking of mobile RFID tags to high precision using cots devices. In: Proc. of ACM MobiCom (2014)
14. Dai, D., An, Z., Gong, Z., Pan, Q., Yang, L.: RFID+: spatially controllable identification of uhf rfids via controlled magnetic fields. In: Proc. of USENIX NSDI, pp. 1351–1367 (2024)
15. Bocanegra, C., Khojastepour, M.A., Arslan, M.Y., Chai, E., Rangarajan, S., Chowdhury, K.R.: RFGO: a seamless self-checkout system for apparel stores using RFID. In: Proc. of ACM MobiCom, pp. 1–14 (2020)
16. Wang, J., Hassanieh, H., Katabi, D., Indyk, P.: Efficient and reliable low-power backscatter networks. Proc. ACM SIGCOMM **42**(4), 61–72 (2012)
17. Ou, J., Li, M., Zheng, Y.: Come and be served: Parallel decoding for cots RFID tags, pp. 500–511. In: Proc. of ACM MobiCom (2015)

FedDEK: Federated Domain-Incremental Learning via Expert Knowledge Construction

Lu Liu[1,2], Juan Li[1,2(✉)], and Tianzi Zang[1]

[1] College of Computer Science and Technology, Nanjing University of Aeronautics and Astronautics, Nanjing, China
{luvluvlu,juanli,zangtianzi}@nuaa.edu.cn
[2] Ministry Key Laboratory for Safety-Critical Software Development and Verification, Nanjing University of Aeronautics and Astronautics, Nanjing, China

Abstract. The rapid expansion of applications of the Internet of Things (IoT) has resulted in a vast amount of distributed data across edge devices. Federated Learning (FL) enables collaborative model training without raw data exchange, making it well-suited for privacy-sensitive IoT environments. However, in domain-incremental scenarios with evolving data distributions, such as lighting shifts in traffic or seasonal changes in agriculture, FL models face catastrophic forgetting. To address this, we propose **FedDEK**, a domain-incremental FL framework based on a pre-trained Vision Transformer (ViT). FedDEK freezes the ViT backbone and incrementally incorporates domain-specific expert modules trained on clients. A global knowledge extractor is designed to select and aggregate relevant expert outputs during inference. We further introduce lightweight adaptation to reduce client communication cost, and apply expert knowledge transfer to accelerate convergence and enhance cross-domain generalization. Extensive experiments on two datasets in different settings demonstrate that FedDEK improves average accuracy, reduces forgetting, and lowers communication overhead across diverse task sequences, outperforming prior federated continual learning methods.

Keywords: Federated Learning · Domain-Incremental Learning · Continual Learning

1 Introduction

In AIoT systems like smart cities and intelligent agriculture, vision-based tasks (e.g., vehicle detection, face recognition, and crop-weed classification) rely on edge-collected image data for model training. However, transmitting raw data from edge devices to a central server for training raises significant privacy concerns, especially in sensitive applications such as healthcare [1] and finance [2]. To address this, Federated Learning (FL) [3] has been widely adopted, enabling distributed devices to collaboratively train models without sharing raw data.

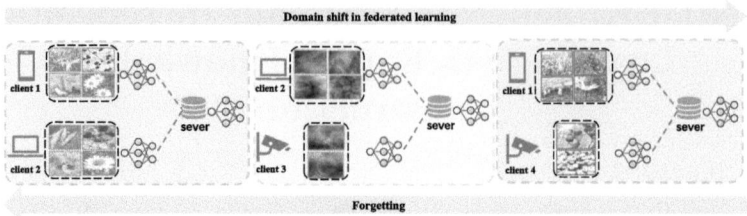

Fig. 1. Illustration of catastrophic forgetting in Federated learning. Multiple clients with non-IID distributions collaboratively train a shared global model through sequential tasks under weather domain shifts (brightness → fog → snow), facing catastrophic forgetting due to parameter overwriting.

Most existing research on Federated Continuous Learning (FCL) focuses mainly on scenarios with incremental classes, commonly referred to as Federated Class-Incremental Learning (FCIL) [4–7]. In FCIL, the model often preserves knowledge of old classes through decision boundary adjustment that maintains class separability. For example, GLFC [26] introduces additional class-imbalance losses, making the model pay more attention to old classes and preventing the decision boundary from tilting excessively towards new classes. However, these methods may struggle in settings where new tasks involve the same set of classes but differ in data distributions due to domain shifts, a scenario known as Federated Domain-Incremental Learning (FDIL) (Fig. 1). Unlike FCIL, FDIL requires more complex global feature adjustments, as new domains can introduce fundamentally different data characteristics, making knowledge retention across domains significantly more challenging.

Current approaches to address catastrophic forgetting in FDIL have shown potential but still face some limitations. **Firstly**, replay-based methods fail to meet privacy requirements. SR-FDIL [8] shows strong performance by selectively retaining and replaying representative samples. However, replay-based methods are not suitable for privacy-sensitive domains such as healthcare or defense , where policy constraints prevent the storage of historical data [1,7,28]. **Secondly**, non-replay methods face a conflict between mitigating catastrophic forgetting and maintaining low storage overhead. PFedDIL [9] achieves domain-incremental tasks by maintaining multiple local models for each client. While this approach mitigates forgetting, it significantly increases local storage requirements. However, naively training a single model for all tasks without data replay leads to catastrophic forgetting when the knowledge from new tasks overlaps with that from old tasks. **Additionally**, FL typically requires exchanging full model parameters between the server and clients in each communication round, leading to significant communication overhead, which is especially problematic for resource-constrained IoT devices [30]. This issue becomes even more pressing with the introduction of continual learning , where the model must continuously adapt to new data distributions over time, thereby demanding more frequent communication . These limitations lead to a critical question: *How can we* *design a communication-efficient federated domain-incremental learning framework that mitigates catastrophic forgetting without data replay, while keeping additional storage costs minimal?*

To address this question, we propose a novel framework, FedDEK (**Fed**erated **D**omain-Incremental Learning via **E**xpert **K**nowledge construction). Inspired by the adapter-based method in NLP [29], which enables efficient knowledge transfer without retraining full models for each task, we hypothesize that it is unnecessary to build a complete model for every domain. Instead, clients can collaboratively train lightweight expert modules to adapt to feature space variations introduced by domain shifts. The Vision Transformer (ViT) encoder offers a natural backbone for this design, as its modular architecture supports the seamless integration of expert components. Thus, based on a pre-trained ViT model, FedDEK progressively constructs and integrates domain-specific expert modules from distributed clients. Additionally, FedDEK introduces a domain knowledge extractor, which efficiently identifies and integrates the most relevant expert knowledge for inference. Our approach improves the model performance in FDIL without data replay, while ensuring minimal additional storage overhead and communication efficiency. The main contributions of our FedDEK are summarized as follows:

- We mitigate catastrophic forgetting in FDIL in a non-replay communication-efficient manner via training domain expert modules across clients based on a pre-trained ViT model and integrating a global knowledge extractor for inference.
- We accelerate the convergence of new tasks and reduce forgetting via expert knowledge transfer- when a new task arrives, the server initializes its parameters using the most relevant historical module.
- We conducted extensive experiments on two benchmark datasets under various settings, demonstrating that FedDEK significantly improves accuracy in FDIL scenarios without data replay, reduces communication costs, and maintains robustness across diverse learning sequences.

2 Related Work

2.1 Continual Learning

Continual Learning (CL), also known as Incremental Learning, enables models to learn sequentially from new tasks without forgetting prior knowledge. The primary variants of incremental learning include: 1) Task-incremental learning: The model learns from distinct tasks with explicit boundaries, where the task identity is known during both training and inference. 2) Class-incremental learning: The model continuously learns new classes without task boundaries, requiring adaptation to new classes while preserving prior knowledge. 3) Domain-incremental learning: The model adapts to new domains with shifting data distributions, maintaining consistent output classes without using domain labels during inference.

Traditional continual learning methods can be broadly categorized into three groups: 1) Regularization-based, typically assess the importance of parameters for previous tasks and use a regularization term to retain important parameters [10,27]. For example, Elastic Weight Consolidation (EWC) [10] selectively penalizes network parameters that are crucial for old tasks. 2) Replay-based, which either implicitly generates or explicitly stores original samples to preserve the model's representation ability for previous tasks [11,12]. However, in FL, both regularization-based and replay-based methods face privacy concerns; : the non-IID nature of data across clients makes it difficult to estimate parameter importance globally without violating privacy constraints; moreover, storing or sharing past data is also impractical due to privacy issues. 3) Parameter isolation-based, which allocates specific model parameters to each task, simplifying subsets of parameters from previous tasks to ensure maximum stability [13,14]. This leads to excessive model expansion in FL, making it infeasible for resource-constrained IoT devices.

Recently, prompt-based CL has gained attention, inspired by ViT-based Pre-training (VPT) [15]. For instance, L2P [16] replaces the replay buffer with a prompt pool, while S-Prompts [17] sequentially learns prompts for different domains, enabling effective knowledge transfer. However, in FL, these methods may struggle to adapt to heterogeneous client data distributions, and merely adjusting prompts does not effectively bridge cross-domain feature gaps.

2.2 Federated Continual Learning

To mitigate the challenge of catastrophic forgetting in Federated Learning, recent studies have attempted to incorporate Continual Learning techniques to retain knowledge across evolving tasks. Some methods assume clients can store and share historical data, simplifying the problem but raising privacy concerns [4,8,19]. Other methods are designed for task-incremental learning. FedWeIT [5] decomposes model weights into global federated parameters and sparse task-specific parameters. FOT [18] operates by extracting the global principal subspace of each layer from previous tasks and ensuring that updates for new tasks are orthogonal to this subspace. However, they are designed for task-incremental learning, assuming clear task boundaries during inference , making them unsuitable for domain-incremental scenarios . GLFC [26] tackles FCIL by introducing a class-aware gradient compensation loss and a class-semantic relation distillation loss to prevent forgetting, but it does not account for domain shifts. Furthermore, some methods are not feasible for resource-constrained scenarios. CFeD [20] employs knowledge distillation with proxy data on both the server and client, which can be computationally intensive and reliant on the quality of proxy data. Target [7] uses a global model to generate synthetic data for knowledge transfer and distills old knowledge on these data, which requires a lot of additional computational resources and storage space.

Most of the aforementioned works were originally designed for FCIL scenarios, while dedicated studies on FDIL remain relatively scarce. A FDIL method, pFedDIL based on adaptive knowledge matching, provides local models for clients but incurs high storage costs and only focuses on local personalized models rather than the global model [9]. In contrast, FedDEK focuses on the global model, which is a communication-efficient approach without data replay, requiring only minimal additional storage (approximately 0.15% of the model size).

3 Preliminaries

3.1 ViT Backbone and Lightweight Expert Layer Injection

The Vision Transformer (ViT) [25] processes images through three core components: embedding layer f_e, Transformer encoder f_r, and classifier φ. First, the embedding layer f_e projects flattened image patches x_p into a d-dimensional embedding space $x_e = f_e(x_p)$, while preserving spatial information through positional embeddings. These embeddings are then processed by the Transformer encoder f_r consisting of L identical blocks, where each block contains: (1) a Multi-Head Self-Attention (MSA) mechanism that captures long-range dependencies through attention weights, and (2) a Multilayer Perceptron (MLP) that applies nonlinear transformations to refine features. The composite feature extractor $f = f_e \circ f_r$ progressively enhances representations through this architecture, combining local patch information with global contextual relationships. Finally, the classifier utilizes the final state of the class token from f's output for prediction.

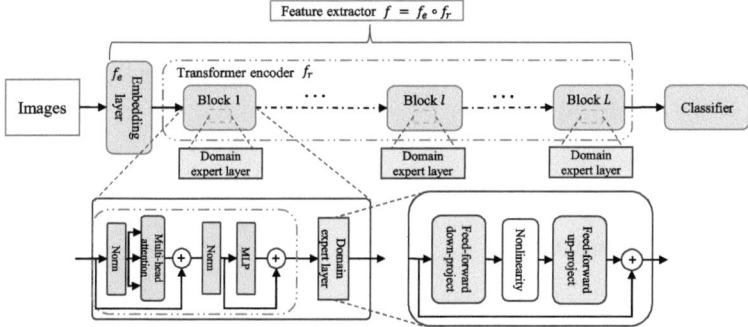

Fig. 2. Illustration of domain expert layers injection into a pre-trained ViT architecture. Expert layers are inserted in a residual manner into each Transformer block to enable domain-specific adaptation.

To enable efficient domain-specific adaptation, we insert lightweight expert modules into each Transformer block of ViT, without modifying the backbone weights of the ViT feature extractor. The domain expert layer adopts the structure of the adapter [29]. It is designed as a bottleneck architecture consisting of a down-projection, a nonlinearity, and an up-projection. This structure enables efficient adaptation by projecting the intermediate features to a lower-dimensional space of size p, where $p \ll d$, significantly reducing the number of trainable parameters on each client. An illustration of this architectural integration is shown in Fig. 2.

3.2 Problem Statement

In a standard Federated Domain-Incremental Learning setting, a central server coordinates a set of distributed clients N to sequentially learn a series of tasks $\{T_1, T_2, \ldots, T_n\}$, each associated with a different data domain. The label space remains fixed across tasks, i.e., $\mathcal{Y}_1 = \mathcal{Y}_t, \forall t \in [1, n]$, While the input distribution evolves over time, i.e., $\mathcal{X}_1 \neq \mathcal{X}_t, \forall t \in [1, n]$, reflecting domain shifts commonly observed in real-world IoT scenarios (e.g., the same object categories captured under varying environmental conditions). During training, each task's data is decentralized, private, and discarded after local updates, making prior task data unavailable for future reference. FDIL thus aims to enable the global model to continually adapt to evolving domains while retaining knowledge of previous ones without access to historical data, all while considering the resource constraints of clients. Formally, FDIL aims to learn a model $\tilde{w}$ that minimizes the following objective, while considering the resource constraints of clients, particularly the need to minimize the volume of transmitted model parameters per round and the additional storage for practical deployment:

$$\min_{\tilde{w}} \sum_{t=1}^{n} \sum_{i \in N} \frac{M_i^t}{M^t} \mathcal{L}(D_i^t; \tilde{w}), \tag{1}$$

where M_i^t and M^t denote the number of training samples from client i and from all clients for task T_t, respectively, D_i^t is the local dataset of client i for task T_t, and $\mathcal{L}(\cdot)$ is the empirical loss.

4 Proposed Method

4.1 Overall Framework

To address the challenge of catastrophic forgetting in Federated Domain-Incremental Learning with an efficient method without data replay, we propose FedDEK, which leverages a pre-trained ViT as the backbone, freezing its base weights, and dynamically expanding and training domain-specific expert modules through collaboration between the server and clients. A detailed procedure is provided in Algorithm 1.

Figure 3 illustrates the FedDEK framework, which operates in eight sequential steps. **Step 1: Initialize global backbone.** The server initializes the backbone w using a frozen pre-trained Vision Transformer (ViT), which serves as a universal feature extractor. **Step 2: Expand model with expert layers.** When a new domain-incremental task T_t arrives, the server expands the model by inserting lightweight domain-specific expert layers θ^t into each Transformer block and a domain-specific classifier φ^t. **Step 3: Distribute backbone to clients.** The server distributes the backbone to clients. During training, each

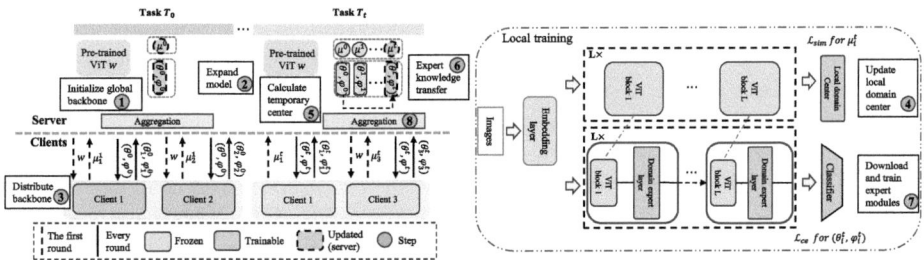

Fig. 3. A global model is extended with domain-specific experts (θ^t, φ^t) for new task T_t based on the pre-trained ViT model on the server. Each client i extracts features, updates local domain center μ_i^t via $\mathcal{L}_{sim}$, and trains experts $(\theta_i^t, \varphi_i^t)$ via $\mathcal{L}_{ce}$. The server aggregates updates and initializes new expert modules via expert knowledge transfer. Dashed lines represent transmission that occurs only during the first communication round. Solid lines indicate continuous transmission that takes place in every round of communication between the client and the server.

Algorithm 1: FedDEK

Input: Pre-trained ViT backbone w; global set of domain centers $\mathcal{M} = \emptyset$
Output: Model $\tilde{w}$ with global expert modules (θ, φ); updated set $\mathcal{M}$

1 **foreach** *task* T_t *with* $t = 1, \ldots, n$ **do**
2   **for** $r = 1$ **to** R **do**
3   Server randomly selects a subset of clients $\mathcal{K}^{tr}$;
4   **Server initialization:**
5   **if** $t > 1$ **and** $r = 1$ **then**
6   Collect $\{\mu_i^t\}_{i \in \mathcal{K}^{tr}}$ and compute $\mu^{t0} \leftarrow \frac{\sum_i D_i^t \mu_i^t}{\sum_i D_i^t}$;
7   Find a^* with Eq. (6), set $\theta^t \leftarrow \theta^{a^*}$, $\varphi^t \leftarrow \varphi^{a^*}$;
8   **Client training (parallel):**
9   **foreach** *client* $k \in \mathcal{K}^{tr}$ **do**
10   **if** μ_i^t *does not exist* **then**
11   **for** $e = 1$ **to** E' **do**
12   Freeze w, update μ_i^t with Eq. (2);
13   Upload μ_i^t;
14   Download θ^t, φ^t;
15   **for** $e = 1$ **to** E **do**
16   Freeze w, update θ_i^t, φ_i^t with Eq. (3);
17   Upload θ_i^t, φ_i^t;
18   **Server aggregation:**
19   Update expert modules:
20         $\theta^t \leftarrow \frac{\sum_i D_i^t \theta_i^{tr}}{\sum_i D_i^t}$, $\varphi^t \leftarrow \frac{\sum_i D_i^t \varphi_i^{tr}}{\sum_i D_i^t}$;
21   **if** $r = R$ **then**
22   Compute μ^t via Eq. (5);
23             $\mathcal{M} \leftarrow \mathcal{M} \cup \mu^t$;

client i receives the pre-trained backbone w only upon its initial participation. The backbone is then reused in all subsequent tasks without re-distribution. **Step 4: Update local domain center.** The domain center refers to the representative feature representation within a given domain. For each task T_t, the local domain center μ_i^t of each client i only needs to be computed once. It is obtained by minimizing a cosine similarity loss $\mathcal{L}_{\text{sim}}$ if it has not been computed, and then it is uploaded to the server. **Step 5: Calculate temporary global domain center.** At the first round of communication of the current task T_t, the server aggregates the local domain centers uploaded by the participating clients and obtains the temporary global domain center μ^{t0}. **Step 6: Expert knowledge transfer.** The server identifies the most similar historical domain by measuring the cosine similarity between the temporary global domain center and historical domain centers. It then transfers the corresponding expert module weights to initialize the expert modules (θ^t, ϕ^t) for task T_t. **Step 7: Download and train expert modules.** Clients download the expert modules (θ^t, ϕ^t) and train them locally using cross-entropy loss $\mathcal{L}_{\text{ce}}$, while keeping the backbone w frozen. **Step 8: Aggregate client updates.** In each training round of the current task T_t, the server aggregates client updates, weighted by local data volume, to refine the global expert modules (θ^t, ϕ^t). In the final round, it updates the global domain center μ^t by aggregating centers received from all clients that participated in T_t.

4.2 Client Side: Local Expert Knowledge Training

Local Domain Center Training. During the training phase of the current incremental domain task T_t, each client i involved in FL establishes a local domain center $\mu_i^t \in \mathbb{R}^d$ aligned with the latent features of the local data. The feature extraction process is based on the Pre-trained ViT feature extractor. Due to the unchanged local data of each client within a task and the freezing of the pre-trained ViT feature extractor, as well as the small number of trainable parameters, updates to the local center are sufficient to converge in one communication round. In order to save computing resources, we update the local center μ_i^t of client i only once during each task T_t. Client i utilizes the feature extractor to extract features from their local data, then updates the local domain center using the following loss function:

$$\mathcal{L}_{sim} = \sum_{j} \left(1 - \cos\left(f\left(x_j\right), \mu_i^t\right)\right), \tag{2}$$

where x_j is the j-th sample, and μ_i^t denotes the local feature center of client i for task T_t. The term $\cos(f(x_j), \mu_i^t) = \frac{f(x_j) \cdot \mu_i^t}{|f(x_j)||\mu_i^t|}$ represents the cosine similarity between the feature vector $f(x_j)$ extracted by the model and the corresponding local center μ_i^t.

Local Domain Expert Modules Training. In the local training process, the base weights of the ViT are frozen, and only the domain-specific expert modules are updated. The expert modules for a domain include both the domain expert layers (which we introduced in Section III.A) and a domain-specific classifier. The objective used for training is the standard cross-entropy loss, defined as:

$$\mathcal{L}_{ce} = \sum_j \sum_c y_{j,c} \log \hat{y}_{j,c}, \qquad (3)$$

where $y_{j,c}$ is the ground truth label for the j-th sample in the c-th class, where $y_{j,c} = 1$ if sample x_j belongs to class c, and $y_{j,c} = 0$ otherwise, and $\hat{y}_{j,c}$ is the predicted probability that sample x_j belongs to class c.

Since the backbone of the ViT remains frozen throughout training, each client i only uploads the parameters of the domain-specific expert layers θ_i^t and classifier φ_i^t to the server during Task T_t. This significantly reduces communication overhead compared to traditional FL methods that transfer the full model.

4.3 Server Side: Global Expert Knowledge Aggregation

When a new domain-incremental task arises, the server dynamically expands and randomly initializes the domain expert modules of the global model.

Expert Knowledge Transfer. Inspired by human learning paradigms in which new knowledge acquisition is usually based on existing related expertise, we propose an expert knowledge transfer mechanism. Assuming that the server's global model already contains domain expert modules from previously trained tasks, when a new domain task T_t arrives, the server transfers parameters from the most relevant historical expert module to accelerate convergence.

Specifically, the server first collects the local domain center μ_i^t uploaded from each client i who participated in the first communication round. The server then computes the temporary domain center for task T_t as $\mu^{t0} = \frac{\sum_i^K D_i^t \mu_i^t}{\sum_i^K D_i^t}$, where K is the number of clients participating in the first communication round of current task T_t, D_i^t is the local data size of client i for T_t. After μ^{t0} is obtained, the server calculates the cosine distance between the current center and the existing domain centers and selects domain a^* with the closest center to initialize parameters of the domain expert modules for the current task T_t. This process can be formulated as follows:

$$a^* = \arg\max_a \cos\left(\mu^{t0}, \mu^a\right) \quad \mu^a \in \mathcal{M}^{t-1}, \qquad (4)$$

where $\mathcal{M}^{t-1}$ is the set of global domain centers on the central server after task T_{t-1}.

Global Knowledge Aggregation. In each round of task T_t, the server receives the locally trained parameters of domain expert modules uploaded by the clients, with the domain expert layer and the classifier of client i denoted by θ_i^t and φ_i^t respectively. The server then aggregates these weights according to the data volume and updates the global weights of expert modules θ^t and φ^t.

To account for the fact that the set of participating clients may vary across training rounds of each task T_t, we design a buffer pool on the server to store the local domain centers μ_i^t from all clients participating in task T_t. In each round, if a client joins T_t for the first time, its local domain center μ_i^t is computed and added to the buffer. When task T_t is completed, the server aggregates all the buffered domain centers to compute the final global domain center μ^t. Assuming there are a total of N clients involved in training task T_t, the global domain center for domain t is calculated as:

$$\mu^t = \frac{\sum_{i=1}^{N} D_i^t \mu_i^t}{\sum_{i=1}^{N} D_i^t}. \tag{5}$$

4.4 Inference: Knowledge Extractor

Since the domain id of an input is x unknown during inference, we employ a global knowledge extractor to dynamically select relevant expert modules. The knowledge extractor first extracts x's features $f(x)$ using the base pre-trained feature extractor. Then $f(x)$ is compared to the domain centers of all domains $\mu^a \in \mathcal{M}$ via cosine similarity, and the ids of top-z closest domain centers are selected into the domain index set A^* by the knowledge extractor (Fig. 4).

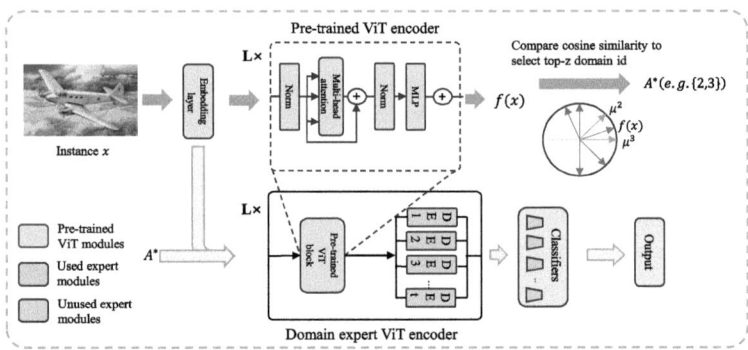

Fig. 4. Inference process with global knowledge extractor. Given an instance x, the knowledge extractor first selects top-z closest domains A^*(e.g.$\{2,3\}$), then aggregates predictions from corresponding experts to produce the final output.

Finally, according to the set of domain indexes selected A^*, the knowledge extractor performs the corresponding branch of the domain expert modules z to generate the final output. Specifically, the output $p(y|x)$ is computed as a weighted combination of the domain-specific expert predictions, where the weights are derived from the exponential squared cosine similarity between the global feature $f(x)$ and each domain center $\mu^a \in A^*$:

$$p(y|x) = \frac{\sum_{a \in A^*} e^{\cos(f(x), \mu^a)^2} \cdot p_a(y|x)}{\sum_{a \in A^*} e^{\cos(f(x), \mu^a)^2}}, \tag{6}$$

where $p_a(y|x)$ represents the prediction from the a-th domain expert for input x. To dynamically emphasize the most relevant domain experts, we employ an exponential squared cosine similarity weighting scheme, a variant of standard cosine similarity. In our experiments, we analyze the impact of different values of z and conclude that the selection of z should be adjusted according to the specific requirements of the application scenario.

5 Experiments

5.1 Experimental Setup

Datasets. We adopted domain-incremental benchmarks, OfficeHome and Sub-DomainNet.

- **OfficeHome** [21]: The OfficeHome dataset consists of images from four domains: Art, Clipart, Product, and Real World. It contains around 15,500 images across 65 object categories.
- **Sub-DomainNet**: The DomainNet dataset [22] consists of images from six domains: Clipart, Infograph, Painting, Quickdraw, Real, and Sketch. It contains around 600,000 images across 345 object categories. To reduce computation cost, we constructed Sub-DomainNet, a subset of DomainNet, by randomly selecting 64 categories that have more than 100 images per domain and reducing the number of images in each category to 1/4 of the original size.

Baselines

- **FedAvg** [3]: FedAvg is a foundational method in federated learning without specific mechanisms to address domain shift.
- **FedEWC**: FedEWC is an extension of the EWC [10] adapted for federated learning. It uses the Fisher information matrix to estimate the importance of parameters and penalizes changes to the parameters deemed important for previous tasks.
- **CFeD** [20]: CFeD divides tasks among clients to handle different objectives. Some clients use knowledge distillation with surrogate datasets to review old tasks, other clients learn new tasks, and the server distills knowledge across rounds.
- **Target** [7]: Target uses the global model to generate synthetic data and uses knowledge distillation to transfer knowledge from previous tasks.
- **FedS-prompts**: FedS-prompts is an extension of the S-prompts [17] adapted for federated learning, which learns prompts for each domain sequentially.

Evaluation Metrics. Following prior work in continual learning [23,24] and considering the communication overhead challenges in federated learning, we adopted the following evaluation metrics:

- **Final Average Accuracy (AN)**: The average test accuracy of the final global model across all N tasks. This metric reflects both the model's learning capability and its resistance to catastrophic forgetting.
- **Final Average Forgetting (FN)**: The average performance degradation across N tasks. This metric specifically quantifies catastrophic forgetting.
- **Local Trainable Parameter Size (PS)**: The number of trainable parameters on the client side, which directly indicates the communication overhead imposed on clients.

Configuration. In our study, all methods are implemented with NVIDIA RTX 3090 GPUs in Pytorch, utilizing pre-trained ViT-B/16 [25] as the model backbone. We set the projection dimension of the domain expert layer as $p = 4$, the number of the total clients as $N = 10$, the client number per round as $K = 5$, and the communication round as $R = 15$ for every task. Like most FL methods on non-IID data distribution, we used the Dirichlet distribution to simulate the data heterogeneity with $\alpha = 1$. For local training, the number of local training epochs of each client is $E = 8$, and the batch size is 32. We used optimizer SGD with a learning rate $lr = 0.001$ for OfficeHome, and $lr = 0.002$ for Sub-DomainNet. We used a momentum of 0.9, and a weight decay of 5×10^{-5} to adjust the learning rate dynamically. We employed a cosine learning rate scheduler with $lrf = 0.01$. During inference, unless specified otherwise, FedDEK dynamically selects the top-z closest experts, with $z = 2$ for the OfficeHome and $z = 1$ for the Sub-DomainNet. For the baseline, we set the EWC penalty coefficient $\lambda = 400$, the distillation temperature τ is 2 for Cfed and 5 for Target, and the prompt length l of FedS-prompts is 10.

5.2 Performance Comparison

Table 1. Performance Comparison on OfficeHome and Sub-DomainNet

Method	OfficeHome			Sub-DomainNet		
	AN(%)	FN(%)	PS	AN(%)	FN(%)	PS
FedAvg	86.78	4.00	85.85 M	73.81	10.57	85.85 M
FedEWC	87.79	2.95	85.85 M	74.87	9.30	85.85 M
Cfed	85.99	4.66	85.85 M	75.33	7.95	85.85 M
Target	87.91	2.99	85.85 M	75.84	8.18	85.85 M
FedS-prompts	86.40	0.86	**0.06 M**	76.53	0.44	**0.06 M**
FedDEK	**88.56**	**0.11**	0.13 M	**78.46**	**0.23**	0.13 M

From the results of Table 1, it was observed that FedDEK consistently achieved the best AN and FN scores on both OfficeHome (with a small domain gap) and

Sub-DomainNet (with a large domain gap), indicating its effectiveness in reducing performance degradation caused by the domain shift in FDIL. In contrast, FedAvg showed a noticeable drop in AN and a rise in FN as the domain gap widened, as mitigating catastrophic forgetting was more critical than model plasticity in domain incremental tasks with significant distribution shifts. For the PS metric, compared to the best-performing FedS-prompts, our method only had a difference of 0.07M in the number of trainable parameters per client locally, but there was a significant improvement in AN.

5.3 Analysis of Our Method

Table 2. Impact of different values of z on AN during the inference phase.

Dataset	z	AN(%)
OfficeHome	1	87.60
	2	**88.46**
	3	88.00
Sub-DomainNet	1	**78.53**
	2	77.78
	3	75.57

Table 3. Three Training Orders on Sub-DomainNet

Order	1	2	3	4	5	6
Order 1	AR	IF	PA	QD	SK	RW
Order 2	PA	RW	IF	AR	SK	QD
Order 3	IF	SK	QD	AR	RW	PA

Note: AR: Art, IF: Infograph, PA: Painting, QD: Quickdraw, SK: Sketch, RW: Real World.

Effect of the Selection of Expert Number z in the Inference Phase. We conducted experiments with different numbers of experts z during inference. The results in Table 2 revealed that on the OfficeHome dataset, where domain distribution differences are relatively small, setting $z = 2$ or $z = 3$ enables more effective utilization of knowledge from similar domains, leading to enhanced performance. In contrast, on the Sub-DomainNet dataset, where domain differences are more pronounced, using $z > 1$ introduces interference between domain-specific knowledge, ultimately degrading model performance. Therefore, we recommend selecting z based on the specific characteristics of the application scenario in IoT systems to optimize performance.

Effect of Task Sequence. To evaluate the robustness of FedDEK, we evaluated its performance under different task sequences by varying the order of incremental domain introductions in the Sub-DomainNet dataset. Three training orders are shown in Table 3. The experimental results in Table 4 demonstrated that FedDEK consistently outperformed baseline methods in terms of AN across all task orders, while maintaining performance stability regardless of domain sequence variations. This confirmed FedDEK's robustness and effectiveness in mitigating catastrophic forgetting. In contrast, traditional methods such as FedAvg and

Table 4. Performance comparison on Sub-DomainNet under three task orders

Method	Order 1		Order 2		Order 3	
	AN(%)	FN(%)	AN(%)	FN(%)	AN(%)	FN(%)
FedAvg	73.81	10.57	59.75	25.28	74.03	9.65
FedEWC	74.87	9.30	66.32	17.82	72.04	9.67
Cfed	75.33	7.95	63.92	20.51	74.90	8.35
Target	75.84	8.18	72.60	12.21	76.37	7.30
FedS-prompts	76.53	0.44	76.29	0.84	76.23	**0.40**
FedDEK	**78.46**	**0.23**	**78.62**	**0.83**	**78.38**	0.78

Cfed exhibited severe instability and were prone to catastrophic forgetting, especially when the distribution of the final domain deviated significantly from the others', with FN reaching as high as 25.2%. Although FedS-prompts demonstrated a slightly lower FN compared to FedDEK in the case of training order 3, their inherent flexibility limitations led to their AN being 2.15% lower than that of FedDEK. The consistently strong results across different domain orders highlighted FedDEK's robustness and adaptability in FDIL scenarios.

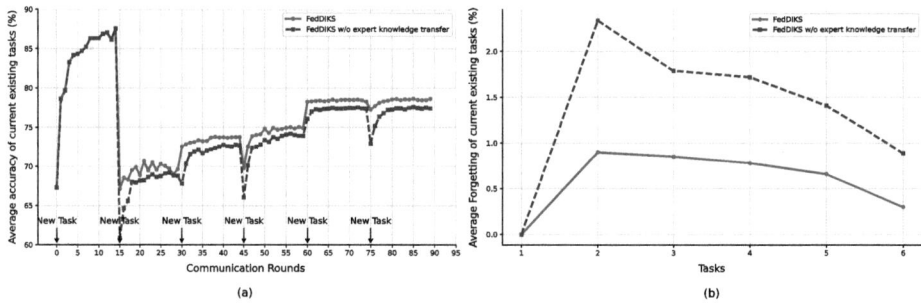

Fig. 5. The ablation study results of expert knowledge transfer: In Figure (a), the average accuracy of current existing tasks is plotted against communication rounds, with arrows indicating the appearance of each new task. Figure (b) illustrates the average forgetting of current existing tasks as new tasks are introduced.

Effect of Expert Knowledge Transfer. We validated the effectiveness of the expert knowledge transfer on the Sub-DomainNet by removing it from the training steps. As shown in Fig. 5(a), FedDEK, by incorporating the expert knowledge transfer, not only accelerated the convergence of the model but also enhanced the overall model performance. Figure 5(b) demonstrated that the inclusion of the expert knowledge transfer reduced forgetting. Thus, the performance improvement could be attributed to the fact that domain expert modules, initialized with knowledge from similar historical experts, possessed stronger capabilities to handle previous domain tasks, thereby reducing catastrophic forgetting. This confirmed the effectiveness of the expert knowledge transfer mechanism in Fed-DEK.

6 Conclusion

We propose FedDEK, a novel framework for Federated Domain-Incremental Learning (FDIL) that leverages a frozen pre-trained ViT backbone while dynamically extending it with domain-specific expert modules and expert knowledge transfer. This design effectively mitigates catastrophic forgetting and significantly reduces communication overhead, making it well-suited for IoT applications. Through extensive experiments on two benchmark datasets, FedDEK demonstrates superior accuracy, lower communication costs, and enhanced robustness across varying task orders in FDIL scenarios.

Acknowledgment. This work was supported by National Natural Science Foundation of China under Grant No. 62202224, Natural Science Foundation of Jiangsu Province under Grant No. BK20220882, China Postdoctoral Science Foundation under Grant No. 2022TQ0154, Open Foundation of Ministry Key Laboratory for Safety-Critical Software Development and Verification (Nanjing University of Aeronautics and Astronautics) under Grant No. NJ2024030, Dual Innovation Doctor Foundation of Jiangsu Province under Grant No. JSSCBS20220213, National Natural Science Foundation of China under Grant 62402215, and China Postdoctoral Science Foundation under Grant No. 2023M741685.

References

1. Ding, M., Li, J., Yi, C., Cai, J.: Federated learning for COVID-19 on heterogeneous CXR images with noise. In: Proceedings of IEEE International Conference on Communications (ICC), pp. 3413–3418 (2023)
2. Mao, Q., Wan, S., Hu, D., Yan, J., Hu, J., Yang, X.: Leveraging federated learning for unsecured loan risk assessment on decentralized finance lending platforms. In: IEEE International Conference on Data Mining Workshops (ICDMW), pp. 663–670 (2023)
3. McMahan, B., Moore, E., Ramage, D., Hampson, S., y Arcas, B.A.: Communication-efficient learning of deep networks from decentralized data. In: Proceedings of the 20th International Conference on Artificial Intelligence and Statistics (AISTATS), pp. 1273–1282 (2017)
4. Dong, J., Li, H., Cong, Y., Sun, X., Liu, Y.: No one left behind: real-world federated class-incremental learning. IEEE Trans. Pattern Anal. Mach. Intell. **46**(4), 2054–2070 (2023)
5. Yoon, J., Jeong, W., Lee, G., Yang, E., Hwang, S.J.: Federated continual learning with weighted inter-client transfer. In: Proceedings of the 38th International Conference on Machine Learning (ICML), pp. 12073–12086 (2021)

6. Shenaj, D., Toldo, M., Rigon, A., Barbantan, I.B., Zanuttigh, P.: Asynchronous federated continual learning. In: Proceedings of the IEEE/CVF Conference on Computer Vision and Pattern Recognition (CVPR), pp. 5055–5063 (2023)
7. Zhang, J., Chen, C., Zhuang, W., Lyu, L.: Target: federated class-continual learning via exemplar-free distillation. In: Proceedings of the IEEE/CVF International Conference on Computer Vision (ICCV), pp. 4782–4793 (2023)
8. Li, Y., Xu, W., Wang, H., Zhang, X., Liu, Y.: SR-FDIL: synergistic replay for federated domain-incremental learning. IEEE Trans. Parallel Distrib. Syst. (2024)
9. Li, Y., Xu, W., Wang, H., Zhang, X., Liu, Y.: Personalized federated domain-incremental learning based on adaptive knowledge matching. In: Proceedings of the European Conference on Computer Vision (ECCV), pp. 127–144 (2024)
10. Kirkpatrick, J., et al.: Overcoming catastrophic forgetting in neural networks. Proc. Natl. Acad. Sci. **114**(13), 3521–3526 (2017)
11. Rebuffi, S.A., Kolesnikov, A., Sperl, G., Lampert, C.H.: iCaRL: incremental classifier and representation learning. In: Proceedings of the IEEE Conference on Computer Vision and Pattern Recognition (CVPR), pp. 2001–2010 (2017)
12. Liu, Y., Schiele, B., Sun, Q.: RMM: reinforced memory management for class-incremental learning. In: Advances in Neural Information Processing Systems, vol. 34, pp. 3478–3490 (2021)
13. Mallya, A., Lazebnik, S.: PackNet: adding multiple tasks to a single network by iterative pruning. In: Proceedings of the IEEE Conference on Computer Vision and Pattern Recognition (CVPR), pp. 7765–7773 (2018)
14. Song, X., He, Y., Dong, S., Liu, Z., Liu, Y.: Non-exemplar domain incremental object detection via learning domain bias. In: Proceedings of the AAAI Conference on Artificial Intelligence (AAAI), pp. 15056–15065 (2024)
15. Jia, M., Tang, L., Chen, B.C., Cardie, C., Belongie, S., Hariharan, B., Lim, S.N.: Visual prompt tuning. In: Proceedings of the European Conference on Computer Vision (ECCV), pp. 709–727 (2022)
16. Wang, Z., et al.: Learning to prompt for continual learning. In: Proceedings of the IEEE/CVF Conference on Computer Vision and Pattern Recognition (CVPR), pp. 139–149 (2022)
17. Wang, Y., Huang, Z., Hong, X.: S-prompts learning with pre-trained transformers: an Occam's razor for domain incremental learning. In: Advances in Neural Information Processing Systems, vol. 35, pp. 5682–5695 (2022)
18. Bakman, D Y.F., Yaldiz, N., Ezzeldin, Y.H., Avestimehr, S.: Federated orthogonal training: Mitigating global catastrophic forgetting in continual federated learning. In: Proceedings of the 12th International Conference on Learning Representations (ICLR) (2024)
19. Li, Y., Li, Q., Wang, H., Zhang, X., Liu, Y.: Towards efficient replay in federated incremental learning. In: Proceedings of the IEEE/CVF Conference on Computer Vision and Pattern Recognition (CVPR), pp. 12820–12829 (2024)
20. Ma, Y., Xie, Z., Wang, J., Liu, Y., Liu, Y.: Continual federated learning based on knowledge distillation. In: Proceedings of the International Joint Conference on Artificial Intelligence, pp. 2182–2188 (IJCAI) (2022)
21. Venkateswara, H., Eusebio, J., Chakraborty, S., Panchanathan, S.: Deep hashing network for unsupervised domain adaptation. In: Proceedings of the IEEE Conference on Computer Vision and Pattern Recognition (CVPR), pp. 5018–5027 (2017)
22. Peng, X., Bai, Q., Xia, X., Huang, Z., Saenko, K., Wang, B.: Moment matching for multi-source domain adaptation. In: Proceedings of the IEEE International Conference on Computer Vision (ICCV), pp. 1406–1415 (2019)

23. Smith, J.S., et al.: Coda-prompt: continual decomposed attention-based prompting for rehearsal-free continual learning. In: Proceedings of the IEEE/CVF Conference on Computer Vision and Pattern Recognition (CVPR) (2023)
24. Wang, Z., et al.: DualPrompt: complementary prompting for rehearsal-free continual learning. In: Proceedings of the European Conference on Computer Vision (ECCV), pp. 631–648 (2022)
25. Dosovitskiy, A., et al.: An image is worth 16x16 words: transformers for image recognition at scale. In: Proceedings of the International Conference on Learning Representations (ICLR) (2021)
26. Dong, J., et al.: Federated class-incremental learning. In: 2022 IEEE/CVF Conference on Computer Vision and Pattern Recognition (CVPR), New Orleans, LA, USA, pp. 10154–10163 (2022)
27. Aljundi, R., Babiloni, F., Elhoseiny, M., Rohrbach, M., Tuytelaars, T.: Memory aware synapses: Learning what (not) to forget. In: Proceedings of the European Conference on Computer Vision, pp. 139–154 (ECCV) (2018)
28. Smith, J., Hsu, Y.-C., Balloch, J., Shen, Y., Jin, H., Kira, Z.: Always be dreaming: a new approach for data-free class-incremental learning. In: Proceedings of the IEEE/CVF International Conference on Computer Vision (ICCV), pp. 9374–9384 (2021)
29. Houlsby, N., et al.: Parameter-efficient transfer learning for NLP. In: Proceedings of the 36th International Conference on Machine Learning, PMLR, vol. 97, pp. 2790–2799 (2019)
30. Han, Z., et al.: Robust privacy-preserving federated learning framework for IoT devices. Int. J. Intell. Syst. **37**(11), 9655–9673 (2022)

An Accurate Indoor Depth Estimation Method Based on Iterative Pose Refinement

Yi Le[✉], Xiang Gao, and Hao Sun

Nanjing Institute of Electronic Engineering, Nanjing, China
lyy2023@whu.edu.cn

Abstract. Monocular depth estimation plays a pivotal role in critical domains such as autonomous driving and robot navigation, holding an extremely high application value. Particularly, research on monocular depth estimation using self-supervised deep learning imposes more flexible and lenient requirements on training datasets, enabling model training without actual depth information. However, compared with outdoor datasets dominated by translational motion, indoor datasets have more rotational motions. This will make it difficult to predict accurate camera poses, thereby affecting the accuracy of depth estimation. To address this problem, we propose an indoor depth estimation method based on iterative pose refinement, thereby improving the rotation processing ability through hierarchical optimization. Using the self-supervised architecture MonoDepth2, we creatively reconstruct the pose estimation module into a cascaded structure. On the one hand, an initial network is designed to predict camera transformations from adjacent frames; on the other hand, a residual network is constructed to iteratively correct rotation drift using photometric reprojection loss. Through this iterative fusion mechanism, the geometric constraints between frames are effectively strengthened. We built a prototype system and conducted experimental evaluations on the KITTI dataset and a self-built indoor dataset rich in rotation scenarios. The experimental results show that compared with the baseline method, our framework effectively reduces the error by 14.5%, significantly improving the robustness to complex rotation scenarios.

Keywords: Monocular Depth Estimation · Pose Refinement · Self-supervised Learning · MonoDepth2

1 Introduction

Depth estimation technology aims to accurately calculate the distance between each pixel point and the camera based on the content of the image, playing an indispensable role in core tasks such as scene reconstruction, target detection, and face recognition. Among them, monocular depth estimation stands out for its characteristic of being able to predict depth values using only a single image. Its remarkable advantage of having low requirements for input data makes it highly adaptable to various practical application scenarios. In recent years, benefiting from the vigorous development of deep learning, monocular depth estimation technology has made remarkable progress. While in the supervised learning framework, each RGB image requires a corresponding depth label, and the acquisition of depth labels often relies on professional devices such as depth cameras or lidar. Additionally, even when depth labels are obtained, they are typically sparse and difficult to precisely match with the original images. In view of this, self-supervised depth estimation methods that do not require depth labels have become a research hotspot in this field.

Most self-supervised monocular depth estimation methods essentially fall into the image generation paradigm, where a target image from one viewpoint is synthesized from a source image of another viewpoint. The core relies on two subnetworks: one is a depth network for predicting the depth value of each pixel, and the other is a pose network for estimating the camera pose variations between adjacent images [1–4]. Through the collaboration of these two subnetworks, new view images can be synthesized based on the source image, and network parameters are optimized according to the photometric reprojection error between the synthesized image and the target image to make them as consistent as possible. Once the image generation network is trained, the depth estimation result of the monocular image can be obtained. Evidently, the accuracy of depth prediction is closely related to the accuracy of pose estimation. However, traditional methods have low accuracy in estimating camera rotation, which directly affects the accuracy of depth estimation. Due to the dense objects and limited depth range in indoor environments, camera rotations are inevitable during image capture, leading to significant pose estimation errors and subsequent depth estimation distortion. If precise estimation of camera motion poses in indoor scenarios can be achieved, the accuracy of indoor depth estimation is expected to improve significantly.

In this paper, we propose an accurate indoor depth estimation method based on iterative pose refinement. As a pioneering framework for self-supervised monocular depth estimation, MonoDepth2 achieved landmark performance on outdoor datasets [5], motivating us to optimize it for exploring the potential of indoor scenarios.

To realize our goals, we encounter two key challenges. The first challenge lies in accurately estimating camera pose within indoor environments. Unlike outdoor scenarios, indoor spaces are cluttered with objects and spatially constrained, necessitating frequent and substantial camera rotations to capture diverse objects and scenes. These pronounced rotations are further accentuated by the limited depth range in indoor settings, posing significant challenges to conventional pose estimation methods. While some researchers have attempted to mitigate rotation effects through data preprocessing, such approaches often compromise flexibility and generalization capabilities, failing to address the root cause of inaccurate pose estimation. To address this, we introduce an iterative pose refinement method that operates directly on raw input data without preprocessing. This module iteratively optimizes the relative camera pose between the target and source images via generating synthetic views based on the current pose estimate and computing residual poses against the target image. By progressively minimizing these residuals, it refines pose estimation accuracy, leading to substantial improvements in both synthetic view quality and depth estimation.

The second challenge lies in how to modify MonoDepth2 to better adapt to depth estimation in indoor scenarios. MonoDepth2 employs ResNet18 as the encoder, which performs excellently in open outdoor scenes. However, the complexity of indoor environments places higher demands on feature extraction. To address this, we replace the encoder with the ResNet50 network. This 50-layer convolutional architecture, through its deeper hierarchical design and bottleneck structures, can extract richer multi-scale contextual features. It exhibits stronger representational abilities particularly for small object details and low-texture regions frequently encountered in indoor scenes. The improved network takes image pairs (including the target image) as input, generates disparity maps at four scales through an encoding-decoding process, and performs reprojection using camera poses estimated by an initial pose network to generate synthetic views. Building upon this, the iterative pose refinement module dynamically optimizes pose parameters, gradually eliminating photometric inconsistencies caused by rotational estimation errors and ultimately achieving more accurate indoor depth estimation.

This paper makes three main contributions. First, we propose an iterative pose refinement method that addresses the problem of inaccurate rotation estimation by iteratively computing residual poses. Second, we improve the classic self-supervised framework by adopting a deeper encoder to fully extract detailed features of indoor scenes, and combining an initial pose module with an iterative pose refinement module to enhance the depth feature extraction. Finally, we built a prototype system and conducted evaluations on public and self-built datasets. Experiments show that our method reduces the absolute relative error and root mean square error of depth prediction by 11.8% and 14.5%, respectively.

2 Related Work

2.1 Indoor Depth Estimation

Depth estimation aims to obtain the distance information between objects in a scene and the camera, and it is widely used in driving scenarios. However, when faced with indoor environments, the accuracy often decreases. First, indoor depth values usually have an irregular span, and it is difficult for the depth network to extract consistent cues. Second, the indoor space is limited, and the camera motion is more arbitrary. Third, moving objects may block the background or other objects, and the camera rotation changes the perspective, making depth estimation need to consider more dynamic factors. To solve the problem of indoor depth estimation, there are already various methods [1,6–12]. Such as traditional algorithms based on structured light or photometric stereo, as well as multi-sensor fusion solutions. Among them, the indoor depth estimation method based on a monocular camera is widely used due to its advantages like low cost, flexible deployment, and strong adaptability to self-supervised learning.

2.2 Monocular Depth Estimation with Deep Learning

The existing learning-based depth estimation methods mainly fall into four categories. The first category is based on unsupervised learning methods. Bian et al. [12] propose an unsupervised monocular depth estimation method that automatically localizes moving objects through geometric consistency loss and self-discovered masks. However, feature extraction in this method is highly challenging. The second category is based on supervised learning, which require large-scale datasets with ground-truth maps collected from active depth sensors, such as infrared sensors [2] or LiDAR [13]. Guizilini et al. [14] mention that although these methods naturally produce metric predictions, they suffer from sparsity and high noise levels in ground-truth data training, as well as limited scalability due to the need for dedicated hardware and calibration. The third category explores the idea of training weakly supervised methods. 3D reconstruction methods are widely adopted, such as using 3D videos [4,15–18] to generate virtual depth data for network training. Spurr et al. [3] propose a semi-supervised and weakly supervised method for hand pose estimation, which can obtain large amounts of data at low cost and has better generalization ability, but may not achieve optimal performance in existing datasets. The fourth category is based on self-supervised depth estimation, which has evolved from early stereo-based methods [19] to monocular video frameworks minimizing photometric loss. Compared with the other three categories, self-supervised depth estimation methods exhibit significant advantages in hardware cost, data acquisition, and model generalizability, making them highly suitable for practical applications. Recent improvements achieve performance comparable to supervised methods through three key extensions: surface normal integration [20], semantic-guided optimization [21], as well as geometric constraint enhancements via improved pose networks [22,23] and direct geometry aids [24]. Although these improvements enhance robustness against low-texture and complex poses from different aspects, they fail to address the fundamental issue of inaccurate rotational pose estimation, resulting in limited robustness.

Overall, taking the typical self-supervised depth estimation method MonoDepth2 as an example, this paper focuses on tackling the problem of the pose network in accurately estimating camera rotation, thereby improving the prediction accuracy of the depth network.

3 System Design

The proposed monocular depth estimation method employs a self-supervised learning paradigm, leveraging the technical framework of MonoDepth2 to optimize the training performance of self-supervised learning. An iterative camera pose refinement mechanism is then introduced to generate synthetic views. The architecture is illustrated in Fig. 1.

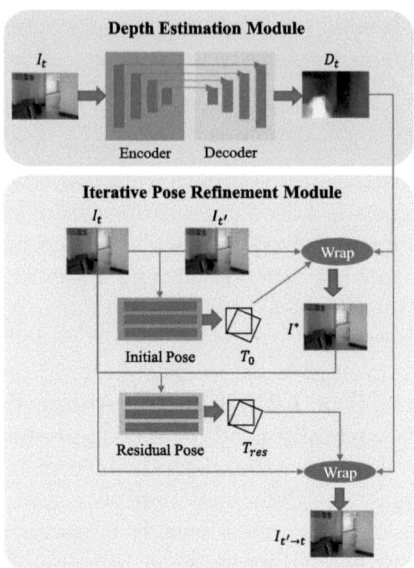

Fig. 1. Architecture of accurate indoor depth estimation

3.1 Depth Estimation Module

In this section, for ease of understanding, we first introduce the self-supervised training and three key techniques of MonoDepth2, and then further present the depth estimation module based on MonoDepth2.

Self-Supervised Training. By inputting a pair of images with parallax (in monocular training, this is a pair of adjacent frames), the pose network can estimate an initial camera transformation pose. After obtaining the camera pose, we can synthesize a virtual view I^* by combining the target image depth map obtained in the first step and the source image.

The self-supervised depth prediction network trains a neural network to predict the appearance of a target image from the perspective of another image (the source image). Given a target image I_t and a source image I'_t from a different viewpoints, it first predicts the relative camera pose $T_{t \to t'}$ from the target image I_t to the source image I'_t. Then, it obtains a dense depth map of the target image D_t through a pair of encoder and decoder. At this point, the learning problem can be formulated as minimizing the photometric reprojection error during training, where the photometric reprojection loss can be constructed as follows:

$$L_p = \sum_{t'} pe(I_{t'}, I_{t' \to t}), \qquad (1)$$

$$I_{t \to t'} = I'_t \langle proj(D_t, T_{t \to t'}, K) \rangle, \qquad (2)$$

where pe represents the photometric reprojection error using the L1 norm and Structural Similarity Index Measure(SSIM) [25]. pe is a weighted combination, as:

$$pe(I_a, I_b) = \frac{\alpha}{2}(1 - SSIM(I_a, I_b)) + (1 - \alpha)\|I_a - I_b\|_1. \quad (3)$$

The transformation function $proj()$ in Eq. 2 is utilized to reproject the image coordinates pt of the target image I_t to the image coordinates of the source image I_t' using the depth map D_t, the camera pose $T_{t \to t'}$ and the intrinsics K of the camera [26]:

$$p_{t'} \sim K T_{t \to t'} D_t(p_t) K^{-1} p_t. \quad (4)$$

The operator $\langle \rangle$ in Eq. 2 represents bilinear sampling. It is a synthesized view of the source image transformed into the target coordinates based on the camera pose and intrinsics according to the depth map of the target image. MonoDepth2 employs edge-aware smoothness [27]:

$$L_s = |\partial_x d_t^*| e^{-|\partial_x I_t|} + |\partial_y d_t^*| e^{-|\partial_y I_t|}, \quad (5)$$

where $d_t^* = \frac{d_t}{\bar{d}_t}$. d_t^* denotes the mean-normalized inverse depth, preventing the contraction of estimated depths.

Key Techniques of MonoDepth2. We use the MonoDepth2 autoencoder as the base network for the depth prediction module, which consists of a pair of encoder and decoder. The encoder can generate the corresponding depth feature information from RGB images. We use the ResNet50 network as its backbone to obtain more in-depth feature information. The decoder then converts these features into the corresponding depth map. In monocular sequence training, the relative pose must be predicted by a pose network [25] for use in the reprojection function. The training process involves solving for the relative camera pose T and depth D to minimize the reprojection error L_p. MonoDepth2 have made three improvements to this self-supervised learning framework: pixel-wise minimum reprojection error loss, automatic filtering of stable pixels, and multi-scale estimation enhancement, achieving high-quality results on KITTI dataset.

Pixel-wise Minimum Reprojection Error Loss: In existing self-supervised depth prediction methods, the reprojection errors computed from multiple source images are typically averaged. However, this averaging operation may lead to errors when pixels visible in the target image are occluded or out of view in some source images. This problem arises in two scenarios: first, pixels located near the image boundaries may move out of the current image's field of view due to inter-frame motion; second, pixels may become occluded due to changes in spatial relationships between frames. When these cases occur, the reprojection error computed between the target image and a source image where the pixel is occluded tends to be large, while other source images without occlusion yield smaller reprojection errors. Averaging these errors results in an overall inflated photometric error. MonoDepth2 addresses this issue by replacing the average with a per-pixel minimum reprojection error, i.e., instead of averaging, it simply takes the minimum reprojection error across source images. Using the minimum value better reflects the actual situation. Therefore, the final per-pixel photometric loss is formulated as:

$$L_p = \min_{t'} pe(I_t, I_{t' \to t}). \tag{6}$$

Automatic Filtering of Stable Pixels: In self-supervised monocular sequence training, effectiveness can be compromised when the camera is stationary or when moving objects appear stationary relative to the camera, leading to incorrect infinite depth predictions. MonoDepth2 addresses this by using an automatic masking technique to filter out unchanged pixels. It applies a per-pixel mask μ to weight pixels, while MonoDepth2 improves upon this by using a binary mask, that is $\mu \in \{0, 1\}$, automatically computed during the forward pass of the neural network. The mask only includes pixels where the reprojection error on the synthesized image $I_{t \to t'}$ is less than that on the source image I_t', and μ is:

$$\mu = \begin{cases} 1 & \text{if } \min_t pe(I_t, I_{t \to t'}) < \min_t pe(I_t, I_{t' \to t}), \\ 0 & \text{otherwise.} \end{cases} \tag{7}$$

Multi-scale Estimation: MonoDepth2 decouples the resolution of the disparity images used for computing reprojection errors from that of the color images. The main approach involves upsampling low-resolution depth maps to the input image size, performing reprojection and resampling at this resolution, and then calculating the errors pe. This process constrains depth maps across all scales, guiding them to reconstruct consistently in one direction.

Combining the above three improvements, the total loss function of MonoDepth2 can be written as:

$$L = \mu L_p + \lambda L_s, \tag{8}$$

where L_p is photometric loss and L_s is smoothness. MonoDepth2 is based on a U-Net structure with an encoder-decoder architecture. The encoder uses ResNet18 initialized with ImageNet weights for efficient feature extraction, while the decoder employs a sigmoid activation function in the output layer and ELU (Exponential Linear Unit) elsewhere, enabling the model to capture both deep-level features and local information effectively. In the process of converting the sigmoid output to depth, the formula is:

$$D = \frac{1}{a\sigma + b}, \tag{9}$$

where σ is the output of sigmoid. The values of a and b represent the predefined minimum and maximum depth values in the environment, respectively. The pose network is constructed using a ResNet18, and the pose estimation follows the design in [28], employing axis-angle representation for rotation prediction. Both translation and rotation are scaled down by a factor of 0.01. The pose network takes a pair of color images as input, totaling 6 channels, and then predicts the relative pose between the images in 6 degrees of freedom.

Depth Estimation Based on MonoDepth2. This module is a network designed to predict the depth value of each pixel. It is based on the common U-Net architecture, which is an autoencoder consisting of an encoder and a decoder. The encoder can extract depth information features from the RGB image input, and the decoder can reconstruct the depth image from depth information features. The encoder is replaced with the ResNet50 network, which is a deeper residual network with 50 layers of fully connected and weighted convolutional layers. It can extract more depth information features, thereby improving the performance of indoor depth prediction.

In each computation of the input, we will obtain a set of images stored in the dataset $inputs["color"]$. First, the encoder extracts the images into feature information. This encoder extracts information at four different scales for use in subsequent steps. Then, this set of feature information is passed through the decoder to obtain a set of 4 scales of disparity maps $outputs["disp"]$. Next, the initial pose network calculates a set of rotation vectors and translation vectors for camera transformation by inputting the image and its adjacent frame. The transformation matrix of the camera can be obtained through Eq. 10 and placed in $outputs["axisangle", "translation", "cam\_T\_cam"]$.

$$\begin{bmatrix} \cos\theta + R_z^2(1-\cos\theta) & R_z(1-\cos\theta) - R_z\sin\theta & R_zR_y(1-\cos\theta) + R_y\sin\theta & 0 \\ R_yR_z(1-\cos\theta) + R_z\sin\theta & \cos\theta + R_y^2(1-\cos\theta) & R_yR_z(1-\cos\theta) + R_z\sin\theta & 0 \\ R_zR_y(1-\cos\theta) - R_y\sin\theta & R_zR_y(1-\cos\theta) + R_z\sin\theta & \cos\theta + R_z^2(1-\cos\theta) & 0 \\ 0 & 0 & 0 & 1 \end{bmatrix}. \tag{10}$$

After obtaining the camera pose, a reprojection operation is performed according to Eq. 2 to obtain a new target image and update the output dictionary $outputs["depth", "translation", "cam\_T\_cam"]$. If the iterative pose refinement module is activated, the algorithm for iterative pose refinement is entered (which will be introduced in the next subsection). Finally, the loss value $losses$ of this is calculated by calling the loss function according to Eq. 5. Then, it returns $losses$ and $outputs$ to the deep learning framework to perform optimization and backpropagation to adjust the model. The corresponding pseudocode is shown in Algorithm 1.

3.2 Iterative Pose Refinement Module

$$L_p = \min_t pe(I_t, I_{t \to t'}) \tag{11}$$

The iterative pose refinement module is the core part of the proposed system, which can effectively improve the prediction accuracy of the pose network, thus improving the quality of the new view and the performance of the depth estimation.

The iterative pose refinement module consists of a standard pose network and a residual pose network. As illustrated in Fig. 2, in the first stage, the target image I_t and the source image $I_{t'_0}$ are input into the pose network to predict the initial pose $T_{t'_0 \to t}$. Using the initial pose $T_{t'_0 \to t}$, the depth map of the target

Algorithm 1. Operations performed for each input

Require: Input dataset: containing a target image;
1: $feature \leftarrow$ encoder encodes the target image;
2: $outputs \leftarrow$ decoder decodes $feature$;
3: $outputs \leftarrow$ initial pose network predicts pose;
4: $I_{t \to t} = I_{t'}(proj(D_t, T_{L \to t'}, K))$; // reproject to generate synthesized view
5: **if** pose iteration is enabled **then**
6:   call pose iteration algorithm to update synthesized view and pose;
7: **end if**
8: $losses \leftarrow \mu L_p + \lambda L_s$;
9: **return** $losses, outputs$;

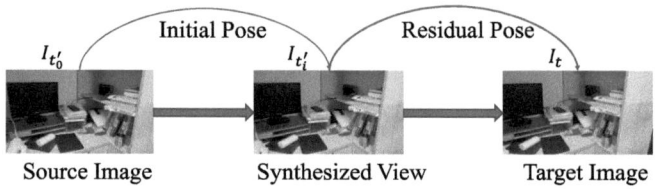

Fig. 2. Illustration of iterative pose refinement

image D_t, and the intrinsic camera parameters K, a synthesized view $I_{t'_0 \to t}$ can be reconstructed according to Eq. 2. This process can be expressed as the following equation:

$$I_{t'_0 \to t} = I'_t \left\langle proj\left(D_t, T^{-1}_{t'_0 \to t}, K\right)\right\rangle. \tag{12}$$

If the camera pose were accurate, then the synthesized view $I_{t'_0 \to t}$ should match the target image I_t. However, as determined in the above analysis, pose prediction is unreliable. Therefore, we also need to continue solving the accurate camera pose through the residual pose network.

Similar to the first stage, in the second stage, the target image I_t and the synthesized view $I_{t'_0 \to t}$ obtained from the first stage are input into the residual pose network $T^{res}_{(t'_0 \to t) \to t}$ to output a residual camera pose, which represents the relative camera transformation from the synthesized view $I_{t'_0 \to t}$ to the target image I_t. At this point, by performing bilinear sampling on the synthesized view again, a new synthesized view can be obtained, thus allowing the next iteration prediction to continue. This process can be defined by the following equation:

$$I_{t'_i \to t} = I_{t_i'} \left\langle proj\left(D_t, T^{res^{-1}}_{t'_i \to t}, K\right)\right\rangle, i = 0, 1, \cdots, \tag{13}$$

where i refers to the number of iterations.

After multiple iterations, an initial pose and several residual poses can be obtained, which means multiple rotation matrices have been acquired. To derive the total pose, all the poses need to be multiplied together [29]:

$$T_{t' \to t} = \prod_i T_{t'_i \to t}, i = \cdots, k, \cdots, 1, 0. \tag{14}$$

The corresponding pseudocode of its iterative algorithm is shown in Algorithm 2, which improves training effects.

Algorithm 2. Iterative pose refinement

Require: Camera intrinsics K, source/target images $\{I_s, I_t\}$ at multiple scales, output dictionary, initial pose T_0, synthesized view I_{syn};
1: **for** each of the 2 source images **do**
2:   // Calculate residual pose
3:   $D' \leftarrow \text{Decode}(I_{\text{syn}})$;
4:   $T_{\text{res}} \leftarrow f_{\text{pose}}(I_s, I_{\text{syn}}; \theta)$;
5:   // Reproject to generate new synthetic view
6:   **for** all 4 scales **do**
7:     $I_{t'_i \to t} = I_{t_i'} \left\langle proj \left(D_t, T_{t'_i \to t}^{\text{res}}{}^{-1}, K\right)\right\rangle, i = 0, 1, \cdots$;
8:   **end for**
9:   Update pose: $T_{t' \to t} \leftarrow \prod_i T_{t'_i \to t}, i = \cdots, k, \cdots, 1, 0$;
10: **end for**

The iterative pose refinement module is the fundamental for addressing the inaccurate rotation estimation. As shown in Fig. 3, the pair of target and source images will be input into the original standard pose network, resulting in two initial pose estimations, the former frame I_{-1} and the latter frame I_{+1}. Firstly, the pair of target and source images are fed into the original standard pose network, yielding two initial pose estimations, $T_{-1 \to 0}$ and $T_{0 \to +1}$ respectively. Then, based on these two initial pose estimations, the target image can be transformed to match the viewpoint of the source images, thereby generating two synthesized views. These are the synthesized view from the current frame to the previous frame I'_{-1} and the synthesized view from the current frame to the next frame I'_{+1}. Next, these two synthesized views serving as the target images in the next iteration, seek the residual pose with respect to the source images. In order to calculate the residual pose, the depth maps of the two synthesized views D'_{-1} and D'_{+1} need to be calculated again.

4 Experiment and Results

4.1 Experiment Setup

Dataset, Hardware and Software. The experiments are conducted on a laptop equipped with an Intel Core i5-11320H processor, using Python 3.7 and

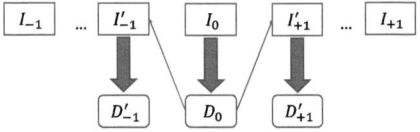

Fig. 3. View synthesis process

PyTorch 1.11.0 as the software environment. To comprehensively evaluate the effectiveness of the proposed method, two datasets are used: the publicly available KITTI dataset and a self-built indoor dataset, which together cover both outdoor and indoor scenarios.

KITTI Dataset: It is collected synchronously using a high-resolution RGB camera and a 3D laser scanner [13]. It includes a variety of outdoor traffic scenes, such as urban, rural, and highway environments. This dataset serves as the training set to verify the effectiveness of the proposed iterative pose refinement module. As KITTI dataset provides accurate ground-truth depth information, it enables a quantitative evaluation of the performance. If the proposed iterative pose refinement module achieves good results on outdoor scenes, it further demonstrates the validity of the improvements described in our work.

Self-Built Indoor Dataset: This self-built dataset consists of video sequences recorded in the public areas of student dormitories. These videos include scenarios such as pure translation, pure rotation, and a combination of both. The videos are recorded at a frame rate of 30 frames per second, and each frame is exported as a sequentially numbered JPG image using Python scripts. In total, more than 3,000 monocular images are obtained and divided into several segments. The resolution of the videos is 640 × 480 pixels. Figure 4 presents several sample images from the self-built dataset.

The camera intrinsics for these two datasets are as follows:

$$\begin{pmatrix} 0.58 & 0 & 0.5 & 0 \\ 0 & 1.92 & 0.5 & 0 \\ 0 & 0 & 1 & 0 \\ 0 & 0 & 0 & 1 \end{pmatrix}, \begin{pmatrix} 1.59 & 0 & 1.17 & 0 \\ 0 & 1.14 & 0.53 & 0 \\ 0 & 0 & 1 & 0 \\ 0 & 0 & 0 & 1 \end{pmatrix}.$$

Evaluation Metrics. We use the error metrics and the accuracy metrics from [30] to evaluate the model results in this data set, including the mean Absolute Relative error (AbsRel), the Root Mean Square error (RMS) and the accuracy at various thresholds (specifically, δ_1, δ_2 and δ_3). The AbsRel metric normalizes the per-pixel error by the actual depth to reduce the impact of large errors, and its calculation formula is:

$$\text{AbsRel} = \frac{1}{N} \sum \frac{|d_i - d_i^*|}{d_i}. \tag{15}$$

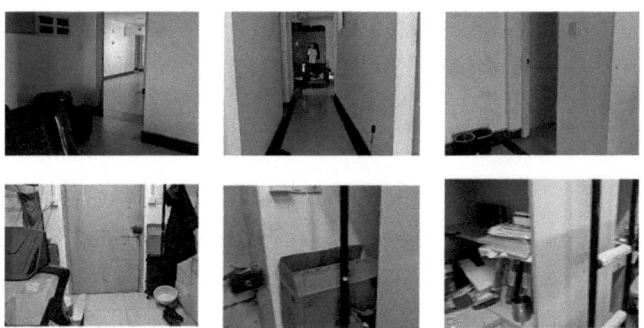

Fig. 4. Self-built indoor dataset samples

RMS, a traditional metric for measuring regression error, is calculated using the following formula:

$$\text{RMS} = \sqrt{\frac{1}{N} \sum |d_i - d_i^*|^2}, \tag{16}$$

where d_i represents the actual depth information, and d_i^* represents the predicted depth value. Both *AbsRel* and *RMS* are error metrics where smaller values indicate lower errors and thus better model performance.

Regarding accuracy, there are three commonly used thresholds $thr = 1.25, 1.25^2, 1.25^3$ for depth estimation, corresponding to δ_1, δ_2, and δ_3, which are defined in [8] as: *Accuracy with threshold* $(\delta < thr)$: % *of* d_i *such that* $\max\left(\frac{d_i}{d_i^*}, \frac{d_i^*}{d_i}\right) < thr, where\ thr = 1.25, 1.25^2, 1.25^3$ That is, select the larger of $\frac{d_i}{d_i^*}$ and $\frac{d_i^*}{d_i}$ such that it is less than the threshold d_i, and calculate the percentage of all such pixel points relative to the total number of pixel points. The closer this percentage is to 1, the better the performance. Therefore, the larger the threshold, the higher the accuracy.

4.2 Experiment Analysis

Performance of Iterative Pose Refinement Module. To evaluate the impact of the iterative pose refinement module, we conduct three training experiments on the KITTI dataset: without enabling the iterative pose refinement module, enabling it once, and enabling it multiple times. The number of training epochs is set to 40, with a learning rate of 10^{-4}. The maximum and minimum depth values in Eq. 9 are set to 100 and 0.01, respectively. This experiment aims to verify the feasibility of the iterative pose refinement module. The KITTI dataset provides accurate depth ground truth, enabling quantitative assessment of the model's performance. As shown in Fig. 5, introducing the iterative pose refinement module significantly reduces *AbsRel* and *RMS*, and improves accuracy across multiple depth thresholds. Repeated training confirms the stability of these results, ruling out random errors and ensuring reliability. However, the results also indicate that increasing the number of iterations beyond one does not lead to further performance improvements; instead, it may cause a slight decline. Therefore, a single iteration of pose estimation is adopted as the final approach.

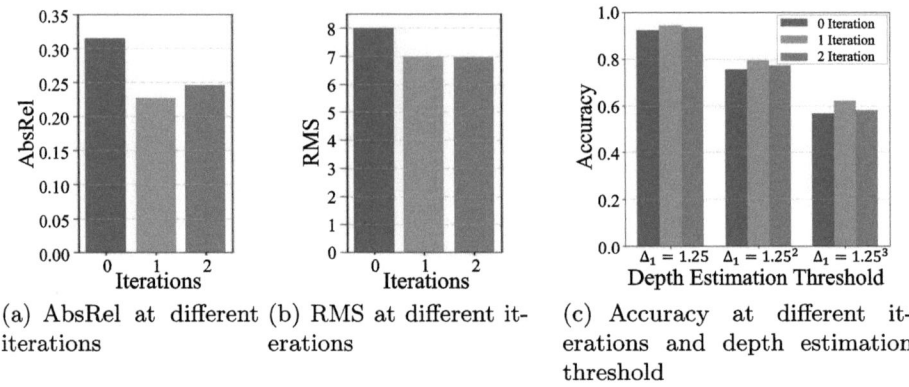

(a) AbsRel at different iterations (b) RMS at different iterations (c) Accuracy at different iterations and depth estimation threshold

Fig. 5. Quantitative evaluation with KITTI dataset

We further conduct training on the self-built indoor dataset with and without the iterative pose refinement module enabled. The training parameters are consistent with the KITTI experiments: 40 epochs and a learning rate of 10^{-4}. Considering the indoor environment, the maximum and minimum depth values in Eq. 9 are set to 10 and 0.01, respectively. The trained models are then tested on sample images to observe depth prediction results. Due to limited computational resources and the relatively small scale of the self-built dataset, the resulting depth maps are not optimal. However, the model that incorporates the iterative pose refinement module consistently outperforms the model without it, demonstrating the effectiveness of the module in enhancing self-supervised monocular depth estimation (Table 1).

Performance of Depth Network. As mentioned in Eq. 11, the proposed method uses a pair of minimum and maximum depth values to constrain the output depth map. Therefore, we set up a set of comparative experiments to verify that reducing the maximum depth value can improve the model's performance in indoor environments. We conducted several training sessions on our self-built indoor dataset, setting different maximum depth constraints at 100, 10, and 5, with the number of epochs set to 20 and the learning rate set to 10^{-4}. We enabled iterative pose refinement once and observed its prediction performance.

Table 1. Comparison of depth estimation results with or without iterative pose refinement module on two datasets

| Input | Self-built dataset | Self-built dataset + Iteration | KITTI | KITTI + Iteration |
|---|---|---|---|---|ування
| | | | | |
| | | | | |

We tested the depth prediction performance of three models, as shown in Table 2. After observing several groups of images, we found that when the original *maxdepth* was set to 100, the resulting depth maps were blurrier, with many details obscured. In contrast, the model with *maxdepth* set to 5 had more details, but its accuracy of depth was slightly worse on a few test images. The model with a *maxdepth* of 10, which was relatively balanced in terms of details and accuracy. Therefore, the setting of *maxdepth* has an impact on the model, and is related to the dataset. When the dataset consists of indoor scenes with shallow depth of field, setting this value lower can be more beneficial for model training.

Experiment Summary. Through the analysis of a series of performance metrics, it is demonstrated that the proposed iterative pose refinement method is effective for depth prediction. Specifically, on the KITTI dataset, the model achieved an absolute relative error of 0.227 and precision on three scales $\delta_1 = 0.622, \delta_2 = 0.976, \delta_3 = 0.944$, both of which are superior to the performance obtained by MonoDepth2 under the same training conditions. It is difficult to obtain an accurate groundtruth in indoor scenes. Therefore, the KITTI dataset is used for testing. By comparing the effects of depth prediction maps, we can demonstrate the results of the model. It was also found that performing the iterative residual pose calculation once can achieve a significant improvement, while multiple iterations do not yield much enhancement and may even lead to a decrease. Furthermore, the impact of the value on indoor depth prediction was verified; a value of 10 is relatively appropriate for indoor environments.

Table 2. Comparison of depth estimation results under different max depth values

Input	Max depth: 5	Max depth: 10	Max depth: 100

5 Conclusion

In traditional self-supervised monocular depth estimation methods, pose estimation and depth estimation are highly coupled. Frequent camera rotations in indoor environments lead to significant pose estimation errors, which in turn cause cumulative errors in depth estimation. To address this challenge, we take the classic framework MonoDepth2 as a baseline and design an iterative pose refinement module to improve rotation estimation accuracy. Additionally, we enhance the depth network to better extract detailed features of indoor scenes, thereby significantly improving the accuracy and robustness of indoor monocular depth prediction. We have developed a prototype system and conducted performance evaluations on public and self-collected datasets, with experimental results fully validating the effectiveness of the proposed method.

References

1. Sun, L., Bian, J.-W., Zhan, H., Yin, W., Reid, I., Shen, C.: SC-depthv3: robust self-supervised monocular depth estimation for dynamic scenes. IEEE Trans. Pattern Anal. Mach. Intell. **46**(1), 497–508 (2023)
2. Silberman, N., Hoiem, D., Kohli, P., Fergus, R.: Indoor segmentation and support inference from RGBD images. In: Fitzgibbon, A., Lazebnik, S., Perona, P., Sato, Y., Schmid, C. (eds.) ECCV 2012. LNCS, vol. 7576, pp. 746–760. Springer, Heidelberg (2012). https://doi.org/10.1007/978-3-642-33715-4_54

3. Spurr, A.: Semi-and Weakly-Supervised Methods for 3D Hand Pose Estimation from Monocular RGB. PhD thesis, ETH Zurich (2023)
4. Cong, P., et al.: Weakly supervised 3D multi-person pose estimation for large-scale scenes based on monocular camera and single lidar. In: Proceedings of the AAAI Conference on Artificial Intelligence, vol. 37, pp. 461–469 (2023)
5. Godard, C., Aodha, O.M., Firman, M., Brostow, G.J.: Digging into self-supervised monocular depth estimation. In: Proceedings of the IEEE/CVF International Conference on Computer Vision, pp. 3828–3838 (2019)
6. Ji, P., Li, R., Bhanu, B., Xu, Y.: Monoindoor: towards good practice of self-supervised monocular depth estimation for indoor environments. In: Proceedings of the IEEE/CVF International Conference on Computer Vision, pp. 12787–12796 (2021)
7. Wu, C.Y., Wang, J., Hall, M., Neumann, U., Su, S.: Toward practical monocular indoor depth estimation. In: Proceedings of the IEEE/CVF Conference on Computer Vision and Pattern Recognition, pp. 3814–3824 (2022)
8. Gui, M., et al.: Depthfm: fast monocular depth estimation with flow matching. arXiv preprint arXiv:2403.13788 (2024)
9. Dong, L., Ren, Q., Shi, J., Zhang, B.: Bigeodepth: leveraging bi-geometric priors for unsupervised monocular depth estimation in indoor environments. IEEE Trans. Consum. Electron. 1–1 (2025)
10. Cheng, A., Yang, Z., Zhu, H., Mao, K.: Gam-depth: self-supervised indoor depth estimation leveraging a gradient-aware mask and semantic constraints. In: 2024 IEEE International Conference on Robotics and Automation (ICRA), pp. 5367–5374 (2024)
11. Wei, C., Yang, M., He, L., Zheng, N.: FS-depth: focal-and-scale depth estimation from a single image in unseen indoor scene. IEEE Trans. Circuits Syst. Video Technol. **34**(11), 10604–10617 (2024)
12. Bian, J.-W., et al.: Unsupervised scale-consistent depth learning from video. Int. J. Comput. Vision **129**(9), 2548–2564 (2021)
13. Geiger, A., Lenz, P., Urtasun, R.: Are we ready for autonomous driving? the kitti vision benchmark suite. In: 2012 IEEE Conference on Computer Vision and Pattern Recognition, pp. 3354–3361. IEEE (2012)
14. Guizilini, V., Vasiljevic, I., Chen, D., Ambruş, R., Gaidon, A.: Towards zero-shot scale-aware monocular depth estimation. In: Proceedings of the IEEE/CVF International Conference on Computer Vision, pp. 9233–9243 (2023)
15. Jiang, X., Jin, S., Lu, L., Zhang, X., Lu, S.: Weakly supervised monocular 3D detection with a single-view image. In: Proceedings of the IEEE/CVF Conference on Computer Vision and Pattern Recognition, pp. 10508–10518 (2024)
16. Peng, L., Yan, S., Wu, B., Yang, Z., He, X., Cai, D.: Weakm3D: towards weakly supervised monocular 3D object detection. arXiv preprint arXiv:2203.08332 (2022)
17. Li, W., et al.: Weakly-supervised 3D building reconstruction from monocular remote sensing images. IEEE Trans. Geosci. Remote Sens. (2024)
18. Han, W., Tao, R., Ling, H., Shen, J.: Weakly supervised monocular 3D object detection by spatial-temporal view consistency. IEEE Trans. Pattern Anal. Mach. Intell. (2024)
19. Garg, R., B.G., V.K., Carneiro, G., Reid, I.: Unsupervised CNN for single view depth estimation: geometry to the rescue. In: Leibe, B., Matas, J., Sebe, N., Welling, M. (eds.) ECCV 2016. LNCS, vol. 9912, pp. 740–756. Springer, Cham (2016). https://doi.org/10.1007/978-3-319-46484-8_45

20. Zhan, H., Weerasekera, C.S., Garg, R., Reid, I.D.: Self-supervised learning for single view depth and surface normal estimation. corr, abs/1903.00112 (2019). arXiv preprint arxiv:1903.00112
21. Zhang, Y., et al.: Cid-sims: complex indoor dataset with semantic information and multi-sensor data from a ground wheeled robot viewpoint. Int. J. Robot. Res. **43**(7), 899–917 (2024)
22. Chen, J., Li, S., Liu, D., Weisheng, L.: Indoor camera pose estimation via style-transfer 3D models. Comput. Aided Civ. Infrastruct. Eng. **37**(3), 335–353 (2022)
23. Tang, S., et al.: Transcnnloc: end-to-end pixel-level learning for 2D-to-3D pose estimation in dynamic indoor scenes. ISPRS J. Photogramm. Remote. Sens. **207**, 218–230 (2024)
24. Zhao, C., et al.: Gasmono: geometry-aided self-supervised monocular depth estimation for indoor scenes. In: Proceedings of the IEEE/CVF International Conference on Computer Vision (ICCV), pp. 16209–16220 (2023)
25. Zhou, T., Brown, M., Snavely, N., Lowe, D.G: Unsupervised learning of depth and ego-motion from video. In: Proceedings of the IEEE Conference on Computer Vision and Pattern Recognition (CVPR) (2017)
26. Li, R., Ji, P., Yi, X., Bhanu, B.: Monoindoor++: towards better practice of self-supervised monocular depth estimation for indoor environments. IEEE Trans. Circuits Syst. Video Technol. **33**(2), 830–846 (2023)
27. Godard, C., Aodha, O.M., Brostow, G.J.: Unsupervised monocular depth estimation with left-right consistency. In: Proceedings of the IEEE Conference on Computer Vision and Pattern Recognition (CVPR) (2017)
28. Wang, C., Buenaposada, J.M., Zhu, R., Lucey, S.: Learning depth from monocular videos using direct methods. In: Proceedings of the IEEE Conference on Computer Vision and Pattern Recognition (CVPR) (2018)
29. Li, R., Ji, P., Yi, X., Bhanu, B.: Monoindoor++: towards better practice of self-supervised monocular depth estimation for indoor environments. IEEE Trans. Circuits Syst. Video Technol. **33**(2), 830–846 (2022)
30. Eigen, D., Puhrsch, C., Fergus, R.: Depth map prediction from a single image using a multi-scale deep network. In: Advances in Neural Information Processing Systems, vol. 27 (2014)

A Dynamic Stress Assessment Framework via Multi-scale Feature Fusion

Kang Yu[1]($\boxtimes$), Wenjing Hu[1], Meng Tian[1,2], Peng Tian[1,2], Jun Zhang[1,2]($\boxtimes$), and Yunfeng Wang[1,2]($\boxtimes$)

[1] Institute of Microelectronics of Chinese Academy of Sciences, Beijing 100029, China
{yukang,zhangjun,wangyunfeng}@ime.ac.cn
[2] University of Chinese Academy of Sciences, Beijing 100049,, China

Abstract. With the accelerating pace of societal development and the escalating competitive pressures, accurate and reliable stress assessment plays a crucial role in the monitoring of mental health and personalized intervention. Electrocardiogram (ECG) signals have emerged as a critical modality for stress assessment due to their non-invasiveness and capacity to convey abundant physiological information. However, existing methods find it difficult to achieve high-precision classification under multi-level stress states. Meanwhile, they are also faced with challenges such as high feature redundancy, weak temporal correlation, and insufficient fusion of multimodal information. In this study, ECG data of 23 subjects under the pressure induced by a mental arithmetic task at three phase were collected. A time-series dataset was constructed through signal preprocessing and segment division. In addition, a multi-scale feature extraction framework was proposed, covering time-domain, frequency-domain, and nonlinear dynamics indices. A CNN-LSTM network was designed to extract deep spatiotemporal features from the original signals. Furthermore, the TabNet network was adopted for model training and prediction to achieve cross-modal feature interaction and stress state classification. The experimental results show that this method achieves an accuracy of 91% in the three phase stress classification task and has good robustness against inter-individual differences. Compared with existing technologies, this study innovatively integrates HRV features with deep learning features, constructs an ECG multi-scale feature fusion framework, and introduces TabNet for efficient feature modeling, which significantly improves the accuracy and reliability of stress assessment. It provides a new idea for real-time stress monitoring by wearable devices.

Keywords: psychological stress · electrocardiogram signal · multi-stage · multi-scale feature · feature fusion

1 Introduction

Psychological stress has become one of the core risk factors threatening the physical and mental health of individuals in modern society. Research shows that long-term or high-intensity stress can trigger cardiovascular diseases, metabolic syndrome, and various mental disorders, leading to obvious pathological changes [1]. However, in clinical practice, it still mainly relies on subjective assessment methods such as questionnaire scales, which are easily affected by recall bias and cognitive limitations, making it difficult to achieve continuous and objective quantitative monitoring of the stress level. Developing non-invasive dynamic stress detection technology based on physiological signals is of great significance for the early warning of stress-related diseases and the optimization of personalized intervention strategies.

The ECG signal directly reflects the activities of the autonomic nervous system, and its indicators such as heart rate variability (HRV) and T-wave morphology are highly correlated with the stress response [2]. Compared with signals such as electroencephalogram (EEG) and electrodermal activity (EDA), ECG is more convenient to collect and has stronger resistance to motion artifacts, so it is more suitable for continuous monitoring in the scenario of wearable devices.[3] However, there are currently two major bottlenecks in the research of stress assessment:

1. Insufficient modeling of the dynamic process: Most experiments only classify a single stress level (such as the stressed/non-stressed), ignoring the continuous evolution process of stress accumulation and release, resulting in limited representational ability of the model in multi-stage stress recognition [4].
2. Challenges in feature fusion: Traditional feature engineering focuses on time-frequency domain features, and there is insufficient exploration of nonlinear dynamic features (such as entropy values, fractal dimensions, etc.) and deep spatiotemporal features. Moreover, redundancy and conflicts often exist among high-dimensional heterogeneous features, reducing the classification robustness.

In addition, although deep learning models (such as CNN and LSTM) have shown potential in physiological signal analysis, they are prone to overfitting due to inter-individual differences under the condition of small samples, and it is difficult to achieve stable generalization across subjects [5].

Aiming at the above challenges, this study proposes a dynamic stress assessment framework based on multi-scale feature fusion [6]: First of all, through a standardized mental arithmetic task, a three-phase stress response is induced to construct an ECG time-series dataset that includes the dynamic evolution process of stress. Secondly, by integrating the complementary advantages of traditional feature engineering and deep learning, time-domain, frequency-domain, and nonlinear features, as well as the deep spatiotemporal features generated by the CNN-LSTM, are synchronously extracted to comprehensively represent the multi-scale characteristics of the stress response. Furthermore, a two-way attention network based on tensor embedding–TabNet is adopted. Through feature space reconstruction and cross-modal interaction optimization, the problems of dimensional mismatch and information conflict in the fusion of heterogeneous features are solved.

This study combines the modeling of the dynamic evolution of stress with the fusion of multimodal ECG features and applies the TabNet architecture. While reducing the interference of individual differences, it enhances the model's ability to identify the temporal patterns of pressure, providing a new method for wearable devices to achieve high-precision pressure classification and detection (Fig. 1).

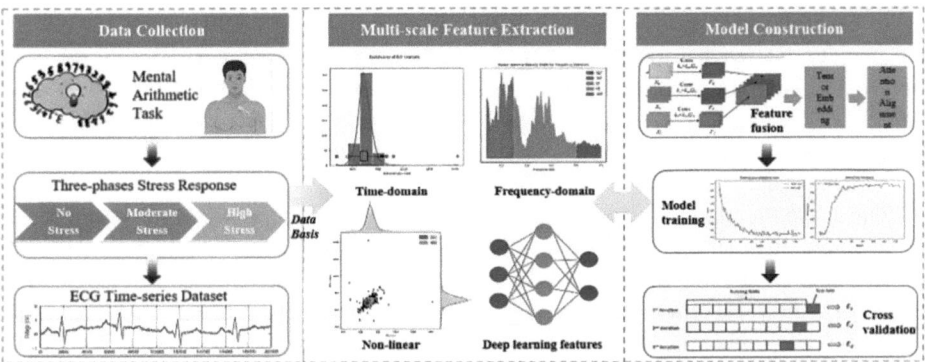

Fig. 1 Overall technical roadmap.

2 Methods

2.1 Experimental Design and Data Collection

Common stress induction methods include: color word test, ice water simulation, public speaking math calculations, watching horror videos, etc. Considering the feasibility and controllability of the induction levels, and based on the Montreal Stress Model Theory, we chose mental arithmetic as the primary stress induction method [7]. The Montreal Stress Model was originally designed by psychologists to assess psychological stress, and it includes three phases: no stress, moderate stress, and high stress induction. The experiment consists of a 5-min rest (no stress) phase, a 5-min moderate stress phase, and a 5-min high stress phase, as show in Fig. 2, with the calculation and answering system designed and implemented using E-Prime software.

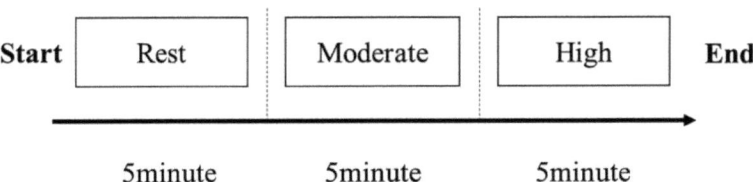

Fig. 2 Three phases of the stress induction experiment.

During the rest phase, no tasks are assigned. Participants are played relaxing music to help them relax as much as possible. This phase is designed to simulate a stress-free situation in the participants' daily lives.

In the moderate stress phase, participants are required to solve 50 simple two-digit arithmetic problems involving addition, subtraction, multiplication, and division. To induce moderate stress, a result would display at the bottom of the screen. Each question is allotted 5 s, and if the time runs out without a response, the system marks the answer as incorrect and automatically proceeds to the next question [8]. After completing the moderate stress task, participants are given a 1-min adjustment time before moving on to the high stress phase.

In the high stress phase, to further induce psychological stress, we introduce reward and penalty. For each correct answer, participants are rewarded 1 RMB, and for each incorrect answer, they are penalized 1 RMB. As shown in Fig. 3, the reward amount is displayed in bold red font in the center of the interface, which increases the participants' focus and creates a sense of pressure. After completing the high stress phase, the experiment ends.

The data collection environment is shown in Fig. 3. During the experiment, a medical-grade single-channel ECG recorder (Model: TKECG-H01) were used to collect participants' ECG physiological data, while a laptop computer was used to run calculation system. The ECG recorder utilized a dual-electrode structure, with one electrode attached below the right end of the left clavicle and the other positioned at a 45-degree downward inclination on the lower left side of the body. The device operated at a sampling rate of 250 Hz and transmitted data to the computer via Bluetooth, meeting the requirements for data collection.

Fig. 3 Data collection scenario.

A total of 23 participants without any cardiovascular or cerebrovascular diseases were recruited for the experiment, including 13 males and 10 females. The participants' ages ranged from 24 to 41 years, with an average age of 28.65 years. It was confirmed that none of the participants had previously participated in similar experiments, and they had no history of smoking or alcohol consumption in the past two days. After obtaining informed consent, they completed the data collection.

2.2 Data Preprocessing

During the acquisition of the original ECG signals, common noises include baseline drift, power frequency interference, and motion artifacts [9]. In order to ensure the accuracy of subsequent feature extraction, this study has designed the following preprocessing procedure (Fig. 4):

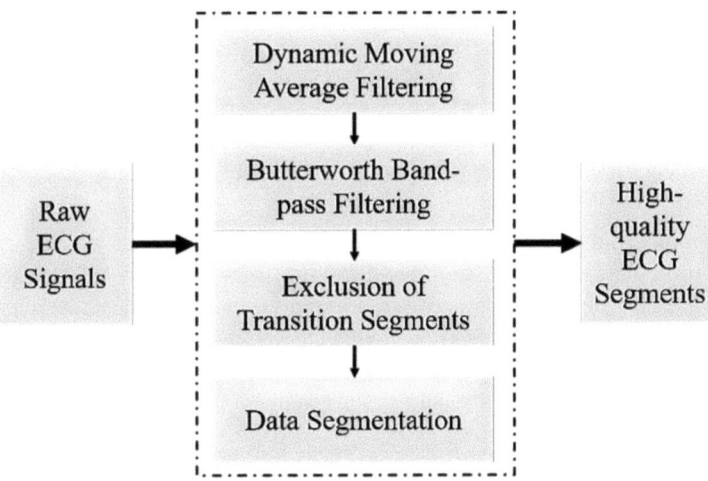

Fig. 4 Data preprocessing pipeline.

1. Dynamic Moving Average Filtering: Aiming at the high-frequency noise in the ECG signals, an improved moving average filter is used for preliminary smoothing. The window length is set to the sampling rate divided by 10 (25 points at a sampling rate of 250 Hz), which can effectively attenuate the high-frequency interference while maximizing the retention of the steep characteristics of the R wave.
2. Butterworth Band-pass Filtering: Based on the smoothed signals, a fourth-order Butterworth band-pass filter (0.5 Hz–40 Hz) is designed to further remove the baseline drift below 0.5 Hz and the high-frequency electromyogram and power frequency artifacts above 40 Hz.
3. Exclusion of Transition Segments: To eliminate the influence of the state transition at the beginning and end of the task for participants, 10 s of data at the front and rear of each recording segment are discarded, and only the middle stable experimental segment is retained to improve the signal-to-noise ratio of subsequent analysis.
4. Data Segmentation: To achieve the temporal modeling of the dynamic stress response, a sliding window strategy is adopted to segment the continuous ECG signals. A fixed-length window of 3 min (45,000 points @ 250 Hz) and a step size of 0.5 min (7500 points @ 250 Hz) are set to ensure that each pressure stage contains at least 2 complete windows [10]. Within each window segment, the position information of the R peak is detected through the NeuroKit2 toolkit, and a data file is generated for each segment to construct a traceable temporal analysis framework.

After the above filtering, clipping and segmentation processes, the data of 23 participants finally obtained 313 segments of high-quality and stable ECG data, providing a structured input for subsequent multi-scale feature extraction.

2.3 Feature Engineering

To comprehensively represent the ECG responses of participants at different stress phase, this study synchronously extracts features from four dimensions: time domain, frequency domain, nonlinear, and deep spatiotemporal [11].

- **HRV Feature Extraction:**

As shown in Table I, the features of HRV include time-domain features, frequency-domain features, and nonlinear features. Temporal features refer to the extraction of statistical features from the obtained R-R interval signals. The time-domain features used in this study are MeanRR, SDNN, RMSSD, NN50, and PNN50 [12].

Table 1 HRV features.

Domain	Feature symbol	Description
Time-Domain	MeanRR	The mean of the RR intervals
	SDNN	The standard deviation of the RR intervals
	RMSSD	The square root of the mean of the squared successive differences between adjacent RR intervals
	NN50	The number of time that RR intervals greater than 50 ms
	pNN50	The percentage of absolute differences in successive RR intervals greater than 50 ms
	TINN	An approximation of the RR interval distribution
Frequency-Domain	LF	The spectral power of low frequencies (0.04 to 0.15 Hz)
	HF	The spectral power of high frequencies (0.15 to 0.4 Hz)
	LFHF	The ratio divided the low frequency power by the high frequency power
	TP	The total spectral power
Non-linear Domain	SampEn1	Sample entropy in scale 1
	SampEn3	Sample entropy in scale 3
	SD1	Index of short-term RR interval fluctuations
	SD2	Index of long-term HRV changes

Time-domain features can reflect the balance state between the sympathetic nerve and the parasympathetic nerve, while frequency-domain features can reflect the activities of the sympathetic nerve and the parasympathetic nerve respectively [13]. For the non-uniformly sampled R-R sequence in this study, the Lomb–Scargle spectral estimation method is used to calculate the power spectral density, covering the standard HRV frequency band of 0.003–0.4 Hz. Through integration, four frequency-domain features are obtained: LF, HF, LFHF, and TP.

The physiological movement of the heart is not a completely regular periodic movement. Therefore, nonlinear features are also required to evaluate the movement pattern of the heart. In this study, SampEn1, SampEn3, SD1, and SD2 are calculated as the nonlinear features.

- **Deep Spatiotemporal Feature Extraction:**

To automatically extract high-dimensional spatiotemporal features from each ECG segment, this study designed a CNN-LSTM hybrid network to extract 128-dimensional features (Fig. 5). Finally, through PCA dimensionality reduction, 20 dimensional deep spatiotemporal features are obtained [14]. The implementation process is as follows:

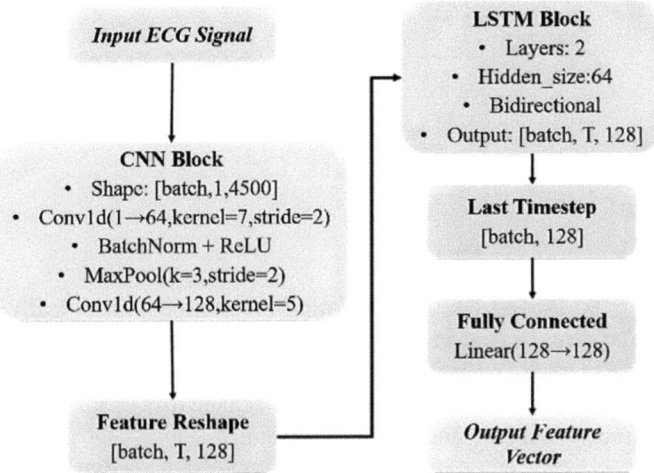

Fig. 5 CNN-LSTM network.

First of all, each segment of the original ECG signal is resampled via FFT or padded with zeros at the end to unify its length to 4500 points. Subsequently, variations in ECG amplitude across subjects and electrode placements can impede the convergence of deep networks due to inconsistent feature scales. Zero-mean and unit-variance normalization removes baseline shifts and rescales signals to a common distribution, stabilizing gradient updates, accelerating convergence, and improving classification accuracy. Finally, a tensor of shape [1, 1, 4500] is generated as the network input. After that, a two-layer CNN network is employed for one dimensional convolution to extract local waveform features. Downsampling with a stride of 2 is carried out to enhance translation invariance. This step effectively captures the local waveform patterns of the ECG signal and outputs feature maps. Then, the feature maps output by the CNN are transposed and input into a two-layer bidirectional LSTM network, which integrates forward and backward context information to represent the global temporal dynamics. Finally, the features are input into a fully connected layer, and the ReLU activation function is used to obtain a 128-dimensional deep feature vector. This vector is then reduced to 20 dimensions through PCA, which serves as the deep spatiotemporal representation of this segment.

2.4 Data Splitting

After the ECG data of the participants have undergone data preprocessing and feature extraction, a total of 313 data samples are generated. Among them, there are 105 samples in the no stress phase, labeled as 0; 108 samples in the moderate stress phase, labeled as 1; and 100 samples in the high stress phase, labeled as 2. Each sample contains 6 time-domain features, 4 frequency-domain features, 4 nonlinear features, and 20 deep learning features, making a total of 34 features, which are used for model training and testing. The data is divided into a training set and a testing set at a ratio of 7:3. The dataset distribution is shown in Table 2.

Table 2 Dataset distribution

Category	Amount	Label	Total
No stress	105	0	313
Moderate stress	108	1	
High stress	100	2	

2.5 Stress Assessment Model

As shown in Fig. 6, TabNet is a deep learning-based model specifically designed for processing tabular data. Its core idea is to handle tabular data using self-attention mechanisms and sparse mask [15]. Unlike traditional neural networks, TabNet employs sparse activation and interpretable feature selection mechanisms, enabling it to efficiently process high-dimensional data and avoid overfitting.

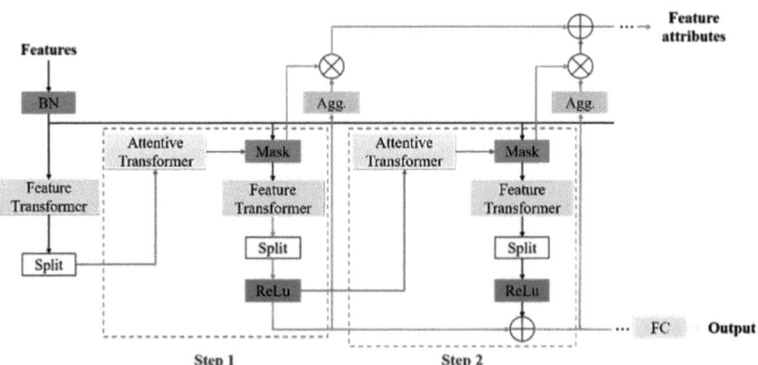

Fig. 6 TabNet architecture. The model do not consider any global feature normalization, but merely apply batch normalization (BN). TabNet use fully-connected (FC) layer to output the results. The feature selection mask provides interpretable information about the model's functionality, and the masks can be aggregated (Agg) to obtain global feature important attribution.

TabNet utilizes self-attention mechanisms to select the most relevant input features [16]. The model could automatically select important features for different tasks during each training iteration. Suppose the input feature set is $X = [x_1, x_2, \ldots, x_n]$, where each input feature x_i is a vector of length d. TabNet computes the weights of the features through the self-attention mechanism. The core operation of the self-attention mechanism can be expressed as Eq. (1):

$$Attention(Q, K, V) = softmax(\frac{QK^T}{\sqrt{d}})V \qquad (1)$$

where Q (query), K (key), and V (value) are matrices derived from the input features. By computing the similarity between these matrices, the model selects the most relevant features for subsequent steps.

TabNet's sparse activation mechanism allows the model to activate only a subset of neurons in each decision step, improving computational efficiency and reducing memory usage. In each training step, only a small number of neurons are activated, and sparse activation strategies are used to avoid overfitting and enhance the model's generalization ability. Let a_i be the activation value of the i feature. The sparse activation computation formula is shown in Eq. (2):

$$a_i = Sparsemax(W \bullet x_i) \qquad (2)$$

where W is the weight matrix, and Sparsemax is an activation function similar to softmax. It enforces sparsity by forcing some activation values to be zero, thereby reducing computation and memory consumption.

In the decision portion, the model consists of multiple decision steps. Each decision step computes by selecting the most relevant features and improving the model's accuracy. The output of each decision step combines the features of the self-attention network and sparse activation. The output of the t decision step z_t can be expressed as Eq. (3):

$$z_t = \sum_{i=1}^{n} a_{t,i} \bullet h_i \qquad (3)$$

where $a_{t,i}$ is the activation value of the t step for feature, and h_i is the representation of feature i. By computing this weighted sum, the model can select the most informative features in different decision steps. After multiple decision steps, TabNet generates the final prediction result through a fully connected layer, shown in Eq. (4):

$$y = sigmoid(W_{out} \bullet z_T) \qquad (4)$$

where W_{out} is the weight matrix of the output layer, z_T is the output of the T decision step, and the *sigmoid* function is used to convert the output into classification probabilities.

TabNet has successfully achieved efficient and interpretable processing of tabular data by introducing a self-attention mechanism and sparse activation. Its self-attention mechanism is capable of selecting the most informative features in different decision-making steps, while the sparse activation mechanism significantly improves computational efficiency. TabNet also has demonstrated excellent performance in multiple tasks. Especially when dealing with data with high-dimensional features, it provides a powerful end-to-end solution.

3 Results and Analysis

3.1 Model Training

To effectively train and evaluate the TabNet model, we optimized several key parameters during the training process while avoiding overfitting. The training process are as follows:

First of all, the maximum number of training epochs is set to 150, ensuring that the model learns effective features within enough training cycles while avoiding overfitting due to excessive training. Secondly, early stopping is set up, where training would terminate early if there is no significant improvement in validation performance for 30 consecutive epochs. Meanwhile, the batch size for each training step is 64 and the drop_last is set to false, meaning training would continue even if the last batch contains fewer samples, to avoid losing useful data. Besides, the cross-entropy loss function is used, which performs well in multi-class problems and effectively measures the difference between the model's output and the true labels. And the Adam optimizer is employed with a learning rate of 0.005. Adam adjusts the learning rate adaptively, ensuring stability and efficiency during the training process. Finally, the hyperparameters of TabNet are optimized using grid search. The final selected hyperparameters are as follows in Table III. After training, the performance of the model is evaluated on the test set. The evaluation metrics include: Accuracy, Precision, Recall, F1-Score, and Area Under The Curve (AUC).

Table 3 Model hyperparameters

Hyperparameters	Description	Value
n_d	Dimension of the decision layer	64
n_a	Dimension of the attention layer	64
n_{steps}	Number of decision steps	5
γ	Sparsity regularization parameter	1.3
l_r	Learning rate	0.005

3.2 Results Analysis

In this section, the performance of the proposed framework in the three-phase stress classification task was deeply analyzed from three aspects: classification performance, ablation experiments, and comparison with other models.

Table 4 Model classification report.

Category	Precision	Recall	F1 Score	Amount
0 (no stress)	0.97	0.95	0.96	41
1 (moderate stress)	0.76	1.00	0.87	26
2 (high stress)	1.00	0.79	0.88	33
Mean	0.91	0.91	0.90	100

As shown in Table IV, the model's classification performance across the three stress phase is generally excellent: For the detection of "no stress", it achieves a precision of 0.97 and a recall of 0.95 (F1 of 0.96), indicating that the model is extremely reliable in distinguishing no stress. For "moderate stress", although the precision is slightly lower at 0.76, the recall reaches 1.00 (F1 of 0.87), which means that all samples actually under moderate stress are correctly captured, with only a few false positives. For "high stress", it shows a precision of 1.00 and a recall of 0.79 (F1 of 0.88), suggesting that the model is very strict in identifying high stress, but there is still about 21% of missed detections.

Overall, the model has an average precision and recall of 0.91, and an average F1 of 0.90, demonstrating a stable and balanced recognition ability for the three-phase stress. The ROC curve and confusion matrix are shown in Fig. 7.

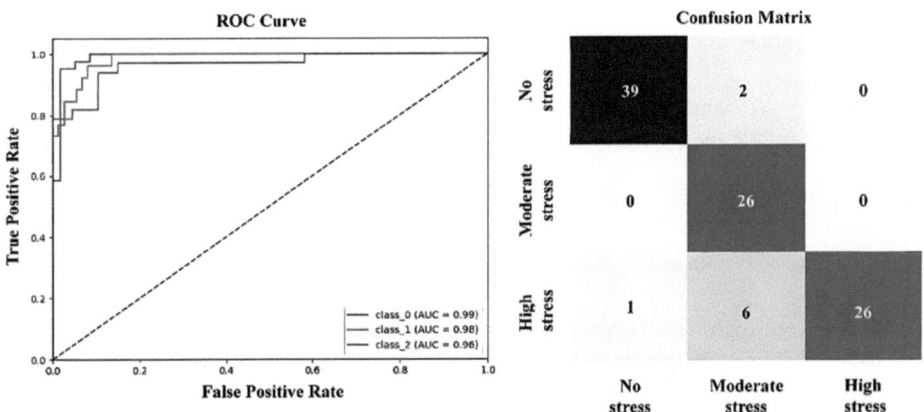

Fig. 7. ROC curve and confusion matrix.

To evaluate the benefits of multi-scale feature fusion, we compared our model (hyperparameters shown in Table 3) with those using only a single feature set (Table 5). "Single-scale" models that use only HRV features or rely solely on deep features generated by CNN-LSTM can both achieve good performance on test set. However, the fused model reaches an accuracy of 91.0%, surpassing the results of any single-scale model, which verifies the complementary advantages of traditional features and deep spatiotemporal features. This ablation experiment shows that the integration of multi-scale and multi-type features can significantly enhance the model's sensitivity to subtle differences in stress assessment.

Table 5 Multi-scale fusion model and single-scale models results.

Model	Accuracy	Amount
HRV Features Model	80.5%	100
DL Features Model	88.4%	100
Multi-scale Fusion Model	**91.0%**	**100**

In the comparison with common machine learning and deep learning baseline models (Table VI), the accuracy of 91.0% of the method proposed in this study is superior to the experimental results of the Random Forest (n_estimators = 100, max_depth = 10), Support Vector Machine (RBF kernel, C = 1.0), and simple CNN models (convolutional layers + pooling + fully connected layers). This demonstrates the advantages of the feature reconstruction and cross-modal interaction mechanism based on TabNet in the multi-stage pressure classification task. Compared with the classical methods, this framework not only improves the classification accuracy but also shows better consistency in reducing the model's sensitivity to individual differences.

Table 6 Comparison of model accuracies.

Model	Accuracy	Amount
Random Forest	86.4%	100
Support Vector Machine	84.2%	100
CNN	89.1%	100
Mortensen et al. [4]	86.2%	100
Song et al. [14]	88.4%	100
Ours	**91.0%**	**100**

We also compared our approach with recent state-of-the-art studies (TABLE VI). Our multi-scale fusion framework achieves 91.0% accuracy, outperforming Mortensen et al. (2023) [86.2%] and Song et al. (2024) [88.4%], representing an improvement of 2.6–4.8 percentage points.

4 Discussion

In this study, by constructing a multi-scale feature fusion and a TabNet architecture, the comprehensive performance of three-phase stress classification has been significantly improved. However, there are still the following points worthy of in-depth discussion:

1) *Balance of classification between moderate and high stress:* Although the model performs well for individual categories, the precision for "moderate stress" is relatively low, and the missed detection rate for "high stress" can be further optimized. This is mainly due to the fact that under the mental arithmetic induction task, the subjective experiences of participants regarding pressure vary greatly, resulting in overlapping feature distributions. To address the lower recall in the high-stress category, future work will incorporate multi-scale entropy and complex network features to better characterize ECG dynamic complexity. Additionally, we will adopt a multi-task learning framework by adding subjective emotion self-report scores as auxiliary labels to enhance discriminative power under high stress.
2) *Value of multi-scale feature fusion:* The ablation experiment shows that the HRV features and the deep spatiotemporal features are highly complementary in information expression. The tensor embedding and attention mechanism of TabNet can automatically learn the weight distribution of each feature subspace, effectively alleviating the redundancy and conflict among high-dimensional heterogeneous features. This idea can be extended to the multi-modal fusion of other physiological signals, such as the joint analysis of EEG, EMG and motion acceleration data.
3) *Sample size and generalization across subjects:* Although our current dataset comprises 23 participants (313 segments) and demonstrates promising results under small-sample conditions, we acknowledge that such a sample size may limit generalizability. And this study was conducted in a controlled laboratory setting without considering motion artifacts, environmental noise, or social interactions. In future work, we will expand data collection to a large sample of population across diverse real-world scenarios (e.g., outdoor exercise, office tasks, social interactions) to assess the model's robustness in real-world applications and cross-population performance.

In summary, the multi-scale feature fusion method based on TabNet not only outperforms single-modal and classical models in terms of accuracy but also provides a new technical path for wearable real-time stress monitoring and clinical intervention. Subsequent research will focus on cross-modal data expansion, personalized adaptation, and real-time deployment to further enhance its application value and promotion potential.

5 Conclusions and Future Work

This study proposed a dynamic stress assessment framework based on multi-scale feature fusion. The time-domain, frequency-domain, nonlinear, and deep spatiotemporal features are uniformly modeled and optimized for fusion through TabNet. In the three-phase stress classification task, the model achieved a classification accuracy of 91% and an AUC value of 0.98, verifying the sensitive capture ability of the CNN-LSTM network for the morphological changes of ECG related to stress, and constructing a dynamic mapping relationship between the parameters of ECG signals and the intensity of stress.

From the application perspective, this model has strong practical value. It can be directly deployed on wearable devices such as smart bracelets and ECG patches to achieve real-time hierarchical early warning of the stress level in the workplace or daily scenarios. In clinical practice, by outputting continuous stress trajectories, it can provide an accurate quantitative basis for personalized interventions for stress-related diseases such as anxiety disorders and hypertension, significantly making up for the subjectivity of traditional scale evaluations and the limitations of breakpoint monitoring.

In the future, the research will focus on promoting the following areas: Firstly, a personalized fine-tuning mechanism will be constructed based on federated learning to alleviate the impact of physiological differences among subjects on the performance of the model. Secondly, data from environmental sensors (acoustic, optical, temperature, humidity, location information) will be integrated to build a more comprehensive multi-modal stress assessment system. Thirdly, the associated modeling of long-term stress exposure and cardiovascular and metabolic risk indicators will be explored to provide new predictive tools for the early warning and intervention of chronic diseases.

6 Disclosure of Interests.

The authors declare no competing interests.

Acknowledgments. This work was supported by the Science and Technology Projects of Xizang Autonomous Region, China (Grant No. XZ202501ZY0060).

References

1. Thielmann, B., Pohl, R., Böckelmann, I.: Heart rate variability as a strain indicator for psychological stress for emergency physicians during work and alert intervention: a systematic review. J. Occup. Med. Toxicol. **16**(1), 24 (2021)
2. Giannakakis, G., Grigoriadis, D., Giannakaki, K., et al.: Review on psychological stress detection using biosignals. IEEE Trans. Affect. Comput. **13**(1), 440–460 (2019)
3. Hantono, B.S., Nugroho, L.E., Santosa, P.I.: Mental stress detection via heart rate variability using machine learning. Int. J. Electr. Eng. Inform. **12**(3), 431–444 (2020)
4. Mortensen, J.A., Mollov, M.E., Chatterjee, A., et al.: Multi-class stress detection through heart rate variability: A deep neural network based study. IEEE Access **11**, 57470–57480 (2023)
5. Savarimuthu, S.R., Karuppannan, S.J.K: CNN based stress detection from ECG: A systematic survey[C]//AIP Conference Proceedings. AIP Publishing **2857**(1) (2023)
6. Yoon, J., Kang, C., Kim, S., et al.: D-vlog: Multimodal vlog dataset for depression detection[C]//Proceedings of the AAAI Conference on Artificial Intelligence. **36**(11): 12226–12234 (2022)
7. Gil-Martin, M., San-Segundo, R., Mateos, A., et al.: Human stress detection with wearable sensors using convolutional neural networks. IEEE Aerosp. Electron. Syst. Mag. **37**(1), 60–70 (2022)
8. Dedovic, K., Renwick, R., Mahani, N.K., Engert, V., Lupien, S.J., Lupien, J.C.: The montreal imaging stress task: Using functional imaging to investigate the effects of perceiving and processing psychosocial stress in the human brain. J. Psychiatr. Neurosci. **30**, 319 (2005)
9. Hasnul, M.A., Ab. Aziz N A, Abd. Aziz A.: Augmenting ECG data with multiple filters for a better emotion recognition system. Arab. J. Sci. Eng. **48**(8): 10313–10334 (2023)
10. Tian, X., Ning, M.: Tracking vigilance fluctuations in real-time: a sliding-window heart rate variability-based machine-learning approach, Sleep, **48**(Issue 2) (February 2025)

11. Lee, S., Hwang, H.B., Park, S., et al.: Mental stress assessment using ultra short term HRV analysis based on non-linear method. Biosensors **12**(7), 465 (2022)
12. Pham, T., Lau, Z.J., Chen, S.H.A., Makowski, D.: Heart rate variability in psychology: A review of HRV indices and an analysis tutorial. Sensors **21**(12), 3998 (2021)
13. Makowski, D., et al.: NeuroKit2: A Python toolbox for neurophysiological signal processing. Behav. Res. Methods **53**(4), 1689–1696 (2021)
14. Song, C.H., Kim, J.S., Kim, J.M., et al.: Stress Classification using ECGs based on a Multi-dimensional Feature Fusion of LSTM and Xception[J]. IEEE Access **12**, 19077–19086 (2024)
15. Arik, S. Ö., Pfister, T.: Tabnet: Attentive interpretable tabular learning[C]//Proceedings of the AAAI conference on artificial intelligence **35**(8): 6679–6687 (2021)
16. Zhang, P., Li, F., Du, L., et al.: Psychological stress detection according to ECG using a deep learning model with attention mechanism. Appl. Sci. **11**(6), 2848 (2021)

Regularized Offline Reinforcement Learning for Energy Efficient Urban Rail Transit System Control

Han Chen[1], Changkai Zhang[1], Jinfeng Ma[2], and Bolei Zhang[2(✉)]

[1] NR Electric Co., Ltd., Nanjing, Jiangsu, China
{chenhan,zhangchangkai}@nrec.com
[2] School of Computer Science, Nanjing University of Posts and Telecommunications, Nanjing, Jiangsu, China
{majinfeng,bolei.zhang}@njupt.edu.cn

Abstract. Urban rail transit (URT) systems have become crucial components of sustainable transportation networks in modern cities. These systems consume substantial amounts of energy, contributing significantly to operational costs and environmental impact. To reduce the energy cost, previous works have adopted reinforcement learning (RL) to optimize the controlling parameters that minimizes the energy cost while ensuring performance. However, RL requires exploration-and-exploitation to interact with the environment, which is costly or even dangerous in real URT systems. This paper proposes a novel model-based offline RL approach for optimizing energy efficiency in URT systems. The basic idea is to build a transition model that can be utilized for policy optimization. To improve the generalization, our method leverages physical laws as regularizers and employs Monte Carlo dropout to quantify model uncertainty. The regularizer and uncertainty are further used to penalize the reward function for conservative policy optimization. Experiments demonstrate that our approach achieves average 20% energy consumption reduction compared to conventional methods while maintaining system stability and performance.

Keywords: Offline reinforcement learning · Physical regularizers · Urban rail transit system

1 Introduction

With the acceleration of the urbanization process, the demand for urban rail transit (URT) systems has been increasing continuously, leading to a growing energy consumption problem. The energy consumption of URT systems mainly comes the following aspects: train traction energy consumption, braking energy

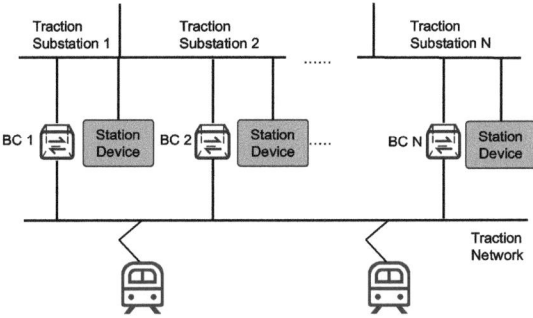

Fig. 1. An illustration of the URT system. The energy consumption of the URT system mainly comes from the traction energy consumption, braking energy consumption, and auxiliary system energy consumption. The regenerative braking energy is collected by the traction network and sent to the bidirectional converter (BC) in the substation.

consumption, and auxiliary system energy consumption. Traction energy consumption is the main energy consumed during the train's operation, used to overcome the train's inertia, friction, and slope resistance. Braking energy consumption is the energy consumed when the train decelerates or stops, and part of the energy can be recovered through regenerative braking technology. Auxiliary system energy consumption includes the energy consumption of equipment such as carriage lighting, air conditioning, and ventilation. As URT systems continue to scale up, the issue of energy consumption has become increasingly prominent, making energy conservation and emission reduction critical challenges that need to be addressed urgently.

When a train generates regenerative braking energy during braking, this energy is collected by the traction network and sent to the **bidirectional converter (BC)** in the substation (illustrated in Fig. 1). After being converted, the energy is fed back to the power grid side or stored in the energy storage device. When the train is in traction, the energy from the power grid side is also converted by these BCs and then supplied to the train. In the traction working mode, the energy feedback device is in a rectifying state, providing voltage and energy to the catenary. In the regenerative braking mode, the excess energy causes the voltage of the traction network to rise. After the system determines it as an inverter mode, the energy will be returned to the AC network side.

The BC can effectively improve the energy efficiency of the URT system, reduce energy consumption, and achieve energy conservation and emission reduction. However, the energy consumption of URT systems is still a critical issue. Previous works have adopted various strategies to reduce energy consumption in URT systems. For example, some works have focused on optimizing the train operation schedule to reduce energy consumption [2,21]. Other works have proposed energy-efficient control strategies for URT systems using supervised learning or model predictive control (MPC) [4,9]. More recently, studies have adopted reinforcement learning (RL) [17] to optimize the BC policies. RL offers a promising data-driven approach to develop optimal control policies for complex systems. However, conventional online RL algorithms that require active interaction with the environment are impractical for URT systems due to safety concerns, operational constraints, and the inability to test potentially risky policies during regular service. Moreover, these methods often require a high-fidelity environment model and are difficult to transfer to the real-world environment. Therefore, to optimize the energy cost of the URT system, it is essential to develop a method that can learn from historical data with minimal online exploration.

In this paper, we propose a model-based offline RL framework for energy-efficient control of BCs in URT systems, with additional ability of fast adapting to online environments. We first formulate URT energy efficiency problem as a Markov Decision Process (MDP). Next, a novel transition model learning approach is introduced, which incorporates physical laws as regularizers to enhance generalization beyond the offline dataset. We propose an uncertainty quantification method combining Monte Carlo dropout with physical constraints to assess model reliability. A policy optimization technique is proposed that penalizes actions associated with high uncertainty, ensuring safe and reliable operation. Empirical evaluation demonstrating significant energy savings compared to conventional control methods.

2 Related Works

Offline reinforcement learning aims to learn effective policies from static datasets without environment interaction, primarily through model-free and model-based approaches. The fundamental challenge lies in addressing distribution shift caused by discrepancies between the offline dataset and potential online environments.

Model-free offline RL methods directly optimize policies using offline data while incorporating conservative mechanisms to prevent overestimation of out-of-distribution actions [6,7,12,19,20]. Notable approaches include CQL [8], which regularizes value estimates to penalize overly optimistic action values, and BCQ [19] that combines imitation learning with Q-learning while constraining the policy to the data distribution via a generative model. While conceptually straightforward, these methods often suffer from limited sample efficiency due to their inability to generate novel transitions.

Model-based offline RL methods first learn environment dynamics from the dataset, then use the learned model for policy optimization [3,11,18,23]. However, these approaches face challenges with extrapolation errors when encountering out-of-distribution states [10], leading to compounding inaccuracies. Recent advances address this through various conservative mechanisms: COMBO [22] employs conservative policy evaluation, RAMBO [14] introduces adversarial model training, and MOPO [23] incorporates uncertainty-based reward penalties. MOREL [5] constructs pessimistic MDPs to constrain transition overestimation. Emerging techniques leverage diffusion models [16,24] to enhance model robustness. Despite these innovations, current methods still struggle to reliably evaluate and utilize out-of-distribution samples, highlighting the need for more informative data collection strategies.

3 Preliminaries

3.1 Markov Decision Process

A Markov Decision Process (MDP) is described by a 6-tuple $M = (\mathcal{S}, \mathcal{A}, \mathcal{T}, \mathcal{R}, \mu_0, \gamma)$, where $\mathcal{S}$ denotes the state space, $\mathcal{A}$ denotes the action space, $\mathcal{R}$ is the reward space, $\mathcal{T}(s'|s,a)$ defines the transition probability of moving to state s' from state s upon executing action a, μ_0 represents the initial state distribution, $\gamma \in (0,1)$ is the discount factor that balances the weight of future rewards with immediate ones'.

The state value function $V^\pi : \mathcal{S} \to \mathbb{R}$ measures the expected cumulative reward received following policy π which is defined as:

$$V^\pi(s) = \mathbb{E}^\pi \left[\sum_{t=0}^{\infty} \gamma^t r_t \,\Big|\, s_0 = s \right] \tag{1}$$

The state-action value function $Q^\pi : \mathcal{S} \times \mathcal{A} \to \mathbb{R}$ captures the expected discounted return from taking action a in state s according to policy π which satisfies:

$$Q^\pi(s,a) = \mathcal{R}(s,a) + \gamma \int_\mathcal{S} V^\pi(s') \mathcal{T}(ds'|s,a) \tag{2}$$

3.2 Model-Based Offline RL

In the offline reinforcement learning paradigm, the learning agent's training is strictly constrained to a fixed dataset $\mathcal{D}_{\text{off}} = \{(s_i, a_i, s'_i, r_i)\}_{i=1}^N$ collected apriori by some *behavior policy* $\pi_\mathcal{D}$. The model-based approach to offline RL proceeds by first learning a parameterized *transition model* $\hat{\mathcal{T}}_\theta(s', r \mid s, a)$ through maximum likelihood estimation:

$$\theta^* = argmin_\theta \mathbb{E}_{(s,a,s',r) \sim \mathcal{D}_{\text{off}}} \left[-\log \hat{\mathcal{T}}_\theta(s', r \mid s, a) \right] \tag{3}$$

where $\hat{\mathcal{T}}_\theta$ provides a probabilistic approximation to the true environment dynamics $P(s', r \mid s, a)$. The transition model is typically parameterized as a multivariate Gaussian distribution:

$$s' \sim \mathcal{N}(\mu_{s'}(s,a;\theta), \Sigma_{s'}(s,a;\theta)) \tag{4}$$

$$r \sim \mathcal{N}(\mu_r(s,a;\theta), \sigma_r^2(s,a;\theta)) \tag{5}$$

where $\mu.$ and $\Sigma.$ denote the mean and covariance functions learned by the neural network.

Following model learning, policy improvement can be performed entirely through interaction with the learned model $\hat{T}_\theta$, circumventing the need for additional environment interactions and thereby achieving superior sample efficiency.

The fundamental challenge in model-based offline RL stems from the distributional shift between. The data distribution induced by $\pi_{\mathcal{D}}$ The state-action visitation distribution $P_\pi(s,a)$ induced by the learned policy π. This mismatch leads to compounding extrapolation error when the model $\hat{T}_\theta$ is queried with out-of-distribution (OOD) inputs $(s,a) \notin \text{supp}(\mathcal{D}_{\text{off}})$. To mitigate these issues, contemporary methods is employed such as policy constraints to ensure $\pi(a|s)$ stays within $\text{supp}(\pi_{\mathcal{D}}(a|s))$, uncertainty quantification to detect OOD queries and pessimistic value estimation to avoid overoptimism about OOD actions

3.3 Soft Actor-Critic

Soft Actor-Critic (SAC) [1] is an off-policy reinforcement learning algorithm specifically designed for continuous control tasks. It aims to strike a balance between exploration and exploitation by jointly maximizing the expected cumulative reward and the entropy of the policy. The inclusion of the entropy term promotes exploration across the action space, helping to prevent the agent from converging prematurely to suboptimal solutions.

SAC employs a stochastic policy network π_φ, along with two Q-value networks (critics). The use of a twin-critic architecture helps reduce overestimation bias by taking the minimum of the two predicted Q-values when computing the target, thereby enhancing training stability.

The critic networks are updated by minimizing the following loss function:

$$\mathcal{L}_{\text{critic}}(\theta_i) = \mathbb{E}_{(s,a,r,s') \sim \mathcal{B}}\left[\left(Q_{\theta_i}(s,a) - \right.\right. \tag{6}$$
$$\left.\left.\left(r + \gamma \mathbb{E}_{a' \sim \pi_\varphi}[\min_{j=1,2} Q_{\bar{\theta}_j}(s',a') - \alpha \log \pi_\varphi(a'|s')]\right)\right)^2\right], i = 1,2$$

where $\mathcal{B}$ denotes the replay buffer containing past transitions, $\bar{\theta}$ represents the target network parameters updated with a delay, and α is a temperature coefficient that scales the entropy term $-\alpha \log \pi_\varphi(a'|s')$. This entropy term encourages the selection of more stochastic actions, facilitating broader exploration.

The policy (actor) is updated to maximize both the expected return and the entropy, according to the following objective:

$$\mathcal{J}_{\text{actor}}(\varphi) = \mathbb{E}_{s \sim \mathcal{B}, \xi \sim \mathcal{N}}\left[\min_{i=1,2} Q_{\theta_i}(s, a_\varphi(s,\xi))\right.$$
$$\left. - \alpha \log \pi_\varphi(a_\varphi(s,\xi)|s)\right] \tag{7}$$

where $a_\varphi(s,\xi) = \tanh(\mu_\varphi(s) + \sigma_\varphi(s) \odot \xi)$ (reparameterized sampling).

This formulation enables SAC to dynamically control the exploration-exploitation trade-off through the entropy coefficient α, which can either remain fixed or be learned during training. By alternately updating the actor and critics, SAC achieves stable and sample-efficient learning, making it effective for high-dimensional control environments.

4 Method

In this section, we present our method in detail. We first describe the problem formulation. Next, we show how to train the model-based offline RL. The last part will introduce method to transfer the policy to online environment.

4.1 Problem Formulation

We consider the problem of energy-efficient control for URT systems. Suppose there are N BCs and M rails in the URT system. The time is discretized into T intervals: $\{1, 2, \ldots, T\}$. The energy consumption of the i-th rail at the t-th interval is denoted as $E_{i,t}$. The total energy consumption of the URT system is the sum of the energy consumption of all the rails, i.e., $E = \sum_{i=1}^{M} \sum_{t=0}^{T} E_{i,t}$. The goal is to adjust the rectification starting voltage and inverter starting voltage of each BC ($V_{i,t}^{rec}, V_{i,t}^{inv}$), $\forall i \in \{1, 2, \ldots, M\}, t \in \{1, 2, \ldots, T\}$ to minimize the total energy consumption of the URT system while maintaining high performance, which can be formulated as the following optimization problem:

$$\min_{V^{rec}, V^{inv}} \sum_{i=1}^{M} \sum_{t=1}^{T} E_{i,t}. \tag{8}$$

Given the problem formulation, we formulate the energy efficiency optimization problem in URT systems as a Markov Decision Process (MDP) defined by the tuple $(\mathcal{S}, \mathcal{A}, \mathcal{T}, \mathcal{R}, \gamma)$. The elements are defined as follows:

- **State space ($\mathcal{S}$):** The state vector encapsulates the complete system status, including: Position, velocity, and passenger load of each train; Passenger counts and waiting queues at each station; Electrical parameters (current, voltage, power) at each converter; Grid connection status and energy storage levels.
- **Action space ($\mathcal{A}$):** The actions consist of the rectified starting voltage and inverter starting voltage for each BC in the network. These voltages control both the direction and magnitude of power flow through the converters, enabling energy recovery during braking and optimal power distribution. Formally, the action at step t can be represented as: $(V_{i,t}^{rec}, V_{i,t}^{inv}), \forall i \in \{1, 2, \ldots, M\}$. As the voltage can only be adjusted at fixed values, the action space is discrete.
- **Transition function ($\mathcal{T}$):** This function describes how the state evolves based on the current state and selected actions, incorporating the physical dynamics of train movement, passenger flow, and electrical power systems.

– **Reward function ($\mathcal{R}$):** The reward function is mathematically defined as the negative summation of absolute power values from all traction substations in the rail transit system, formally expressed as:

$$R = -\sum_{i=1}^{N} |P_i| \qquad (9)$$

where P_i represents the power consumption of the i-th traction substation. This formulation explicitly establishes an optimization objective that drives the control policy to minimize the total traction energy consumption throughout the rail network.

4.2 Transition Model

Given the offline dataset $\mathcal{D}_{off}$, we first build a supervised transition model $\hat{\mathcal{T}}_\theta(s', r | s, a)$, that predicts the next state and reward based on current state and action. Learning an accurate transition model from offline data is crucial but also challenging. On one hand, as the policy is optimized by interacting with the transition model, accurate predictions are essential to avoid compounding errors. On the other hand, the model may overfit to the training data, leading to poor generalization to unseen states. This is particularly problematic in offline RL, where the model is trained on a limited dataset that may not cover the entire state-action space. As a result, the model may fail to accurately predict transitions for states that are not well-represented in the training data, leading to suboptimal policies and poor performance in real-world scenarios.

To address this, we propose a novel approach that incorporates physical laws as regularizers during model training. Our transition model $\mathcal{T}_\theta(s', r | s, a)$ is implemented as a neural network with parameters θ, trained to minimize the following loss function:

$$\mathcal{L}(\theta) = \mathcal{L}_{data}(\theta) + \lambda_0 \mathcal{R}_{phys}(\theta), \qquad (10)$$

where $\mathcal{L}_{data}(\theta)$ is the standard prediction error on the offline dataset:

$$\mathcal{L}_{data}(\theta) = \mathbb{E}_{(s,a,s',r) \sim \mathcal{D}}[||(s', r) - \mathcal{T}_\theta(s, a)||^2], \qquad (11)$$

and $\mathcal{R}_{phys}(\theta)$ is a physics-based regularization term that penalizes violations of known physical constraints. In particular, we introduce three parts of regularizers that can be formulated as follows:

– **Kinematic Constraints Regularizer:** The train's motion must satisfy kinematic constraints with bounded acceleration. For the i-th train, we enforce:

$$\mathcal{R}_{\text{kin}} = \lambda_v \sum_t \max\left(0, |v_{i,t+1} - v_{i,t}| - a_{\max} \Delta t\right)^2 \qquad (12)$$

where λ_v is the regularization weight, $a_{\max}$ denotes the maximum allowable acceleration ($3.0 \, \text{m/s}^2$ for typical metro systems), and Δt is the discretization time step. The max operator ensures soft constraint enforcement.

- **Jerk Minimization Regularizer:** To ensure passenger comfort and mechanical safety, we minimize the jerk (time derivative of acceleration):

$$\mathcal{R}_{\text{jerk}} = \lambda_j \sum_t \left(\frac{a_{i,t+1} - 2a_{i,t} + a_{i,t-1}}{\Delta t^2} \right)^2 \quad (13)$$

where λ_j controls regularization strength. This discrete formulation approximates the third-order time derivative of position, with typical values kept below $1.0 \, \text{m/s}^3$ for comfort.
- **Power Balance Regularizer:** The energy conservation constraint enforces real-time power equilibrium:

$$\mathcal{R}_{\text{power}} = \lambda_p \sum_t \left(P_{\text{grid},t} - \sum_{i=1}^{N_t} \eta_i P_{\text{train},i,t} - P_{\text{ESS},t} \right)^2 \quad (14)$$

where λ_p is the penalty coefficient, $\eta_i \in [0,1]$ represents the i-th train's energy conversion efficiency, and $P_{\text{ESS},t}$ denotes energy storage system power (positive for discharging). The formulation accounts for regenerative braking through sign conventions ($P_{\text{train},i,t} < 0$ when generating).

Therefore, the physics-based regularization term can be formulated as:

$$\mathcal{R}_{phys}(\theta) = \mathcal{R}_{kin} + \mathcal{R}_{jerk} + \mathcal{R}_{power}, \quad (15)$$

The first term penalizes large changes in velocity, the second term penalizes large changes in acceleration, and the third term penalizes power imbalance. By incorporating these physical constraints into the transition model training process, we can improve the model's generalization ability and ensure that it adheres to the underlying physical laws governing the system.

4.3 Policy Optimization

We employ a conservative policy optimization approach that accounts for model uncertainty. The key idea is to modify the reward function to penalize actions associated with high uncertainty:

$$\tilde{r}(s,a) = r(s,a) - \beta u(s,a), \quad (16)$$

where β is a non-negative hyperparameter controlling the penalty strength, and $u(s,a)$ is the uncertainty quantification of the transition model. To quantify the uncertainty in our transition model, we employ Monte Carlo dropout (MC Dropout) as a Bayesian approximation technique. MC Dropout is a technique used to estimate uncertainty in neural networks. It is based on the idea of treating dropout, a regularization method, as a way to approximate a Bayesian posterior distribution. The uncertainty can be computed as follows:

1) Sampling model parameters: We sample K independent sets of transition model parameters $\{\theta_k\}_{k=1}^K$ by dropping activations with probability p.

2) Performing forward passes: For each set of sampled parameters θ_k, we perform a forward pass. Let $\mu_k, \sigma_k = \mathcal{T}_{\theta_k}(s,a)$ be the output of the k-th forward pass. The output is a Gaussian distribution with mean μ_k and standard deviation σ_k.
3) The uncertainty consists of aleatoric uncertainty and epistemic uncertainty. Aleatoric uncertainty is the uncertainty inherent in the data, while epistemic uncertainty is the uncertainty in the model parameters. The aleatoric uncertainty can be estimated as the variance of the predicted means:

$$u_{alea}(s,a) = \frac{1}{K}\sum_{k=1}^{K}\sigma_k, \tag{17}$$

while the epistemic uncertainty can be estimated as the variance of the predicted variances:

$$u_{epist}(s,a) = \frac{1}{K}\sum_{k=1}^{K}(\mu_k - \mu(s,a))^2 + \lambda_{phys}\mathcal{R}_{phys}(s,a). \tag{18}$$

where λ_{phys} is a hyperparameter that controls the strength of the physical regularization term. The total uncertainty can be computed as the sum of the two terms. In addition, we incorporate domain knowledge by assigning higher uncertainty to predictions that violate physical constraints, creating a physically-informed uncertainty quantification method. The final uncertainty can be formulated as:

$$u(s,a) = u_{alea}(s,a) + u_{epist}(s,a). \tag{19}$$

The structure of the transition model and the policy model are illustrated in Fig. 2. Both transition network and the policy network have two layers of gated recurrent unit layers (GRUs) with 256 hidden units, followed by a multi-layer perceptron layer (MLP) with 128 hidden units. The output of the transition model is a Gaussian distribution, while the output of the policy network is a softmax layer. Each GRU layer is followed by a ReLU layer which is not presented in the figure.

4.4 Online Fine-Tuning

In the offline phase, we train conservative policy by penalizing the reward with the estimated uncertainty. However, the offline data may not cover all possible states and actions, leading to distribution shift when deploying the policy in an online environment. To address this issue, we further propose to improve the policy with online samples, to bridge the gap between offline transition model and online environment. The online fine-tuning process consists of two steps: 1) Collecting online samples by interacting with the environment; 2) Updating the policy using the collected samples.

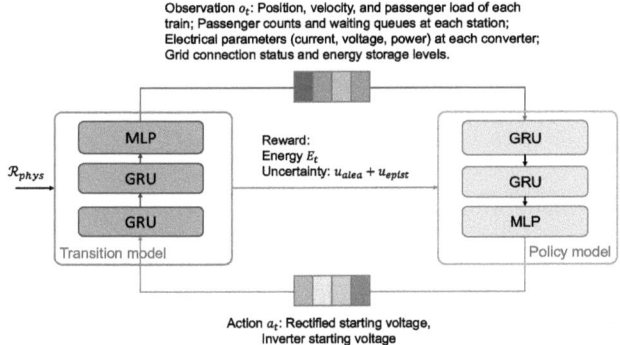

Fig. 2. This figure illustrates the structural design of the transition model and the policy model. Both models feature two stacked Gated Recurrent Unit (GRU) layers (256 hidden units each), followed by a Multi-Layer Perceptron (MLP) layer (128 hidden units).

When collecting the samples, our main motivation is to focus on the regions where the model is uncertain. We can achieve this by sampling actions from the policy with high uncertainty. When collecting online samples, we guide the policy to explore the uncertain regions by sampling actions from the policy with high uncertainty. In particular, we use the epistemic uncertainty since it can be interpreted as a measure of how much the model does not know about the environment. Therefore, we can use it to guide the exploration process. The exploration process can be formulated with Boltzman distribution as follows:

$$\pi(a|s) = \frac{\exp(Q(s,a) + \lambda_u u_{epist}(s,a))}{\sum_{a'} \exp(Q(s,a') + \lambda_u u_{epist}(s,a'))}, \tag{20}$$

where λ_u is a temperature parameter that controls the strength of the uncertainty term. The higher the uncertainty, the more likely the action will be selected. This encourages the policy to explore actions that are associated with high uncertainty, allowing it to gather more informative samples.

The detailed algorithm is summarized in Algorithm 1, named as Regularized Model-based Offline Policy Optimization (RMOPO). In the algorithm, we first train the transition model using the offline dataset. Then, we optimize the policy using the transition model and the uncertainty-guided reward function. Finally, we collect online samples by interacting with the environment and update the policy using the collected samples.

5 Experiments

We designed our experiments to answer the following questions: What is the performance of RMOPO in the URT system? Would the physical regularizers be beneficial for transition model and policy optimization? What is the performance of RMOPO in the online environment? To answer these questions, we conducted

---
Algorithm 1. Regularized Model-based Offline Policy Optimization (RMOPO)

1: **INPUT**: Offline dataset $\mathcal{D}_{off}$, learning rate α, hyperparameters β, λ;
2: **OUTPUT**: Policy π_{on};
3: Randomly initialize parameters φ, ϕ, θ;
4: # *offline stage:*
5: Train transition model $\hat{T}_\theta(s', r|s, a)$ with $\mathcal{D}_{off}$;
6: Compute uncertainty $u(s_i, a_i)$ with Eq. 19;
7: Optimize policy π_φ with PPO.
8: # *online stage:*
9: **for** *epoch* $= 0, 1, 2, \ldots$ **do**
10:    Sample state s_0 from the online environment;
11:    **for** $t = 0, 1, 2, \ldots$ **do**
12:      Sample action according to Eq. 20;
13:      Execute a_t, get reward r_t and next state s_{t+1};
14:      Update policy parameters φ with PPO;
15:    **end for**
16: **end for**
17: **return** π_φ;
---

experiments on a tide calculation simulated URT system with 15 BCs, 15 stations and 23 rails.

5.1 Experiment Settings

Data Collection. The dataset contains the train operation data, including position, velocity, power consumption, passengers, electrical parameters, grid connection, etc. We used this dataset to train our transition model and evaluate our method. There are three kinds of collection policies: a random one that explores many possible actions, a conservative one that seldom explores, and an expert one that balances exploration and exploitation. We denote the three policies as: random, conservatism, and expert. The data is collected every minute for 30 days. The dataset is divided into training set and test set with a ratio of 8:2. The performance of the policy is also evaluated in the simulation system.

Experimental Implementation. We primarily follow the PPO [15] and MOPO [23] settings for basic hyperparameters, while the other adjusted hyperparameters mainly include the online data ratio f and the weight adjustment position t_0. We ran RMOPO on 3 random seeds for each task. In each run, we conducted 50k steps of online interaction and the weight adjustment rate k is set to 0.001. Next, considering fairness and consistency, we conducted 1M steps of policy training according to convention. We evaluated the policy every 1k steps.

Hyperparameters. By default, the hyperparameters are set as follows: the learning rate is set to 1e-3, the batch size is set to 256, the discount factor γ is set to 0.99, the temperature parameter α is set to 2, and the number of Monte Carlo samples K is set to 10.

The hyperparameters for the physical regularizers are set as follows: λ_0 and λ_{phys} is set to 0.1. λ_u is set as 1.0. λ_1, λ_2, and λ_3 are set to 0.1, 0.1, and 0.5 respectively. Similar to MOPO, the hyperparameter β for the uncertainty term is set as 1.0.

5.2 Results

Table 1. Results of ablation experiment.

Collecting Policy	RMOPO(ours)	MOPO	COMBO	AWAC	MORel	RAMBO
random	**48.3 ± 1.2**	33.6	31.9	42.3	21.5	39.4
conservatism	**42.1 ± 0.7**	25.4	21.3	25.5	27.3	25.9
expert	**78.5 ± 2.3**	63.3	52.4	55.9	73.5	65.7

Performance of RMOPO. In the first experiment, we compared RMOPO with offline-to-online RL algorithm AWAC [13], as well as several model-based baseline algorithms, including MOPO [23], COMBO [22], MORel [5], RAMBO [14], and AWAC [13]. Since RMOPO is built upon MOPO with an added online fine-tuning process, comparing it to MOPO can be seen as an ablation study. The comparison results are presented in Table 1. We report the normalized performance averaged over 3 seeds. ± captures the standard deviation over seeds. We bold the highest mean.

As shown in Table 1, RMOPO performs the best among different data collecting policies. The performance of RMOPO is significantly better than those of the other algorithms, indicating that RMOPO can effectively learn from offline data and adapt to the online environment. The performance of MOPO is significantly worse than RMOPO. This indicates that the physical-regularization and online fine-tuning process are beneficial for improving the performance of the policy. The performance of COMBO and MORel is relatively worse, indicating that they are not suitable for this task. AWAC and RAMBO also have high performance. The performance of the random policy is better than that of the conservatism policy, since the random policy can explore more informative samples. The performance of the expert policy is the best, meaning that the expert policy can provide more informative samples for training.

The Role of Physical Regularizers. In the second experiment, we validate the effectiveness of the physical regularizers. We compare the performance of RMOPO with different weights of the physical regularizers. First, we conduct experiments w.r.t. different values of the hyperparameter λ_0, which controls the strength of the physical regularization term for the transition model. The results are shown in Fig. 3. As presented, the performance of RMOPO is significantly improved with the physical regularizers. The performance of RMOPO is better

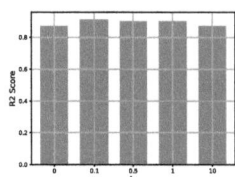
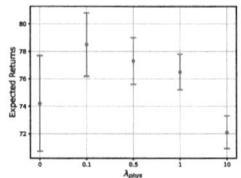
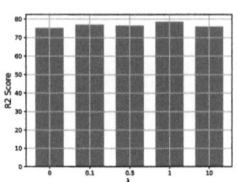

Fig. 3. The expected return w.r.t different values of λ_0

Fig. 4. The expected return w.r.t different values of λ_{phys}

Fig. 5. The expected return w.r.t different values of λ_u

than those of the baseline algorithm, indicating that the physical regularizers are beneficial for the transition model and policy optimization. The optimal value for λ_0 is 0.1. When this value increases, the performance may decrease due to over-regularization.

Figure 4 shows the performance of RMOPO with different values of λ_{phys}, which controls the strength of the physical regularization term for the policy. The performance of RMOPO is significantly improved with the physical regularizers. The optimal value for λ_{phys} is also 0.1.

In Fig. 5, we show the performance of RMOPO with different values of λ_u, which controls the strength for the policy during online learning. As presented, the optimal value for λ_u is 1.0. When this value is too small, the performance may decrease since the collected samples are not informative. On the other hand, large values of λ_u may lead to over-exploration, which can also degrade the performance.

Online Fine-Tuning. In this experiment, we validate the effectiveness of the online fine-tuning. In particular, we compare the performance of RMOPO with and without the online fine-tuning process. The results are shown in Table 2. The version of RMOPO without fine-tuning is denoted as RMOPO w/o ol. As presented, the performance of RMOPO with the online fine-tuning process is significantly better than that of RMOPO without the online fine-tuning process. This indicates that the online fine-tuning process can effectively improve the performance of the policy. Among the collecting policies, the conservative policy improves the most, since the original collected data has very limited coverage.

Table 2. Results of ablation experiment.

Collecting Policy	RMOPO	RMOPO w/o ol
random	**48.3 ± 1.2**	39.7
conservatism	**42.1 ± 0.7**	30.6
expert	**78.5 ± 2.3**	75.3

6 Conclusion

In this work, we study the problem of energy-efficient control for URT systems. We propose a novel model-based offline RL algorithm, RMOPO, which incorporates physical regularizers into the transition model and policy optimization. The proposed method is able to learn from offline data and adapt to the online environment. We conduct extensive experiments on a tide calculation simulated URT system with 15 BCs, 15 stations and 23 rails. The experimental results show that RMOPO outperforms several state-of-the-art algorithms in terms of energy efficiency. In addition, we validate the effectiveness of the physical regularizers and the online fine-tuning process.

The proposed method can be easily extended to other energy-efficient control problems in transportation systems. In the future, we will explore more advanced physical regularizers and online fine-tuning methods to further improve the performance of the proposed method.

References

1. Haarnoja, T., Zhou, A., Abbeel, P., Levine, S.: Soft actor-critic: off-policy maximum entropy deep reinforcement learning with a stochastic actor. In: International Conference on Machine Learning, pp. 1861–1870. PMLR (2018)
2. Huang, Y., Yang, L., Tang, T., Cao, F., Gao, Z.: Saving energy and improving service quality: bicriteria train scheduling in urban rail transit systems. IEEE Trans. Intell. Transp. Syst. **17**(12), 3364–3379 (2016)
3. Janner, M., Fu, J., Zhang, M., Levine, S.: When to trust your model: model-based policy optimization. In: Advances in Neural Information Processing Systems, vol. 32 (2019)
4. Jia, Z., Jiang, J., Lin, H., Cheng, L.: A real-time MPC-based energy management of hybrid energy storage system in urban rail vehicles. Energy Procedia **152**, 526–531 (2018)
5. Kidambi, R., Rajeswaran, A., Netrapalli, P., Joachims, T.: Morel: model-based offline reinforcement learning. In: Advances in Neural Information Processing Systems, vol. 33, pp. 21810–21823 (2020)
6. Kostrikov, I., Nair, A., Levine, S.: Offline reinforcement learning with implicit q-learning. arXiv preprint arXiv:2110.06169 (2021)
7. Kumar, A., Fu, J., Soh, M., Tucker, G., Levine, S.: Stabilizing off-policy q-learning via bootstrapping error reduction. In: Advances in Neural Information Processing Systems, vol. 32 (2019)
8. Kumar, A., Zhou, A., Tucker, G., Levine, S.: Conservative q-learning for offline reinforcement learning. In: Advances in Neural Information Processing Systems, vol. 33, pp. 1179–1191 (2020)
9. Liu, X., Dabiri, A., Wang, Y., De Schutter, B.: Modeling and efficient passenger-oriented control for urban rail transit networks. IEEE Trans. Intell. Transp. Syst. **24**(3), 3325–3338 (2022)
10. Liu, Y., Swaminathan, A., Agarwal, A., Brunskill, E.: Provably good batch off-policy reinforcement learning without great exploration. In: Advances in Neural Information Processing Systems, vol. 33, pp. 1264–1274 (2020)

11. Lu, C., Ball, P., Teh, Y.W., Parker-Holder, J.: Synthetic experience replay. In: Advances in Neural Information Processing Systems, vol. 36 (2024)
12. Lyu, J., Ma, X., Li, X., Lu, Z.: Mildly conservative q-learning for offline reinforcement learning. In: Advances in Neural Information Processing Systems, vol. 35, pp. 1711–1724 (2022)
13. Nair, A., Gupta, A., Dalal, M., Levine, S.: AWAC: accelerating online reinforcement learning with offline datasets. arXiv preprint arXiv:2006.09359 (2020)
14. Rigter, M., Lacerda, B., Hawes, N.: Rambo-RL: robust adversarial model-based offline reinforcement learning. In: Advances in Neural Information Processing Systems, vol. 35, pp. 16082–16097 (2022)
15. Schulman, J., Wolski, F., Dhariwal, P., Radford, A., Klimov, O.: Proximal policy optimization algorithms. arXiv preprint arXiv:1707.06347 (2017)
16. Sun, J., Jiang, Y., Qiu, J., Nobel, P., Kochenderfer, M.J., Schwager, M.: Conformal prediction for uncertainty-aware planning with diffusion dynamics model. In: Advances in Neural Information Processing Systems, vol. 36 (2024)
17. Wang, D., Wu, J., Wei, Y., Chang, X., Yin, H.: Energy-saving operation in urban rail transit: a deep reinforcement learning approach with speed optimization. Travel Behav. Soc. **36**, 100796 (2024)
18. Wang, J., Li, W., Jiang, H., Zhu, G., Li, S., Zhang, C.: Offline reinforcement learning with reverse model-based imagination. In: Advances in Neural Information Processing Systems, vol. 34, pp. 29420–29432 (2021)
19. Wu, Y., Tucker, G., Nachum, O.: Behavior regularized offline reinforcement learning. arXiv preprint arXiv:1911.11361 (2019)
20. Xu, H., Jiang, L., Jianxiong, L., Zhan, X.: A policy-guided imitation approach for offline reinforcement learning. In: Advances in Neural Information Processing Systems, vol. 35, pp. 4085–4098 (2022)
21. Yang, X., Li, X., Ning, B., Tang, T.: A survey on energy-efficient train operation for urban rail transit. IEEE Trans. Intell. Transp. Syst. **17**(1), 2–13 (2015)
22. Yu, T., Kumar, A., Rafailov, R., Rajeswaran, A., Levine, S., Finn, C.: Combo: conservative offline model-based policy optimization. In: Advances in Neural Information Processing Systems, vol. 34, pp. 28954–28967 (2021)
23. Yu, T., et al.: Mopo: model-based offline policy optimization. In: Advances in Neural Information Processing Systems, vol. 33, pp. 14129–14142 (2020)
24. Zhou, G., et al.: Diffusion model predictive control. arXiv preprint arXiv:2410.05364 (2024)

P2MFDS: A Privacy-Preserving Multimodal Fall Detection System for Elderly People in Bathroom Environments

Haitian Wang[1,2], Yiren Wang[2], Xinyu Wang[2], Yumeng Miao[2], Yuliang Zhang[2], Yu Zhang[1(✉)], and Atif Mansoor[2]

[1] School of Computer Science, Northwestern Polytechnical University, No. 1 Dongxiang Road, Chang'an District, Xi'an 710129, China
zhangyu@nwpu.edu.cn
[2] Department of Computer Science and Software Engineering, The University of Western Australia, 35 Stirling Hwy, Crawley, WA 6009, Australia

Abstract. By 2050, people aged 65 and over are projected to make up 16% of the global population. As aging is closely associated with increased fall risk, particularly in wet and confined environments such as bathrooms where over 80% of falls occur. Although recent research has increasingly focused on non-intrusive, privacy-preserving approaches that do not rely on wearable devices or video-based monitoring, these efforts have not fully overcome the limitations of existing unimodal systems (e.g., WiFi-, infrared-, or mmWave-based), which are prone to reduced accuracy in complex environments. These limitations stem from fundamental constraints in unimodal sensing, including system bias and environmental interference, such as multipath fading in WiFi-based systems and drastic temperature changes in infrared-based methods. To address these challenges, we propose a Privacy-Preserving Multimodal Fall Detection System for Elderly People in Bathroom Environments. First, we develop a sensor evaluation framework to select and fuse millimeter-wave radar with 3D vibration sensing, and use it to construct and preprocess a large-scale, privacy-preserving multimodal dataset in real bathroom settings, which will be released upon publication. Second, we introduce P2MFDS, a dual-stream network combining a CNN–BiLSTM–Attention branch for radar motion dynamics with a multi-scale CNN–SEBlock–Self-Attention branch for vibration impact detection. By uniting macro- and micro-scale features, P2MFDS delivers significant gains in accuracy and recall over state-of-the-art approaches. Code and pretrained models are available at https://github.com/HaitianWang/P2MFDS-A-Privacy-Preserving-Multimodal-Fall-Detection-Network-for-Elderly-Individuals-in-Bathroom.

Keyword: Fall detection

1 Introduction

By 2050, people aged 65 and above are projected to constitute 16% of the global population, up from 10% in 2022 [1]. A growing number of elderly people prefer to live independently, yet this increases their vulnerability to medical emergencies such as falls, which remain one of the leading causes of injury-related morbidity and mortality among elderly adults [2,3]. Bathrooms, with their slippery surfaces, are particularly hazardous; studies indicate that over 80% of elderly falls occur in these environments [4]. Given that falls can cause severe injuries or deaths, a real-time and accurate solution to detect falls in the bathroom is essential for the elderly [5,6].

Existing fall detection solutions for elderly people can be categorized into three primary approaches. The first involves in-home caregivers who provide continuous monitoring and immediate assistance in the event of a fall [7,8]. Although effective, this solution is cost-prohibitive and lacks scalability [9]. The second approach utilizes wearable devices, such as smart watches and electrocardiogram (ECG) monitors. These devices continuously track physiological parameters such as heart rate, blood pressure, and vibration [10]. However, these devices face adoption challenges, particularly among elderly users who may find them uncomfortable, forget to wear them, or struggle to use them in wet environments such as bathrooms [11]. The third approach integrates smart home technologies, including video-based and audio-based fall detection systems [12,13]. While these systems can achieve high detection accuracy, their deployment in private spaces, such as bathrooms, raises significant privacy concerns, limiting user acceptance. Given these limitations, there is a critical need for a non-intrusive, real-time fall detection system that ensures both accuracy and privacy in bathroom environments [14].

Recent research has explored non-intrusive fall detection methods, focusing on privacy-preserving solutions that do not require wearable devices or video-based monitoring [15,16]. Palipana [12] et al. proposed FallDeFi, a WiFi-based system that detects falls by analyzing signal disturbances. While effective in controlled environments, its accuracy degrades in complex spaces like bathrooms due to multipath interference and environmental noise. Similarly, Zigel et al. [17] developed a vibration- and sound-based detection system that addresses privacy concerns but is highly sensitive to background noise, leading to false positives from non-fall events such as dropped objects. Akash et al. [6] explored millimeter-wave (mmWave) radar for fall detection, utilizing its high-resolution motion tracking capabilities.

However, its effectiveness depends on precise calibration and can be influenced by bathroom wall tiles and flooring [18,19]. These approaches predominantly rely on unimodal sensor data, which undergoes preprocessing, feature extraction, and classification to detect falls [20,21]. However, unimodal systems face fundamental limitations: the performance is approaching theoretical bounds, leaving minimal room for improvement through feature engineering alone [22,23]. Moreover, single-sensor data is highly susceptible to system bias and environmental interference. For instance, infrared sensors are affected by temperature

Fig. 1. System overview of the P2MFDS pipeline. First, a multidimensional evaluation framework scores candidate sensing modalities (e.g., mmWave radar, vibration sensor, RGB and thermal cameras) to select the optimal combination. Next, multimodal data are collected in a realistic bathroom setup. The P2MFDS Network then fuses long-term motion features extracted by a 1D CNN–BiLSTM–Attention stream with short-term impact signatures captured by a Multi-Scale CNN–SEBlock–Self-Attention stream. Finally, the fused representation is classified into fall or non-fall events.

variations, while vibration sensors can misinterpret background noise or minor disturbances as falls, leading to increased false alarm rates. These challenges underscore the need for a multimodal, sensor-fusion approach to enhance accuracy and robustness in real-world applications [24,25].

To solve this issue, we propose a Privacy-Preserving Multimodal Fall Detection System for Elderly People in Bathroom Environments. This system comprises two parts (as shown in Fig. 1)). The first part is a systematic sensor evaluation framework that guides the selection and fusion of millimeter-wave radar with 3D vibration sensing, enabling the collection and preprocessing (smoothing, filtering, fusion) of a large-scale, privacy-preserving multimodal dataset in real bathroom settings; this dataset will be released upon publication. The second part is P2MFDS, a dual-stream network combining a CNN–BiLSTM–Attention branch for capturing long-term motion dynamics from radar point clouds and a multi-scale CNN–SEBlock–Self-Attention branch for detecting short-term impact signatures in vibration data. By uniting macro- and micro-scale features, P2MFDS achieves 95.0% accuracy, 87.9% recall, and a 91.3% F1-score, substantially outperforming state-of-the-art unimodal approaches. The complete implementation and pretrained models are available at P2MFDS Github Repository.

2 Offline Multimodal Sensors Selection

To develop a robust non-intrusive fall detection system suitable for bathroom environments, we propose a comprehensive Multimodal Sensor Evaluation Framework (as shown in Table 1). This framework systematically assesses various sensing modalities based on their feasibility, performance, and deployability in real-world conditions. The evaluation criteria include accuracy, non-intrusiveness, energy efficiency, computational complexity, deployability, recall rate, availability, and cost-effectiveness.

Table 1. Proposed evaluation framework for selecting optimal sensing modalities in privacy-preserving fall detection.

Selection Criteria	Description	Rating	Weight
Target Relevance	Relevance for fall detection	1–3 scale	3
Non-Intrusiveness	Level of privacy preservation	1–5 scale	2
Energy Efficiency	Power consumption (AC/DC)	1–5 scale	3
Comp Complexity	Processing demands	1–5 scale	2
Deployability	Ease of integration	1–5 scale	2
Recall Rate	Ability to detect falls	1–5 scale	3
Availability	Reliability in real conditions	1–5 scale	2
Cost-Effectiveness	Economic feasibility	1–5 scale	1

Table 2. Evaluation of 14 sensing modalities across nine key factors for privacy-preserving fall detection in bathrooms. TR: Target Relevance, CE: Cost Effectiveness, Rec: Recall, EE: Energy Efficiency, CC: Computational Complexity, Dep: Deployability, NI: Non-Intrusiveness, Avail: Availability, Score: Usability Score.

Method	TR	CE	Rec	EE	CC	Dep	NI	Avail	Score
Wi-Fi	3	3	4	2	4	3	5	3	61
Infrared	3	2	3	3	5	4	5	1	61
Bluetooth	3	4	4	2	4	3	5	2	64
MMW	3	3	5	4	5	4	5	3	**73**
Ultrasonic	3	3	4	2	4	4	4	3	62
Audio	3	4	4	3	3	4	2	4	65
Thermal Imaging	3	1	5	5	3	3	4	1	62
3D Vibration	3	4	4	4	3	3	5	4	**68**
Light Sensors	1	5	4	1	2	1	5	5	47
Door Sensors	1	5	4	2	2	5	5	4	62
Furniture Sensors	1	5	2	1	1	5	5	3	51
Weight Sensors	1	5	3	1	1	5	5	4	57
Electricity Usage	1	2	1	1	2	2	5	4	63
Water Usage	1	2	3	1	2	3	5	4	54

Each sensing modality is assigned a weighted score for the eight evaluation dimensions, contributing to an overall usability score (OU) using the following equation: $OU_m = \sum_{i=1}^{N} w_i \times S_{m,i}$, where OU_m represents the overall usability of modality m, $S_{m,i}$ denotes the score for modality m in the i-th dimension, and w_i is the weight assigned to the i-th evaluation criterion. Higher OU_m values indicate superior suitability for bathroom fall detection.

Using the Multimodal Sensor Evaluation Framework, we evaluate 14 widely used sensing modalities across multiple categories. The evaluation results (as shown in Table 2) indicate that mmWave radar and a three-axis vibration achieve the highest usability scores, making them the most suitable choices for privacy-preserving fall detection in bathroom environments. The mmWave radar offers high-precision motion tracking that remains effective in varying lighting conditions while maintaining privacy, whereas the three-axis vibration provides real-time vibration sensing with low power consumption, ensuring reliable detection. The combination of these two modalities enables a complementary approach, where the mmWave radar captures overall motion patterns, and the vibration sensor detects localized ground vibrations. This multimodal integration enhances detection accuracy, minimizes false positives, and improves robustness against environmental noise, making the system well-suited for real-world deployment.

3 Privacy-Preserving Multimodal Fall Detection System Network

Privacy-Preserving Multimodal Fall Detection System (P2MFDS) network fuses mmWave radar and triaxial vibration streams to exploit their complementary sensing characteristics. After targeted low-pass filtering to remove high-frequency noise while preserving fall-related signatures, the radar branch captures macro-scale motion attributes (e.g., velocity, distance, signal energy) and the vibration branch extracts micro-scale impact signatures. Two independent processing streams—one employing a CNN–BiLSTM–Attention pipeline for radar point clouds and the other using a Multi-Scale CNN–SEBlock–Self-Attention pipeline for vibration data—then distill salient features. Finally, these embeddings are concatenated and fed into a lightweight classifier for fall vs. non-fall prediction (Fig. 2).

3.1 Signal Preprocessing

Prior to feature extraction, each modality undergoes targeted low-pass filtering to suppress high-frequency noise while preserving fall-related signatures. For vibration data, a Moving Average Filter (MAF) smooths transient disturbances by computing

$$y_{\text{vib}}(t) = \frac{1}{N} \sum_{i=0}^{N-1} x_{\text{vib}}(t-i), \tag{1}$$

where N is chosen to balance noise reduction against responsiveness. For radar data, a first-order Exponential Low-Pass Filter (ELPF) adapts to rapid motion changes:

$$y_{\text{radar}}(t) = \alpha\, x_{\text{radar}}(t) + (1-\alpha)\, y_{\text{radar}}(t-1), \tag{2}$$

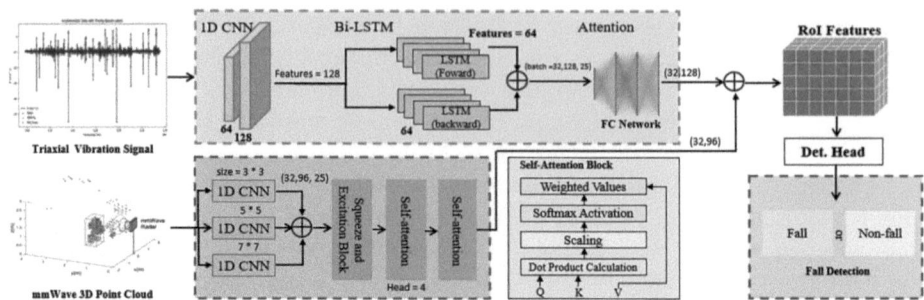

Fig. 2. Overview of the P2MFDS network architecture. The upper pipeline employs a 1D CNN–BiLSTM–Attention sequence to extract global motion features from mmWave 3D point clouds, while the lower pipeline utilizes multi-scale CNN, SE block, and self-attention to capture localized vibration signals. Both streams are fused to enable robust, privacy-preserving fall detection in bathroom environments.

with $\alpha \in (0,1)$ controlling the trade-off between stability and sensitivity. This dual-filtering strategy mitigates environmental artifacts (e.g., electrical interference, multipath reflections) and enhances the robustness of subsequent feature extraction.

3.2 3D Vibration-Based Impact Feature Extraction

To capture fine-grained temporal dynamics and transient impacts related to fall events, the triaxial vibration signal initially undergoes feature extraction through a 1D Convolutional Neural Network (1D CNN) and a Bidirectional Long Short-Term Memory (Bi-LSTM) network. Specifically, the input vibration data is first convolved by multiple CNN kernels to produce intermediate temporal-spatial embeddings, which are subsequently fed into a Bi-LSTM composed of forward and backward sequences. The Bi-LSTM effectively integrates long-term dependencies from historical and future contexts, yielding comprehensive temporal embeddings as follows:

$$\overrightarrow{h_t} = \text{LSTM}_{\text{fw}}(X_t, \overrightarrow{h_{t-1}}), \quad \overleftarrow{h_t} = \text{LSTM}_{\text{bw}}(X_t, \overleftarrow{h_{t+1}}) \tag{3}$$

where $\overrightarrow{h_t}$ and $\overleftarrow{h_t}$ represent forward and backward hidden states at time step t, respectively. These embeddings are concatenated to form the final hidden state $h_t = [\overrightarrow{h_t}, \overleftarrow{h_t}]$, effectively capturing bidirectional temporal information.

To highlight critical temporal segments indicative of potential falls, an attention mechanism is subsequently applied. Given the Bi-LSTM output $H = \{h_1, h_2, \ldots, h_T\}$, the attention weights α_t are calculated by:

$$\alpha_t = \frac{\exp(h_t W_a)}{\sum_{i=1}^{T} \exp(h_i W_a)}, \quad F_{\text{vibration}} = \sum_{t=1}^{T} \alpha_t h_t \tag{4}$$

where W_a denotes learnable parameters, and $F_{\text{vibration}}$ represents the refined impact feature embedding emphasizing the most informative segments of vibration data.

3.3 mmWave Radar-Based Motion Feature Extraction

The mmWave radar stream focuses on capturing macro-scale motion patterns within the 3D point cloud data through a multi-scale convolutional strategy enhanced by channel-wise and temporal attention mechanisms. Initially, radar signals are processed by parallel 1D CNN branches with varying kernel sizes (3, 5, and 7) to capture diverse motion scales. These multi-scale features are merged into a unified embedding for subsequent recalibration.

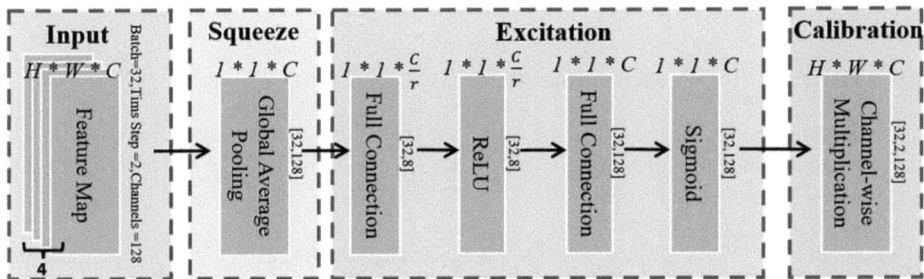

Fig. 3. Architecture of the SE block. Channel-wise statistics are extracted via global average pooling, passed through two fully connected layers with a bottleneck structure, and used to generate attention weights via sigmoid activation. The input feature map is then recalibrated through channel-wise multiplication.

To further enhance critical channel-level characteristics and suppress redundant features, we incorporate a Squeeze-and-Excitation (SE) block [26]. The SE block contributes to improved generalization by reducing the risk of overfitting through dynamic channel reweighting, which is particularly beneficial under data-limited training conditions. The SE block first applies global average pooling (GAP) to aggregate global spatial information across each channel, forming a channel descriptor z_c:

$$z_c = \frac{1}{H \times W} \sum_{i=1}^{H} \sum_{j=1}^{W} x_c(i,j) \tag{5}$$

where $x_c(i,j)$ denotes the feature map value at position (i,j) of channel c (as shown in Fig. 3). Subsequently, channel-wise recalibration weights are learned through fully connected layers with a bottleneck architecture:

$$s_c = \sigma\left(W_2 \operatorname{ReLU}(W_1 z_c)\right), \quad \hat{x}_c = s_c \cdot x_c \tag{6}$$

where W_1 and W_2 represent learnable weights of fully connected layers, σ denotes the sigmoid activation function, and $\hat{x}_c$ is the recalibrated channel-wise feature map. This recalibration adaptively emphasizes significant motion-related features across channels.

Finally, a self-attention mechanism utilizes these recalibrated features to enhance temporal sensitivity and motion dynamics (as shown in Fig. 2). Query (Q), Key (K), and Value (V) matrices are constructed, and the self-attention output is computed as:

$$F_{\text{radar}} = \text{softmax}\left(\frac{QK^{\text{T}}}{\sqrt{d_k}}\right)V \tag{7}$$

Here, d_k is the dimensionality of keys, and the resulting embedding F_{radar} captures the most salient temporal motion characteristics relevant to accurate fall detection.

3.4 Multimodal Feature Fusion and Classification

After the mmWave 3D point cloud and vibration streams independently extract motion and impact features, respectively, these two embeddings are concatenated to form a joint representation:

$$F_{\text{fusion}} = \sigma\big(W_f[F_{\text{radar}}; F_{\text{acc}}] + b_f\big), \tag{8}$$

where W_f and b_f are learnable parameters, and σ is the activation function. This fused feature map (labeled "RoI Features" in Fig. 2) is then fed into a lightweight *detection head* (Det Head)—a final classification module comprising fully connected layers and a confidence function. The Det Head transforms the high-level fused representation into discrete predictions (fall vs. non-fall) [27,28].

4 Experiment

This section presents the experimental design and setup used to evaluate the proposed P2MFDS Network in real-world bathroom scenarios. We first describe the testing environment and sensor placement, followed by the methodology for collecting multimodal data under various fall and non-fall conditions. Finally, we present the metrics and benchmarking procedures used to validate our system's effectiveness and robustness.

4.1 Experiment Setup

To ensure the reliability, reproducibility, and real-world applicability of our proposed fall detection system, we conducted experiments in a controlled environment designed to replicate a typical residential bathroom (Fig. 4 shows the Top-down floor plan). The experimental space measure 2.5 m (L) × 1.1 m (W) × 2.2 m (H), forming a 6.05 m² enclosed testing area. The bathroom features ceramic tiled

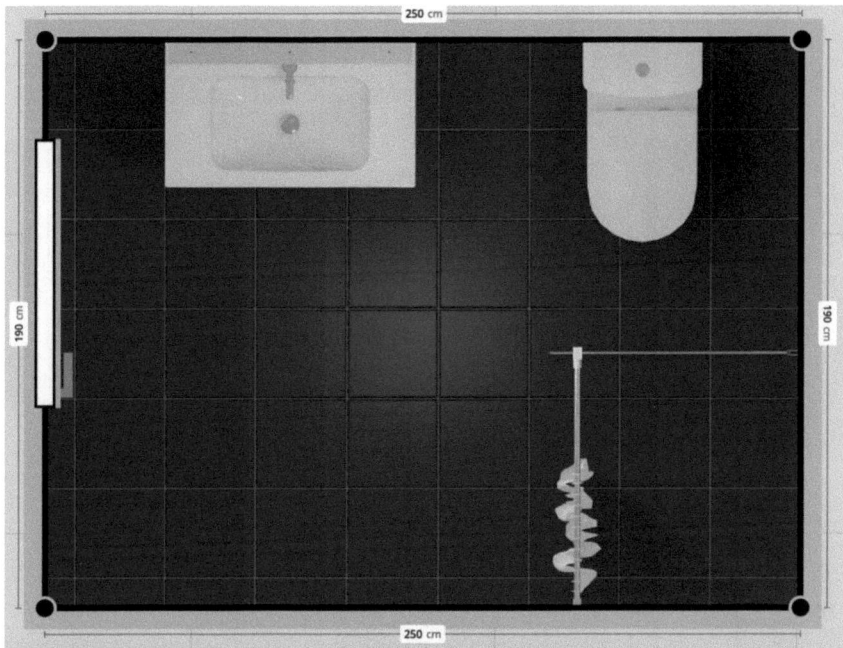

Fig. 4. Top-down floor plan of the experimental bathroom (2.50 m × 1.90 m), showing the mmWave radar and vibration sensor mounting positions, annotated coverage areas, and key dimensions.

walls, along with a glass partition and a shower curtain that resemble a typical residential bathroom. The floor consists of anti-slip ceramic tiles, with standard bathroom utilities including a showerhead, drain, and storage shelves containing toiletries such as soap and shampoo (as shown in Fig. 5).

To achieve optimal sensor coverage while minimizing occlusion and environmental interference, two sensor nodes were strategically deployed within the test area. The triaxial vibration sensor node comprising an ADXL345 3-axis vibration, an ESP32-C3 microcontroller, and a rechargeable battery was securely placed 0.5 m from the wall beneath the showerhead, a location frequently occupied during bathing activities. A waterproof case was used to protect the sensor from moisture exposure. The mmWave radar node consisting of a C4001 mmWave radar sensor, an ESP32-C3 MCU, and a battery unit was mounted on a metal beam at a height of 2.2 m, ensuring unobstructed coverage of the entire bathroom space.

4.2 Experimental Design

The evaluation protocol comprises nine representative bathroom scenarios (Fig. 6) chosen to span a spectrum of non-fall and fall events, namely: (1) empty room (no motion), (2) light object drop (e.g., soap), (3) heavy object drop (e.g.,

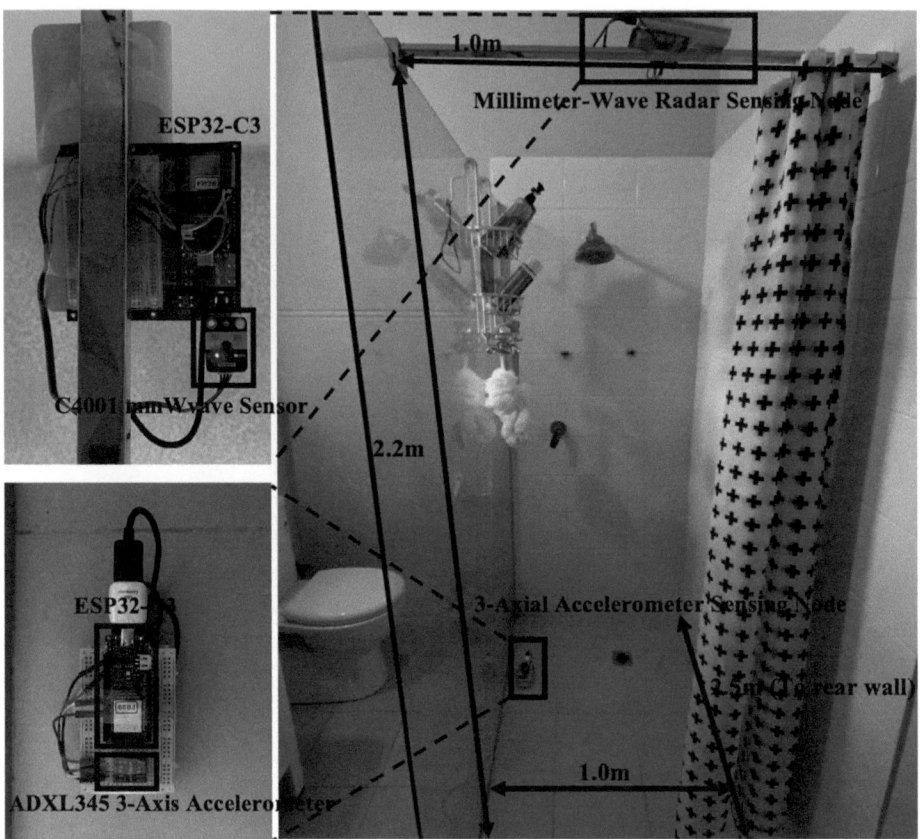

Fig. 5. Illustration of the controlled bathroom environment and sensor placements. The mmWave radar node is mounted at 2.2 m for broad coverage, while the ADXL345 vibration sensor is installed near the shower area to capture ground impacts. This setup closely replicates typical residential bathroom conditions for realistic fall detection experiments.

mop), (4) normal walking, (5) bent-posture walking, (6) wall-supported walking, (7) static standing, (8) squatting, and (9) intentional falls. These scenarios were selected to challenge the system with variations in motion intensity, impact signature, and postural change, thereby testing P2MFDS's ability to distinguish true falls from everyday disturbances and fall-like activities.

Each scenario was conducted in multiple trials by all participants, resulting in a dataset comprising 20 minutes per scenario, with a total duration of 3 h. The mmWave radar operated at a sampling rate of 10 Hz, generating 3D point clouds at 10 frames per second, capturing motion dynamics over time. Simultaneously, the triaxial vibration recorded 3D vibration data at 100 Hz, ensuring high temporal resolution for detecting impact forces associated with falls. The

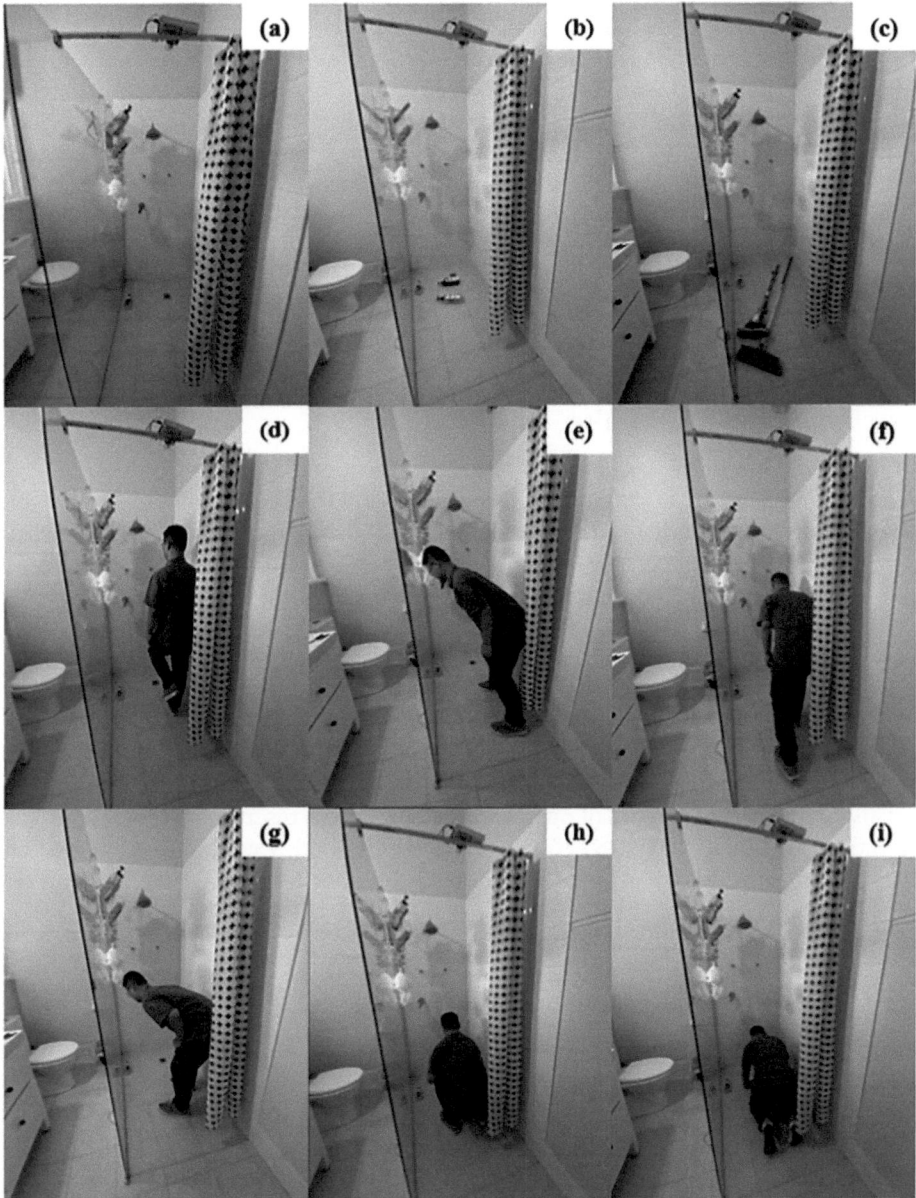

Fig. 6. Overview of the eight experimental scenarios simulating typical bathroom conditions:(a) empty bathroom (no motion),(b) lightweight object drop,(c) heavy object drop,(d) normal walking,(e) bent-posture walking,(f) assisted walking with wall support,(g) static standing or squatting,(h)squat down and (i) fall down. These scenarios collectively capture a wide range of non-fall and fall-like events for robust real-world detection analysis.

dataset included over 120,000 vibration data points and 18,000 mmWave frames, allowing for robust multimodal feature extraction.

This experimental design ensures the system can effectively distinguish genuine falls from environmental disturbances, such as object drops, while accounting for variations in human movement patterns.

The P2MFDS Network achieves an overall accuracy of 95.0%, precision of 94.6%, recall of 87.8%, and F1-score of 91.3% across eight representative bathroom scenarios (Table 3). Scenario-wise F1-scores exceed 90% in six activities, peaking at 97.9% for normal walking and 97.3% for squatting. The confusion matrices in Fig. 7 confirm a balanced distribution of true positives and true negatives, with squatting yielding zero false negatives and static standing maintaining a 93.9% F1-score despite potential environmental noise.

4.3 Experimental Results

By fusing macro-scale motion features from the mmWave radar with micro-impact signatures from 3D vibration sensing, P2MFDS effectively suppresses false alarms in non-fall events (e.g., 97.5% F1 in an empty bathroom) while retaining high fall detection recall under complex disturbances (e.g., 94.6% recall in light object drops). This robust performance across both quantitative metrics and confusion-matrix analyses underscores the method's resilience to noise and its practical viability for privacy-preserving fall monitoring in confined bathroom environments.

4.4 Comparison

To validate the effectiveness of our proposed Privacy-Preserving Multimodal Fall Detection System (P2MFDS) Network, we conducted a comprehensive performance comparison with 16 state-of-the-art fall detection models, as summarized

Table 3. P2MFDS performance across eight bathroom scenarios, reporting accuracy (Acc), recall (Rec), precision (Pre), and F1-score (F1) for non-fall and fall detection.

Scenario	Non-Fall (%)				Fall (%)			
	Acc	Rec	Pre	F1	Acc	Rec	Pre	F1
Empty Bathroom	96.0	96.1	96.1	96.1	96.0	95.9	95.9	95.9
Light Object Drop	95.0	94.3	96.2	95.2	95.0	95.7	93.8	94.7
Heavy Object Drop	90.4	91.4	89.8	90.6	90.4	89.5	91.1	90.1
Normal Walking	94.7	95.7	93.8	94.7	94.7	93.8	95.7	94.8
Bent Posture Walk	89.2	81.8	88.2	84.9	89.2	93.6	89.7	91.6
Wall-Supported Walk	85.3	89.1	84.5	86.7	85.3	80.9	86.4	83.5
Static Standing	92.6	94.6	92.9	93.7	92.6	90.0	92.3	91.1
Squatting	97.0	96.2	98.1	97.1	97.0	97.9	95.8	96.8
Total	**95.0**	**98.0**	**95.1**	**96.5**	**95.0**	**94.6**	**87.8**	**91.3**

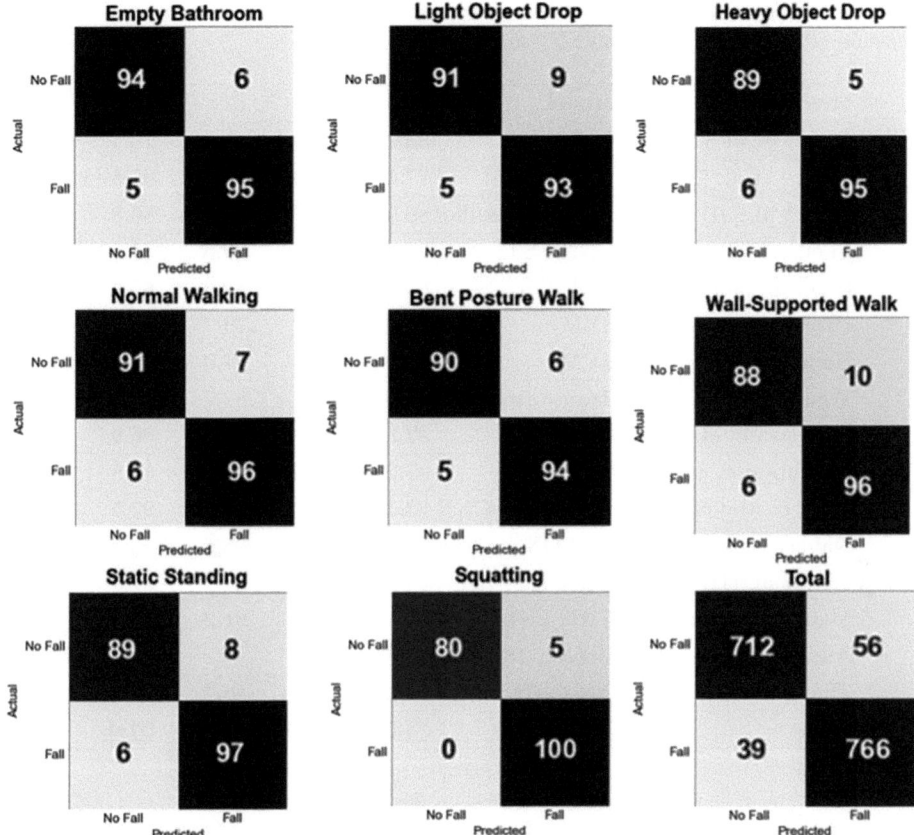

Fig. 7. Confusion matrices of the P2MFDS Network across eight experimental scenarios and the overall total. Each matrix shows classification counts for fall and non-fall events, illustrating high true positive and true negative rates with minimal misclassification.

in Table 4. These methods span a diverse range of sensing modalities—including mmWave radar, UWB radar, Wi-Fi CSI, vibration sensors, and multimodal fusion—and model architectures such as CNNs, LSTMs, attention mechanisms, and SVMs.

Among radar-only methods, Rezaei et al. [29], He et al. [33], and Yao et al. [31] employed traditional CNN and RBF-based models, achieving moderate recall (82.8%–88.9%) but relatively lower precision, particularly in cluttered environments. Transformer-based [36] and Bi-LSTM+CNN [39] architectures showed improved performance due to better temporal modeling, yet still relied on a single sensing modality.

Several UWB radar-based solutions [30,37,38] demonstrated high recall (up to 98.8%) but suffered from limited precision or overall accuracy, likely due to lower spatial resolution and vulnerability to occlusion. Similarly, vision-assisted

Table 4. Comparison of P2MFDS with state-of-the-art methods on accuracy (Acc), precision (Pre), recall (Rec), and F1-score (F1).

Methods	Model	Acc (%)	Pre (%)	Rec (%)	F1 (%)
Rezaei et al. [29]	CNN	88.7	69.9	85.5	76.9
Maitre et al. [30]	CNN-LSTM	77.0	86.0	89.0	87.5
Yao et al. [31]	RVAM Classifier	86.7	84.7	88.9	86.8
Li et al. [32]	CNN-LSTM	**96.8**	84.6	89.6	87.0
He et al. [33]	RBF NN	85.9	88.3	82.8	85.4
Clemente et al. [34]	SVM	91.2	74.2	73.7	73.9
Hanif et al. [35]	SVM	89.0	88.0	87.0	87.5
Wang et al. [36]	Transformer	94.5	94.2	86.7	90.3
Sadreazami et al. [37]	CNN	82.9	92.8	81.2	86.6
Swarubini et al. [38]	ResNet-50	67.2	50.6	**98.8**	66.8
Yang et al. [39]	Bi-LSTM+CNN	93.1	93.8	91.3	92.5
Dai et al. [40]	YOLOX	93.4	94.6	95.6	95.1
Sun et al. [41]	Att-GRU	94.2	92.4	83.8	87.9
Meng et al. [42]	CNN-LSTM-Att	93.1	94.3	91.2	92.7
Alkhaldi et al. [43]	ResNet-18	93.0	91.4	90.2	90.8
Zhang et al. [44]	Att-CNN-LSTM	93.5	90.2	86.7	88.4
Ours	**P2MFDS**	95.0	**94.6**	87.8	**91.1**

fusion [40] and Wi-Fi CSI-based fusion [42] achieved high recall but may raise privacy concerns in sensitive environments like bathrooms.

Compared to these approaches, P2MFDS achieves competitive accuracy (95.0%), superior precision (94.6%), and robust recall (87.8%) through effective multimodal integration of mmWave radar and 3D vibration sensing. This fusion enables the model to simultaneously capture macro-scale motion dynamics and micro-scale impact signals, enhancing robustness against noise and confounding activities. Notably, our method outperforms other multimodal fusion networks such as Sun et al. [41] and Zhang et al. [44], both in terms of accuracy and precision. These results underscore the superiority of the P2MFDS architecture in balancing detection performance with privacy preservation in challenging, confined environments such as residential bathrooms.

4.5 Ablation Study

To assess the impact of individual components in the Privacy-Preserving Multimodal Fall Detection System (P2MFDS) Network, we conducted an ablation study, as summarized in Table 5. The results demonstrate that multimodal sensor fusion significantly improves fall detection performance. Using only the vibration, the CNN+LSTM model achieves 92.3% accuracy, while mmWave radar alone

Table 5. Ablation study showing the impact of model components (Attention, CNN, LSTM) and sensor modalities (Vibration, Radar) on P2MFDS performance (Acc, Rec, Pre, F1, AUC-ROC).

Model Components			Sensors		Acc	Rec	Pre	F1	AUC
Att	CNN+LSTM	CNN	Vib	Rad	(%)	(%)	(%)	(%)	(%)
✓	✓		✓		92.3	80.6	83.4	82.0	86.2
✓	✓			✓	87.2	88.8	64.3	74.7	75.8
✓	✓		✓	✓	98.3	76.7	96.4	85.4	97.5
✓		✓	✓		84.5	51.2	70.0	59.1	72.1
✓		✓		✓	80.2	84.8	66.3	74.5	68.5
✓		✓	✓	✓	93.4	83.1	91.4	87.1	87.4
✓	✓	✓	✓		95.9	91.4	93.7	92.5	88.3
✓	✓	✓		✓	87.2	89.3	77.7	83.2	78.3
✓	✓	✓	✓	✓	**95.0**	**94.6**	**87.8**	**91.1**	**96.4**

yields 87.2% accuracy. When both modalities are combined, accuracy increases to 98.3%, confirming the advantages of multimodal integration.

In terms of model architecture, the full P2MFDS Network (CNN+LSTM+Attention) achieves the highest performance, outperforming CNN- or LSTM-only configurations. The results highlight the complementary strengths of CNN for spatial feature extraction, LSTM for temporal modeling, and Attention for adaptive weighting, leading to robust and privacy-preserving fall detection.

5 Conclusion

Our experimental results confirm that the Privacy-Preserving Multimodal Fall Detection System (P2MFDS) Network effectively fuses mmWave 3D point cloud and vibration data, achieving high accuracy under various activities such as object drops, squatting, and bent-posture walking. This fusion addresses limitations found in unimodal systems—particularly noise susceptibility and poor recall in cluttered bathroom settings. Furthermore, the attention mechanisms within the CNN-BiLSTM-Attention and Multi-Scale CNN-SEBlock-Self-Attention modules enhance the model's adaptability to different motion patterns while preserving user privacy.

Despite these promising outcomes, environmental factors—such as temperature fluctuations or partial occlusions—may still pose challenges. Additionally, large-scale, long-term deployments in real homes with elderly people are needed to further validate robustness and usability. Future work will focus on exploring adaptive calibration, extending the dataset across diverse bathroom layouts, and improving noise reduction to strengthen overall fall detection performance.

Acknowledgement. This research was funded by the UWA - Department of Computer Science and Software Engineering Research Project Fund under the "Non-Invasive Fall Detection System for Elderly in Bathroom" initiative. Dr. Atif Mansoor served as project leader and supervisor, overseeing the design and implementation to ensure adherence to scientific, ethical, and methodological standards. We also thank all participants for their valuable contributions to the experimental evaluation.

References

1. Leong, D., et al.: World population prospects 2019. Department of economic and social affairs population dynamics. World **73**(7) (2018)
2. Strini, V., Schiavolin, R., Prendin, A.: Fall risk assessment scales: a systematic literature review. Nurs. Rep. **11**(2), 430–443 (2021)
3. Wang, X., Ellul, J., Azzopardi, G.: Elderly fall detection systems: a literature survey. Front. Robot. AI **7**, 71 (2020)
4. Singh, A., Rehman, S.U., Yongchareon, S., Chong, P.H.J.: Sensor technologies for fall detection systems: a review. IEEE Sens. J. **20**(13), 6889–6919 (2020)
5. Yazar, A., Erden, F., Cetin, A.E.: Multi-sensor ambient assisted living system for fall detection. In: Proceedings of the IEEE International Conference on Acoustics, Speech, and Signal Processing (ICASSP 2014), pp. 1–3 (2014)
6. Akash, M.S.R., Shahria, M.N., Morshed, M.A., Rodsee, S.S., Hannan, N., Imam, M.H.: Elderly patient monitoring and fall detection using mmwave fmcw radar system. In: 2023 26th International Conference on Computer and Information Technology (ICCIT), pp. 1–6. IEEE (2023)
7. Chen, W., Jiang, Z., Guo, H., Ni, X.: Fall detection based on key points of human-skeleton using openpose. Symmetry **12**(5), 744 (2020)
8. Tun, S.Y.Y., Madanian, S., Mirza, F.: Internet of things (IoT) applications for elderly care: a reflective review. Aging Clin. Exp. Res. **33**, 855–867 (2021)
9. Cardoso, G.P., Damaceno, D.G., Alarcon, M.F.S., Marin, M.J.S.: Care needs of the elderly who live alone: an intersectoral perception. Rev. Rene **21**, 44395 (2020)
10. Guiñón, J.L., Ortega, E., García-Antón, J., Pérez-Herranz, V.: Moving average and savitzki-golay smoothing filters using mathcad. Papers ICEE **2007**, 1–4 (2007)
11. Perumal, T., et al.: A review on fall detection systems in bathrooms: challenges and opportunities. Multimed. Tools Appl. 1–29 (2024)
12. Palipana, S., Rojas, D., Agrawal, P., Pesch, D.: Falldefi: ubiquitous fall detection using commodity wi-fi devices. Proc. ACM Interact. Mob. Wearable Ubiquit. Technol. **1**(4), 1–25 (2018)
13. Riquelme, F., Espinoza, C., Rodenas, T., Minonzio, J.G., Taramasco, C.: ehome-seniors dataset: an infrared thermal sensor dataset for automatic fall detection research. Sensors **19**(20), 4565 (2019)
14. Wagner, J., Mazurek, P., Morawski, R.Z.: Non-invasive monitoring of elderly persons: systems based on impulse-radar sensors and depth sensors. Springer, Cham (2022)
15. Mavaddat, A., Armaki, S.H.M., Erfanian, A.R.: Millimeter-wave energy harvesting using 4×4 microstrip patch antenna array. IEEE Antennas Wirel. Propag. Lett. **14**, 515–518 (2014)
16. Gerstmair, M., Melzer, A., Onic, A., Stuhlberger, R., Huemer, M.: Highly efficient environment for fmcw radar phase noise simulations in if domain. IEEE Trans. Circuits Syst. II Express Briefs **65**(5), 582–586 (2018)

17. Zigel, Y., Litvak, D., Gannot, I.: A method for automatic fall detection of elderly people using floor vibrations and sound–proof of concept on human mimicking doll falls. IEEE Trans. Biomed. Eng. **56**(12), 2858–2867 (2009)
18. Neipp, C., Hernndez, A., Rodes, J., Mrquez, A., Belndez, T., Belndez, A.: An analysis of the classical doppler effect. Eur. J. Phys. **24**(5), 497 (2003)
19. Taha, A.: Intelligent ensemble learning approach for phishing website detection based on weighted soft voting. Mathematics **9**(21), 2799 (2021)
20. Chen, Y., Du, R., Luo, K., Xiao, Y.: Fall detection system based on real-time pose estimation and SVM. In: 2021 IEEE 2nd International Conference on Big Data, Artificial Intelligence and Internet of Things Engineering (ICBAIE), pp. 990–993. IEEE (2021)
21. Hashim, H.A., Mohammed, S.L., Gharghan, S.K.: Accurate fall detection for patients with Parkinson's disease based on a data event algorithm and wireless sensor nodes. Measurement **156**, 107573 (2020)
22. Appeadu, M.K., Bordoni, B.: Falls and fall prevention in the elderly. In: StatPearls [Internet]. StatPearls Publishing (2023)
23. Too, J., Abdullah, A.R., Zawawi, T.T., Saad, N.M., Musa, H.: Classification of EMG signal based on time domain and frequency domain features. Int. J. Hum. Technol. Interact. (IJHaTI) **1**(1), 25–30 (2017)
24. Khan, S.R., Pavuluri, S.K., Cummins, G., Desmulliez, M.P.: Wireless power transfer techniques for implantable medical devices: a review. Sensors **20**(12), 3487 (2020)
25. Kim, J., Lee, W., Lee, S.H.: A systematic review of the guidelines and delphi study for the multifactorial fall risk assessment of community-dwelling elderly. Int. J. Environ. Res. Public Health **17**(17), 6097 (2020)
26. Hu, J., Shen, L., Sun, G.: Squeeze-and-excitation networks. In: Proceedings of the IEEE Conference on Computer Vision and Pattern Recognition, pp. 7132–7141 (2018)
27. Kiranyaz, S., Avci, O., Abdeljaber, O., Ince, T., Gabbouj, M., Inman, D.J.: 1D convolutional neural networks and applications: a survey. Mech. Syst. Signal Process. **151**, 107398 (2021)
28. Alzubaidi, L., et al.: Review of deep learning: concepts, CNN architectures, challenges, applications, future directions. J. Big Data **8**, 1–74 (2021)
29. Rezaei, A., et al.: Unobtrusive human fall detection system using mmwave radar and data driven methods. IEEE Sens. J. **23**(7), 7968–7976 (2023). https://doi.org/10.1109/JSEN.2023.3245063
30. Maitre, J., Bouchard, K., Gaboury, S.: Fall detection with UWB radars and CNN-LSTM architecture. IEEE J. Biomed. Health Inform. **25**(4), 1273–1283 (2021). https://doi.org/10.1109/JBHI.2020.3027967
31. Yao, Y., et al.: Fall detection system using millimeter-wave radar based on neural network and information fusion. IEEE Internet Things J. **9**(21), 21038–21050 (2022). https://doi.org/10.1109/JIOT.2022.3175894
32. Li, W., et al.: Real-time fall detection using mmwave radar. In: ICASSP 2022 - 2022 IEEE International Conference on Acoustics, Speech and Signal Processing (ICASSP), pp. 16–20 (2022). https://doi.org/10.1109/ICASSP43922.2022.9747153
33. He, C., et al.: A noncontact fall detection method for bedside application with a mems infrared sensor and a radar sensor. IEEE Internet Things J. **10**(14), 12577–12589 (2023). https://doi.org/10.1109/JIOT.2023.3251980
34. Clemente, J., Li, F., Valero, M., Song, W.: Smart seismic sensing for indoor fall detection, location, and notification. IEEE J. Biomed. Health Inform. **24**(2), 524–532 (2020). https://doi.org/10.1109/JBHI.2019.2907498

35. Hanifi, K., Karsligil, M.E.: Elderly fall detection with vital signs monitoring using CW doppler radar. IEEE Sens. J. **21**(15), 16969–16978 (2021). https://doi.org/10.1109/JSEN.2021.3079835
36. Wang, J., Lin, L.H., Wang, M., Xu, W.: Non-contact fall detection system based on mmwave radar and transformer. IEEE Sens. J. **23**(20), 24002–24009 (2023). https://doi.org/10.1109/JSEN.2023.3313270
37. Sadreazami, H., Bolic, M., Rajan, S.: Compressed domain contactless fall incident detection using UWB radar signals. IEEE Trans. Biomed. Eng. **69**(1), 244–256 (2022). https://doi.org/10.1109/TBME.2021.3094044
38. Swarubini, P., Ganapathy, N.: Radar-based elderly fall detection using spwvd and resnet network. Curr. Directions Biomed. Eng. **10**(4), 498–501 (2024). https://doi.org/10.1515/cdbme-2024-2122
39. Yang, B., Guo, L., Zhang, X., Yang, J., Ren, Y., Li, X.: An intelligent fall detection method based on millimeter-wave radar and hybrid deep learning model. Sensors **22**(21), 8176 (2022)
40. Dai, J., Ji, Z., Xie, Z., Wang, C., Wu, J.: Multimodal fusion for indoor fall detection using millimeter-wave radar and vision sensor. Expert Syst. Appl. **231**, 120770 (2023)
41. Sun, Y., Li, G., Zhu, Z., Huang, C., Hou, X., He, Y.: Attention-GRU based multimodal fusion for elderly fall detection. Sensors **21**(2), 1–16 (2021)
42. Meng, X., Liu, Y., Zhou, B., Wang, P., Song, J., Zhang, D.: Multimodal sensor fusion for elderly fall detection using wi-fi CSI and accelerometer data. IEEE Internet Things J. **9**(21), 21247–21257 (2022)
43. Alkhaldi, H., Faruque, M.A.A.: Fall detection using radar and deep learning for elderly safety. In: Proceedings of the IEEE Sensors Applications Symposium (SAS), pp. 1–6. IEEE (2022). https://doi.org/10.1109/SAS54614.2022.9764038
44. Zhang, Y., Wang, L., Li, M., Liu, X.: A privacy-preserving radar-based system for elderly fall detection using attention-enhanced CNN-LSTM architecture. In: Proceedings of the 45th Annual International Conference of the IEEE Engineering in Medicine and Biology Society (EMBC), pp. 1234–1238. IEEE (2023). https://doi.org/10.1109/EMBC46164.2023.10345123

Enhancing Indoor Trajectory Tracking with XGBoost-Based Classification on mmWave Radar Point Clouds

Yuru Lu[1], Zhanjun Hao[1(✉)], Yuejiao Wang[2], Guowei Wang[2], and Xiangyu Wang[1]

[1] Northwest Normal University, Lanzhou 730000, China
haozhj@nwnu.edu.cn
[2] Lanzhou University, Lanzhou 730000, China
{wyuejiao2024,wanggw2024}@lzu.edu.cn

Abstract. This paper tackles the challenge of achieving high-accuracy classification of human trajectories in complex indoor environments using millimeter-wave radar point clouds, while preserving privacy. After enhancing raw point-cloud motion signals via median and MTI filtering, we extract multi-dimensional physical features—including position, velocity, and radar reflection intensity—to form robust trajectory descriptors. An XGBoost-based classifier is then employed to discriminate between different motion patterns with over 96% overall accuracy. Finally, a Perceiver IO model captures spatio-temporal dependencies of classified trajectories to predict short-term target positions. Experiments demonstrate that the proposed system delivers both precise classification and sub-5 cm prediction error, validating its effectiveness and robustness for real-time indoor monitoring.

Keywords: Point cloud · Indoor human trajectory classification · Millimeter-wave radar · XGBoost · Short-Term trajectory prediction

1 Introduction

Millimeter-wave (mmWave) radar has emerged as a promising technology for human sensing applications due to its robustness in challenging environmental conditions, such as fog, rain, and poor lighting [1]. It can provide high-resolution point clouds that capture 3D information of the environment, making it suitable for various applications, including autonomous driving [2], human activity recognition [3], and indoor monitoring systems [4]. The ability of mmWave radar to penetrate obstacles such as smoke and dust further enhances its applicability in real-world settings. However, despite its potential, mmWave radar point clouds are typically sparse and irregular, which poses challenges for precise target classification and trajectory analysis.

In recent years, the focus has primarily been on trajectory tracking and prediction using mmWave radar [5], with several approaches demonstrating success in dynamic environment monitoring. However, the classification of different trajectories based on radar point clouds has received limited attention. This gap is critical, as the ability to accurately classify trajectories can significantly enhance applications such as behavior recognition, object identification, and autonomous decision-making systems. Existing methods for trajectory classification, often relying on traditional sensors like Wi-Fi [6] or cameras [7], struggle to adapt to the unique characteristics of mmWave radar, such as data sparsity and sensor limitations. While some methods have employed machine learning techniques to improve classification accuracy [8], the exploitation of physical layer features like position, velocity, and reflection intensity for mmWave radar data remains underexplored.

To address the challenges of multi-target tracking in millimeter-wave radar point cloud data, we designed a millimeter-wave radar-based trajectory classification and tracking prediction system. To solve the target occlusion problem in multi-person trajectory recognition, we utilize XGBoost [9], which integrates multiple decision trees and employs a gradient boosting algorithm. In each iteration, the model updates the prediction results and minimizes the loss function to optimize the model. This approach effectively handles noise and missing data, and regularization is used to prevent overfitting, thus improving the accuracy of trajectory classification. Additionally, we use the Perceiver IO model [10] for accurate tracking and prediction of target trajectories. We constructed a walking trajectory dataset based on real-world scenarios and successfully trained and evaluated the system. We introduce a high-precision XGBoost-based classification method that leverages multi-dimensional physical-layer features from mmWave radar point clouds to improve accuracy in complex indoor, multi-target scenarios, seamlessly integrated with a Perceiver IO module for short-term trajectory prediction to form a real-time monitoring and forecasting system; extensive experiments on real-world indoor datasets confirm classification accuracy above 96%, prediction errors below 5 cm, and strong robustness under varied conditions.

2 Related Work

The use of millimeter-wave (mmWave) radar for human sensing and trajectory prediction has gained significant attention due to its robustness in adverse weather conditions and its ability to penetrate occlusions such as smoke and fog. While mmWave radar provides sparse point cloud data, recent advancements in signal processing and machine learning techniques have enabled its application in diverse fields such as target classification, human activity recognition, and tracking. Several studies have proposed innovative methods to address the challenges associated with radar data, including data sparsity, noise, and the integration of machine learning models to improve performance in real-world scenarios.

2.1 Classification and Recognition

A major challenge in radar-based classification is the sparsity of point clouds, which limits the detection of fine-grained features. Fan et al. [11] tackled this by incorporating visual-inertial supervision to enhance radar point clouds, improving classification accuracy. Models like XGBoost have also been applied to classify radar point clouds by handling noise and missing data effectively. Additionally, Xie et al. [12] employed a Lift-and-Deform Module (LDM) to combat sparsity and generate high-quality 3D human models, boosting performance in tasks like classification.

In human activity recognition, mmWave radar has been used to capture motion and behavioral patterns. Lin et al. [13] proposed a multi-activity classification system that fuses time-domain features with PCANet, improving recognition accuracy across diverse human activities. These feature fusion techniques make mmWave radar promising for gesture detection and biometric applications where privacy and non-intrusiveness are vital. Furthermore, human pose estimation is emerging as a key application for classifying body parts and postures. An et al. [14] developed a scalable framework using meta-learning to process sparse point clouds, enabling rapid adaptation to new scenarios and improving pose estimation performance.

2.2 Tracking

Traditional tracking often uses point target tracking (PTT), treating each target as a single point. However, mmWave radar captures multiple reflections per target, raising the need for extended target tracking (ETT). Jiang et al. [15] proposed a robust multi-target tracking method combining extended Kalman filters with group tracking to handle dense environments.

Beyond tracking, radar-based gait recognition has been adapted to open-set scenarios where unseen individuals must be identified. Mazzieri et al. [16] addressed this using a neural network combining supervised classification with unsupervised sparse point cloud reconstruction, enhancing robustness and accuracy in open-set gait recognition.

Further developments include Wang et al. [17], who improved ETT via a group association model to better process sparse radar data. Li et al. [18] introduced MCGait, a multiuser gait system using micro-Doppler calibration to boost identification accuracy in crowds, achieving 95.45% accuracy when tracking up to four users. Moreover, Prabhakara et al. [19] presented RadarHD, which uses deep learning to enhance radar point cloud resolution, significantly improving tracking precision for applications like autonomous navigation and multi-target tracking.

3 Method

We performed effective preprocessing on the millimeter-wave radar point cloud data to improve the accuracy of target classification and tracking. First, we removed zero-velocity points and applied median filtering to eliminate pulse noise, smooth the data, and reduce environmental noise. Next, we used a Moving Target Indication (MTI) filter to exclude stationary objects, keeping only dynamic targets to ensure data accuracy. We then applied DBSCAN clustering to the denoised point cloud data to identify target groups. Finally, we applied Kalman filtering to the cluster centers for trajectory tracking and smoothing, ensuring accurate tracking of target movement. These steps effectively cleaned and enhanced the data, providing a reliable foundation for subsequent target detection and tracking.

3.1 Target Classification with XGBoost

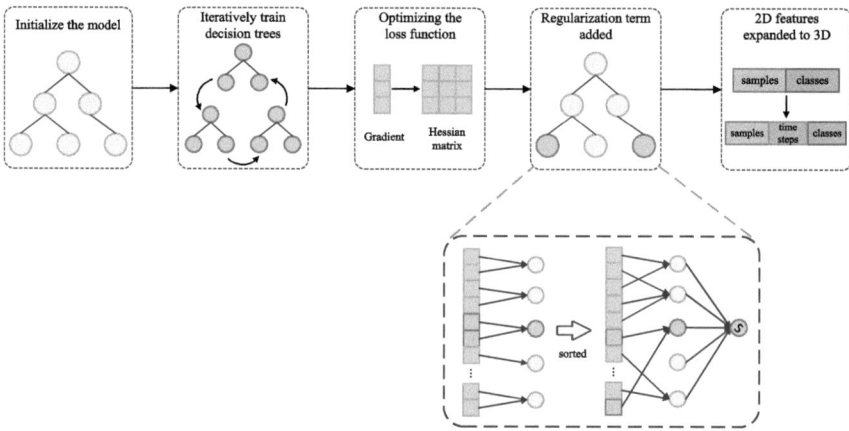

Fig. 1. XGboost model architecture.

We employ XGBoost (Extreme Gradient Boosting) as the trajectory classifier. Its core idea is to iteratively add multiple weak regression trees, where each new tree corrects the residual errors of the previous ensemble. As illustrated in Fig. 1, the model consists of T regression trees. Each tree is grown based on the residuals and the negative gradient of the overall objective function from the previous iteration, with leaf nodes outputting residual correction values. The process equates to iteratively minimizing the loss function along its negative gradient, with a regularization term to control tree complexity and prevent overfitting.

Next, We extract spatial coordinates (x, y, z) and Doppler velocity from each frame of the raw millimeter-wave radar point cloud, and compute statistical and geometric features over a sliding time window of length W frames. We calculate the mean, standard deviation, skewness, and kurtosis of range; mean and variance of Doppler velocity; the three eigenvalues of the 3D coordinate covariance matrix; and a point cloud density metric (total number of points in the window). To eliminate scale differences, all features are standardized via z-score normalization and concatenated into a 10-dimensional feature vector, which serves as the input to XGBoost. At iteration t, model predictions are updated by:

$$F_t(x_i) = F_{t-1}(x_i) + \eta \cdot h_t(x_i) \tag{1}$$

where $F_{t-1}(x_i)$ is the cumulative prediction from the previous iteration, $h_t(x_i)$ is the current regression tree's output, and η (learning rate) controls each tree's contribution.

The overall optimization objective is:

$$L = \sum_{i=1}^{n} l(y_i, F_{t-1}(x_i) + h_t(x_i)) + \Omega(h_t) \tag{2}$$

where the logistic loss is defined as:

$$l(y, \hat{y}) = -[y \cdot \ln(\sigma(\hat{y})) + (1 - y) \cdot \ln(1 - \sigma(\hat{y}))], \tag{3}$$

and the regularization term is:

$$\Omega(h) = \gamma \cdot T + \frac{1}{2} \cdot \lambda \cdot \sum_{j=1}^{T} w_j^2, \tag{4}$$

which penalizes both the number of leaves T and the leaf weights w_j.

To optimize efficiently, XGBoost applies a second-order Taylor expansion of the loss, computing each sample's first derivative g_i and second derivative h_i:

$$h_i = \frac{\partial^2}{\partial \hat{y}^2} l(y_i, \hat{y}) \Big|_{\hat{y}=F_{t-1}(x_i)} \tag{5}$$

The optimal weight for each leaf node is:

$$w_j = -\frac{\sum_{i \in \text{leaf}_j} g_i}{\sum_{i \in \text{leaf}_j} h_i + \lambda} \tag{6}$$

This formula shows that the leaf weight equals the negative sum of gradients divided by the sum of Hessians plus the regularization term λ, ensuring smooth and bounded predictions.

When evaluating candidate splits, XGBoost computes the gain:

$$\text{Gain} = \frac{1}{2} \left[\frac{(\sum_L g)^2}{\sum_L h + \lambda} + \frac{(\sum_R g)^2}{\sum_R h + \lambda} - \frac{(\sum g)^2}{\sum h + \lambda} \right] - \gamma \tag{7}$$

where $\sum_L$ and $\sum_R$ are the sums of gradients and Hessians for samples in the left and right child nodes, respectively. A higher gain indicates a larger reduction in loss relative to model complexity; γ penalizes splits to avoid unnecessary tree growth. Features and thresholds are ranked by gain, and the best split is chosen until maximum depth or minimum sample constraints are met.

Key hyperparameters are tuned via five-fold cross-validation: 100 trees, maximum depth of 6, learning rate of 0.1, subsample and column sampling ratios of

0.8, and regularization parameters γ and λ both set to 1. These settings ensure sufficient model capacity while effectively controlling overfitting. In the testing phase, the model automatically identifies single-target and two-target trajectories from unlabeled radar point cloud sequences, labels them as "Track 1" and "Track 2", and saves the spatial coordinates and time frame information of each segment for subsequent tracking prediction modules.

3.2 Tracking and Prediction with Perceiver IO

In the previous subsection, we covered data preprocessing and trajectory classification. Next, we will illustrate the trajectory tracking and prediction workflow using the Perceiver IO model, its pseudocode is presented in Algorithm 1.

Algorithm 1 Perceiver IO for Trajectory Prediction

Require: Training set $X = \{x_t\}_{t=1}^{T}$, where each x_t denotes the position at time t
Ensure: Predicted outputs Y
1: Initialize latent representation $Z_0 \in R^{M \times D}$ {M: latent space size; D: latent vector dimension}
2: **for** $t = 1$ to T **do**
3:    Apply cross-attention mechanism:
$$\text{Attention}(Q, K, V) = \text{softmax}\left(\frac{QK^\top}{\sqrt{d_k}}\right) V$$
4:    Compute the loss:
$$\text{Loss} = \frac{1}{K} \sum_{k=1}^{K} \|\hat{y}_k - y_k\|^2$$
5:    Optimize the loss and update model parameters
6: **end for**
7: **return** $Y = \text{Decoder}(Z_T)$

First, a data-embedding layer transforms the input traffic tensor over T time steps into embeddings suitable for the model. These embeddings are then combined with a randomly initialized latent array and fed into a multi-head attention mechanism, which is split into three parallel components each capturing distinct spatiotemporal features. The three attention outputs are concatenated along the channel dimension, passed through a feed-forward network, and then added back via a residual connection to form the encoder's output. Six such encoder layers are stacked, enabling the latent representation to progressively aggregate long-range dependencies and multimodal features from the history.

For millimeter-wave point-cloud data, we first extract each object's coordinates to construct input vectors. These vectors are then projected into a high-dimensional feature space by an affine mapping, as follows:

$$z_i = W_{\text{in}} \cdot x_i + b_{\text{in}}, \tag{8}$$

where x_i is the input vector for the ith frame, W_{in} denotes the weight matrix for this mapping, and b_{in} is the bias. The generated embedding z_i is then passed on to the latent array.

The model then applies multi-head self-attention to capture long-range dependencies. We use 6 encoder layers ($L = 6$) and 8 attention heads. In each layer, we compute queries Q, keys K, and values V from the current latent and embedding representations, and update the latent.

After attention, the layer output passes through a feed-forward network, and is added residually to the previous latent:

$$z_{l+1} = \text{LayerNorm}(z_l + \text{Attention}(z_l) + \text{FeedForward}(z_l)) \tag{9}$$

By stacking multiple layers in a recursive manner, the model effectively learns long-range dependencies. The use of residual connections helps address the vanishing gradient issue. After the encoding process is completed, the latent representation is passed to the decoder for further refinement. The decoder executes six iterative steps, projecting the latent features back into the original space to generate trajectory predictions. The output is a two-dimensional vector, representing the predicted x and y coordinates.

4 Experiment and Evaluation

4.1 Experiment Setting

For data collection, we use a commercial IWR1843 millimeter-wave radar. The radar offers a distance resolution of 4.4 cm, a 9 m detection range, 0.13 m/s radial velocity resolution, and a 10 fps sampling rate. It is mounted 1.2 m above ground to ensure stable data acquisition. Experiments are conducted in three indoor scenarios: an office (3.5×4 m), a classroom (4×5 m), and a hall (5×7 m). For each scenario, multiple trajectory paths are designed, including straight, L-shaped, and S-shaped paths. In the classroom, a predefined path is established, leading from the radar's initial position to a specific table. Each scene collects 600 sets of data, totaling 30,000 frames. The high-precision GPS system records real-time motion trajectories, including latitude, longitude, speed, and time, ensuring the authenticity and reliability of the data.

The dataset is split into training and testing sets with an 80:20 ratio. All models are trained for 100 epochs with a batch size of 32. For classification tasks, the evaluation metrics include Accuracy and F1 Score, which assess the classification effectiveness of the models. Additionally, Processing Time is recorded to evaluate the computational efficiency. For the tracking and prediction tasks, the models are evaluated using Mean Squared Error (MSE), Mean Absolute Error (MAE), and Coefficient of Determination (R^2), which measure prediction accuracy, stability, and overall model fit. These metrics are used to assess both the classification and tracking performance, ensuring the system's robustness and efficiency across different conditions.

4.2 Trajectory Classification Evaluation

To evaluate the classification performance and robustness of the system's trajectory classification module, a series of experiments were conducted using four widely adopted machine learning models: Support Vector Machine (SVM), Random Forest, Long Short-Term Memory (LSTM), and eXtreme Gradient Boosting (XGBoost). These models were evaluated based on classification accuracy, F1 score, and processing time, reflecting both their effectiveness and computational efficiency.

As illustrated in Fig. 2, a comparison of the four models revealed that XGBoost demonstrated superior classification performance. In single-target scenarios, it attained an accuracy of 98.3%, while in dual-target scenarios, it recorded an F1 score of 92.0%. Although LSTM also demonstrated strong performance, with an accuracy of 94.1% and an F1 score of 85.0%, its average processing time was relatively high at 15.2 ms. XGBoost, in comparison, maintained a significantly lower processing time of 4.8 ms while delivering superior classification metrics. Random Forest and SVM achieved faster processing times, with averages of 3.5 ms and 2.1 ms respectively. However, their classification results were comparatively lower. Random Forest reached an accuracy of 92.4% and an F1 score of 83.0%, while SVM achieved an accuracy of 88.7% and an F1 score of 79.0%. Overall, the results indicate that XGBoost provides the most balanced combination of classification performance and computational efficiency, making it well-suited for real-time gesture recognition in smart home applications.

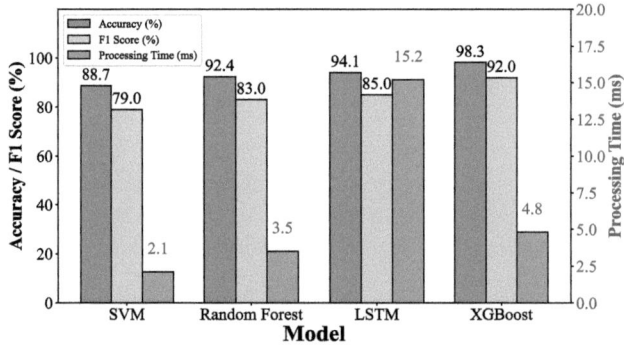

Fig. 2. Comparison of classification performance and efficiency across different models

Further experiments focused specifically on the selected XGBoost model to assess its robustness across different scenarios. As shown in Fig. 3, the model's accuracy varied depending on the experimental environment and the number of targets. In terms of environmental impact, the Hall environment generally yielded the highest accuracy for both single-target (98.3%) and dual-target (91.0%) scenarios, which can be attributed to its open layout and reduced signal interference. The Office environment followed with relatively strong performance, achieving 97.5% for single-target and 89.2% for dual-target cases. As anticipated, the Classroom environment exhibited slightly lower accuracy due to increased signal occlusion, with results of 93.8% for single-target and 84.5% for dual-target situations.

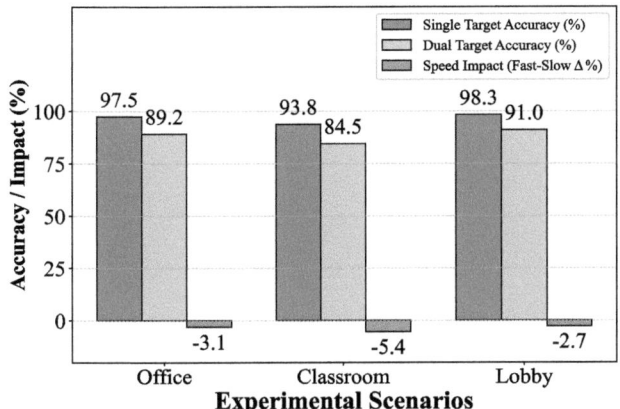

Fig. 3. Comparison of XGboost overall performance in different environments, target counts, and speeds

The number of targets also had a notable impact on classification accuracy. Across all environments, single-target scenarios consistently outperformed dual-target ones. For example, in the Office environment, accuracy decreased from 97.5% to 89.2% when moving from a single to a dual target. This reduction can be attributed to increased signal overlap and interference when multiple individuals are present. Additionally, the influence of varying walking speeds was evaluated by analyzing the percentage point change in accuracy between fast and slow walking speeds. As illustrated by the 'Speed Impact' bars in Fig. 3, the effect of speed variation was relatively minor. The accuracy differences ranged from -2.7% in the Lobby to -5.4% in the Classroom. These results suggest that while walking speed variations may slightly affect performance, the XGBoost model retains a reasonable level of robustness across different pace conditions.

In conclusion, the trajectory classification experiments confirm XGBoost's superiority as a classifier for the proposed system. It delivers high accuracy and F1 scores with computational efficiency. Furthermore, it demonstrates acceptable robustness across various indoor environments, different numbers of targets, and changes in walking speed, establishing it as a reliable and practical component for indoor trajectory tracking applications.

4.3 System Performance Evaluation

To evaluate the robustness of the proposed system, experiments were conducted across diverse indoor environments, including an office, classroom, and lobby.

Table 1. Prediction Performance Indicators for Different Experimental Scenarios

Environment	MSE (m^2)	MAE (m)	R^2
Office	0.0094	0.0595	0.9405
Classroom	0.0117	0.0977	0.8799
Lobby	0.0079	0.0391	0.9746

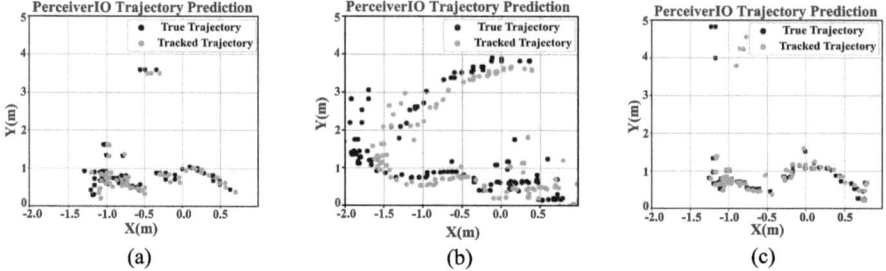

Fig. 4. Comparison of the prediction results of different environment(a) office (b) classroom (c) lobby

These tests encompassed both single-target and dual-target tracking scenarios under varying walking speeds. In terms of environmental influence, the system demonstrated consistent performance across all settings. As shown in Table 1 and Fig. 4, the lobby environment yielded the highest accuracy, featuring the lowest mean squared error (MSE) of 0.0079 m^2, a mean absolute error (MAE) of 0.0391 m, and the highest coefficient of determination (R^2) of 0.9746. These results highlight the system's robustness in spacious environments with minimal signal interference. The office environment followed closely, achieving an R^2 value of 0.9405, while the classroom environment showed slightly reduced performance

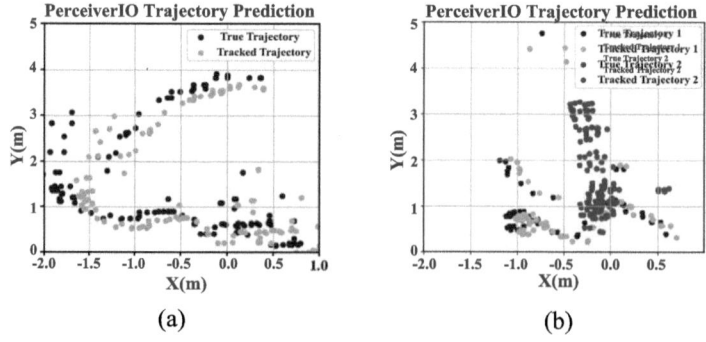

Fig. 5. Comparison of the prediction results of different number of people tracking (a) single person (b) double person

with $R^2 = 0.8799$, likely due to higher levels of signal occlusion caused by denser layouts and obstacles.

Additional experiments were conducted to assess system performance under different target densities. As summarized in Table 2 and illustrated in Fig. 5, the tracking accuracy in single-target scenarios remained high, with $R^2 = 0.9405$ and an MAE of 0.0595 m. In contrast, dual-target conditions introduced moderate degradation in tracking quality. The R^2 dropped to 0.7287, and the MAE increased to 0.0825 m, primarily due to signal overlap and interference between targets. Nevertheless, the system's performance under dual-target conditions remained within acceptable limits for practical deployment in real-world applications.

To evaluate the robustness of this system, we conducted experiments in diverse indoor scenarios including an office, classroom, and lobby, addressing both single and dual-target tracking under varying walking speeds. In terms of environmental impact, the system showed stable performance across all settings. As shown in Table 1 and Fig. 4, the lobby environment yielded the highest accuracy, achieving a mean squared error (MSE) of 0.0079 m^2, a mean absolute error (MAE) of 0.0391 m, and a coefficient of determination R^2 of 0.9746, demonstrating the system's robustness in a spacious, low-interference setting. The office environment followed closely, with an R^2 value of 0.9405, while the classroom showed slightly reduced accuracy, reaching an R^2 of 0.8799, likely due to increased signal occlusion.

Further experiments evaluated the system's performance under single-target and dual-target conditions. As presented in Table 2 and Fig. 5, the tracking performance for a single target maintained high precision, with an R^2 of 0.9405 and an MAE of 0.0595 m. However, under dual-target scenarios, signal overlap resulted in a moderate decline in performance, with R^2 decreasing to 0.7287 and MAE increasing to 0.0825 m. Despite this reduction, the system maintained accuracy within acceptable limits for practical applications.

Table 2. Prediction Performance Indicators for Single and Dual Tracking

Number of Participants	MSE (m^2)	MAE (m)	R^2
Single	0.0094	0.0595	0.9405
Double	0.0097	0.0825	0.7287

These results suggest that our system provides consistent and reliable trajectory tracking and prediction across various environments, target conditions, and walking speeds. The system's performance is particularly strong in environments with fewer obstacles and lower signal interference, while still maintaining acceptable accuracy in more challenging conditions.

5 Conclusion

This paper proposes a trajectory classification approach based on millimeter-wave radar point clouds, with an emphasis on addressing the challenges of multi-target classification in indoor environments. The core classification module adopts the XGBoost algorithm, which outperforms other machine learning models in both accuracy and computational efficiency. Experiments demonstrate that the module achieves 92% accuracy in dual-target scenarios within the Classroom setting, while maintaining a low processing latency of 4.8 ms, making it suitable for real-time applications. Furthermore, the classifier exhibits robust performance across varying indoor environments, target counts, and walking speeds. The classification module is designed to complement a trajectory tracking pipeline built on the Perceiver IO model, which provides position data with high spatial precision. This model achieves a single-target tracking error of 3.67 cm and approximately 5 cm for two targets. By enabling reliable discrimination between trajectory patterns, the classification process enhances the interpretability and utility of the tracking results, especially in complex indoor scenarios. However, challenges remain under conditions of severe occlusion or high target density, due to the intrinsic resolution limitations of millimeter-wave radar. Future work will explore multimodal sensing and multi-radar collaboration to further improve both tracking accuracy and the effectiveness of classification under such constraints.

Acknowledgments. This work was supported by the National Natural Science Foundation of China (Grant 62262061), Major Science and Technology Projects of Gansu (23ZDGA009), the Central Government Guides Local Funds (25ZYJA007), and Gansu Provincial Department of Education: Industry Support Program Project (2022CYZC-12).

References

1. Ji, Q., Cheng, X., Fu, S.: Research on identification method based on point cloud sequence with millimeter-wave radar. In: 9th International Conference on Intelligent Computing and Signal Process (ICSP), pp. 758–762 (2024)
2. Han, Z., et al.: 4D millimeter-wave radar in autonomous driving: a survey. arXiv preprint arXiv:2306.04242 (2023)
3. Kim, Y., Alnujaim, I., Oh, D.: Human activity classification based on point clouds measured by millimeter wave MIMO radar with deep recurrent neural networks. IEEE Sensors J. **21**(12), 13522–13529 (2021)
4. Shen, Z., Nunez-Yanez, J., Dahnoun, N.: Advanced millimeter-wave radar system for real-time multiple-human tracking and fall detection. Sensors **24**(11), 3660 (2024)
5. Zhao, P., et al.: Human tracking and identification through a millimeter wave radar. Ad Hoc Netw. **116**, 102475 (2021)
6. Li, W., Bocus, M.J., Tang, C., Piechocki, R.J., Woodbridge, K., Chetty, K.: On CSI and passive Wi-Fi radar for opportunistic physical activity recognition. IEEE Trans. Wirel. Commun. **21**(1), 607–620 (2021)

7. Singh, S., Shekhar, C., Vohra, A.: FPGA-based real-time motion detection for automated video surveillance systems. Electronics **5**(1), 10 (2016)
8. Li, Z., Du, L., Yu, Z.: Point cloud features-based random forest for road user classification via millimeter wave radar. In: 2021 CIE International Conference on Radar (Radar), pp. 1463–1466 (2021)
9. Chen, T., Guestrin, C.: XGBoost: a scalable tree boosting system. In: 22nd ACM SIGKDD International Conference Knowledge Discovery and Data Mining, pp. 785–794 (2016)
10. Jaegle, A., et al.: Perceiver IO: a general architecture for structured inputs & outputs. arXiv preprint arXiv:2107.14795 (2021)
11. Fan, C., Zhang, S., Liu, K., Wang, S., Yang, Z., Wang, W.: Enhancing mmWave radar point cloud via visual-inertial supervision. In: 2024 IEEE International Conference on Robotics and Automation (ICRA), pp. 9010–9017 (2024)
12. Xie, Q., et al.: mmPoint: dense human point cloud generation from mmWave. In: British Machine Vision Conference (BMVC), pp. 194–196 (2023)
13. Wang, Y., Su, J., Murakami, H., Tonouchi, M.: PointNet++ based concealed object classification utilizing an FMCW millimeter-wave radar. J. Infrared Millim. Terahertz Waves **45**(11), 1040–1057 (2024)
14. Lin, Y., Li, H., Faccio, D.: Human multi-activities classification using mmWave radar: feature fusion in time-domain and PCANet. Sensors **24**(16), 5450 (2024)
15. An, S., Ogras, U.Y.: Fast and scalable human pose estimation using mmwave point cloud. In: 59th ACM/IEEE Design Automation Conference, pp. 889–894 (2022)
16. Jiang, M., Guo, S., Luo, H., Yao, Y., Cui, G.: A robust target tracking method for crowded indoor environments using mmWave radar. Remote Sens. **15**(9), 2425 (2023)
17. Mazzieri, R., Pegoraro, J., Rossi, M.: Open-set gait recognition from sparse mmWave radar point clouds. arXiv preprint arXiv:2503.07435 (2025)
18. Li, J., Li, B., Wang, L., Liu, W.: Passive multiuser gait identification through micro-Doppler calibration using mmWave radar. IEEE Internet Things J. **11**(4), 6868–6877 (2023)
19. Prabhakara, A., et al.: High resolution point clouds from mmWave radar. In: 2023 IEEE International Conference on Robotics and Automation (ICRA), pp. 4135–4142 (2023)

Towards Large-Scale Wireless Sensing in Smart Buildings Using LoRa Signals

Xinyu Xue[1](✉), Zhaoxin Chang[1], Xujun Ma[1], Pei Wang[1], Fusang Zhang[2], Badii Jouaber[1], and Daqing Zhang[1,3]

[1] SAMOVAR, Telecom SudParis, Institut Polytechnique de Paris, 91120 Palaiseau, France
`xinyu.xue@telecom-sudparis.eu`
[2] Beihang University and Institute of Software, Chinese Academy of Sciences, Beijing 100191, China
[3] Peking University, Beijing 100871, China

Abstract. With the increased need for intelligent functions in smart buildings, the ability to sense the states of human subjects becomes essential. In recent years, wireless signals have demonstrated strong capability for contactless sensing. However, most wireless sensing systems currently focus on room-level scenarios. The deployment challenges and solutions in large-scale scenarios have not been sufficiently investigated. In this paper, we take the first step to explore the feasibility of utilizing LoRa signals for large-scale sensing, leveraging their advantages in wide-area sensing capabilities. However, given the fixed deployment in buildings, the sensing coverage of each device is likely to mismatch with the desired sensing area of interest (AoI). To address this challenge, we first investigate the factors affecting sensing coverage. Then, we propose to control the sensing coverage by adjusting hardware parameters, enabling human presence detection within the desired area. The effectiveness of the proposed method is validated through benchmark experiments and two case studies in real-world environments.

Keywords: Smart Building · Wireless Sensing · LoRa Signal · Sensing Coverage · Sensing-signal-to-noise Ratio

1 Introduction

The rapid evolution of IoT technologies has facilitated the realization of smart buildings. In such environments, sensing the presence and activities of human subjects is essential for enabling a wide range of intelligent functions, such as the automatic control of appliances, security control, space utilization analytics, and human-device interaction applications. To support these intelligent functions, smart buildings typically rely on a variety of sensing technologies, including infrared sensors and cameras. However, infrared-based solutions suffer from high false-alarm rates under environmental disturbances and offer limited coverage due to their narrow sensing range and Line-of-Sight (LoS) constraints.

On the other hand, camera-based systems can provide profound visual information, while posing serious privacy concerns and inevitably consuming significant computational and storage resources.

In recent years, wireless signals have demonstrated the capability of contactless sensing, offering an ubiquitous, non-intrusive, and privacy-preserving sensing solution. Various types of wireless signals, e.g., Wi-Fi [1–3], mmWave [4–6], LoRa [7–9], and UWB [10–12], have been employed for different sensing applications, including presence detection [13–15], vital sign monitoring [16–18], gesture recognition [19–21], activity recognition [22–24], and tracking [25–27]. The principle of wireless sensing lies in that the movement of human targets affects signal propagation, resulting in signal fluctuation. Therefore, sensing objectives can be achieved by interpreting the changes in wireless signals. While promising in many applications, existing research has mainly focused on room-scale deployments, such as bedrooms, offices, or laboratories, where the sensing area is relatively limited and the number of involved devices remains small. Currently, few studies explore the system deployment requirements and practical sensing solutions for large-scale smart building environments. Such environments typically involve numerous spatially distributed sensing zones, posing new challenges in achieving scalable, accurate, and efficient sensing with minimal infrastructure.

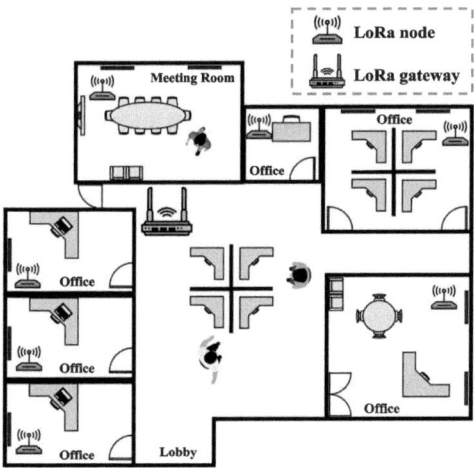

Fig. 1. Scenario of large-scale wireless sensing using LoRa.

In this work, we present an initial investigation of enabling large-scale wireless sensing in smart buildings by exploring the system design considerations, deployment requirements, and practical solutions. We propose to leverage LoRa signals for building-scale sensing. LoRa is a wireless technology designed for low-power and long-range communication. It offers several key advantages that make it suitable for large-scale deployment in smart buildings, including low power consumption, cost-effectiveness, and strong capabilities in long-range and wide-area sensing [7,8].

Figure 1 demonstrates the envisioned building-scale sensing scenarios by using LoRa signals, involving a centralized LoRa gateway as the receiver (Rx) and a large number of LoRa nodes as transmitters (Tx). Our objective is to deploy one LoRa node in a specific AoI. For example, a LoRa node placed in an office room can be used to detect human presence within this room. This one-to-one mapping between each device and target sensing area enables location-aware sensing while maintaining system scalability. Thus, it is important to control the sensing coverage of each device. Prior studies [28,29] have shown that the sensing coverage can be controlled by adjusting the distance between transceivers. However, for large-scale sensing system deployment, the placement of LoRa devices is fixed. For LoRa nodes located close to the gateway, the sensing coverage tends to be large, potentially extending beyond the intended sensing area and causing significant interference issues. In contrast, nodes positioned farther from the gateway may exhibit a much smaller sensing coverage, insufficient to fully cover the designated sensing area. Therefore, in large-scale smart building environments, the key challenge lies in how to control the sensing coverage of each LoRa node, making it align with the physical boundary of the intended sensing area.

To address this challenge, we first investigate the factors that influence the sensing coverage. Our analysis reveals that the effective sensing coverage is primarily determined by the distance between the LoRa nodes and the LoRa gateway, as well as the transceiver gain. Given the fact that it is not flexible to adjust the distance between devices in a layout-fixed building scenario, we propose to adjust the gain of the transceiver pairs such that the coverage of each sensing area could be independently controlled. We first conduct benchmark experiments in a real-world environment to validate the effectiveness of gain control method for sensing coverage manipulation. Then, we showcase the use of the proposed method with two case studies, which are presence detection in an office room and a public area, respectively. Experiment results show that our prototyped system can accurately detect human presence in both scenarios, moving a first step towards large-scale wireless sensing in smart building environments.

2 Preliminary

In this section, we first introduce the principle behind LoRa sensing. We then present the challenges associated with deploying large-scale LoRa sensing systems.

2.1 Principle of LoRa Sensing

LoRa is a technology designed for long-range communication between IoT devices. During the sensing process, LoRa signals transmitted by a LoRa node propagate to a LoRa gateway through multiple paths. As shown in Fig. 2, the received signals can be grouped into two distinct categories: static and dynamic components. Among these, the static components originate from signal propagation along the LoS path between transceivers and reflection paths from stationary objects. Meanwhile, the dynamic component is related to the moving target-reflected signal, which changes over time. These multipath signals are superimposed at the receiver. Therefore, the received signal can be represented as:

$$s(t) = H_s + H_d(t) = H_s + a(t)e^{-j\frac{2\pi f_c d(t)}{c}}, \tag{1}$$

where H_s and $H_d(t)$ are the static and dynamic components, respectively, $a(t)$ and $d(t)$ are the amplitude and the path length of the target-reflected signal, respectively, f_c is the carrier frequency, and c is the speed of light. According to Eq. 1, both the amplitude and phase of the received signal $s(t)$ vary with respect to the movement of the human target. Thus, by analyzing the amplitude and phase change of $s(t)$, target motion can be estimated for sensing purposes.

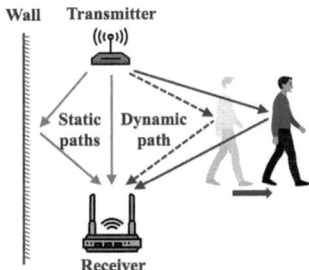

Fig. 2. Illustration of LoRa signal propagation.

2.2 Large-Scale Sensing

In smart building scenarios, LoRa is particularly well-suited for large-scale deployments due to its low cost and low power consumption. For example, as illustrated in Fig. 1, LoRa signals can be utilized for human presence detection within specific AoIs, such as private rooms and public spaces. A large number of low-cost LoRa nodes can be distributed in these AoIs, while a smaller number of LoRa gateways are centrally deployed to receive signals transmitted by LoRa nodes. Intuitively, each node should be bound to a specific AoI. Thus, it is essential to control the sensing coverage of each pair of transceivers.

Previous studies have demonstrated that sensing coverage can be controlled by changing the distance between transceivers [28,29]. However, in the case of large-scale deployment in smart buildings, it is not practical to adjust the distance between LoRa nodes and gateways due to the constraints of fixed building layout, especially when there are many AoIs. Therefore, how to effectively control the sensing coverage of each pair of transceivers becomes a key challenge for large-scale sensing.

3 Methodology

In this section, we first analyze the factors that may influence the size of the sensing coverage. Then we propose a method for human presence detection within a specific area by adjusting the sensing coverage of the transceiver pair.

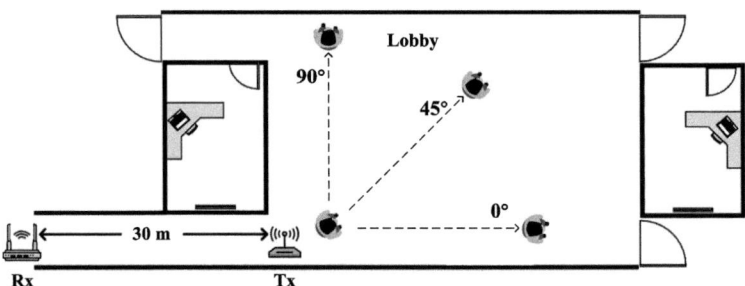

Fig. 3. Benchmark experiment setup.

Sensing coverage is defined as an area in which target movement can induce detectable signal fluctuations for sensing. In essence, whether an object's reflection signal at the receiver can be detected depends on its power. If the power of the reflection signal is lower than the noise power, the object's movement-induced signal variation will be overwhelmed by the noise. Thus, it is crucial to analyze the power of the subject-reflected signal at the receiver. Previous studies have demonstrated that sensing-signal-to-noise ratio (SSNR) can be employed to quantify whether the reflection signal of an object can be detected at the receiver [28,29]:

$$SSNR = \frac{P_d}{P_n} = \frac{\lambda^2 P_T G_R G_{AT} G_{AR} \sigma}{(4\pi)^3 (d_T d_R)^2 P_n}, \quad (2)$$

where P_d and P_n stand for the target signal power and the noise power at the baseband output, respectively. P_T represents the power of the Tx, G_R denotes the gain of Rx, G_{AT} and G_{AR} represent the gains of the transmitting and receiving antennas, respectively, σ represents the effective reflection area of the object, d_T and d_R are the distance from the human subject to the Tx and the Rx, respectively. According to Eq. 2, it can be observed that the device placement (i.e., d_T and d_R) affects the SSNR of the received signal. Denote the minimum SSNR of the detectable object-reflected signal as $SSNR_{min}$. The function of the boundary of the sensing coverage can be represented as:

$$(d_T d_R)_{max} = (\frac{\lambda^2 P_T G_R G_{AT} G_{AR} \sigma}{(4\pi)^3 P_n SSNR_{min}})^{\frac{1}{2}}. \quad (3)$$

It can be observed that the control of sensing boundary can also be achieved by adjusting the transceiver gains (P_T, G_R, G_{AT}, and G_{AR}). Specifically, reducing the transceiver gains results in a smaller sensing coverage. Note that in large-scale scenarios, the deployment of LoRa devices is fixed. Thus, it is impossible to control the sensing coverage by changing the placement of LoRa transceivers. To this end, in this work, we propose to adjust transceiver gains.

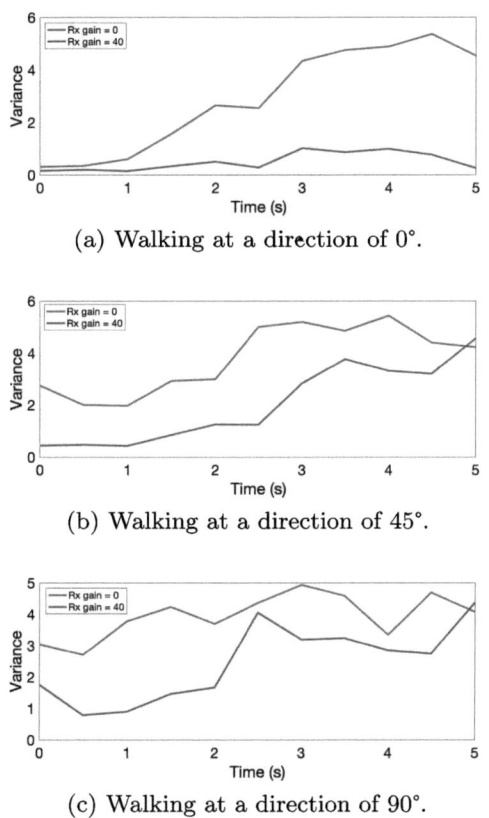

(a) Walking at a direction of 0°.

(b) Walking at a direction of 45°.

(c) Walking at a direction of 90°.

Fig. 4. Results of walking in different directions.

To investigate the impact of transceiver gains on sensing coverage, we conduct benchmark experiments in a real-world environment as shown in Fig. 3. The objective is to validate that lower transceiver gain leads to a reduced SSNR, thereby shrinking the effective sensing coverage. To this end, we measure the SSNR of the received signal while a human subject performs identical activities under different transceiver gain settings. To quantify the SSNR, which is the ratio of the power of the target-reflected signal to that of the noise, we first compute the difference between adjacent samples to remove the static component ($\Delta s(t)$). Then, we notice a fundamental distinction between the target-reflected and noise components, which is that the dynamic component generally exhibits low-frequency, smooth variations over time, whereas the noise component is characterized by high-frequency, stochastic fluctuations that lack temporal coherence. Thus, if the signal is dominated by the target-reflected component (i.e., high SSNR case), the phase difference between adjacent samples of $\Delta s(t)$ is subtle. In contrast, a lower SSNR corresponds to a stronger noise interference, leading to more irregular phase difference fluctuations. To this end, we propose to first compute the phase difference of $\Delta s(t)$ as $\Delta(\angle \Delta s(t))$, and then calculate its variance to quantify SSNR.

Figure 3 shows the setup of the benchmark experiment. The subject walks along three predefined trajectories away from the Tx, including at 0°, 45°, and 90° with respect to the LoS direction, respectively. In this experiment, the distance between Tx and Rx is fixed at 30 m, and the speed of walking is approximately 1 m/s. We tune the transceiver gain by setting the Rx gain as 0 dB and 40 dB, respectively.

The results are presented in Fig. 4. Across all trajectories, a lower Rx gain results in higher signal variance, indicating a reduced SSNR. Moreover, as the subject moves farther from the Tx, the SSNR gradually decreases. This indicates that a lower transceiver gain not only reduces the overall SSNR but also makes it more difficult to detect human activity at longer distances. As a result, the effective sensing coverage shrinks. These findings validate that adjusting the transceiver gain provides a practical mechanism to control the sensing coverage.

4 Implementation

To evaluate the effectiveness of our sensing coverage control approach, we implement the proposed method in real-world environments. Tx is implemented using a LoRa node equipped with a commercial Semtech SX1276 chip. The LoRa node is configured to transmit signals at a carrier frequency of 915 MHz with a bandwidth of 125 kHz. LoRa signals are received by a LoRa gateway realized by a USRP B210, which supports signal reception with two antennas. A laptop is connected to the LoRa gateway via a USB cable to collect packet samples for signal processing. Note that due to the unsynchronized LoRa transceivers, raw signals received by each antenna are distorted by phase offsets, including carrier frequency offset (CFO) and sampling frequency offset (SFO), hindering accurate phase extraction. To this end, we first calculate the ratio between the signals received by two antennas for phase offset elimination [7]. Then, the preprocessed signals can be used for various applications.

5 Evaluation

In this part, we evaluate our proposed method using two case studies to demonstrate the effectiveness of sensing coverage control in large-scale LoRa sensing in a smart building. In the following sections, we present the detailed designs for both use case studies.

5.1 Case Study A: Presence Detection in an Office Room

In this section, we present our design for room-level presence detection. By deploying a LoRa node in an office room, human activities within the room can be monitored for presence detection.

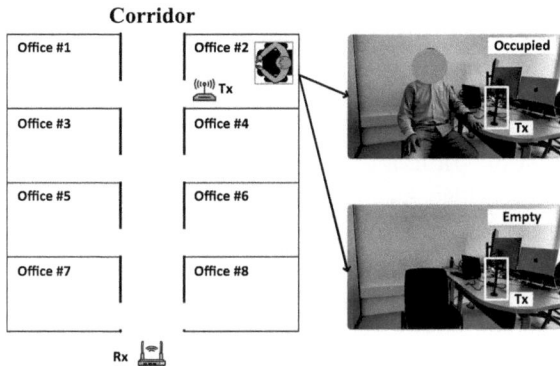

Fig. 5. Experiment setup for presence detection in an office room.

As shown in Fig. 5, a LoRa node is placed in the office room. The LoRa gateway is positioned in the corridor to receive signals. During the experiments, the human subject is asked to perform a series of activities. To avoid the interference induced by human movement in other rooms or the corridor, the Rx gain is fine-tuned to ensure the sensing coverage is within the room. A camera is used to record the video for ground truth information of human activities.

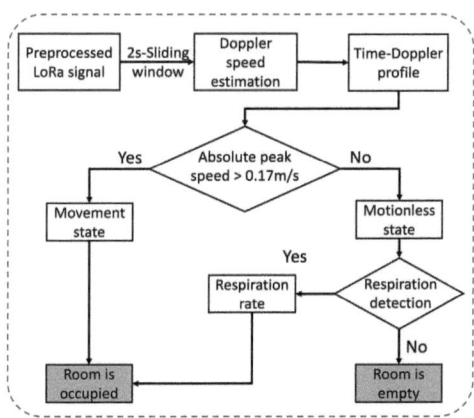

Fig. 6. Signal processing pipeline of presence detection in an office room.

Figure 6 shows the signal processing pipeline for in-room presence detection. The core idea is to detect human movement, including both large-scale activities such as walking and subtle physiological motions like respiration. We conduct an experiment in which different human activities are performed within a period of 190 s, including in-room walking, leaving the room, entering the room, and staying stationary. Figure 7a shows the amplitude of the signal after preprocessing and the ground truth of human activities. We first apply the short-time Fourier transform (STFT) on the received signal to analyze the Doppler speed of the target. Note that a 2-second sliding window is applied to generate the time-Doppler profile using STFT as shown in Fig. 7b. Figure 7c shows the absolute peak Doppler speed extracted from the profile. It is clear that when the human subject is in the motion state, a distinct speed distribution appears in the time-Doppler profile, from which an absolute peak speed greater than a predefined threshold (i.e., 0.17 m/s) can always be extracted. In this case, the room is identified as occupied by a human subject. When the human body is stationary or the office is empty, no obvious Doppler speed distribution can be observed, and the corresponding peak Doppler speed is close to zero. As shown in Fig. 7d, when the room is empty, the signal is stable, while it exhibits periodic fluctuation when a static human exists. To this end, we further perform FFT on the signal using a 30-s window. The presence of a spectral peak within the typical respiration frequency range (i.e., 0.1 – 0.5 Hz) indicates that a person is still present in the room.

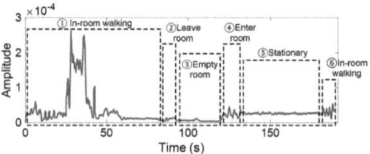

(a) Amplitude of the signal under different states.

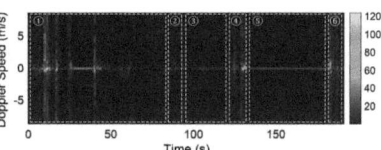

(b) Time-Doppler profile.

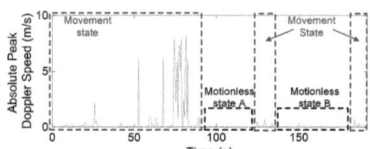

(c) Absolute peak Doppler speed over time.

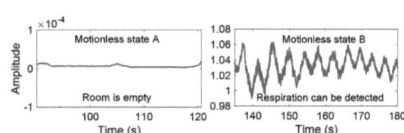

(d) Waveforms of two motionless states.

Fig. 7. Results of presence detection in an office room.

Fig. 8. Experiment setup of presence detection in a public area.

5.2 Case Study B: Presence Detection in a Public Area

In this section, we present a case study of human presence detection in a public area within a smart building (i.e., a lobby scenario). The system deployment is shown in Fig. 8. The LoRa node is positioned in the corner of the lobby, while the LoRa gateway is placed 30 m away from the Tx. Three human subjects are asked to walk within the entire lobby area. A 5 m × 2 m area is set as the sensing coverage. During the experiment, three subjects walk sequentially through the lobby over a total duration of 200 s. Figure 9 illustrates the signal processing pipeline. The core objective is to measure the SSNR of the received signal to detect human presence, specifically within the designated sensing coverage. Since Doppler speed can always be observed regardless of the person's location, relying solely on Doppler measurements does not allow for accurate presence detection within the target area. To this end, we use the method proposed in Sect. 3 to quantify the SSNR of the received signal. We first compute the difference between each signal sample (i.e., $\Delta s(t)$), and then the phase difference of it (i.e., $\Delta(\angle \Delta s(t))$). Finally, we calculate the variance of $\Delta(\angle \Delta s(t))$ as a metric

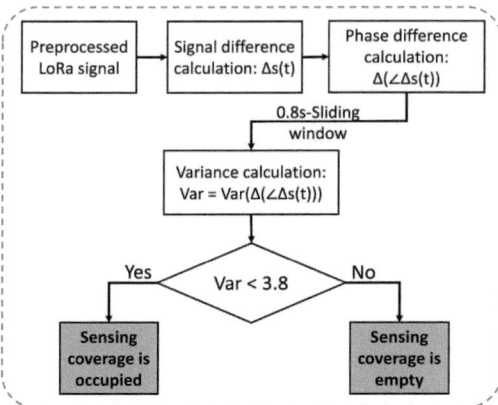

Fig. 9. Signal processing pipeline of presence detection in a public area.

to quantify the SSNR. Note that a 0.8-second sliding window is applied for variance calculation, and Fig. 10a shows the result. We set a threshold (i.e., 3.8) to determine whether a human target is walking in the sensing coverage. Figure 10b shows the presence detection result and the ground truth status. It can be observed that our detection result is highly similar to the ground truth.

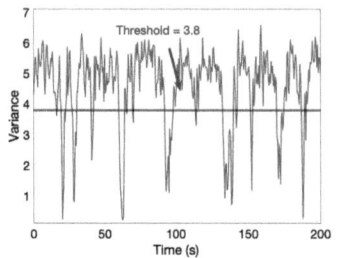
(a) Variance of the phase difference.

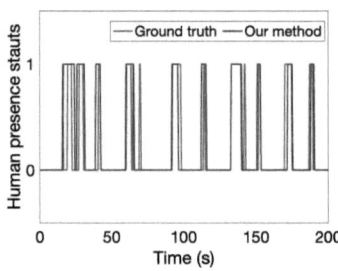
(b) Comparison between our method and ground truth.

Fig. 10. Results of presence detection in a public area.

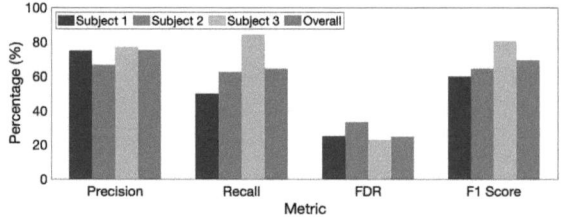

Fig. 11. Presence detection performance in a public area.

To evaluate the generalization of our system, we further present statistics on the presence detection performance across three subjects. Specifically, we calculate the precision, recall, false discovery rate (FDR), and F1 score as matrices. As illustrated in Fig. 11, the system achieves good performance for presence detection, with the precision, recall, FDR, and F1 score of 75.3%, 64.4%, 24.6%, and 70.1%. The performance between different subjects is similar, demonstrating the generalizability and reliability of the system.

6 Related Work

6.1 LoRa Sensing

Due to the long communication range, low cost, and low power consumption, the use of LoRa signals for wireless sensing has attracted increasing research attention [7–9, 29–35]. Zhang et al. [7] first propose the sensing model with LoRa signals and utilize the signal ratio method to achieve long-range through-wall sensing. Xie et al. [9] further improve the sensing distance beyond 100 m by adopting a chirp concentration scheme. Zhang et al. [31] propose to utilize beamforming techniques to achieve multi-target sensing.

6.2 Sensing Coverage

The sensing coverage of wireless sensing has been well studied. Yang et al. [36] investigate the impact of transceiver distance on the coverage of bi-static radar sensing systems. Wang et al. [28] theoretically model the coverage of Wi-Fi sensing system and utilize it for interference mitigation. Xie et al. [29] further explore the effect of walls on the sensing coverage model and provide insights on how to plan device placement to expand sensing coverage and mitigate interference.

7 Conclusion

In this work, we explore the feasibility of scalable wireless sensing in large smart building environments by using LoRa technology. We provide a novel perspective for adapting the sensing coverage of LoRa nodes, which involves hardware-based adjustments. We demonstrate the capability of the proposed framework in two real-world case studies, showing that this approach can effectively manage large-scale deployments and provide reliable presence detection.

Acknowledgment. This work was supported in part by the European Union through the Horizon EIC Pathfinder Challenge project SUSTAIN (Grant No. 101071179), and in part by the Innovative Medicines Initiative 2 Joint Undertaking under the IDEA-FAST project (Grant No. 853981).

References

1. Pu, Q., Gupta, S., Gollakota, S., Patel, S.: Whole-home gesture recognition using wireless signals. In: Proceedings of the 19th Annual International Conference on Mobile Computing & Networking, pp. 27–38 (2013)
2. Wang, H., et al.: Human respiration detection with commodity WiFi devices: do user location and body orientation matter? In: Proceedings of the 2016 ACM International Joint Conference on Pervasive and Ubiquitous Computing, pp. 25–36 (2016)
3. Zeng, Y., Wu, D., Xiong, J., Yi, E., Gao, R., Zhang, D.: FarSense: pushing the range limit of WiFi-based respiration sensing with CSI ratio of two antennas. Proc. ACM Interact., Mob., Wearable Ubiquit. Technol. **3**(3), 1–26 (2019)
4. Wei, T., Zhang, X.: mTrack: High-precision passive tracking using millimeter wave radios. In: Proceedings of the 21st Annual International Conference on Mobile Computing and Networking, pp. 117–129 (2015)

5. Palipana, S., Salami, D., Leiva, L.A., Sigg, S.: Pantomime: mid-air gesture recognition with sparse millimeter-wave radar point clouds. Proc. ACM Interact., Mob., Wearable Ubiquit. Technol. **5**(1), 1–27 (2021)
6. Chang, Z., Zhang, F., Xiong, J., Chen, W., Zhang, D.: MSense: boosting wireless sensing capability under motion interference. In: Proceedings of the 30th Annual International Conference on Mobile Computing and Networking, pp. 108–123 (2024)
7. Zhang, F., et al.: Exploring LoRa for long-range through-wall sensing. Proc. ACM Interact., Mob., Wearable Ubiquit. Technol. **4**(2), 1–27 (2020)
8. Chen, L., et al.: WideSee: towards wide-area contactless wireless sensing. In: Proceedings of the 17th Conference on Embedded Networked Sensor Systems, pp. 258–270 (2019)
9. Xie, B., Cui, M., Ganesan, D., Chen, X., Xiong, J.: Boosting the long range sensing potential of LoRa. In: Proceedings of the 21st Annual International Conference on Mobile Systems, Applications and Services, pp. 177–190 (2023)
10. Zheng, T., Chen, Z., Cai, C., Luo, J., Zhang, X.: V2iFi: in-vehicle vital sign monitoring via compact RF sensing. Proc. ACM Interact., Mob., Wearable Ubiquit. Technol. **4**(2), 1–27 (2020)
11. Zhang, F., Xiong, J., Chang, Z., Ma, J., Zhang, D.: Mobi2Sense: empowering wireless sensing with mobility. In: Proceedings of the 28th Annual International Conference on Mobile Computing and Networking, pp. 268–281 (2022)
12. Zhang, F., et al.: Embracing consumer-level UWB-equipped devices for fine-grained wireless sensing. Proc. ACM Interact., Mob., Wearable Ubiquit. Technol. **6**(4), 1–27 (2023)
13. Zhou, Z., Yang, Z., Wu, C., Shangguan, L., Liu, Y.: Towards omnidirectional passive human detection. In: Proceedings IEEE INFOCOM, pp. 3057–3065. IEEE (2013)
14. Wu, C., Yang, Z., Zhou, Z., Liu, X., Liu, Y., Cao, J.: Non-invasive detection of moving and stationary human with WiFi. IEEE J. Sel. Areas Commun. **33**(11), 2329–2342 (2015)
15. Xin, T., et al.: FreeSense: a robust approach for indoor human detection using Wi-Fi signals. Proc. ACM Inter., Mob., Wearable Ubiquit. Technol. **2**(3), 1–23 (2018)
16. Patwari, N., Wilson, J., Ananthanarayanan, S., Kasera, S.K., Westenskow, D.R.: Monitoring breathing via signal strength in wireless networks. IEEE Trans. Mob. Comput. **13**(8), 1774–1786 (2013)
17. Liu, X., Cao, J., Tang, S., Wen, J.: Wi-Sleep: contactless sleep monitoring via WiFi signals. In: IEEE Real-Time Systems Symposium, pp. 346–355. IEEE (2014)
18. Wang, P., et al.: SlpRof: improving the temporal coverage and robustness of RF-based vital sign monitoring during sleep. IEEE Trans. Mob. Comput. **23**(7), 7848–7864 (2023)
19. Abdelnasser, H., Youssef, M., Harras, K.A.: WiGest: a ubiquitous WiFi-based gesture recognition system. In: IEEE Conference on Computer Communications (INFOCOM), pp. 1472–1480. IEEE (2015)
20. Zheng, Y., et al.: Zero-effort cross-domain gesture recognition with Wi-Fi. In: Proceedings of the 17th Annual International Conference on Mobile Systems, Applications, and Services, pp. 313–325 (2019)
21. Wu, D., et al.: FingerDraw: sub-wavelength level finger motion tracking with WiFi signals. Proc. ACM Inter., Mob., Wearable Ubiquit. Technol. **4**(1), 1–27 (2020)

22. Wang, W., Liu, A. X., Shahzad, M., Ling, K., Lu, S.: Understanding and modeling of WiFi signal based human activity recognition. In: Proceedings of the 21st Annual International Conference on Mobile Computing and Networking, pp. 65–76 (2015)
23. Wang, H., Zhang, D., Wang, Y., Ma, J., Wang, Y., Li, S.: RT-Fall: a real-time and contactless fall detection system with commodity WiFi devices. IEEE Trans. Mob. Comput. **16**(2), 511–526 (2016)
24. Hu, Y., Zhang, F., Wu, C., Wang, B., Liu, K.R.: DeFall: environment-independent passive fall detection using WiFi. IEEE Internet Things J. **9**(11), 8515–8530 (2021)
25. Li, X., Li, S., Zhang, D., Xiong, J., Wang, Y., Mei, H.: Dynamic-music: Accurate device-free indoor localization. In: Proceedings of the 2016 ACM International Joint Conference on Pervasive and Ubiquitous Computing, pp. 196–207 (2016)
26. Li, X., et al.: IndoTrack: device-free indoor human tracking with commodity Wi-Fi. Proc. ACM Inter., Mob., Wearable Ubiquit. Technol. **1**(3), 1–22 (2017)
27. Qian, K., Wu, C., Zhang, Y., Zhang, G., Yang, Z., Liu, Y.: Widar2. 0: passive human tracking with a single Wi-Fi link. In: Proceedings of the 16th Annual International Conference on Mobile Systems, Applications, and Services, pp. 350–361 (2018)
28. Wang, X., et al.: Placement matters: understanding the effects of device placement for WiFi sensing. Proc. ACM Interact., Mob., Wearable Ubiquit. Technol. **6**(1), 1–25 (2022)
29. Xie, B., Cui, M., Ganesan, D., Xiong, J.: Wall matters: rethinking the effect of wall for wireless sensing. Proc. ACM Interact., Mob., Wearable Ubiquit. Technol. **7**(4), 1–22 (2024)
30. Xie, B., Xiong, J.: Combating interference for long range LoRa sensing. In: Proceedings of the 18th Conference on Embedded Networked Sensor Systems, pp. 69–81 (2020)
31. Zhang, F., et al.: Unlocking the beamforming potential of LoRa for long-range multi-target respiration sensing. Proc. ACM Interact., Mob., Wearable Ubiquit. Technol. **5**(2), 1–25 (2021)
32. Xie, B., Yin, Y., Xiong, J.: Pushing the limits of long range wireless sensing with LoRa. Proc. ACM Interact., Mob., Wearable Ubiquit. Technol. **5**(3), 1–21 (2021)
33. Xie, B., Ganesan, D., Xiong, J.: Embracing LoRa sensing with device mobility. In: Proceedings of the 20th ACM Conference on Embedded Networked Sensor Systems, pp. 349–361 (2022)
34. Chang, Z., Zhang, F., Xiong, J., Ma, J., Jin, B., Zhang, D.: Sensor-free soil moisture sensing using LoRa signals. Proc. ACM Interact., Mob., Wearable Ubiquit. Technol. **6**(2), 1–27 (2022)
35. Song, Z., Tong, S., Wang, J.: LoSense: integrated long-range sensing and communication with LoRa signals. In: 2023 IEEE 31st International Conference on Network Protocols (ICNP), pp. 1–11. IEEE (2023)
36. Yang, Q., He, S., Chen, J.: Energy-efficient area coverage in bistatic radar sensor networks. In: IEEE Global Communications Conference (GLOBECOM), pp. 280–285. IEEE (2013)

Wi-CLIP: Toward Zero-Shot Air Gesture Recognition Based on RF-Text Foundation Model

Haoyu Zhang, Yifan Guo, Zhu Wang(✉), Zhuo Sun, Bin Guo, and Zhiwen Yu

Northwestern Polytechnical University, Xi'an 710072, Shaanxi, China
wangzhu@nwpu.edu.cn

Abstract. Wi-Fi-based gesture recognition, driven by deep learning, holds significant promise for privacy-preserving and all-weather sensing. However, current methods typically rely on large amounts of labeled data, and Wi-Fi signals vary significantly across gestures, leading to severe performance degradation when models encounter unseen gestures. To address these challenges, we explore the potential of transferring knowledge from large pre-trained language models to improve the generalization of Wi-Fi-based gesture recognition systems. To this end, we propose a zero-shot gesture recognition framework, named *Wi-CLIP*. Inspired by the vision-language pre-training model CLIP, our method constructs a cross-modal radio frequency-text model centered on aligning Wi-Fi signals with textual semantics. Specifically, we develop a novel Wi-Fi signal encoder and a BERT-based text encoder, aligning the two modalities within a shared semantic space using contrastive learning. Our framework achieves an average recognition accuracy of 89.12% across 6 gestures. Notably, when trained on only 5 gestures, *Wi-CLIP* demonstrates a remarkable zero-shot recognition accuracy of 78.79% on the sixth, previously unseen gesture. This highlights its strong generalization capability and effectiveness in cross-modal representation learning.

Keywords: Wireless Sensing · Gesture Recognition · Zero Shot Learning · Vision Language Model

1 Introduction

With the continuous development of smart technology, people's daily life is increasingly connected with various electronic devices, and human gesture recognition has become one of the core technologies in the fields of smart home control, contactless interaction, health monitoring, and security. Since the 1970s, researchers have achieved many important results in this field, and traditional methods mainly rely on sensing modules such as cameras [1] and wearable devices [2]. However, these methods are gradually being replaced by more advanced technologies due to the high risk of privacy leakage, the need to wear additional sensors, and the limited sensing range. Early wireless communication systems

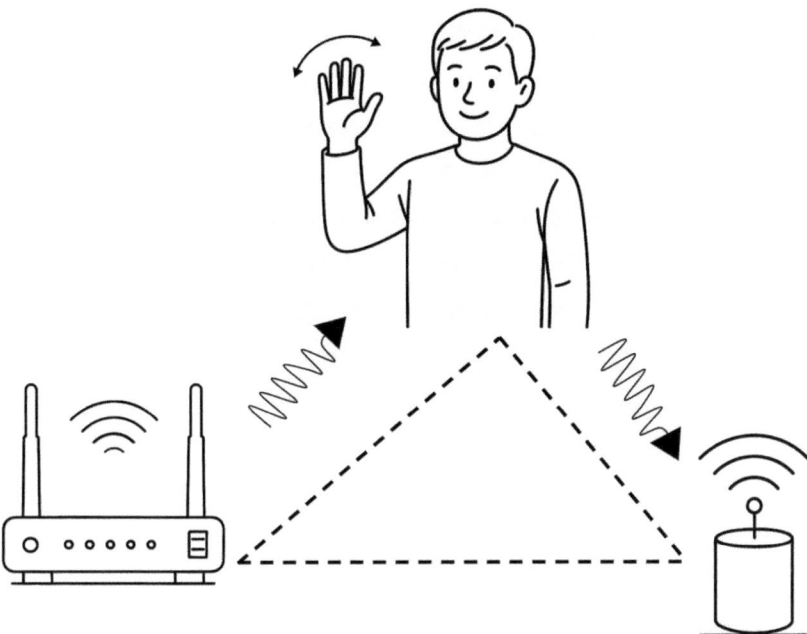

Fig. 1. The process of WiFi gesture detection.

had data transmission as their core objective, but with the rapid development of the Internet of Things (IoT) and smart home technologies, researchers have begun to explore how to utilize the existing Wi-Fi infrastructure to achieve vital signs monitoring [3–6], material detection [7,8], and gesture recognition [9,10], as shown in Fig. 1. Compared with the traditional visual gesture recognition, Wi-Fi gesture recognition shows great potential in the field of gesture recognition by virtue of its non-invasiveness and good privacy preservation ability.

In recent years, the field of Wi-Fi gesture recognition has witnessed a surge in innovative research aimed at bolstering the cross-domain generalization capabilities of recognition models. For instance, cutting-edge studies [11] have demonstrated that fusing acoustic signals with Wi-Fi signals can enhance gesture recognition performance, effectively overcoming the limitations inherent in relying on a single modality. Another notable approach [12] involves employing a subcarrier selection strategy based on fluency and discrimination, coupled with a novel gesture coding method that enables position-independent gesture recognition using only a pair of transceivers. While these advancements have indeed improved the generalization abilities of the models to some extent, existing classifiers still face the challenge of limited generalization domains. They are primarily optimized for specific, narrowly defined classification tasks. Given the virtually infinite variety of gestures, each with varying levels of complexity, this limitation poses a significant obstacle to further enhancing the generalization capabilities of Wi-Fi gesture recognition.

In recent years, artificial intelligence has made significant progress, especially in natural language processing and computer vision. These advances are driven by large-scale data mining and the effective use of hyperscale models. Scaling laws show that combining large datasets with large models significantly improves generalization, enabling strong performance in complex tasks. This raises a key question: Can knowledge from large AI models trained on text and images be transferred to Wi-Fi-based gesture recognition to boost generalization?

To explore this problem, we propose the *Wi-CLIP* framework, which implements zero-shot gesture recognition based on Wi-Fi signals with the help of knowledge embedded in a large language model. The core idea lies in aligning the high-level representation space of Wi-Fi signals with the textual semantic space obtained from training a large language model (LLM), so as to predict unseen activities by leveraging the generalization ability of the language model. However, accurately aligning these two embedding spaces requires a large amount of signal-text pairing data corresponding to different gestures, while directly adopting the raw Wi-Fi data is not only a huge amount of data, which leads to a slower training process, but also suffers during feature extraction. Therefore, we chose the publicly available dataset widar3.0 [13] as the experimental data, and extracted a domain-independent feature from it, the Body Coordinate Velocity Profile (BVP), which describes the power distribution of the body parts involved in the gesture action at different velocities. This not only shortens the model training time and improves the training efficiency, but also achieves an improvement in accuracy.

Meanwhile, it is often difficult to achieve a satisfactory generalization effect by directly aligning radio frequency (RF) data with original text labels. The reason is that Wi-Fi signal belongs to continuous time-series data, while text is discrete symbolic data, and there are significant differences in data distribution and feature representation between the two, making direct alignment extremely challenging. In addition, the training process relies on high-quality pairs of positive and negative samples, whereas the direct adoption of RF data tends to introduce noise, which reduces the quality of the data paired with Wi-Fi and text, and thus leads to unstable model convergence. For this reason, we design a novel data encoder. Specifically, this encoder is based on BVP data, and first extracts the spatial relational features of the signals through convolutional neural network (CNN), then captures the temporal dependencies of the data using recurrent neural network (RNN), and finally captures the global dependencies with the help of the self-attention mechanism in the Transformer architecture, which dynamically pays attention to the importance of each element in the input sequence. In this way, RF data and text data can be effectively aligned at the feature level.

To sum up, we summarize the contributions of this paper as follows:

- We propose and open-source the *Wi-CLIP* framework[1], which aligns Wi-Fi signals with the text embedding space for unseen activity recognition based

---

[1] https://github.com/yanbanliu/Wi-CLIP.

on Wi-Fi signals, where the model is never trained on unseen activity Wi-Fi data or exact text labels.
- We design a new Wi-Fi signal encoder that utilizes CNNs, RNNs, and Transformer architecture to extract spatial relational features, local and global temporal dependencies on the RF data, and effectively align the RF data embedding space with the text embedding space.
- We develop a prototype system based on the *Wi-CLIP* framework, which achieve an average accuracy of 78.79% in the zero-shot classification task, highlighting the effectiveness of the *Wi-CLIP* framework in zero-shot Wi-Fi gesture recognition activities.

2 Related Work

2.1 Wi-Fi-Based Gesture Recognition

Wi-Fi-based gesture recognition methods can be broadly categorized into two types: handcrafted feature-based methods and deep learning-based methods. The former relies on manually designed features to distinguish different gesture patterns, while the latter leverages machine learning techniques to automatically learn signal patterns for gesture recognition. In early research, WiGest [14] detected changes in Wi-Fi access point (AP) received signal strength and used received signal strength indicator (RSSI) to map hand movements to specific gestures. Although this method is inspiring, its applicability is limited due to the coarse granularity of RSSI and its susceptibility to environmental variations. WiDraw [15] utilized the angle-of-arrival (AoA) information of Wi-Fi signals to capture hand movement trajectories and even recognize air-drawn characters. However, this approach requires a dense deployment of Wi-Fi devices, typically needing at least 12 signal sources to achieve high tracking accuracy. QGesture [16] inferred gesture motion characteristics by analyzing changes in channel state information (CSI) caused by hand movements. While it offers high measurement accuracy and environmental adaptability, it requires gestures to contain specific preambles, imposing certain constraints on its application.

Deep learning-based methods, on the other hand, rely on large-scale data-driven pattern learning. Before 2019, Wi-Fi gesture recognition primarily adopted traditional machine learning approaches. With the rise of deep learning, researchers began applying it to this field. For instance, the Wi-Do system [17], introduced in 2020, incorporated an attention mechanism and bidirectional GRU network to classify motion feature data, thereby improving recognition accuracy. Widar3.0 [13]introduced a novel domain-independent feature to capture human posture changes. This feature significantly enhanced cross-domain generalization in gesture recognition, even under low signal levels. Building on this, WiHF [18] further extracted arm gesture motion patterns, incorporating individual execution styles to enhance recognition precision. WiSGP [19] designed an intelligent wireless data augmentation technique to address cross-domain issues in wireless signals, thereby improving model generalization. However, these methods generally assume that test samples belong to the same gesture categories as training samples, overlooking the complexity of gestures in real-world applications. As a result, they struggle to adapt to open-set gesture recognition scenarios.

2.2 Vision-Language Foundation Models

Contrastive Language-Image Pre-training (CLIP) [20] aligns visual and textual representations through large-scale contrastive learning, demonstrating powerful zero-shot transfer capabilities. In recent years, CLIP's cross-modal properties have been extended to various domains. For example, AudioCLIP [21] expanded CLIP by integrating the ESResNeXt audio model, enabling it to process audio, text, and image data for environmental sound classification tasks. ViFi-CLIP [22] fine-tuned CLIP with video data, adapting it for video-related tasks. mmCLIP [23] applied similar techniques to millimeter-wave radar (mmWave Radar), aligning signals with textual descriptions to enhance zero-shot human activity recognition (HAR) performance in radar-based sensing. However, transferring CLIP's vision-language knowledge to Wi-Fi signals remains an open challenge. Due to the substantial modality gap between Wi-Fi CSI data and images/text, directly applying CLIP is difficult. Therefore, new adaptation methods tailored to Wi-Fi data characteristics are required to fully leverage the knowledge embedded in CLIP's pre-trained model.

2.3 Wi-Fi-Text Foundation Models

Despite the outstanding performance of large models in vision and language tasks, their application in wireless sensing tasks, such as Wi-Fi-based gesture recognition, remains underexplored. For instance, RF-CM [24] leveraged extensive Wi-Fi data to assist in training mmWave radar-based HAR models, improving sensing capabilities. RF-Diffusion [25] employed a diffusion model to generate high-quality temporal wireless RF data, enhancing the accuracy of wireless sensing systems. Unlike these studies, our work is the first to explore transferring CLIP's vision-language alignment capabilities to Wi-Fi gesture recognition. We propose a modality adaptation approach that enables the model to effectively utilize CLIP's pre-trained knowledge, improving both zero-shot generalization and zero-shot learning capabilities. This research not only introduces new possibilities for Wi-Fi sensing technology but also lays the foundation for applying large models to wireless signal processing.

3 *Wi-CLIP* System Overview

In this study, we propose *Wi-CLIP*, a cross-modal understanding framework designed to achieve semantic interpretation and generalized recognition of complex human actions by aligning low-cost Wi-Fi signals with textual semantics. As shown in Fig. 2, the system architecture comprises four core modules: Wi-Fi signal pre-processing, Wi-Fi feature extraction, text encoding, and cross-modal alignment.

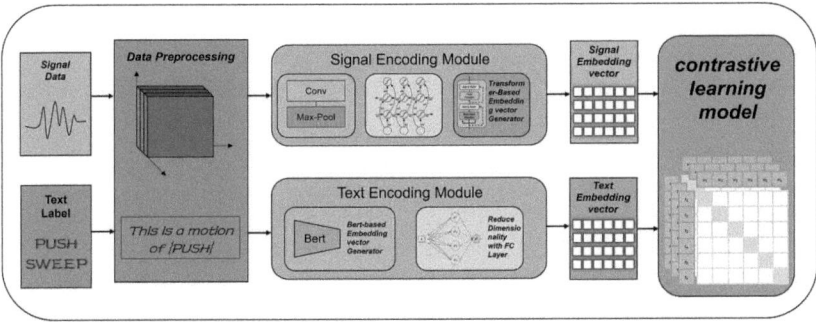

Fig. 2. Overview of the *Wi-CLIP* framework.

Using the Widar3.0 dataset, the system converts raw CSI data into BVPs as the Wi-Fi input modality. These profiles are derived from multipath variations caused by human body movements. BVPs effectively capture the spatial motion characteristics of the human body, facilitating the encoder's extraction of temporal dynamic information.

To extract spatio-temporal features from Wi-Fi BVP signals, we design a multilayer hybrid encoder that combines CNNs for local perception, RNNs for sequential modeling, and Transformers for global attention, enabling multi-scale action feature modeling. The encoder produces fixed-dimensional embedding vectors that represent the semantic features of Wi-Fi-based actions. For each human action, a pre-trained BERT model is employed to extract embedded features from its corresponding semantic description. The text encoder generates embeddings with the same dimensionality as the Wi-Fi encoder, enabling effective feature alignment.

Leveraging the contrastive learning mechanism of the CLIP framework, the system optimizes the similarity between Wi-Fi and text features by constructing positive and negative sample pairs, thereby achieving cross-modal alignment and semantic association. Specifically, in each training batch, the system constructs RF-Text pairs, maximizes the similarity of positive pairs, and minimizes that of negative pairs using the *InfoNCE* loss function, ultimately achieving semantic alignment in a shared embedding space.

3.1 Wi-Fi Signal Pre-processing

Widar3.0 [13], developed by Tsinghua University in 2020, is a representative large-scale dataset for Wi-Fi-based action recognition in the field of wireless sensing. It contains more than 1,000,000 samples that encompass six basic action types, such as pushing and pulling, sweeping, applauding, swiping, drawing circles, and drawing jagged lines, as well as a variety of more complex and semantically rich gestures. Data collection was carried out in three typical indoor environments (office, hallway, and living room) and involved multiple participants with various body types.

The key innovation of Widar3.0 lies in its pioneering implementation of omnidirectional motion recognition using commercial Wi-Fi devices. It captures CSI through a multiple-input multiple-output (MIMO) system composed of six transmitterreceiver antenna pairs. Additionally, inertial measurement unit (IMU) data is synchronously recorded as auxiliary annotation, and detailed timefrequency domain features are provided for further analysis.

CSI captures both the magnitude and phase characteristics of each propagation path in a multipath wireless channel. It can be mathematically modeled as follows:

$$\hat{H}(f,t) = \left(\sum_{l=1}^{L} \alpha_l(f,t) e^{-j2\pi f \tau_l(f,t)} \right) e^{j\epsilon(f,t)} \tag{1}$$

Here, L denotes the number of propagation paths, while α_l and τ_l represent the attenuation and time delay of the l path, respectively. $\epsilon(f,t)$ denotes the phase error.

When body movement alters the reflection path, it induces a Doppler Frequency Shift (DFS), which can be expressed as:

$$\hat{H}(f,t) = \left(H_s(f) + \sum_{l \in P_d} \alpha_l(t) e^{j2\pi \int_{-\infty}^{t} f_{D_l}(u) du} \right) e^{j\epsilon(f,t)} \tag{2}$$

Here, the constant H_s represents the sum of all static signal components—such as line-of-sight signals—with zero Doppler shift, while P_d denotes the set of dynamic signal components, including those reflected from the target, that exhibit non-zero Doppler shift.

Significant multipath components with non-zero DFS are preserved through a dual-antenna CSI conjugate product computation, combined with out-of-band noise filtering and quasi-static offset cancellation. A single application of the Short-Time Fourier Transform (STFT) then produces power distributions across both the time and Doppler frequency domains.

Human gesture movements induce site-specific DFS, generating a DFS profile that contains gesture-related information at the receiver, although it is significantly affected by environmental interference. A local coordinate system is defined with the human body's position as the origin, and the positive x-axis aligned with the body's orientation. For link i, the DFS contribution from the transmitter position $\bm{l}_t^{(i)} = \left(x_t^{(i)}, y_t^{(i)} \right)$ and receiver position $\bm{l}_r^{(i)} = \left(x_r^{(i)}, y_r^{(i)} \right)$, with velocity $\bm{v} = (v_x, v_y)$, is given by:

$$f^{(i)}(\bm{v}) = a_x^{(i)} v_x + a_y^{(i)} v_y \tag{3}$$

where the coefficients $a_x(i)$ and $a_y(i)$ are determined by geometric relations:

$$a_x^{(i)} = \frac{1}{\lambda} \left(\frac{x_t^{(i)}}{\left\| l_t^{(i)} \right\|_2} + \frac{x_r^{(i)}}{\left\| l_r^{(i)} \right\|_2} \right)$$
$$a_y^{(i)} = \frac{1}{\lambda} \left(\frac{y_t^{(i)}}{\left\| l_t^{(i)} \right\|_2} + \frac{y_r^{(i)}}{\left\| l_r^{(i)} \right\|_2} \right) \quad (4)$$

where λ represents the wavelength of the Wi-Fi signal.

Based on the coefficients $a_x^{(i)}$ and $a_y^{(i)}$ depend solely on the position of the i-th link, the allocation matrix $A_{F \times N^2}^{(i)}$ is defined to map the BVP velocity component v_k to the Doppler frequency sampling point f_j:

$$A_{j,k}^{(i)} = \begin{cases} 1, & \text{if } f_j = f^{(i)}(v_k) \\ 0, & \text{otherwise} \end{cases} \quad (5)$$

Thus, the relationship between the DFS profile of link i and the BVP can be formulated as:

$$D^{(i)} = c^{(i)} A^{(i)} V \quad (6)$$

where $c^{(i)}$ is the scaling factor accounting for the propagation loss of the reflected signal.

Due to the limited number of links, BVP estimation is an underdetermined problem. By exploiting the sparsity of the BVP, since human actions involve only a few significant velocity components, the BVP estimation problem can be formulated as the following optimization problem, solved using a compressed sensing approach:

$$\min_V \sum_{i=1}^{M} \left| \text{EMD}(A^{(i)} V, D^i) \right| + \eta \| V \|_0 \quad (7)$$

where M is the number of Wi-Fi links.

The first term, Earth Mover's Distance (EMD), quantifies the difference between the estimated DFS profile $A^{(i)}$ and the measured D^i, while the second term, $\eta \| V \|_0$, enforces BVP sparsity by minimizing the number of non-zero velocity components. The significant velocity components are preferentially retained through an iterative optimization algorithm, such as orthogonal matching pursuit.

3.2 Wi-Fi Feature Extraction

BVP data can be regarded as a sequence of images, where each BVP instance represents the power distribution of physical velocity over a short time window. Consecutive BVPs capture the temporal evolution of this distribution corresponding to a specific human action. To efficiently extract multilevel spatiotemporal features from such data, this paper proposes a hybrid encoder architecture that integrates CNN, Bidirectional Gated Recurrent Unit (GRU) network, and Transformers in a cascaded structure for hierarchical feature modeling.

Spatial Feature Extraction Module. CNNs are widely used for spatial feature extraction and data compression. In this study, two consecutive two-dimensional convolutional layers (Conv2D) are used to extract spatial features from the BVP data. The input BVP sequence is represented as $V \in \mathbb{R}^{N \times N \times T}$, where $N \times N$ indicates the spatial resolution of each BVP snapshot, and T denotes the total number of time frames. For the t -th sampled BVP snapshot, the input tensor V_{in} is processed through two Conv2D layers to extract local spatial features, producing an intermediate output F_{local}. Subsequently, max-pooling is applied to perform spatial downsampling, generating F_{pooled}. Finally, the output is flattened into a one-dimensional feature vector, $F_{\text{flattened}}$, which serves as the input for the subsequent temporal modeling module.

Temporal Feature Extraction Module. To model the temporal dynamics of gestures in BVP data, this paper utilizes a recurrent neural network (RNN) to capture the intricate temporal dependencies within the time series. Various types of RNN units exist, including the Simple-RNN, the LSTM, and the gated recurrent unit (GRU). Compared with basic RNNs, both LSTM and GRU are more effective at learning long-term dependencies.In this study, a bidirectional GRU (Bi-GRU) network is employed for temporal modeling, which enables the model to capture contextual information from both past and future time steps. Specifically, the output from the spatial feature extraction module is fed into the Bi-GRU network to produce a vector $h_{(LSTM)}$. A dropout layer is then applied to prevent overfitting, as expressed by:

$$\hat{X} = \text{Dropout}(h_{\text{GRU}}, p) \tag{8}$$

Global Relationship Modeling Module. Building on the modeling of temporal dynamics, this paper further incorporates the Transformer architecture to capture global dependencies across time steps. The Transformer is a sequence modeling architecture based on a self-attention mechanism, enabling parallel input processing and efficient capture of long-range global dependencies. In comparison to traditional recurrent networks, the Transformer offers more efficient and expressive feature modeling by leveraging a multi-head attention mechanism alongside feed-forward neural networks.

Specifically, this study employs a three-layer Transformer encoder, with eight attention heads in each layer, to model the global dependency structures in the time series. To retain information about the sequential order of the input, sinusoidal positional encoding is added to the input embeddings prior to Transformer processing. The positional encoding $\mathbf{PE} \in \mathbb{R}^{L \times d}$ is defined as follows:

$$\begin{aligned} \text{PE}_{(pos,\,2i)} &= \sin\left(\frac{pos}{10000^{\frac{2i}{d}}}\right) \\ \text{PE}_{(pos,\,2i+1)} &= \cos\left(\frac{pos}{10000^{\frac{2i}{d}}}\right) \end{aligned} \tag{9}$$

where *pos* denotes the position index and i is the dimension index. This positional encoding is added element-wise to the input embeddings to preserve the temporal order.

Finally, the extracted features are projected into a uniform vector space via a fully connected layer, yielding the final embedded representation of the Wi-Fi signal.

3.3 Text Encoding

A text encoder is a crucial module in natural language processing that converts discrete language units (such as words or sentences) into continuous vector representations (embeddings), enabling computers to understand and process them. Several well-established pre-trained text encoders exist, including BERT-base proposed by Google and RoBERTa-base developed by Facebook. In this study, we select the BERT-base-uncased model, which uses the [CLS] token as the semantic representation of an entire sentence and maps it to a unified vector space via a projection layer. Given that the BERT model comprises approximately 110 million parameters, all of which are frozen during training, only the projection layer is trained to substantially improve training efficiency and bolster the stability of multimodal alignment.

3.4 Cross-Modal Alignment

For the Wi-Fi signal embedding $z_i^{(w)}$ and text embedding $z_i^{(t)}$ obtained in the previous section, an effective fusion mechanism is required to integrate the representations of these two modalities.In this study, we employ a contrastive learning approach to achieve multimodal embedding alignment.Contrastive learning is a self-supervised technique that derives discriminative feature representations by quantifying the similarities and differences among samples.

In contrastive learning frameworks, a variety of loss functions can be employed for optimization.Common loss functions in contrastive learning include Contrastive Loss, Triplet Loss, InfoNCE Loss, and Decoupled Contrastive Loss (DCL). In this study, the InfoNCE loss function is adopted as the optimization objective for contrastive learning. This loss function is founded on the principle of mutual information maximization, achieving multimodal feature alignment by reducing the representation distance between pairs of positive samples (e.g., signals and text embeddings corresponding to the same activity) while increasing the distance between pairs of negative samples (e.g., embeddings corresponding to different activities).

Specifically, for each Wi-Fi encoding $z_i^{(w)}$, we aim to maximize its similarity with the corresponding text embedding $z_i^{(t)}$, while minimizing its similarity with all other negative text samples. The corresponding loss function is defined as follows:

$$\mathcal{L}_{w \to t} = \frac{1}{N} \sum_{i=1}^{N} -\log \left(\frac{\exp\left(\text{sim}(z_i^{(w)}, z_i^{(t)})\right)}{\sum_{j=1}^{N} \exp\left(\text{sim}(z_i^{(w)}, z_j^{(t)})\right)} \right) \quad (10)$$

In turn, for each text encoding $z_i^{(t)}$, we also want it to maximize the similarity with its corresponding Wi-Fi encoding $z_i^{(w)}$. This objective is defined by the following contrastive loss:

$$\mathcal{L}_{t \to w} = \frac{1}{N} \sum_{i=1}^{N} -\log \frac{\exp\left(\text{sim}(z_i^{(t)}, z_i^{(w)})\right)}{\sum_{j=1}^{N} \exp\left(\text{sim}(z_i^{(t)}, z_j^{(w)})\right)} \quad (11)$$

Take the average of both as the final training loss:

$$\mathcal{L}_{\text{InfoNCE}} = \frac{1}{2} \left(\mathcal{L}_{w \to t} + \mathcal{L}_{t \to w} \right) \quad (12)$$

4 Performance Evaluation

4.1 Dataset Processing

Experiments were conducted using the Widar 3.0 dataset in three typical indoor environments: a classroom, a hall, and a furnished office. Six common hand gestures were selected: push-pull, sweep, clap, slide, circle, and zigzag. The dataset provides detailed annotations of position and orientation variations. In total, 12,000 samples were collected from 16 participants across five locations and five body orientations, with each gesture repeated five times for consistency.

4.2 Overall Accuracy

After considering multiple scenarios and influencing factors, the system achieves an overall gesture recognition accuracy of 89.12%. Specifically, data from Rooms 1, 2, and 3 are divided by 90% used for training and 10% for testing. Figure 3a presents the confusion matrix for the six gestures in Dataset 1. The *Wi-CLIP* framework achieves more than 84% accuracy for all gestures, demonstrating robust and consistent recognition performance.

Additionally, the *Wi-CLIP* framework demonstrates strong performance in zero-shot gesture recognition. Six gesture categories are used in the experiment, with five for model training and one reserved as an unseen category for testing. Few-shot learning results are also used as a comparative reference. Specifically, introducing 1 to 5 samples of the target gesture into the training set yields only a slight improvement in test accuracy.

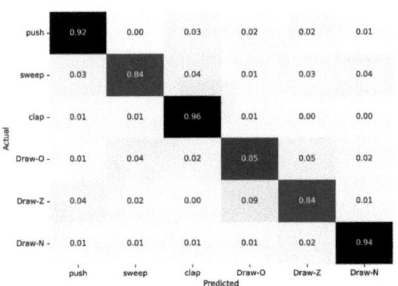 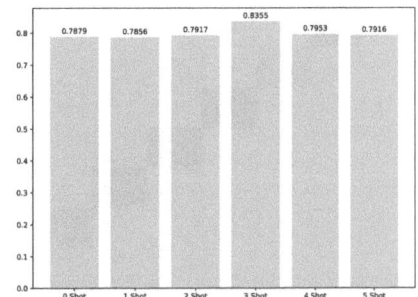

(a) Confusion Matrix for Six Gesture Recognition.

(b) Accuracy comparison of gesture recognition under different few-shot settings.

Fig. 3. Hit rate after overall training and hit rate with zero sample training.

As shown in Fig. 3b, the zero-shot recognition accuracy for unseen gestures is 78.79%, whereas the few-shot accuracy ranges from 78.56% to 83.55%. These results indicate that the *Wi-CLIP* framework exhibits strong generalization in modeling the cross-modal alignment between gestures and semantics. It achieves high recognition accuracy even without samples from the target category, demonstrating robust cross-modal transfer capabilities and making it well-suited for recognizing gestures from unseen categories.

However, adding a small number of samples results in only marginal improvement in recognition performance. This limitation may stem from the small number of gesture categories in the experiment (only six), which may not adequately capture the diversity of gesture types. Given that each gesture category contains approximately 2,000 samples, the performance gap between few-shot and zero-shot recognition remains relatively small.

4.3 Cross-Cutting Assessments

This section evaluates the overall performance of the proposed framework under various domain factors. For each individual factor, all other conditions are kept constant, and evaluation is conducted using Leave-One-Out Cross-Validation (LOOCV) on the dataset. System performance is assessed using a confusion matrix, with the results presented in Fig. 4.

Location Independence. In this experiment, BVP data from four locations, five orientations, and eight participants were randomly selected for training in Room 1. Data collected from the remaining location in the same room was used for testing. The experimental results show that all six gestures achieved recognition rates above 80%. Specifically, the sweep gesture, which had the lowest recognition rate, reached 84%, while the average recognition rate was 90.6%. The push-pull and circle-drawing gestures achieved the highest performance, with recognition rates of approximately 94%.

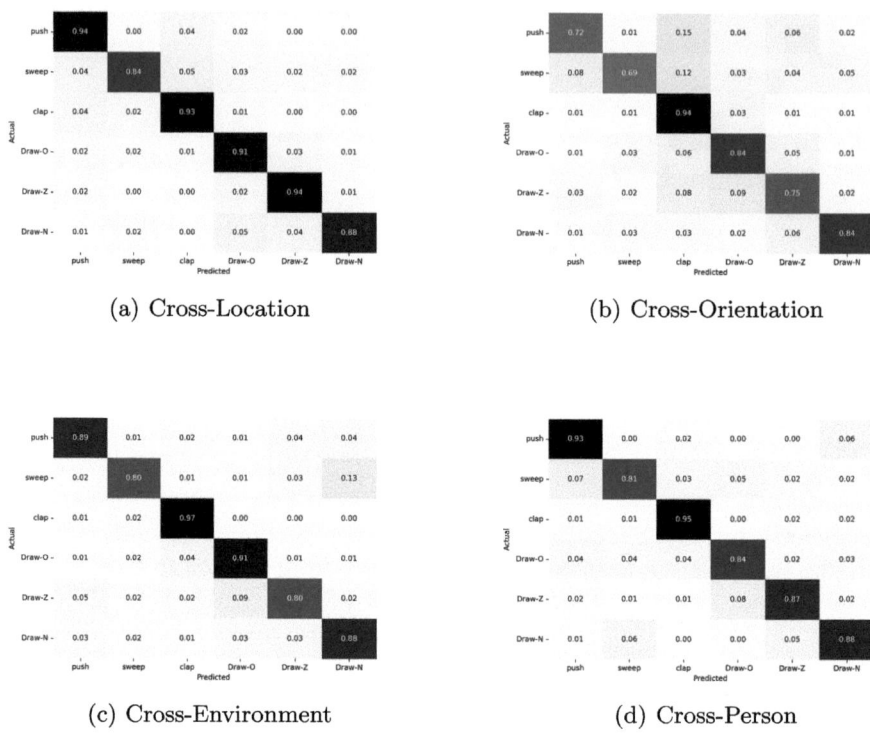

Fig. 4. Confusion matrices of WiFi-CLIP in four generalization scenarios.

Orientation Sensitivity. The experimental data included five orientations, with one selected as the target domain and the remaining four as the source domain. The test results show that the first, second, and fifth gestures exhibited relatively low recognition rates, while the third gesture achieved the highest accuracy at 94%. The overall average recognition rate was 80%. This performance degradation may be attributed to occlusion of gesture movements by the human body, which reduces the number of active wireless links contributing to the generation of BVP signals.

Environment Diversity. In this experiment, data from Room 1 and Room 2 were used as the training set, while data from Room 3 served as the test set. The experimental results indicate that the second and fourth gestures had relatively low recognition rates, though both exceeded 80%. The third gesture achieved the highest recognition rate of 97%. These results demonstrate that the proposed framework maintains robust gesture recognition performance across different experimental environments.

Person Variety. This experiment involved data from 14 participants. However, due to individual differences in behavior patterns, the data collected from different participants may exhibit some variation. During the testing phase, data from "user3" was selected as the test set, while data from the other participants were used for training. The experimental results show that the recognition rates of all gestures exceeded 80%, with two gestures achieving recognition rates of 90%, and the highest reaching 95%.

5 Future Work

At present, the *Wi-CLIP* framework still has a lot of room for improvement. In this section, we introduce several possible future directions worthy of further study.

First, extensions of multimodal encoders for multi-source wireless sensing data are explored. Current models are primarily based on Wi-Fi CSI. However, in practical applications, single-modal data is highly susceptible to environmental interference and occlusion, leading to limited stability and generalization performance of the system. Therefore, in the future, a more versatile multimodal sensing encoder framework can be developed to handle multiple wireless signal sources, such as millimeter-wave radar, Bluetooth RSSI/BLE signals, infrared/wearable sensor data, and multi-antenna array information. Strategies like modal adaptive mechanisms, cross-modal alignment modules, and unified embedding spaces can be integrated to enable joint modeling and fusion of multi-source data, significantly enhancing the robustness and scalability of the model in complex dynamic environments.

At present, additional efforts are required to enhance performance in more challenging zero-shot scenarios. Currently, the zero-shot recognition capability of Wi-CLIP is primarily demonstrated in the "5 training classes + 1 test class" setting. However, when evaluated under a "4 training classes + 2 test classes" configuration, the performance declines significantly, with accuracy dropping below 60%. This performance gap may result from limited training data, low sample diversity, or architectural constraints in representational capacity. Future research can address these limitations by enlarging the dataset, increasing the diversity of gesture samples, and enhancing model generalization through more effective encoding strategies and cross-modal alignment mechanisms.

Third, task design and model expansion for complex semantic interactions are explored. The current model primarily focuses on matching static gestures with brief text semantics, and the task scale is relatively small, covering only six basic gestures. Future research can extend this work to include richer, dynamic, and semantically diverse interaction tasks, including but not limited to the following directions:

1. Diversified gesture recognition tasks, such as multi-stage sequential actions, micro-gesture recognition, and two-handed cooperation;
2. Fine-grained semantic parsing, introducing multilevel linguistic expressions (like "raise your right hand" vs. "slowly raise your right hand");

3. Context-aware and multi-round interaction modeling, integrating contextual information such as time, space, and the user's historical behaviors to enhance the naturalness and coherence of human-computer dialogues;
4. Self-supervised and cross-domain generalization mechanisms, utilizing large-scale unlabeled data for multimodal pre-training to improve the model's adaptability to different scenarios and individual user differences.

6 Conclusion

In this paper, we propose a zero-shot gesture recognition framework named *Wi-CLIP* for Wi-Fi signals. which is inspired by the contrastive pre-training model CLIP and aligns Wi-Fi data with textual semantics within a shared semantic space. After being trained on five gesturetext pairings, the model is evaluated on previously unseen gestures. Experimental results demonstrate that *Wi-CLIP* exhibits strong generalization ability and effective cross-modal modeling capabilities.

References

1. Gkioxari, G., Girshick, R., Dollár, P., He, K.: Detecting and recognizing human-object interactions (2018). https://arxiv.org/abs/1704.07333
2. Guan, Y., Plötz, T.: Ensembles of deep LSTM learners for activity recognition using wearables. Proc. ACM Interact. Mob. Wearable Ubiquitous Technol. **1**(2) (2017). https://doi.org/10.1145/3090076
3. Song, W., et al.: FinersSnse: a fine-grained respiration sensing system based on precise separation of Wi-Fi signals. IEEE Trans. Mob. Comput. (2024)
4. Ren, Z., et al.: Characterizing the through-wall sensing mechanism of Wi-Fi signals with a refraction-aware Fresnel zone model. IEEE Trans. Mob. Comput. (2024)
5. Wang, H., et al.: Human respiration detection with commodity WiFi devices: do user location and body orientation matter? In: Proceedings of the 2016 ACM International Joint Conference on Pervasive and Ubiquitous Computing, pp. 25–36 (2016)
6. Wang, Z., et al.: Size matters: characterizing the effect of target size on Wi-Fi sensing based on the Fresnel zone model. Proc. ACM Interact., Mob., Wearable Ubiquit. Technol. **8**(4), 1–22 (2024)
7. Feng, C., et al.: WiMi: target material identification with commodity Wi-Fi devices. In: IEEE 39th International Conference on Distributed Computing Systems (ICDCS), pp. 700–710. IEEE (2019)
8. Wang, Z., et al.: GrainSense: a wireless grain moisture sensing system based on Wi-Fi signals. Proc. ACM Interact., Mob., Wearable Ubiquit. Technol. **8**(3), 1–25 (2024)
9. Niu, K., Zhang, F., Wang, X., Lv, Q., Luo, H., Zhang, D.: Understanding WiFi signal frequency features for position-independent gesture sensing. IEEE Trans. Mob. Comput. **21**(11), 4156–4171 (2021)
10. Liu, Y., et al.: UniFi: a unified framework for generalizable gesture recognition with Wi-Fi signals using consistency-guided multi-view networks. Proc. ACM Interact., Mob., Wearable Ubiquit. Technol. **7**(4), 1–29 (2024)

11. Li, M., Wang, W.: Hybrid zone: bridging acoustic and Wi-Fi for enhanced gesture recognition. In: IEEE INFOCOM 2024 - IEEE Conference on Computer Communications, pp. 981–990 (2024)
12. Yu, X., Jiang, T., Ding, X., Yao, Z., Zhou, X., Zhong, Y.: Towards position-independent gesture recognition based on WiFi by subcarrier selection and gesture code. In: 2023 IEEE Wireless Communications and Networking Conference (WCNC), pp. 1–6 (2023). https://api.semanticscholar.org/CorpusID:258641622
13. Zheng, Y., et al.: Zero-effort cross-domain gesture recognition with Wi-Fi. In: Proceedings of the 17th Annual International Conference on Mobile Systems, Applications, and Services, pp. 313–325 (2019)
14. Abdelnasser, H., Youssef, M., Harras, K.A.: WiGest: a ubiquitous WiFi-based gesture recognition system. In: 2015 IEEE Conference on Computer Communications (INFOCOM), pp. 1472–1480. IEEE (2015)
15. Sun, L., Sen, S., Koutsonikolas, D., Kim, K.H.: WiDraw: enabling hands-free drawing in the air on commodity WiFi devices. In: Proceedings of the 21st Annual International Conference on Mobile Computing and Networking (MobiCom), pp. 77–89. ACM, New York (2015)
16. Yu, N., Wang, W., Liu, A.X., Kong, L.: QGesture: quantifying gesture distance and direction with WiFi signals. Proc. ACM Interact., Mob., WearableUbiquit. Technol. **2**(1), 1–23 (2018)
17. Hao, Z., Qiao, Z., Dang, X., Zhang, D., Duan, Y.: Wi-do: a robust human motion perception model using WiFi signals. J. Comput. Res. Dev. **59**(2), 463–477 (2022). chinese
18. Li, C., Liu, M., Cao, Z.: WiHF: gesture and user recognition with WiFi. IEEE Trans. Mob. Comput. **21**(2), 757–768 (2022)
19. Liu, S., Chen, Z., Wu, M., Wang, H., Xing, B., Chen, L.: Generalizing wireless cross-multiple-factor gesture recognition to unseen domains. IEEE Trans. Mob. Comput. **23**(5), 5083–5096 (2024)
20. Radford, A., et al.: Learning transferable visual models from natural language supervision (2021). https://arxiv.org/abs/2103.00020
21. Guzhov, A., Raue, F., Hees, J., Dengel, A.: AudioCLIP: extending clip to image, text and audio (2021). https://arxiv.org/abs/2106.13043
22. Rasheed, H., Khattak, M.U., Maaz, M., Khan, S., Khan, F.S.: Fine-tuned clip models are efficient video learners (2023). https://arxiv.org/abs/2212.03640
23. Cao, Q., et al.: mmCLIP: boosting mmWave-based zero-shot HAR via signal-text alignment. In: Proceedings of the 22nd ACM Conference on Embedded Networked Sensor Systems, ser. SenSys '24, pp. 184–197. Association for Computing Machinery, New York (2024) https://doi.org/10.1145/3666025.3699331
24. Wang, X., Liu, T., Feng, C., Fang, D., Chen, X.: RF-CM: cross-modal framework for RF-enabled few-shot human activity recognition. Proc. ACM Interact. Mob. Wearable Ubiquit. Technol. **7**(1) (2023). https://doi.org/10.1145/3580859
25. Chi, G., et al.: RF-Diffusion: radio signal generation via time-frequency diffusion. In: Proceedings of the 30th Annual International Conference on Mobile Computing and Networking, ser. ACM MobiCom '24, pp. 77–92. Association for Computing Machinery, New York (2024). https://doi.org/10.1145/3636534.3649348

Object Size Classification in Garbage Disposal Sensing System Using Monocular Depth Estimation

Takashi Ito[1,2], Wenhao Huang[1], Yin Chen[3], and Jin Nakazawa[1,4(✉)]

[1] Graduate School of Media and Governance, Keio University, Minato, Japan
{takito,gerhua,jin}@sfc.keio.ac.jp
[2] INTEC Inc, Tokyo, Japan
[3] Faculty of Engineering, Reitaku University, Kashiwa, Japan
ychen@reitaku-u.ac.jp
[4] Faculty of Environment and Information Studies, Keio University, Minato, Japan

Abstract. Deep learning-based object detection is widely used in urban sensing, enabling tasks such as pedestrian, pothole, and waste detection. Automotive sensing with dashcams facilitates large-scale, real-time detection across urban environments. However, existing studies primarily focus on detection without estimating object size, which is crucial for event classification. Conventional size estimation methods rely on RGB-D cameras, multiple cameras, or LIDAR, making them unsuitable for large-scale automotive sensing with single RGB dashcams. Monocular depth estimation provides relative depth but does not yield absolute size measurements. To address this limitation, we propose a novel approach that combines monocular depth estimation with a reference object of known size. By comparing the detected object's pixel dimensions with those of the reference object, its physical size can be estimated. To validate our approach, we developed an automotive sensing platform that detected and quantified household garbage bags using footage from the rear-view camera of garbage trucks. The truck body serves as the reference object, ensuring reliable size estimation. Experiments conducted with real-world data collected using an NVIDIA Jetson TX2 demonstrated the effectiveness of our method. The proposed approach achieves size estimation accuracy with mean squared errors (MSEs) of 20.02 for width and 18.68 for height while maintaining an end-to-end processing rate of 19.21 frames per second (FPS) for detection, tracking, and size estimation.

Keywords: automotive sensing · monocular depth estimation · object size estimation

1 Introduction

The development of deep learning-based object detection has led to various advancements in urban sensing, such as pedestrian detection [2], pothole detection [19], and waste detection and counting [14,18,22]. By applying these models to video data from surveillance or mobile cameras, it becomes possible to rapidly identify urban events, significantly improving urban management and planning. Automotive sensing, which utilizes in-vehicle cameras as video sources, is emerging as a promising technology for large-scale urban sensing, enabling continuous monitoring across wide areas through vehicle-mounted systems [5,10,16].

Despite extensive research, most studies focus primarily on detecting objects' locations and categories while neglecting critical sub-information such as physical dimensions. However, object size plays a crucial role in various applications:

- In pedestrian detection, distinguishing between children and adults is necessary for safety measures.
- Potholes require repair only if they exceed a predefined size threshold.
- Garbage bag size is essential for accurate waste estimation.

Since physical dimension estimation is essential for decision-making in such scenarios, developing an effective and efficient size estimation method is key to advancing automotive sensing technology.

Several computer vision methods exist for estimating object size in camera images, including those employing RGB-D cameras [1,4,9], multi-camera setups [7,8], and LIDAR-based methods [12,23,25]. These approaches estimate an object's distance using specialized devices or multiple cameras, allowing pixel measurements to be converted into physical dimensions. However, such approaches are unsuitable for most automotive sensing systems, which primarily rely on a single RGB camera. To address this limitation, we employ monocular depth estimation [6,13,15,20,27], a technique that estimates the relative size of objects using a single RGB camera. However, this technique alone is insufficient for automotive sensing applications, where absolute size measurements are required.

A potential solution to this problem is to first identify a reference object, referred to as a ruler object, within the camera's field of view. This object must have known or measurable physical dimensions. By combining the ruler object's dimension information with the relative size estimates of detected objects, absolute size estimation becomes feasible. The selected ruler object must remain stationary relative to the camera to ensure consistency. While this concept is intuitive, its application to automotive sensing has not been reported in the literature. To investigate its practicality, this paper presents a case study on a household waste sensing platform. The proposed system, DeepCounter, is designed to detect and count garbage bags using video data captured by a rear-mounted camera on garbage trucks in Japan [14,22]. The primary goal of DeepCounter is to enable fine-grained waste discharge analysis in Japanese cities. Previous implementations of the system only detected the category and number of garbage bags. However, as shown in Table 1, garbage bags come in four different sizes, with the largest being eight times the size of the smallest. This makes size estimation a critical component for achieving accurate waste discharge assessment. The key contributions of this research are as follows:

- We redesigned the DeepCounter system to incorporate a size estimation function, modeling the problem as a classification task based on garbage collection system requirements.
- We proposed three approaches that leverage the ruler object's length, its distance from the camera, or a combination of both.
- We conducted a performance evaluation with real-world garbage collection video data and compared the effectiveness of the three estimation strategies.

The remainder of this paper is structured as follows. Section 2 discusses related work. Section 3 provides an overview of the system and problem formulation. Section 4 details the proposed size estimation technology. Section 5 presents experimental evaluations. Section 6 discusses the results, and finally Sect. 7 concludes the paper.

2 Related Work

Despite the high accuracy of size estimation using RGB-D cameras [1,4,23], multi-camera systems [7,8], and LIDAR [12,25], these approaches are impractical for most existing automotive sensing systems using one single camera. To address the challenge of acquiring depth information with a single camera, monocular depth estimation has been explored. [27] proposed a domain adaptation framework that improves depth models by re-weighting image transformations, enhancing performance in nighttime conditions. [20] introduced a pruned depth model optimized for IoT devices, enabling real-time sensing with low power consumption. GAN-based methods, such as Mini-Unet GAN by [13], leveraged adversarial training for accurate and efficient depth estimation in low-computation environments. Additionally, [24] demonstrated a size-based estimation approach that detects objects of known size for depth inference, proving effective in structured environments.

Despite these advancements, several challenges remain, including domain generalization, handling texture less regions, computational efficiency in high-resolution settings, and real-time inference on embedded devices. To enhance robustness and reduce latency, recent studies have introduced innovative depth estimation schemes. [3] presented ZoeDepth, which pre-trains on 12 datasets for relative depth and fine-tunes on two for metric depth, achieving strong generalization. [17] proposed a multi-objective learning framework that optimizes cross-domain consistency, enabling zero-shot transfer across five datasets. [26] introduced depth anything, a self-supervised method that utilizes large-scale unlabeled data to reduce dependence on labeled datasets. In this study, we compare the performance of these models, with results discussed in the following section.

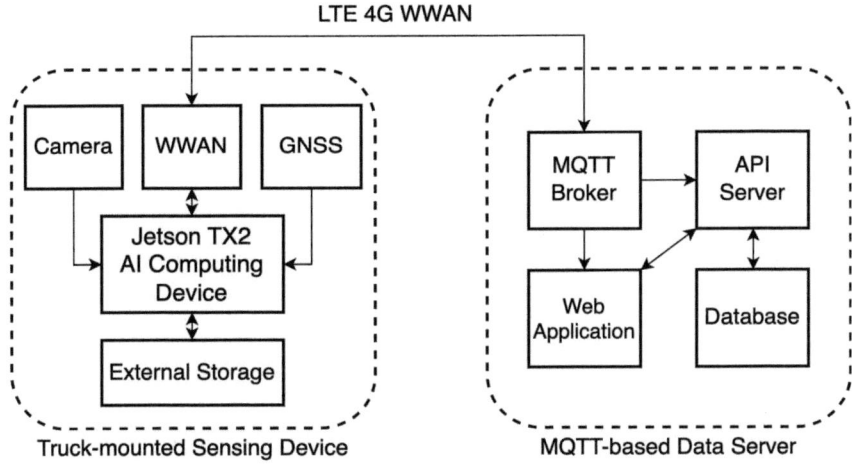

Fig. 1. Real-Time Sensing System Architecture

3 Garbage Bag Sensing System with Size Estimation

This section describes how we modified the DeepCounter system [14,22] to incorporate size estimation for garbage bags, leveraging detection results to enhance accuracy.

3.1 Overview

Figure 1 shows the real-time garbage bag sensing system architecture, which has been enhanced with size estimation functionality. The system's core component is an artificial intelligence (AI) computing device based on the NVIDIA Jetson TX2, operating in a heterogeneous CPU-GPU mode to run YOLOv5 [11] for real-time, fine-grained garbage bag sensing.

The device is equipped with several peripherals, including a rear camera, a USB video capture module, a WWAN module, a GNSS module, and external storage. The rear camera captures video footage of the garbage collection process, while the WWAN module enables Internet connectivity. The GNSS module provides crucial data such as the truck's location, speed, and timestamp. Additionally, external storage serves as a repository for historical sensing results and recorded collection process videos.

3.2 Sensing Algorithm

The video footage of the garbage collection process is processed by an embedded AI computing device using an algorithm that extends the detection, tracking, and counting (DTC) method originally proposed in our previous work [22]. This extended algorithm, referred to as DSTC, consists of four key processes: garbage

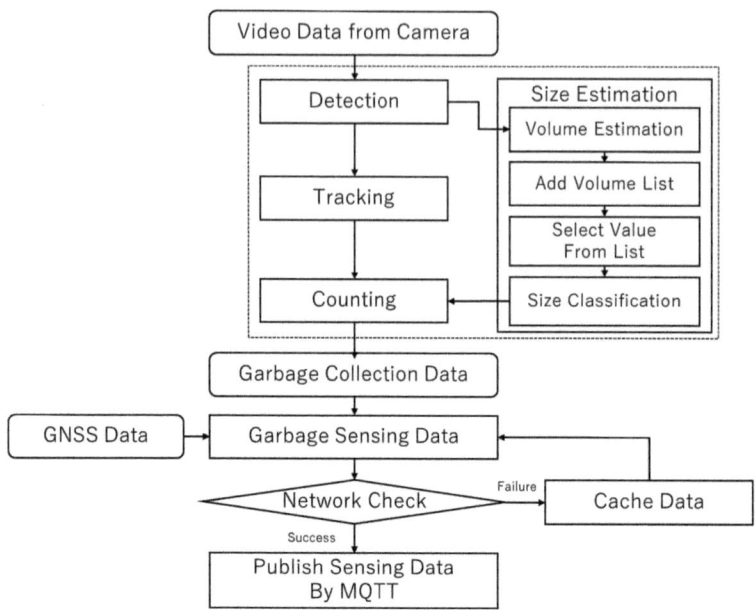

Fig. 2. Workflow in Sensing Device

bag detection, garbage bag tracking, counting, and size estimation. Figures 2 and 3 represent the workflow of the embedded sensing device and the algorithm in action. Notably, the sensing architecture itself does not include the display of images. The DSTC algorithm operates as follows:

1. It detects the location of garbage bags in each frame.
2. Followed by size estimation for each detected bag.
3. It tracks the detected bags by analyzing their relative positions in successive frames.
4. A garbage bag is counted as collected when its tracked location surpasses a predefined threshold line at the bottom of an image frame.

The size estimation process is implemented separately from object detection. It takes as input the original image information and two-dimensional coordinate information, including the start and end points of the bounding box identified during object detection. The output is the estimated volume information corresponding to the bounding box. Separating size estimation from object detection enhances system maintainability and facilitates the integration of multiple machine learning models, such as attribute estimation and attribute verbalization of detected objects.

3.3 Problem Formulation

In the cities where this sensing system is currently deployed, residents are required to dispose of their waste using designated garbage bags sold by the

Fig. 3. DSTC Algorithm in Action

Table 1. Size of Designated Garbage Bags

S	M	L	LL
5 liters	10 liters	20 liters	40 liters
40 × 18 cm	50 × 26 cm	60 × 34 cm	75 × 45 cm

local government in four standardized volumes: S, M, L, and LL. Table 1 provides a classification of these sizes. Given that the majority of these bags are assumed to be filled with waste, the system must accurately classify each collected bag into one of these four categories to ensure precise waste quantification and analysis.

Time constraints must also be taken into account. In previous work, garbage bag counting was performed at 15 frames per second (FPS), matching the camera's frame rate and enabling real-time execution. The new system, which integrates size estimation, must maintain this performance to ensure seamless operation. However, processing every frame within $1/15$ s is not strictly necessary. Since most of the garbage collection process involves driving between pick-up points, the computational load is concentrated during moments when bags are thrown into the truck. Consequently, if a frame takes longer than $1/15$ s to process, incoming frames can be buffered and handled concurrently while the truck is in transit, optimizing computational efficiency without disrupting real-time detection.

4 Estimating Size of Garbage Bags

4.1 Volume Calculation of Detected Objects

The implementation of volume detection begins with object detection using the YOLOv5 algorithm. During the detection phase, the number of pixels within the detected bounding box is utilized as input for size estimation, preceding the counting process in the existing DTC algorithm. Since the shape of a garbage bag deforms throughout the loading process, determining its precise volume in each frame is challenging. Consequently, the object is approximated as an ellipsoid for volume calculation. The volume is determined using the following formula, where a, b, and c are the width, height, and depth of the garbage bag, respectively.

$$V = \frac{4}{3}\pi abc \qquad (1)$$

The width and height of a garbage bag are approximated based on the number of vertical and horizontal pixels within its detected bounding box, respectively. The depth c is estimated as the smaller of a and b, considering that garbage bags are flattened due to centrifugal forces when thrown into the truck.

To obtain physical dimensions, the pixel values a, b, and c must be converted into real-world measurements. This conversion is conducted by leveraging the garbage truck's garbage input table, which has a known physical width of 1400 mm. The table appears in the lower 1/10 of the images, as shown in Fig. 4a. Additionally, Fig. 4b provides an overview of the physical dimensions of typical garbage trucks in the study's target region. If the garbage input table occupies N horizontal pixels in the image, the conversion ratio is calculated as 1,400/N. This ratio is then applied to the detected pixel values to estimate the actual width, height, and depth of garbage bags.

4.2 Actual Length Calculation Strategy

There are three strategies for estimating the length of an unknown object from a monocular image. The first strategy, known as the *distance strategy*, utilizes the pixel size of the object along with its distance from the camera. This approach necessitates prior training data on objects similar to the target, including their actual size, distance from the camera, and corresponding pixel size. However, variations in object distance and pixel size across different cameras and lenses necessitate accurately collected training data for each specific camera setup to ensure accuracy.

The second strategy, known as the *ruler strategy*, estimates the target object's size by comparing its pixel dimensions to those of a reference object with a fixed and known physical size, such as a manhole. This approach assumes that the target and reference objects are physically close to each other and appear within the same frame. Unlike the distance strategy, it does not require prior training data for unknown objects. However, the accuracy of this method is highly dependent on the proximity between the target and reference objects; as the distance

(a) Standard Length of Garbage Input Table in Camera Image

(b) Standard Length of Garbage Truck for Size Calculation

Fig. 4. Standard Lengths of Garbage Trucks

between them increases, so does the estimation error. This limitation makes the ruler strategy less suitable for scenarios where the positional relationship between the target and reference objects changes dynamically.

The third strategy, known as the *distance + ruler strategy*, combines elements of both the distance strategy and the ruler strategy. It first adjusts for the target object's distance from the camera using training data and then compares the target object's pixel size to that of a known reference object. This method requires information on the physical size of the reference object, its distance from the camera, and the corresponding pixel size recorded by the camera. Similar to the ruler strategy, it relies on the reference object being present in the frame at the time of computation. However, by incorporating distance correction through training data, it is more robust against dynamically changing positional relationships between the target and reference objects, improving estimation accuracy in varying conditions.

4.3 Usage of Monocular Depth Estimation Model

In practice, the pixel size of objects captured by a monocular camera is inversely proportional to the distance between the camera and the object, though this relationship depends on the characteristics of the camera lens. Therefore, to estimate the volume of an object from monocular camera information, it is necessary to determine its distance from the camera. Although the physical dimensions of a garbage bag remain constant, the number of pixels representing the bag in video frames changes as it is transported into the truck. Consequently, the tentative size estimates calculated using the previously described method must be adjusted based on the bag's distance from the camera.

To achieve this, a monocular depth estimation model is utilized to generate a relative depth map of the camera image. We compared the execution speed and estimation accuracy of four monocular depth estimation models with different weight file sizes: MiDaS [17], ZoeDepth [3], and versions 1 and 2 of Depth-Anything [26]. This evaluation was conducted on a PC equipped with an NVIDIA RTX 4080, with the average performance calculated after processing 200 images containing a garbage bag. The results, presented in Table 2, show that the estimated depth is generated as a matrix corresponding to the pixel size of the input image. In this representation, the farthest point from the camera is assigned the smallest value, while the closest point is assigned the largest value. As indicated in the table, the depth maps vary across models and weight files, and their accuracy differs depending on the scene. Based on these findings, we use MiDaS dpt_swin2_tiny_256 in this study due to its superior processing speed. Notably, Depth-Anything could not be used in combination with YOLOv5 in our implementation. However, since the size estimation mechanism is modular and separate from other parts of the system (as described in Sect. 3.2), alternative monocular depth estimation models can be integrated as needed.

4.4 Distance Calculation from Camera to Object

The depth map generated by the monocular depth estimation model provides only relative depth values, meaning that the estimated depth cannot be directly interpreted as the actual distance between the camera and the target object. To address this limitation, we use the body of the garbage truck as a reference for distance estimation. In particular, we extract three key values from the depth map, as illustrated in Fig. 5:

The depth value D_{std} is determined as the lesser of the right or left end of the "garbage input table." This approach accounts for potential obstructions, as a garbage bag being thrown into the truck may block one end of the table, resulting in an erroneous depth reading. As illustrated in Fig. 4b, the distance from the camera to the garbage input table, denoted as H_{std}, is approximately 980 mm, while the table's width, W_{std}, is approximately 1400 mm for typical garbage trucks, including those utilized in our experiments. It is important to note that the depth estimation between the camera and the garbage input table can be approximated as a linear relationship. Based on this assumption, the distance from the camera to the detected object, denoted as L_{obj}, can be calculated using the following equation.

$$L_{obj} = H_{std}\frac{D_{max} - D_{obj}}{D_{max} - D_{std}} \qquad (2)$$

4.5 Pixel Size Correction Based on Camera-to-Object Distance

The pixel size of a garbage bag varies depending on its distance from the camera. To account for this variation, we constructed a mathematical formula to estimate the number of pixels an average-sized garbage bag occupies in an image. This formula is particularly useful in scenarios where the trash bag appears blurred due to camera limitations or is partially obscured by other objects. To build the formula, we created garbage bags corresponding to the four predefined sizes listed in Table 1 and recorded 4,757 images of these bags positioned at distances

Table 2. Comparison of Monocular Depth Estimation Model

MiDaS			
Weight File	dpt_beit_large_512	dpt_swin2_large_384	midas_swin2_tiny_256
Time [ms]	156.103 ± 59.882	35.566 ± 0.105	7.986 ± 0.221
Depth Map			

ZoeDepth			
Weight File	zoe_k	zoe_nk	zoe_n
Time [ms]	480.012 ± 34.797	120.506 ± 3.015	116.735 ± 2.774
Depth Map			

Depth-Anything			
Weight File	depth_anyting_vitb	depth_anyting_vitl	depth_anyting_vits
Time [ms]	6.196 ± 0.21	8.118 ± 0.223	5.758 ± 0.212
Depth Map			

Depth-Anything v2			
Weight File	depth_anyting_v2_vitb	depth_anyting_v2_vitl	depth_anyting_v2_vits
Time [ms]	45.561 ± 0.608	118.304 ± 2.335	22.789 ± 0.364
Depth Map			

of 30 cm, 50 cm, 100 cm, 150 cm, 200 cm, 250 cm, and 300 cm from the camera. The height and width of the bounding boxes detected using YOLOv5 in the recorded images were averaged for each distance from the camera and plotted on two curves: $px_{height} = \alpha_{height} \frac{1}{L_{obj}} + \beta_{height}$ and $px_{width} = \alpha_{width} \frac{1}{L_{obj}} + \beta_{height}$

Fig. 5. Reference Points in Depth Map

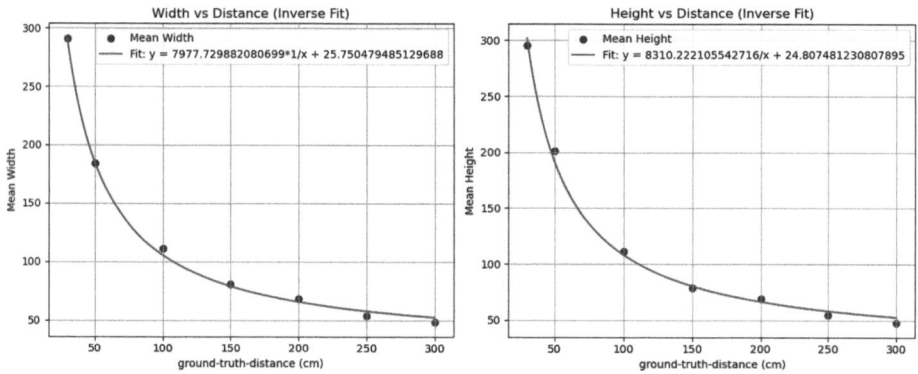

Fig. 6. Inverse Approximation Curve of Pixel Size vs Distance

using the curve fit shown in Fig. 6. px_{height} and px_{width} in this equation refer to the pixel size of the garbage bags used to create the equation.

The functions for estimating the length of garbage bags are shown below. These functions use the previously derived approximate curve equation, incorporating L_{obj} (the distance from the camera to the garbage bag), the distance from the camera to the "input table" of the truck body, and the number of pixels corresponding to a bag. In the following equations, px_{height} and px_{width} represent the pixel dimensions of the garbage bag, while $\alpha_{heigth}, \beta_{heigth}, \alpha_{width}$, and β_{width} are constants determined by the approximate formula.

$$a = W_{std} * \frac{px_{height}}{px_{std}} * \frac{\dfrac{\alpha_{height}}{H_{std}} + \beta_{height}}{\dfrac{\alpha_{height}}{L_{obj}} + \beta_{height}} \quad (3)$$

$$b = W_{std} * \frac{px_{width}}{px_{std}} * \frac{\dfrac{\alpha_{width}}{H_{std}} + \beta_{width}}{\dfrac{\alpha_{width}}{L_{obj}} + \beta_{width}} \quad (4)$$

The dimensions of an object, including the depth of a garbage bag represented by c, can be expressed using the following equations.

$$c = \min(a, b) \quad (5)$$

4.6 Selection Algorithm

In the video of the garbage collection process, a garbage bag remains visible across multiple frames before being collected and eventually disappearing. Initially, the YOLOv5 algorithm may struggle to detect the bag accurately due to insufficient feature acquisition, particularly when the object is more than 2 m away, as its pixel size does not reliably reflect its actual size. In the final frames of the counting algorithm, the bag is often more than half obscured, making size estimation more challenging. To address this, the study calculates volume and distance for all frames from the point of detection to disappearance, storing them in memory along with a Tracking ID. When a bag becomes hidden, the appropriate volume estimate is selected from past frames, with the algorithm prioritizing the last frame in which the estimated distance is approximately 1 m.

4.7 Size Classification

The estimated volume is discretized into four classes, corresponding to the sizes shown in Table 1. While the aforementioned estimation logic provides a means to determine volume, the resulting data is often too granular for practical applications. In addition, certain frames may contain anomalies, such as overlapping garbage bags or partially hidden bags in the back of the truck, leading to unreliable estimations. The discretization process improves the system's robustness by mitigating the impact of such irregularities.

Table 3. Comparison of Resource Utilization

	CPU	GPU	RAM	Frame Rate
DTC	7.64	40.81	43.36	24.19 FPS
DSTC	12.59	47.30	55.60	19.21 FPS

5 Experiment Evaluation

5.1 Resource Utilization

Table 3 summarizes the resource utilization during the execution of the sensing algorithm described in Sect. 3.2. These measurements were obtained on a Jetson Orin Nano while processing 6,145 frames from a pre-recorded garbage collection video spanning 309 s. The table presents a detailed comparison of computational resource utilization for two distinct schemes: the DTC of garbage bags (upper row) and the same process with the addition of size estimation (lower row). The results indicate that incorporating size estimation using MiDaS increases the computational load by approximately 20%. It is important to note that detection, tracking, counting, and size estimation processes are executed only when a garbage bag is present in the frame. Consequently, absolute resource utilization depends on the contents of the frame. However, the relative increase in computational load due to size estimation remains consistently approximately 20%.

5.2 Performance Evaluation

To ensure the practical execution performance of the implemented function on edge devices, we measured the end-to-end frame rate (FPS). This measurement reflects the time required to process a frame through the complete pipeline, including detection, tracking, size estimation using the corresponding models, and MQTT message creation. The evaluation was carried out by processing a pre-recorded garbage collection video on the Jetson Orin Nano, the same hardware used in the truck-mounted sensing device. In the naive implementation, where every frame is processed through the entire pipeline, a significant performance reduction was observed, resulting in an approximate frame rate of 1.0 FPS. To reduce the computational load on edge devices while maintaining functionality, we optimized the implementation to perform size estimation only when a garbage bag is detected. The upper graph in Fig. 7 depicts the execution performance, showing an average of 19.21 FPS. This significant improvement highlights the efficacy of event-driven execution for size estimation. For reference, the lower graph in Fig. 7 presents the execution performance without size estimation, with an average of 24.19 FPS. When a garbage bag is detected in a frame, the frame rate decreases regardless of whether size estimation is applied. While the inclusion of size estimation reduced the instantaneous minimum processing performance to approximately 1 FPS, this impact is mitigated by skipping size

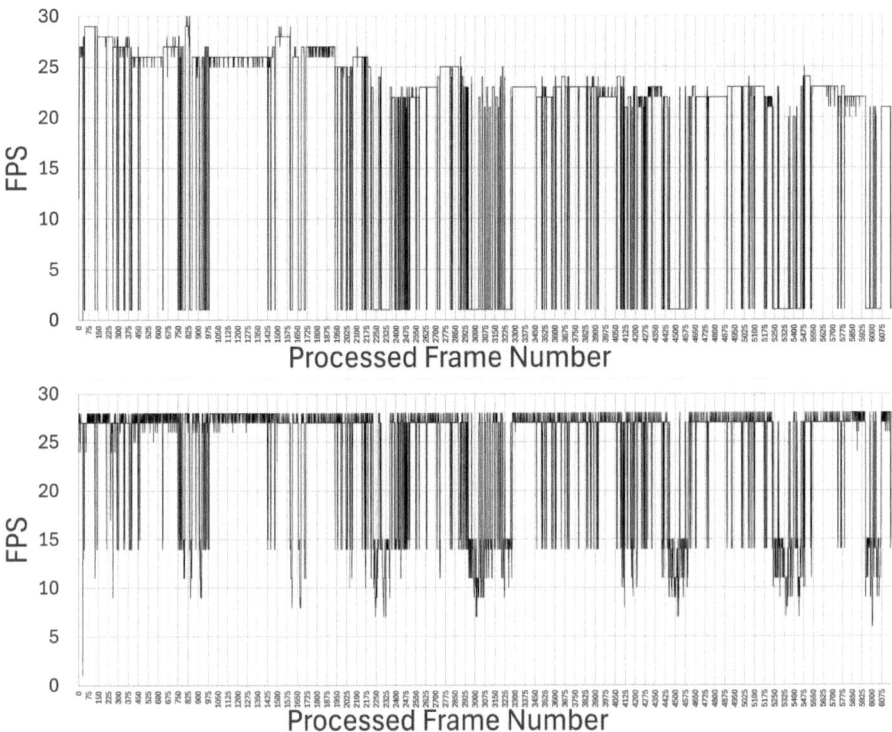

Fig. 7. Execution Performance with (upper) and without (lower) Size Estimation

estimation when no garbage bag is detected. A comparison of the average frame rates indicates that incorporating size estimation leads to an approximate 20% reduction, from 24.19 FPS to 19.21 FPS. It is important to note that the extent of this reduction depends on the frequency of frames requiring size estimation. In scenarios where a substantial proportion of frames contain garbage bags, real-time processing on edge devices may become impractical. In such cases, an alternative scheme, such as task offloading [21], would be necessary.

5.3 Accuracy Evaluation

Size Estimation. We conducted a comparison of the *ruler strategy* and the *distance + ruler strategy*, as described in Sect. 4.2, by evaluating their length estimates against actual values when estimating the height and width of sample garbage bags. In this study, the garbage input table, located approximately 1 m from the camera, served as the reference ruler. As described in Sect. 4.6, the frames selected for size estimation contained a garbage bag positioned at a similar distance of approximately 1 m from the camera. For evaluation, we used an M-size garbage bag measuring 27 cm in height and 22 cm in width, recorded with a rear camera mounted on a garbage truck at distances of 30 cm, 50 cm, 100 cm,

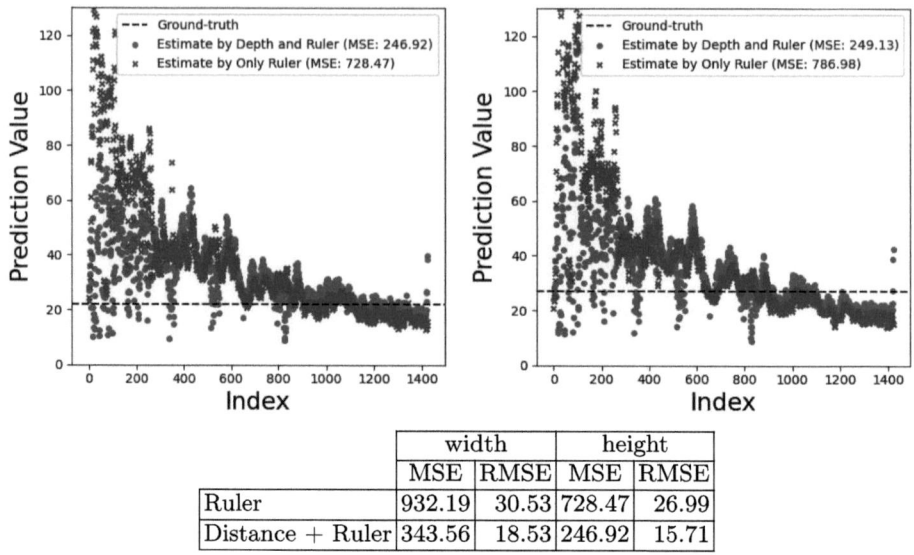

	width		height	
	MSE	RMSE	MSE	RMSE
Ruler	932.19	30.53	728.47	26.99
Distance + Ruler	343.56	18.53	246.92	15.71

Fig. 8. Width (left) and Height (right) Estimations at Ranging Distance

150 cm, 200 cm, 250 cm, and 300 cm from the camera. Since size estimation is expected to be most accurate at a distance of 100 cm, this distance serves as a benchmark for comparison.

Figure 8 presents the size estimation results across all distances. The blue and red dots represent estimations using the *ruler strategy* and the *distance + ruler strategy*, respectively, with their distributions approximated by the curve shown in Fig. 6. The results indicate that the *distance + ruler strategy* tends to produce estimates closer to the ground truth, particularly at shorter distances (30 cm and 50 cm). Figure 9 further examines the estimation results at 100 cm, where the *ruler strategy* outperformed the *distance + ruler strategy*. The results exhibit systematic errors of approximately 20 cm, fluctuating in both positive and negative directions. These fluctuations are attributed to intentional vertical movement of the garbage bag during recording. Notably, red dots (representing the distance + ruler strategy) show a tendency to underestimate the size when the garbage bag is positioned near the upper center of the frame and to overestimate it when the bag is near the left or right edges of the screen.

Size Classification. We evaluated the accuracy of the size classification function using two 10-minute videos recorded during real-world garbage collection. These videos contain 46 garbage bags that were visually classified into four size categories: S, M, L, and LL. Tables 4 and 5 present the correspondence between the estimated and ground truth size classes when using the *ruler strategy* and the *distance + ruler strategy*, respectively. With the *ruler strategy*, the classification accuracy was 43%, with all objects being classified as LL. In contrast,

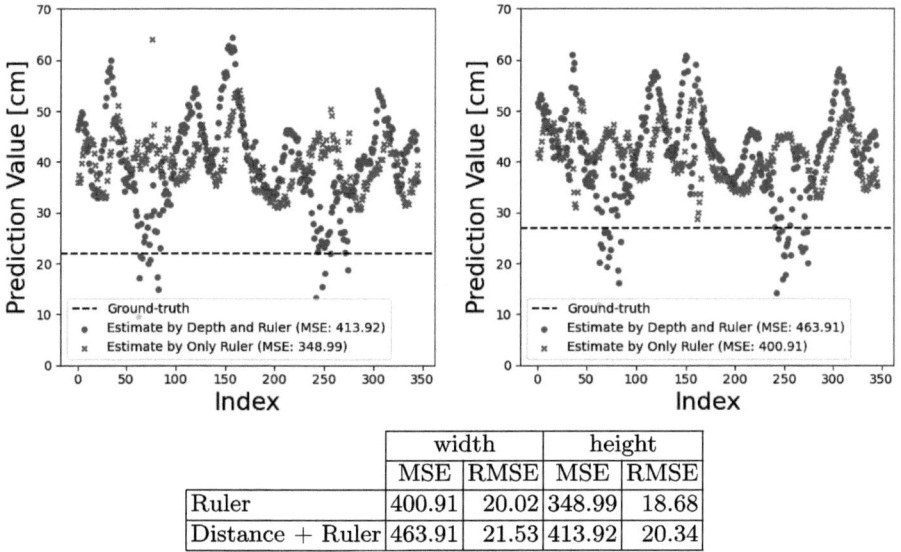

	width		height	
	MSE	RMSE	MSE	RMSE
Ruler	400.91	20.02	348.99	18.68
Distance + Ruler	463.91	21.53	413.92	20.34

Fig. 9. Width (left) and Height (right) Estimations at 100 cm

Table 4. Classification Results by Ruler Strategy

	precision	recall	f1-score	support
S	0.00	0.00	0.00	1
M	0.40	0.22	0.29	9
L	0.22	0.13	0.17	15
LL	0.62	0.76	0.68	21
accuracy			0.43	46

Table 5. Classification Results by Distance + Ruler Strategy

	precision	recall	f1-score	support
S	0.00	0.00	0.00	1
M	1.00	0.33	0.50	9
L	0.55	0.40	0.46	15
LL	0.62	0.54	0.56	21
accuracy			0.55	46

the *distance + ruler strategy* achieved an accuracy of 55%. Additionally, in 37 out of 46 cases (80%), the classification was either correct or off by only one size category (e.g., S misclassified as M).

6 Discussions

6.1 Characteristics of Strategies

In the experiment, the *ruler strategy* provided more stable estimations and higher accuracy, demonstrating its effectiveness in comparing a target object with a known reference. The *distance + ruler strategy* performed slightly worse, suggesting the need for improvements in the correction scheme based on the curves in Fig. 6. These curves represent the average size of various garbage bags, and more precise estimations could potentially be achieved by creating size-specific curves instead of using an averaged approach. Due to the limited availability of training data for different-sized garbage bags, results using only the *distance strategy* are not included in this paper. Future work will focus on refining the size classification process to further evaluate and enhance its accuracy.

6.2 Performance of Monocular Depth Estimation

It was observed, though not explicitly visible in the figures, that the accuracy of size estimation decreased at specific collection points. For example, when the back of the vehicle was positioned against a wall and workers, along with garbage bags, were closer to the vehicle than at other collection sites, accuracy declined. At this location, the estimated relative depth value of the garbage bags was lower than expected, causing the system to perceive them as closer to the camera than their actual distance. As a result, bags were classified as smaller than their true size. This issue was noted with MiDaS; however, it is possible that other advanced monocular depth estimation models may not exhibit the same problem. Since the system is designed with a modular structure, future improvements will focus on integrating alternative monocular depth estimation models to enhance accuracy.

6.3 Robustness of Size Estimation

Our size estimation and classification depend on the pixel size of the bounding box detected by YOLOv5, making them susceptible to errors when detections are inaccurate. For instance, if YOLOv5 detects a garbage bag along with an additional object as a single entity, the bounding box will be larger than the actual bag, leading to overestimation. Similarly, multiple overlapping bags may be identified as one, further inflating size estimates. While our pixel size correction scheme effectively reduced errors using the *distance + ruler strategy* (Fig. 8), replacing the bounding box approach with an edge detection model could further improve accuracy and robustness.

7 Conclusion

We proposed a size estimation method for urban object detection using a ruler object with monocular depth estimation. Our approach estimates physical dimensions without LIDAR or stereo cameras, making it suitable for large-scale automotive sensing with single RGB dashcams. We validated it on a platform detecting household garbage bags from rear-view camera footage, using the truck body as a reference. Experiments on real-world data showed its effectiveness, achieving MSEs of 20.02 (width) and 18.68 (height) at 19.21 FPS, confirming its feasibility for automotive sensing applications.

Acknowledgment. These research results were obtained from NICT Grant Number 23602 and JST Grant Number JPMJPF2111.

References

1. Baisa, N.L., Al-Diri, B.: Mushrooms detection, localization and 3D pose estimation using RGB-D sensor for robotic-picking applications. arXiv preprint: arXiv:2201.02837 (2022)
2. Belhadi, A., Djenouri, Y., Belbachir, A.N., Michalak, T., Srivastava, G.: Knowledge guided visual transformers for intelligent transportation systems. IEEE Trans. Intell. Transp. Syst. (2025)
3. Bhat, S.F., Birkl, R., Wofk, D., Wonka, P., Müller, M.: ZoeDepth: zero-shot transfer by combining relative and metric depth. https://github.com/isl-org/ZoeDepth (2023). https://doi.org/10.48550/arXiv.2302.12288. Accessed 21 Sep 2023
4. Brachmann, E., Rother, C.: Visual camera re-localization from RGB and RGB-D images using DSAC. IEEE Trans. Pattern Anal. Mach. Intell. **44**(9), 5847–5865 (2021)
5. Chen, Y., Nakazawa, J., Yonezawa, T., Tokuda, H.: Cruisers: an automotive sensing platform for smart cities using door-to-door garbage collecting trucks. Ad Hoc Netw. **85**, 32–45 (2019)
6. Feng, C., Zhang, C., Chen, Z., Hu, W., Ge, L.: Real-time monocular depth estimation on embedded systems. In: 2024 IEEE International Conference on Image Processing (ICIP), pp. 3464–3470. IEEE (2024)
7. Guo, X., Chen, Z., Li, S., Yang, Y., Yu, J.: Deep eyes: binocular depth-from-focus on focal stack pairs. In: Pattern Recognition and Computer Vision: Second Chinese Conference, PRCV 2019, Xi'an, China, 8–11 November 2019, Proceedings, Part III 2, pp. 353–365. Springer (2019)
8. Hazarika, A., Vyas, A., Rahmati, M., Wang, Y.: Multi-camera 3D object detection for autonomous driving using deep learning and self-attention mechanism. IEEE Access **11**, 64608–64620 (2023)
9. Hu, T., Wang, W., Gu, J., Xia, Z., Zhang, J., Wang, B.: Research on apple object detection and localization method based on improved YOLOx and RGB-D images. Agronomy **13**(7), 1816 (2023)
10. Huang, W., Mikami, K., Chen, Y., Nakazawa, J.: Real-time image-based automotive sensing: a practice on fine-grained garbage disposal. In: Proceedings of the 13th International Conference on the Internet of Things, pp. 138–145 (2023)
11. Jocher, G., et al.: YOLOv5: Pytorch implementation of YOLO object detector (2020). https://github.com/ultralytics/yolov5
12. Li, Z., Wang, F., Wang, N.: Lidar R-CNN: an efficient and universal 3D object detector. In: Proceedings of the IEEE/CVF Conference on Computer Vision and Pattern Recognition, pp. 7546–7555 (2021)

13. Mebtouche, N.E.D., Baha, N., Bekhelifi, O.: mini-Unet GAN: Optimized GAN for monocular depth estimation. In: 2024 6th International Conference on Pattern Analysis and Intelligent Systems (PAIS), pp. 1–7. IEEE (2024)
14. Mikami, K., Chen, Y., Nakazawa, J., Iida, Y., Kishimoto, Y., Oya, Y.: DeepCounter: using deep learning to count garbage bags. In: 2018 IEEE 24th International Conference on Embedded and Real-Time Computing Systems and Applications (RTCSA), Hakodate, Japan, pp. 1–10 (2018). https://doi.org/10.1109/RTCSA.2018.00010
15. Ming, Y., Meng, X., Fan, C., Yu, H.: Deep learning for monocular depth estimation: a review. Neurocomputing **438**, 14–33 (2021)
16. Pandharipande, A., et al.: Sensing and machine learning for automotive perception: a review. IEEE Sens. J. **23**(11), 11097–11115 (2023)
17. Ranftl, R., Lasinger, K., Hafner, D., Schindler, K., Koltun, V.: Towards robust monocular depth estimation: mixing datasets for zero-shot cross-dataset transfer. IEEE Trans. Pattern Anal. Mach. Intell. **1**(1), 99 (2020). https://doi.org/10.1109/TPAMI.2020.3019967
18. Sayem, F.R., et al.: Enhancing waste sorting and recycling efficiency: robust deep learning-based approach for classification and detection. In: Neural Computing and Applications, pp. 1–17 (2024)
19. Tsung, C.K., Kristiani, E., Chiu, C.K., Liu, J.C., Yang, C.T.: HPPH: computer vision-based service for high-performance pavement health recognition. IEEE Internet Things J. (2025)
20. Tu, X., et al.: Efficient monocular depth estimation for edge devices in internet of things. IEEE Trans. Industr. Inf. **17**(4), 2821–2832 (2020)
21. Wang, C., Eicher, O., Han, Q.: Sharing the edge: System status aware object recognition task offloading. In: 2024 IEEE International Conference on Smart Computing (SMARTCOMP), pp. 77–84. IEEE (2024)
22. Huang, W., Mikami, K., Chen, Y., Nakazawa, J.: Real-time image-based automotive sensing: a practice on fine-grained garbage disposal. In: IoT 2023: 13th International Conference on the Internet of Things, Nagoya, Japan, pp. 138–145 (2023). https://doi.org/10.1145/3627050.3627058
23. Wisultschew, C., Mujica, G., Lanza-Gutierrez, J.M., Portilla, J.: 3D-LIDAR based object detection and tracking on the edge of IoT for railway level crossing. IEEE Access **9**, 35718–35729 (2021)
24. Wu, Y., Ying, S., Zheng, L.: Size-to-depth: a new perspective for single image depth estimation. arXiv preprint: arXiv:1801.04461 (2018)
25. Wu, Y., Wang, Y., Zhang, S., Ogai, H.: Deep 3d object detection networks using lidar data: a review. IEEE Sens. J. **21**(2), 1152–1171 (2020)
26. Yang, L., Kang, B., Huang, Z., Xu, X., Feng, J., Zhao, H.: Depth anything: unleashing the power of large-scale unlabeled data. In: Proceedings of the IEEE/CVF Conference on Computer Vision and Pattern Recognition, pp. 10371–10381 (2024)
27. Zhao, C., Tang, Y., Sun, Q.: Unsupervised monocular depth estimation in highly complex environments. IEEE Trans. Emerg. Top. Comput. Intell. **6**(5), 1237–1246 (2022)

Memory-Aware Structured Pruning for DL with Joint Optimization of L1-norm and Peak Memory

Yu Gong and Ling Wang(✉)

Department of Computer Science and Technology, Harbin Institute of Technology, Harbin, China
gongyu@stu.hit.edu.cn, wangling@hit.edu.cn

Abstract. Efficient deep neural networks are essential for deployment on resource-constrained platforms. In this paper, we propose an improved structured pruning framework that jointly considers L1-norm sparsity and peak memory usage to effectively reduce both computational cost and runtime memory overhead. Our method introduces a memory-aware criterion that iteratively selects channel configurations by minimizing a combined loss of L1-norm and peak memory consumption. We validate our approach on benchmark datasets such as CIFAR-10 using VGG16 with Batch Normalization (VGG16-BN), ResNet, and Transformer architectures. To ensure practical applicability, we implement true channel removal along with synchronized BatchNorm alignment and dynamic reconstruction of the inference graph. Experimental results demonstrate that the proposed method significantly reduces peak memory usage while maintaining high accuracy, making it suitable for deployment on edge devices. Additionally, the framework supports an optional memory loss toggle, enabling flexible analysis of the trade-off between memory consumption and model accuracy.

Keywords: Structured Pruning · Memory-Aware Compression · Channel Sparsity · Edge Deployment · Peak Memory

1 Introduction

Deep neural networks (DNNs) have achieved remarkable success across various domains such as image classification, object detection, and semantic segmentation. However, their substantial computational and memory demands often render them impractical for deployment on resource-constrained platforms like mobile devices, embedded systems, and edge devices [1]. These platforms necessitate models that can deliver real-time performance while maintaining energy efficiency. Consequently, reducing both the computational load and memory footprint of DNNs is essential for practical deployment in such environments [2].

Structured pruning has emerged as a promising technique for compressing deep networks by removing entire channels or filters, facilitating acceleration on standard hardware. Traditional pruning methods often rely on sparsity-inducing criteria like the L1-norm of convolutional weights to identify and eliminate less important network components [3,4]. While effective in reducing parameters and floating-point operations (FLOPs), these methods tend to overlook critical runtime memory characteristics, which significantly influence the real-world performance of models, particularly in memory-constrained environments.

Recent work has proposed advanced frameworks for efficient DNN deployment. For example, BitTrain [15] leverages activation sparsity with bitmap compression to reduce memory during training. Qu et al. [24] proposed the framework GETA, which automatically and efficiently performs joint structured pruning and quantization-aware training on any DNNs. AutoML-based solutions like APQ [25] co-optimize network architecture, pruning, and quantization strategies tailored to resource-constrained hardware.

In this paper, we propose a novel memory-aware structured pruning framework that jointly optimizes both L1-norm sparsity and peak memory usage to better balance model compression and deployment efficiency. Our method introduces a unified pruning criterion that evaluates each channel's importance by considering both its contribution to L1-norm sparsity and its associated memory cost. This dual-objective approach addresses peak memory bottlenecks, which are especially crucial for deployment on edge devices.

To accurately assess memory consumption, we implement a C++-based profiling tool that measures runtime memory usage, providing precise evaluations beyond what dynamic graph estimations offer. Additionally, we implement true physical channel removal and dynamically reconstruct the inference graph to ensure that the pruned model not only has fewer parameters but also results in practical runtime benefits. To maintain model stability, we synchronize Batch Normalization layers with the pruned configurations. Our framework also includes an optional memory loss toggle that allows for flexible exploration of the trade-offs between memory usage and model accuracy.

We evaluate our memory-aware structured pruning method on common architectures such as VGG16-BN, ResNet, and Transformer models using the CIFAR-10 dataset. Our experimental results demonstrate that the proposed method significantly reduces peak memory consumption while maintaining competitive accuracy, making it ideal for deployment in resource-constrained environments (Fig. 1).

The key contributions of this work are as follows:

- We propose a memory-aware structured pruning framework that explicitly introduces runtime memory as a key factor in pruning decisions, enabling compression strategies better aligned with deployment constraints.
- We define a resource-aware ranking function that combines normalized L1-norm sparsity and estimated memory usage through a tunable weighting scheme, providing a principled and flexible metric for channel selection.
- We validate the proposed method across diverse architectures— including VGG16-BN, ResNet50, and Transformer— on the CIFAR-10 benchmark, demonstrating consistent memory reduction and strong accuracy retention.

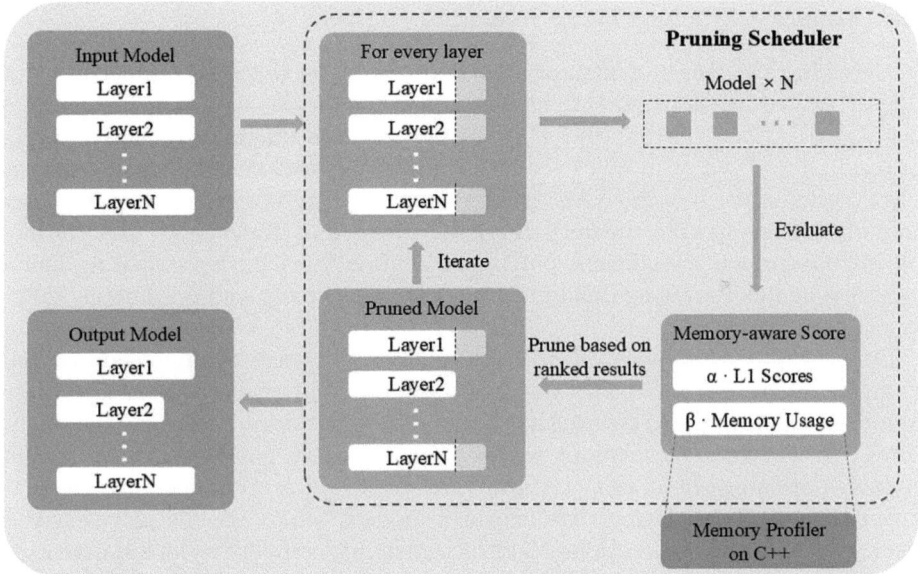

Fig. 1. Architecture of the Memory-Aware Pruning framework.

2 Related Work

2.1 Structured Pruning

Structured pruning has become a widely adopted approach for compressing and accelerating deep neural networks by removing entire filters, channels, or blocks in a structured manner. Unlike unstructured pruning, which generates irregular sparsity patterns that are challenging to exploit on hardware, structured pruning produces models that can benefit from practical speedup and memory reduction. Notable methods, such as Network Slimming [3], ThiNet [4], and Deep Compression [5], use L1-norm or channel statistics to rank and eliminate less important channels. He et al. [6] further proposed an automated pruning approach based on reinforcement learning to find optimal sparsity patterns. Molchanov et al. [7] introduced a Taylor expansion-based criterion for evaluating the importance of network parameters. Li et al. [8] proposed pruning filters based on their L1-norm, effectively reducing computation while maintaining accuracy. However, these methods focus primarily on reducing FLOPs or parameter count, often neglecting runtime memory constraints.

Recent advancements have introduced more versatile frameworks capable of pruning various architectures without manual intervention. For instance, the Structurally Prune Anything (SPA) framework [9] leverages standardized computational graphs to enable architecture-agnostic pruning. Additionally, torque-based structured pruning methods have been proposed to enhance pruning efficiency without requiring complex gradient computations [10].

2.2 Memory-Efficient Deep Learning

The growing challenge of memory bottlenecks in deep models, especially for edge or embedded deployment, has garnered significant attention. Techniques such as activation checkpointing [11], mixed-precision inference [12], and operator fusion have been proposed to reduce runtime memory usage. Memory-centric neural architecture search (NAS) frameworks [13] aim to optimize architectures under hardware-specific memory budgets. Moreover, frameworks like Hermes [14] introduce memory-efficient pipeline execution mechanisms, reducing memory usage by incorporating dynamic memory management and minimizing inference latency.

Other approaches focus on optimizing memory usage during training. For example, BitTrain [15] exploits activation sparsity and proposes a novel bitmap compression technique to reduce the memory footprint during training. Similarly, TinyTL [16] reduces activations rather than trainable parameters for efficient on-device learning. Reformer [17] and Linformer [18] are representative efforts in building memory-efficient Transformer models, which use locality-sensitive hashing or low-rank projections to reduce memory complexity in self-attention layers.

2.3 Joint Optimization Objectives in Pruning

To better align model compression with real-world deployment requirements, several works have explored multi-objective pruning strategies.He et al. [19] proposed an optimization framework based on Alternating Direction Method of Multipliers (ADMM) to jointly prune and quantize the DNNs, which incorporates latency and precision into the pruning objective. He et al. [26] proposed an AutoML-based approach that learns layer-wise pruning policies with reinforcement learning, targeting specific resource constraints. Similarly, NetAdapt [27] performs iterative pruning guided by hardware-specific feedback. Guo et al. [20] proposed a differentiable pruning framework that simultaneously minimizes loss and hardware cost by leveraging gate sparsity and differentiable polarization.

Joint pruning and quantization frameworks, such as PQ-PIM [21] and HAQ [28], have been proposed for efficient inference in energy or latency budgets. Explicit incorporation of peak memory usage, which is crucial for edge devices, has received limited attention until recently. Most existing strategies still assume FLOPs or latency as primary constraints, which may not hold in memory-bound systems.

Recent studies have begun to address this gap. For example, joint pruning and mixed precision channelwise quantization methods [22] target the optimization of the memory footprint, achieving significant size reductions in isoaccuracy. Additionally, frameworks such as SwapNet [23] enable efficient swapping for DNN inference on edge AI devices beyond the memory budget, facilitating the execution of large models within limited memory constraints.Nevertheless, these works either rely on heuristic memory control (e.g., via model size) or address memory constraints indirectly. In contrast, our work explicitly models the peak inference memory as a formal constraint in the pruning optimization process. By jointly ranking channels using both sparsity and memory-aware metrics, we provide a unified framework that directly targets practical memory bottlenecks during inference.

3 Proposed Method

3.1 Problem Description

We formalize structured pruning with memory constraints as a global channel selection problem under a resource budget. Let $\mathcal{C} = \{1, 2, \ldots, N\}$ denote the index set of all prunable output channels across the convolutional layers in a deep neural network. For each channel $i \in \mathcal{C}$, we define an importance score $s_i \in \mathbb{R}_{\geq 0}$, derived from both sparsity and memory-related criteria. The goal is to prune a subset of channels such that the total importance score of pruned channels is minimized, while the peak inference memory usage remains below a specified threshold.

We introduce a binary pruning vector $\mathbf{z} = [z_1, z_2, \ldots, z_N] \in \{0, 1\}^N$, where $z_i = 1$ indicates that channel i is selected for pruning. The constrained optimization problem is formulated as:

$$\min_{\mathbf{z} \in \{0,1\}^N} \sum_{i=1}^{N} z_i s_i \quad (1)$$

$$\text{s.t.} \quad \text{Mem}(f_{\Theta, \mathbf{z}}) \leq \text{Mem}_{\text{target}}$$

where $f_{\Theta,\mathbf{z}}$ denotes the resulting network after pruning the selected channels, and $\text{Mem}(\cdot) : \mathcal{F} \to \mathbb{R}_+$ is a profiling-based function that measures peak inference memory.

Each pruning score s_i is a convex combination of a normalized sparsity term and a normalized memory-saving estimate:

$$s_i = \alpha \cdot l_i + \beta \cdot (1 - m_i) \quad (2)$$

$$l_i = \frac{\|w_i\|_1}{\max_{j \in \mathcal{C}} \|w_j\|_1} \quad (3)$$

$$m_i = \frac{m_i^{\text{raw}}}{\max_{j \in \mathcal{C}} m_j^{\text{raw}}} \quad (4)$$

where $\alpha, \beta \in [0, 1]$ are weighting coefficients satisfying $\alpha + \beta = 1$, and:

- $\mathcal{C}$: Index set of all prunable output channels.
- $z_i \in \{0, 1\}$: Binary indicator of whether channel i is pruned.
- $w_i \in \mathbb{R}^{C_{\text{in}} \times K \times K}$: Convolutional filter corresponding to channel i.
- $\|w_i\|_1$: L1-norm of weights, used as a sparsity metric.
- m_i^{raw}: Estimated raw memory saving from pruning channel i.

- $l_i, m_i \in [0, 1]$: Normalized sparsity and memory terms.
- s_i: Hybrid pruning score for ranking channels.
- $\text{Mem}(f)$: Peak memory usage of network f during inference.
- $\text{Mem}_{\text{target}}$: Target memory budget for post-pruning model.

3.2 Memory Peak Estimation

To more accurately monitor memory usage during pruning, we employ a C++-level memory measurement method, which provides more detailed and precise data compared to PyTorch's built-in memory tracking. This approach allows us to capture exact memory allocations during model execution, giving us a more accurate picture of memory consumption, especially in complex pruning scenarios. The steps are as follows:

1. **Model Saving for Analysis** The model is saved in a traced format using `torch.jit.trace` for memory measurement.
2. **C++-Level Memory Measurement** A custom C++ script runs the model and captures memory usage during execution, providing more precise data than PyTorch's memory functions.
3. **Input and Output Tensors** The memory usage is calculated based on the input tensor $[1, 3, 32, 32]$(VGG16-BN) and the output tensor during the forward pass.
4. **Peak Memory Calculation** The C++ script tracks the maximum memory allocated during the forward pass to calculate peak memory usage.
5. **Memory Optimization** After each pruning step, memory usage is re-evaluated using the C++-level measurement to assess the reduction in memory footprint.

3.3 Pruning Process

To reduce the memory footprint while preserving performance, we design a structured pruning pipeline applicable to deep neural networks. The overall procedure is formalized as Algorithm 1.

The pipeline begins by deploying a pre-trained model onto the target device, followed by identifying the set of prunable layers. For each candidate layer, the importance of every output channel is assessed using a hybrid score that combines the L1-norm of the channel weights and its estimated contribution to peak memory usage. Channels with lower scores are considered less important.

The pruning process is conducted in an iterative manner. At each step, the channel with the lowest importance score is removed, and the model is fine-tuned for a small number of epochs to restore any potential performance loss. This process continues until the model satisfies both the target memory constraint and a minimum accuracy threshold. Finally, the pruned and retrained model is returned.

Algorithm 1. Structured Channel Pruning Pipeline.
Input: Pre-trained model $\mathcal{M}$; memory constraint $\mathcal{M}_{\text{target}}$; minimum accuracy $\mathcal{A}_{\min}$.
Output: Pruned model $\mathcal{M}_{\text{pruned}}$.

1: Deploy $\mathcal{M}$ onto the designated compute device
2: Identify all prunable layers $\mathcal{L} \leftarrow \{l_1, l_2, \ldots, l_n\}$
3: **for** each layer $l \in \mathcal{L}$ **do**
4:   Compute L1-norms $\|W_c\|_1$ for each channel c in l
5:   Estimate per-channel peak memory usage μ_c
6:   Compute hybrid importance score $s_c \leftarrow f(\|W_c\|_1, \mu_c)$
7: **end for**
8: **while** $\text{Mem}(\mathcal{M}) > \mathcal{M}_{\text{target}} \wedge \text{Acc}(\mathcal{M}) \geq \mathcal{A}_{\min}$ **do**
9:   Identify the channel c^* with minimal score s_{c^*} in $\mathcal{L}$
10: Prune c^* from l^* in model $\mathcal{M}$
11: Retrain $\mathcal{M}$ for T epochs to recover accuracy
12: **end while**
13: Evaluate and return $\mathcal{M}_{\text{pruned}} \leftarrow \mathcal{M}$

4 Experiment Results

4.1 Experimental Settings

All experiments are conducted on a high-performance workstation equipped with dual NVIDIA RTX 3090 GPUs (each with 24GB VRAM), an AMD Ryzen 9 5950X 16-core processor, and 64GB of DDR4 system memory. The models are implemented using PyTorch 2.3.0 and accelerated with CUDA 12.1. We evaluate all models on the CIFAR-10 dataset, which comprises 50,000 training images and 10,000 test images spanning 10 distinct classes.

For training, we adopt standard data augmentation including random horizontal flipping and 32×32 random cropping with 4-pixel padding. Models are optimized using stochastic gradient descent (SGD) with a momentum of 0.9 and weight decay of 5×10^{-4}. Unless otherwise stated, the initial learning rate is set to 0.01 and decayed by a factor of 0.1 every 30 epochs.

Each pruning step is followed by a brief fine-tuning phase to recover accuracy. Peak memory usage is measured through a C++-level inference tool that executes traced TorchScript models and records the maximum allocated memory via native CUDA APIs (`cudaMemGetInfo`). This approach provides fine-grained insight into actual memory usage during inference, beyond PyTorch's dynamic estimates.

We evaluate pruning performance based on two key metrics: classification accuracy on the test set and peak memory usage during inference. These metrics are recorded after each pruning iteration to analyze both compression efficiency and model robustness. To accelerate and stabilize the pruning process, each iteration selects a group of $k = 32$ channels from the same layer for evaluation and pruning.

4.2 Pruning VGG16-BN

We evaluate our pruning framework on VGG16 with Batch Normalization (VGG16-BN) using the CIFAR-10 dataset. Given the model's redundancy, it is well-suited for structured channel pruning. We adjust the classifier layer for 10-class output and apply iterative channel pruning guided by our hybrid score, which balances normalized L1-norm and peak memory reduction.

In each pruning iteration, channels with the lowest L1-norms are shortlisted, and their impact on memory and accuracy is assessed individually. A hybrid score is computed for each candidate, and the best is removed. Memory profiling is conducted via traced TorchScript models and C++ scripts for accurate peak memory measurement. The model is fine-tuned periodically to recover potential accuracy degradation.

We compare L1-only and memory-aware(Hybird L1-norm memory usage). The target is to reduce memory to 60% of the baseline peak usage.

The memory-aware method consistently achieves better memory reduction while maintaining accuracy, confirming its suitability for deployment in constrained environments. Channel dependencies and fully connected dimensions are adjusted dynamically to ensure correctness after pruning. These results validate that our method enables efficient compression without architectural redesign (Figs. 2 and 3).

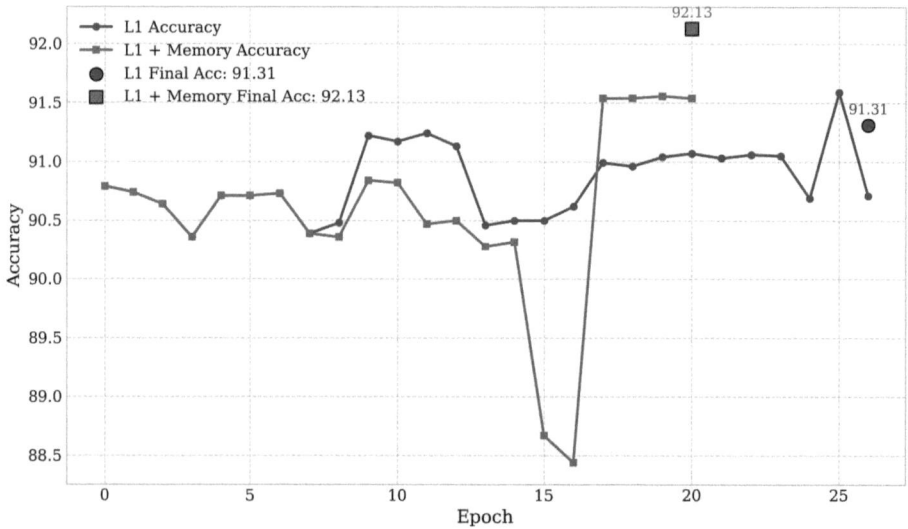

Fig. 2. Accuracy curves over pruning epochs for VGG16-BN. The memory-aware method consistently achieves higher final accuracy.

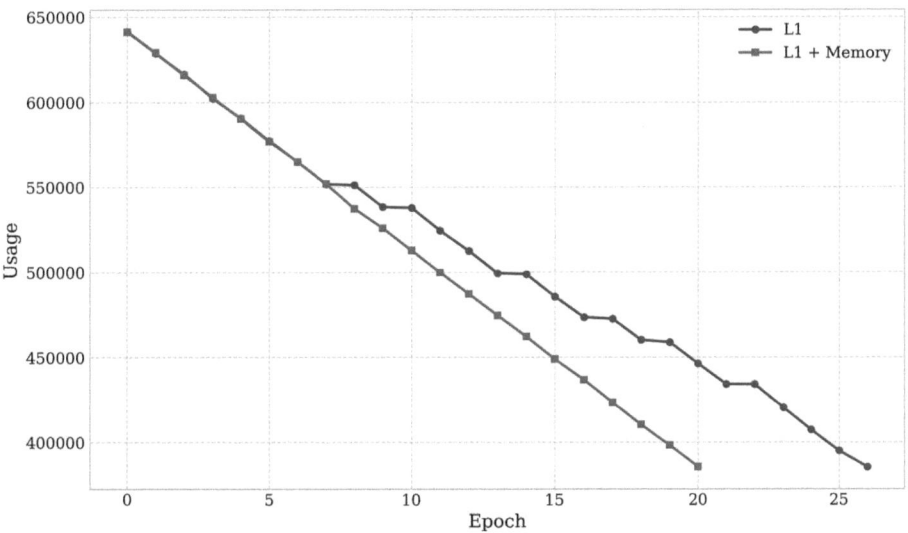

Fig. 3. Peak memory usage during pruning. The Memory-aware method reduces memory more effectively and rapidly.

4.3 Pruning ResNet50

We further evaluate the proposed pruning strategy on ResNet50 to verify its generality across deeper architectures. Given the residual design and higher parameter count, ResNet50 presents unique challenges for structured pruning, especially under strict memory constraints.

Due to the significant variation in channel widths and feature map resolutions across different layers of ResNet50, we revise our original pruning strategy and adopt a combined scheme based on proportional pruning and average L1-norm selection. This adjustment ensures that pruning remains structurally balanced across stages and adapts to the heterogeneous scaling of the network, avoiding over-pruning in shallow layers with fewer parameters or feature maps.

Figures 4 and 5 present the accuracy and memory usage trajectories during pruning. The L1-only baseline shows significant instability, with frequent accuracy collapses and poor recovery. In contrast, the memory-aware method yields markedly improved accuracy retention, achieving a final test accuracy of 87.52%, compared to 82.91% under L1-only pruning.

In terms of memory reduction, the hybrid-aware method maintains a consistently steeper decline, ultimately saving 20% of peak memory relative to the unpruned model, as shown in Fig. 5. This validates the effectiveness of integrating runtime memory profiling in deeper architectures, where intermediate activations dominate memory footprint.

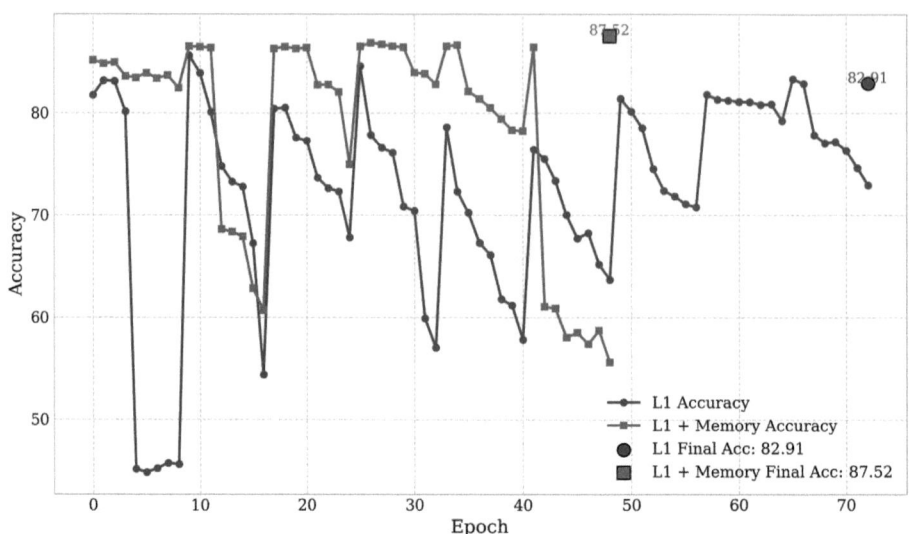

Fig. 4. Accuracy during pruning for ResNet50. The memory-aware strategy preserves accuracy more effectively.

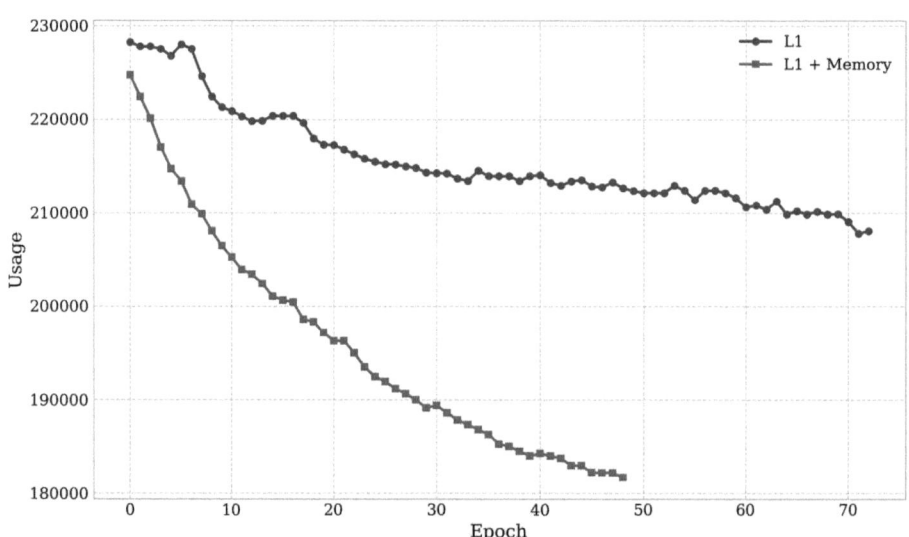

Fig. 5. Peak memory usage comparison during ResNet50 pruning. Memory-aware shows significantly better memory reduction.

These results confirm that memory-guided pruning generalizes well to deeper and more complex architectures like ResNet50, achieving better trade-offs between accuracy and deployment cost.

4.4 Pruning Transformer

To evaluate the generality of our framework beyond convolutional architectures, we conduct pruning experiments on a standard Transformer encoder model. Due to architectural constraints inherent to attention mechanisms, we restrict pruning to the feed-forward (FFN) layers. Specifically, we do not prune the multi-head attention layers, since the embedding dimension must remain constant to maintain shape compatibility across query (Q), key (K), and value (V) projections.

As a result, the achievable memory reduction is naturally limited, especially given that attention layers account for a substantial portion of parameter count and activation memory. Nonetheless, this setting serves to evaluate the pruning strategy's robustness and behavior under restricted flexibility.

To ensure comparability, we prune for a fixed number of steps—stopping once memory reaches approximately 90% of the initial peak—rather than targeting a hard threshold. Figures 6 and 7 show the accuracy and memory usage across pruning epochs. While both L1-only and memory-aware variants yield relatively modest memory reductions (less than 5%), the accuracy of the memory-aware approach remains consistently higher, achieving a final accuracy of 83.08% versus 82.90% for L1-only pruning.

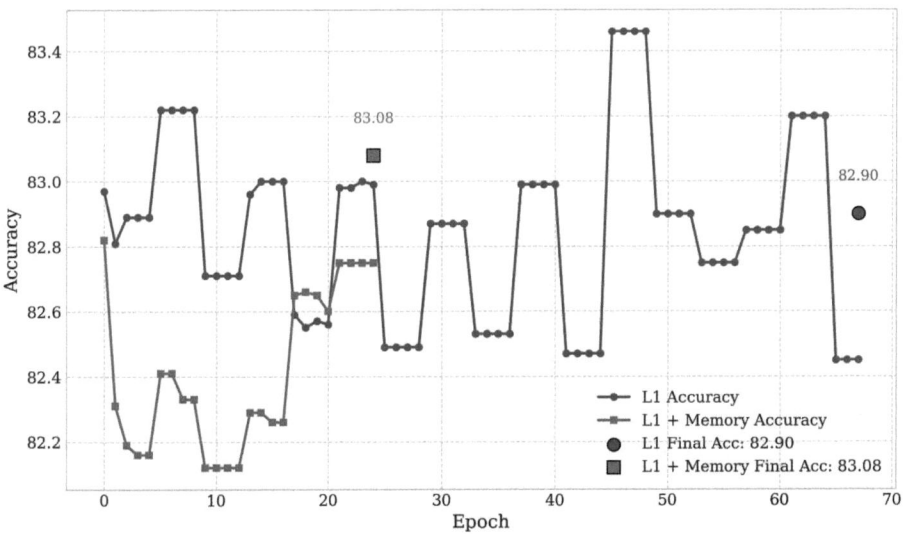

Fig. 6. Accuracy during Transformer pruning. Memory-aware pruning consistently yields higher accuracy.

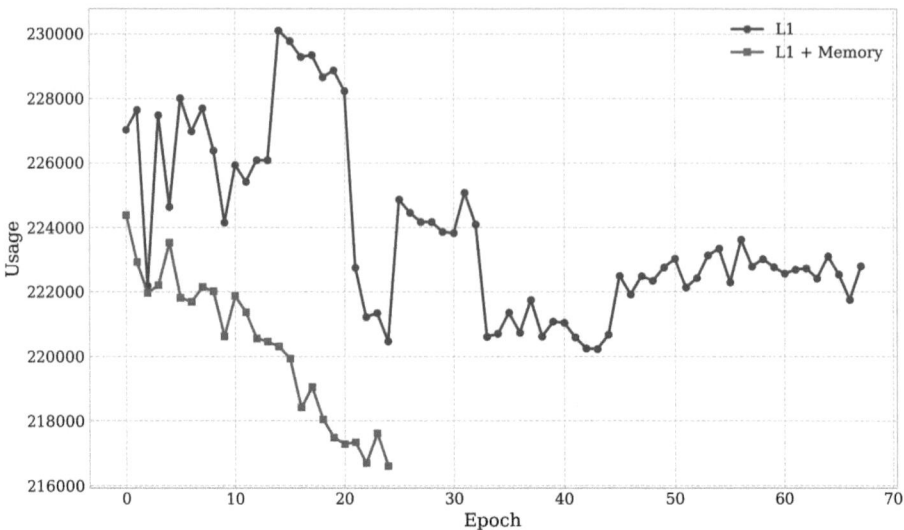

Fig. 7. Peak memory usage for Transformer model. Only FFN layers are pruned; thus, the overall reduction is limited.

These findings highlight a limitation of structured pruning in Transformer architectures—namely, that attention blocks cannot be sparsified without architectural modifications. Nonetheless, the hybrid-aware pruning method proves effective even under such constrained settings, offering a robust solution for partial model slimming.

4.5 Performance Analysis

To assess the effectiveness of our pruning strategy, we summarize classification accuracy, peak memory reduction, and accuracy drop across VGG16-BN, ResNet50, and Transformer models in Table 1. Compared to the L1-only baseline, our memory-aware method consistently achieves better accuracy-memory trade-offs.

On VGG16-BN, which exhibits high channel redundancy, the memory-aware strategy achieves 40% memory reduction with only 1.16% accuracy drop—demonstrating the benefit of jointly considering sparsity and memory usage.

For ResNet50, pruning leads to 20% memory savings and a notable accuracy improvement, suggesting regularization effects in deeper networks. However, deeper residual architectures are more sensitive to aggressive pruning.

In the Transformer case, pruning is limited to feed-forward layers, resulting in modest memory gains (3%) but well-preserved accuracy. This indicates that careful module selection is essential when applying structured pruning to non-convolutional models.

Overall, the results validate that our hybrid pruning framework effectively balances memory efficiency and accuracy across diverse architectures, supporting practical deployment in memory-constrained environments.

Table 1. Summary of Pruning Results on Different Architectures

Model	Method	Final Acc. (%)	Mem. Red. (%)	Acc. Drop (%)
VGG16-BN	L1-only	91.31	40	1.98
	Memory-Aware Pruning	92.13	40	1.16
ResNet50	L1-only	82.91	10	6.46
	Memory-Aware Pruning	87.52	20	1.85
Transformer	L1-only	82.90	3	0.84
	Memory-Aware Pruning	83.08	10	0.64

5 Conclusion

In this paper, we proposed a memory-aware structured pruning framework that jointly optimizes for both L1-norm sparsity and peak memory usage. By introducing a hybrid importance criterion and employing precise C++-level runtime memory profiling, our method effectively guides channel pruning toward solutions that are not only compact but also practical for deployment on memory-constrained devices.

We validated our approach on VGG16-BN, ResNet50, and Transformer models using the CIFAR-10 dataset. Experimental results demonstrate that the proposed hybrid strategy consistently outperforms conventional L1-based pruning in terms of memory reduction, while maintaining or even improving classification accuracy in certain scenarios. In particular, our method achieves 40% memory savings on VGG16-BN with minimal accuracy degradation, and offers stable performance across different model families.

The proposed framework is model-agnostic, hardware-friendly, and does not require specialized layers or retraining protocols, making it well-suited for real-world applications such as edge AI and embedded inference. In future work, we aim to extend this framework to support structured pruning in attention modules and explore its integration with automated architecture search under memory constraints.

Acknowledgment. This work is partly supported by the Key Research and Development Program of Heilongjiang Province No. 2024ZX01A07.

References

1. Ramadan, M.N.A., Ali, M.A.H., Khoo, S.Y.: Federated learning and TinyML on IoT edge devices: challenges, advances, and future directions. ICT Express (2025)
2. Lin, Z., et al.: AdaOper: energy-efficient and responsive concurrent DNN inference on mobile devices. In: Proc. ACM MobiSys (2024)
3. Liu, Z., Li, J., Shen, Z., Huang, G., Yan, S., Zhang, C.: Learning efficient convolutional networks through network slimming. In: Proc. ICCV (2017)
4. Luo, J., Wu, J., Lin, W.: ThiNet: a filter level pruning method for deep neural network compression. In: Proc. ICCV (2017)
5. Han, S., Mao, H., Dally, W.J.: Deep compression: compressing deep neural networks with pruning, trained quantization and Huffman coding. In: Proc. ICLR (2016)
6. He, Y., Zhang, X., Sun, J.: Channel pruning for accelerating very deep neural networks. In: Proc. ICCV (2017)
7. Molchanov, P., Tyree, S., Karras, T., Aila, T., Kautz, J.: Pruning convolutional neural networks for resource efficient inference. In: Proc. ICLR (2017)
8. Li, H., Kadav, A., Durdanovic, I., Samet, H., Graf, H.P.: Pruning filters for efficient convnets. In: Proc. ICLR (2017)
9. Wang, X., Rachwan, J., Günnemann, S.: Structurally prune anything: any architecture, any framework, any time (2024). arXiv:2403.18955
10. Gupta, A., Agrawal, S., Yadav, S., Kumar, R.: Torque based structured pruning for deep neural network. In: Proc. WACV (2024)
11. Chen, T., Xu, B., Zhang, C., Guestrin, C.: Training deep nets with sublinear memory cost (2016). arXiv:1604.06174
12. Micikevicius, P., et al.: Mixed precision training. In: Proc. ICLR (2018)
13. Cai, H., Gan, C., Wang, T., Zhang, Z., Han, S.: Once-for-all: train one network and specialize it for efficient deployment. In: Proc. ICLR (2020)
14. Han, X., et al.: hermes: memory-efficient pipeline inference for large models on edge devices (2024). arXiv:2409.04249
15. Hosny, M., Abdelhalim, H.S.: BitTrain: efficient deep neural network training via activation bitmaps. In: Proc. ICCV (2021)
16. Cai, H., Wang, T., Zhang, Z., Han, S.: TinyTL: reduce memory, not parameters for efficient on-device learning. In: Proc. NeurIPS (2020)
17. Kitaev, N., Kaiser, Ł., Levskaya, A.: ReFormer: the efficient transformer. In: Proc. ICLR (2020)
18. Wang, S., Li, B., Khabsa, M., Fang, H., Ma, H.: LinFormer: self-attention with linear complexity. In: Proc. NeurIPS (2020)
19. Yang, H., Gui, S., Zhu, Y., Li, B.: Automatic neural network compression by sparsity-quantization joint learning: a constrained optimization-based approach. In: Proc. CVPR (2020)
20. Guo, Y., Yuan, H., Tan, J., Huang, T.: GDP: stabilized neural network pruning via gates with differentiable polarization. In: Proc. ICCV (2021)
21. Zhang, Y., Pan, J., Wang, H., Li, C., Yang, H.: PQ-PIM: a pruning–quantization joint optimization framework for ReRAM-based processing-in-memory DNN accelerator. J. Syst. Archit. **127**, 102531 (2022)
22. Motetti, B.A., Pudipeddi, S.R., Sharma, A., Kumar, N.: Joint pruning and channel-wise mixed precision quantization for efficient deep neural networks. IEEE Trans. Comput. **73**(11), 2619–2633 (2024)

23. Wang, T., Cao, J., Li, Z., Kumar, S., Zhang, L.: SwapNet: efficient swapping for DNN inference on edge AI devices beyond the memory budget (2024). arXiv:2401.16757
24. Qu, X., Zhang, Y., Chen, L., Zhao, J., Liu, M.: Automatic joint structured pruning and quantization for efficient neural network training and compression (2025). arXiv:2502.16638
25. Wang, S., Shelhamer, E., Vinyals, O., Darrell, T.: APQ: joint search for network architecture, pruning and quantization policy. In: Proc. CVPR (2020)
26. He, Y., Lin, J., Liu, Z., Wang, H., Li, L., Han, S.: AMC: AutoML for model compression and acceleration on mobile devices. In: Proc. ECCV (2018)
27. Yang, T.-J., Howard, A., Chen, B., Zhang, X., Wang, J.: NetAdapt: platform-aware neural network adaptation for mobile applications. In: Proc. ECCV (2018)
28. Wang, K., Liu, Z., Lin, Y., Lin, J., Han, S.: HAQ: hardware-aware automated quantization with mixed precision. In: Proc. CVPR (2019)

SAULC: Semantic-Driven Adaptive Method for UAV-LEO Satellite Communication

Liangwei Qin[1,3], Dongbo Li[1,2,3](✉), Chongrong Li[1,3], Yibo Hou[1], Jiahe Gao[1], and Bo Yin[1,3]

[1] Faculty of Computing, Harbin Institute of Technology, Harbin 150001, China
ldb@hit.edu.cn
[2] Department of Information Engineering, The Chinese University of Hong Kong, Hong Kong SAR 999077, China
[3] State Key Laboratory of Smart Farm Technologies and Systems, Harbin Institute of Technology, Harbin 150001, China

Abstract. The collaboration between Unmanned Aerial Vehicles (UAVs) and Low Earth Orbit (LEO) satellites represents an emerging edge computing paradigm that holds significant potential for enabling ubiquitous connectivity in the Internet of Things (IoT). However, highly dynamic UAV-satellite links lead to high Doppler shift, high path loss, and other issues, which cause serious semantic distortion. Meanwhile, LEO satellite spectrum resources are extremely limited. Thus, realizing end-to-end reliable UAV-LEO satellite communications with lower communication bandwidth in poor channel environments presents a significant challenge. To address these challenges, this paper proposes the Semantics-driven Adaptive UAV-LEO Satellite Cooperative Communication System (SAULC), establishing the first UAV-LEO satellite semantic communication architecture. The system systematically incorporates time-varying channel characteristics, including Doppler shift, path loss, and atmospheric attenuation, while employing a low-complexity Flatten Transformer architecture with integrated channel feedback mechanisms. Experiments show that the proposed model significantly outperforms conventional schemes under time-varying UAV-satellite channels.

Keywords: Semantic communication · UAV-LEO collaboration · Edge computing · Time-varying channels

1 Introduction

The deep integration of Artificial Intelligence of Things (AIoT) [1] and edge computing [2,3], particularly UAVs, and the development of edge computing technologies related to LEO satellites have greatly expanded air-space communication links. Various air-ground cooperative systems have been proposed, significantly improving the demand for high-bandwidth and low-latency image transmission [4] in dynamic environments for air-space cooperative systems. However, the current shortage of satellite channel resources means that high-bandwidth image transmission occupies a large amount of spectrum resources, hindering large-scale deployment [5]. Simultaneously, the high dynamics of UAV-LEO satellite channels cause significant Doppler shift, drastically impacting image transmission quality. Traditional wireless communication schemes, while achieving better image transmission under abundant bandwidth conditions, struggle to adapt to time-varying channels [6,7] with limited bandwidth and harsh conditions. Achieving stable UAV-LEO satellite communication under limited bandwidth and harsh conditions has become an urgent problem.

Semantic communication as a new generation of communication technology [8,9], aims to focus on the semantic information behind the transmission of information, the core of which lies in the use of large models to compress the data as large as possible without loss, to achieve semantic level of data comprehension [10], which can greatly reduce the size of the data to be transmitted, and the current semantic communication has been applied in the field of car networking, virtual reality and other fields [11–13]. Semantic communication comprehension that is the compression of the characteristics of the data to get a higher dimension of the extraction [14], very suitable for resource-limited satellite channels.

Current semantic communication systems, though successful in terrestrial applications [15], suffer significant semantic distortion in environments with time-varying Doppler frequency shifts [16], intermittent line-of-sight transmissions, and other issues [17] in air-space links. Existing semantic coders are almost exclusively trained under static ground link conditions and perform well in static links [18–20], especially achieving far better end-to-end image transmission quality than traditional communication forms at low signal-to-noise ratios. However, when facing highly dynamic and harsh UAV-satellite channels, model performance remains a concern. The lower bandwidth of UAV-satellite channels also challenges semantic communication, as existing models lack the ability to adapt to the spatio-temporal correlation characteristics of satellite-aircraft collaboration scenarios and inadequately address the dual challenges of semantic fidelity and adaptive dynamic channels [21].

To address these challenges, we propose the SAULC method. The principal innovations include:

- This paper proposes SAULC, an end-to-end semantic communication method designed for UAV-satellite dynamic channels. We develop a kinematic model for UAV-satellite systems that uniquely integrates time-varying Doppler shift, free-space path loss, and atmospheric attenuation in air-space links. The model incorporates real-time channel state feedback and demonstrates through numerical simulations the superior performance of semantic communications under harsh time-varying channel conditions.
- SAULC introduces the Flatten Transformer architecture, demonstrating dual advantages over conventional Swin Transformer-based DeepJSCC architectures: reduced computational complexity and enhanced image reconstruction fidelity.

- We conducted comprehensive experiments under time-varying UAV satellite channel modeling. The results show that the proposed SAULC semantic communication architecture significantly improves the Peak Signal-to-Noise Ratio (PSNR) and Structural Similarity (SSIM) index at each Doppler-compensation rate compared to traditional approaches.

2 System Model

This section develops a semantic communication architecture for UAV-LEO satellite systems, as illustrated in Fig. 1. We first formulate a novel channel modeling methodology that explicitly accounts for Doppler shift dynamics in these hybrid systems. Following this, we delineate the structural relationships governing data mapping within the semantic communication paradigm.

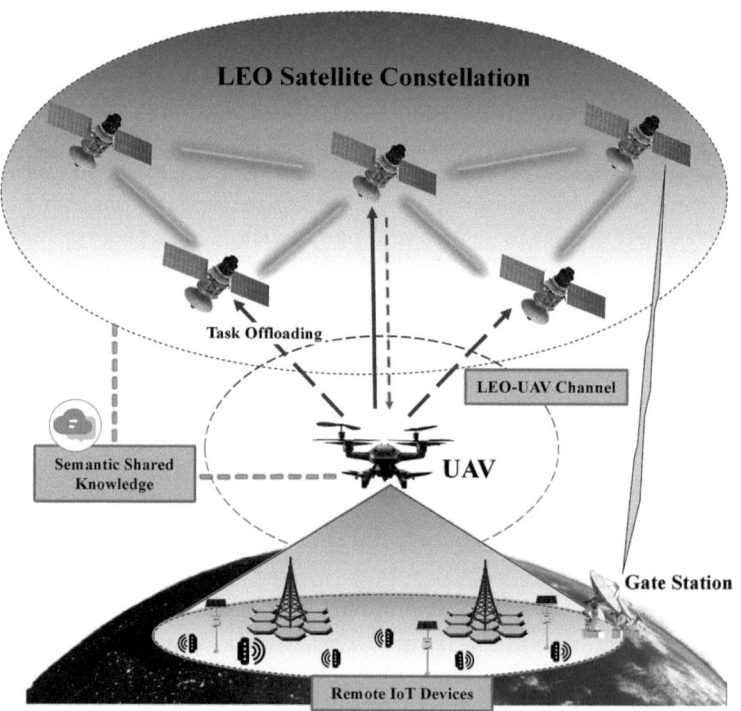

Fig. 1. UAV-LEO Satellite Semantic Communication Architecture

2.1 LEO Satellite Constellation

This paper adopts the Walker Delta constellation architecture to build a seamless network based on multi-satellite cooperative networking technology. The constellation is characterized by three core orbital parameters (T, P, F) to fully characterize satellite spatial distribution:

$$W = (T, P, F) \quad \text{s.t.} \quad \begin{cases} T = P \times S \\ F = \left\lfloor \dfrac{T}{P} \cdot \dfrac{\Delta \Omega}{2\pi} \right\rfloor \end{cases}, \tag{1}$$

where T denotes the total number of satellites, P is the number of orbital planes, and F is the phase factor.

2.2 Channel Model

This paper establishes the UAV-LEO satellite channel from three dimensions: physical relationship, channel characteristic quantification, and environmental parameter correction.

Physics Principles. Based on the Walker Delta constellation, the ascending node of the mth satellite in the mth orbital plane in ascending longitude Ω_m and latitudinal magnitude u_{mn} is determined by the following kinematic equations:

$$\begin{cases} \Omega_m = \Omega_0 + \dfrac{2\pi m}{P} \\ u_{mn} = \dfrac{2\pi n}{S} + \dfrac{2\pi F}{T} m. \end{cases} \tag{2}$$

The position vector of the satellite in the geocentric inertial (ECI) coordinate system can be expressed as:

$$r_{\text{sat}}^{(k)} = \mathbf{R} \cdot r_{\text{orb}}^{(k)}. \tag{3}$$

Define the UAV ECI position vector as r_{UAV}, and its instantaneous slant distance from the satellite as:

$$d = \left\| r_{\text{sat}}^{(k)} - r_{\text{UAV}} \right\|, \tag{4}$$

where the orbital coordinate system position vector:

$$r_{\text{orb}} = R_{\text{obr}} \begin{bmatrix} \cos u_{mn} \\ \sin u_{mn} \\ 0 \end{bmatrix}, \tag{5}$$

the rotation matrix is

$$\mathbf{R} = \mathbf{R}_Z(\Omega_m) \cdot \mathbf{R}_X(i) \cdot \mathbf{R}_Z(u_{mn}). \tag{6}$$

Doppler Shift. The relative velocity along the signal propagation direction can be expressed as:

$$v_{\text{los}} = (\boldsymbol{v}_{\text{sat}}^{(k)} - \boldsymbol{v}_{\text{UAV}}) \cdot \boldsymbol{e}, \tag{7}$$

where the unit vector is defined as:

$$\boldsymbol{e} = \frac{\boldsymbol{r}_{\text{sat}}^{(k)} - \boldsymbol{r}_{\text{UAV}}}{d}. \tag{8}$$

This leads to the derivation of the Doppler shift expression:

$$f = f_0 \cdot (1 + \frac{v_{\text{LOS}}}{c}), \tag{9}$$

where f_0 is the satellite communication carrier frequency and c is the speed of light.

Free Path Loss. The free space path loss (FSPL) expression derived based on the Friis transport equation is:

$$\text{FSPL}_{\text{dB}} = 20 \log_{10} \left(\frac{4\pi d f_0}{c} \right), \tag{10}$$

where d is the transmission distance.

Atmospheric Attenuation. Total atmospheric attenuation is determined through the ITU-R P.676-11 recommendation model as follows:

$$A_{\text{dB}} = \frac{\gamma_0 h_0 - \gamma_\omega h_\omega}{\sin \varphi}, \tag{11}$$

where γ_0 and γ_ω are dry air and water vapor attenuation rates, h_0 and h_ω are dry air and water vapor equivalent heights, and φ is the channel elevation angle.

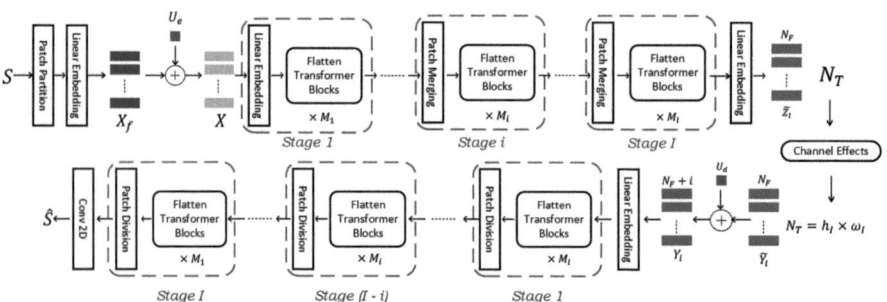

Fig. 2. SAULC semantic communication approach

Received Signal and Capacity. Considering each channel effect, the received power expression is:

$$P_r = P_t + G_t + G_r - \text{FSPL}_{\text{dB}} - A, \qquad (12)$$

where G_T and G_R are the transmitter and receiver antenna gains, respectively. From this, the receiver-side signal-to-noise ratio and instantaneous capacity are derived:

$$\begin{cases} \gamma = P_r - \sigma^2 \\ R = \max_{k \in \mathcal{V}} B \log_2(1+\gamma), \end{cases} \qquad (13)$$

where σ^2 is the noise power, B is the channel bandwidth, and $\mathcal{V}$ is the current set of visible satellites.

2.3 Semantic Communication Model

The semantic communication model consists of the following four main components:

Adaptive Encoder. The adaptive encoder is defined as:

$$f_e : S \times U \to Z, \qquad (14)$$

where $S \in \mathbb{R}^{C \times H \times W}$ is the input image, $U \in \mathbb{R}^{1 \times 4}$ is the channel state parameter, and $Z \in \mathbb{R}^{N_F \times N_T}$ is the feature tensor mapped by the encoder.

Channel Adaptive Strategy. For different bandwidth ratios, the method performs dynamic resizing of the features:

$$\tilde{Z}_l = \tilde{Z}[:n_f,:] \in \mathbb{R}^{n_f \times N_T}, \qquad (15)$$

$$\rho_l = \frac{R}{R_{\max}}, \qquad (16)$$

where ρ_l is the bandwidth ratio and R is the current channel capacity, where $N_F N_T = 2\rho_L N$ and $n_f = \lfloor 2\rho_l N / N_T \rfloor$.

Channel Transmission. The $\tilde{Z}_l$ is transmitted over the drone-satellite channel $h : \mathbb{R}^{n_f N_T} \to \mathbb{C}^{\rho_l N}$, which is mathematically characterized as:

$$s_l = \mathcal{Q}(\tilde{Z}_l) \in \mathbb{C}^{\rho_l N}, \qquad (17)$$

where $\mathcal{Q}(\cdot)$ is a joint quantization-modulation operation in the complex domain. After channel transmission, the symbols are deformed to obtain the damaged signal $\tilde{s}_l$, which is later demodulated-decoded to the feature signal $y_l \in \mathbb{C}^{\rho_l N}$.

Adaptive Decoder. The received signal y_l is reconstructed into a dimensionally matched tensor $\tilde{Y}_l = \mathcal{P}(y_l)$ by transformations, after which it is mapped by a decoder to produce a reconstructed image:

$$f_d : \tilde{Y}_l \times U \to Y_l, \tag{18}$$

where $Y_l \in \mathbb{R}^{C \times H \times W}$ is the recovered image.

This paper aims to optimize the encoder-decoder function (f_e, f_d) to improve image end-to-end transmission metrics and minimize the model's end-to-end computational complexity under this channel.

3 Proposed Method

In this section, the SAULC method is proposed for time-varying UAV-LEO satellite channels. It utilizes Flatten Attention as a more efficient encoder/decoder backbone and proposes an adaptive coding mechanism for UAV-LEO satellite channels, reducing model complexity while guaranteeing performance.

3.1 Flatten Block

We employ the Flatten Attention module [22] as the core component of the semantic encoder. The scheme significantly improves computational efficiency while maintaining expressive power compared to the traditional Swin Transformer [23].

In Vision Transformers, the standard self-attention mechanism operates through pairwise similarity computation between all query-key pairs. For an input token $x \in \mathbb{R}^{N \times C}$ where N denotes sequence length and C channel dimension, the attention computation per head is expressed as:

$$Q = xW_Q, \quad K = xW_K, \quad V = xW_V,$$
$$O_i = \text{Softmax}\left(\frac{QK^\top}{\sqrt{d}}\right) V, \tag{19}$$

where $W_Q, W_K, W_V \in \mathbb{R}^{C \times d}$ are the projection matrices. Explicitly constructing the $N \times N$ attention matrix $QK^\top$ leads to quadratic computational complexity $\mathcal{O}(N^2 d)$.

To adapt to resource-constrained edge scenarios, the focused linear attention module proposed in this paper reduces computational complexity to $\mathcal{O}(Nd^2)$ while maintaining model representation capability. As illustrated in Fig. 2, the computational process is reorganized into three key components:

Firstly, the non-linear mapping function below is introduced to reshape feature orientation without changing feature size:

$$\phi_p(x) = f_p(\text{ReLU}(x)), \quad f_p(x) = \frac{\|x^{\circ p}\|}{\|x\|} x^{\circ p}, \tag{20}$$

where $x^{\circ p}$ represents element-wise exponentiation. This transformation enhances proximity between similar features and increases separation among dissimilar ones, inducing a manifold concentration effect.

Secondly, the standard attention computation can be reconstructed as:

$$O_{\text{linear}} = \frac{\phi_p(Q)\left(\sum_{j=1}^{N}\phi_p(K_j)\right)}{\phi_p(Q)\left(\phi_p(K)^\top V\right)}. \tag{21}$$

Thirdly, the proposed architecture incorporates depthwise convolution(DWC) module to enhance semantic saliency detection while preserving feature diversity:

$$O = O_{\text{linear}} + \text{DWC}(V). \tag{22}$$

Based on the above Flatten Attention, we form the Flatten Transformer block module of the semantic coder.

3.2 Semantic Communication Data Flow

As illustrated in Fig. 2, this paper adopts the Flatten block module as the backbone network of the SAULC semantic communication method, which includes image codec, channel role, and channel feedback mechanism, aiming to realize adaptive semantic compression under dynamic UAV satellite channels and ensure efficient and accurate image transmission.

Encoder Data Stream. The input image $S \in \mathbb{R}^{C \times H \times W}$ is processed by 2×2 non-overlapping block segmentation with a linear embedding layer to generate the initial feature tensor $X_f \in \mathbb{R}^{c \times H/2 \times W/2}$. Enhanced features are constructed by fusing channel state information:

$$\begin{cases} u = \text{MLP}([A_{\text{dB}}, \text{FSPL}, f, \rho_l]) \\ X = X_f \oplus U_e \\ U_e[:, i, j] = u, \end{cases} \tag{23}$$

where $\oplus$ denotes channel splicing and $U_e \in \mathbb{R}^{n_u \times H/2 \times W/2}$ is the edge information matrix after repeated expansion. The augmented feature X is subsequently processed by the M_1-level Flatten Swin Transformer block:

$$X_{l+1} = \text{FlattenBlock}(X_l). \tag{24}$$

The inter-stratum is down-sampled by the patch merge module, converting $c \times h_i \times w_i$ features to $c \times h_{i+1} \times w_{i+1}$ ($h_{i+1} = h_i/2$), and finally outputs the features $\tilde{Z} \in \mathbb{R}^{N_F \times N_T}$. For the bandwidth ratio ρ_l, the method dynamically adjusts n_f to achieve continuous rate adaptation.

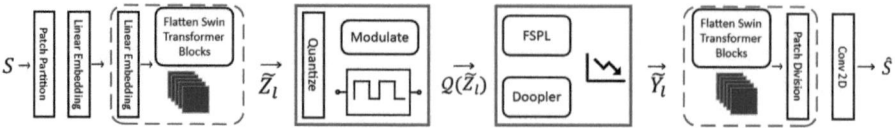

Fig. 3. Overview of the channel action model

Channel Action Model. As illustrated in Fig. 3, the feature tensor $\tilde{Z}_l$ generated by the encoder undergoes initial modulation to produce channel symbol s_l. Following channel transmission, these symbols experience distortion, resulting in the impaired signal $\tilde{s}_l$. This impaired signal is then demodulated and decoded to obtain the received feature signal y_l. Given the significant Doppler effects inherent in LEO satellite channels, uncompensated experimental implementations are infeasible, necessitating the implementation of Doppler shift compensation measures [24].

The specific channel transmission model pseudo-code action mechanism is shown in Algorithm 1.

Decoder Data Stream. The received signal y_l is reconstructed into a dimensionally matched tensor $\tilde{Y}_l$ by a real-domain transformation and zero-value filling operation. The decoding side input is constructed by splicing the same edge information u as the encoding side:

$$Y_l = \tilde{Y}_l \oplus U_d \quad U_d[:, i, j] = u. \tag{25}$$

The decoder uses a mirror-symmetric structure and undergoes I stages of processing containing pixel-mixing up-sampling cascaded with a Flatten Transformer block. Each stage up-samples $c \times h_i \times w_i$ features to $c \times 2h_i \times 2w_i$, ultimately reconstructing the original image resolution.

Based on the above semantic communication method, we design the following simulation experiments, and the results show that the proposed model achieves good performance at high computational speeds.

4 Experiment

This section evaluates three communication methods for UAV-LEO satellite image transmission under varying Doppler compensation rates: the SAULC method based on Flatten Transformer (SAULC-F), the Swin Transformer-based SAULC-S method, and the conventional BPG-LDPC-QAM for UAV-LEO Satellite Communication (BLQULC) baseline. Performance metrics including PSNR and SSIM values were systematically compared.

Experimental results show 12–14% PSNR enhancement and 11–40% SSIM improvement over baseline methods under varying Doppler compensation rates. The following is the experimental implementation procedure:

Algorithm 1. The compensation part of SAULC

Input: Encoded feature tensor $\tilde{Z}_l \in \mathbb{R}^{n_f \times N_T}$, Channel parameters $\Theta = \{h_d, \alpha_p, \alpha_a, \tau\}$
Output: Decoded feature tensor $\tilde{Z} \in \mathbb{R}^{N_F \times N_T}$
   \# Step1: converting a 3D tensor to a bitstream
1: $B \leftarrow [b_1, \ldots, b_n] \leftarrow$ unpackBits$(\tilde{Z}_l \in \mathbb{R}^{n_f \times N_T})$;
   \# Step2: applied channel effect
2: Compute b_i for $1 \leq i \leq n$ according to Eq. (9) and (11);
3: Compute b_i for $1 \leq i \leq n$ with random positively Gaussian noise and uniformly phase noise;
   \# Step3: applying channel compensation mechanisms
4: **for** each $i \in [1, n]$ **do**
5:   Doppler compensation based on the time-varying Brug method and Doppler shift compensation rate for b_i;
6:   Compensate for free path loss computationally;
7: **end for**
   \# Step4: converting the bitstream back to a 3D tensor
8: $\tilde{Z} \in \mathbb{R}^{N_F \times N_T} \leftarrow$ packBits(B);
9: **return** Decoded feature tensor $\tilde{Z} \in \mathbb{R}^{N_F \times N_T}$

4.1 Constellation and Channel Simulation

Based on the UAV-LEO satellite kinematic model described in the SAULC-F method, we constructed a Walker satellite operating at an altitude of 1,400 km with an orbital inclination of 60°. The architecture comprises 8 distinct orbital planes, each populated with 3 evenly distributed satellites, resulting in a total of 24 satellites in the constellation. Within this satellite network configuration, we randomly generated UAV positions and velocities, then computed end-to-end communication link parameters, including Doppler shift, path loss, and atmospheric attenuation, between the UAV and nearest satellites. Additionally, we incorporated random amplitude and phase noise within specified ranges (Table I). Using these parameters, we developed time-varying UAV-satellite channel environments for the SAULC-F, SAULC-S, and BLQULC methods. The experimental validation procedures are detailed below (Table 1 and Fig. 4).

Table 1. Satellite Channel Parameters

Parameter	Value
Carrier Frequency	20 GHz
Doppler Shift	-40 kHz ~ 40 kHz
FSPL	137 dB $\sim$ 143 dB
A_{dB}	28 dB $\sim$ 32 dB
Amplitude Noise	0 dB $\sim$ 5 dB
Phase Noise	-23 dBc/Hz

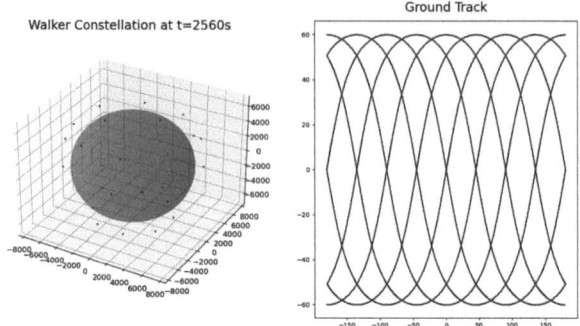

Fig. 4. Map of satellite status and sub-star point trajectories

4.2 Comparison

The STL-10 dataset was employed to train the SAULC-F and SAULC-S methods, with 100 training iterations per Doppler-compensation rate. For the BLQULC method under conventional configuration, the parameters were set as follows: code rate = 0.5, codeword length = 1296 bits, decoding iterations = 50, and modulation scheme = 16-ary Quadrature Amplitude Modulation (16-QAM). Each method was evaluated by transmitting 10 batches of RGB images. SSIM and PSNR metrics were recorded for each test batch and mapped as follows (Figs. 5 and 6):

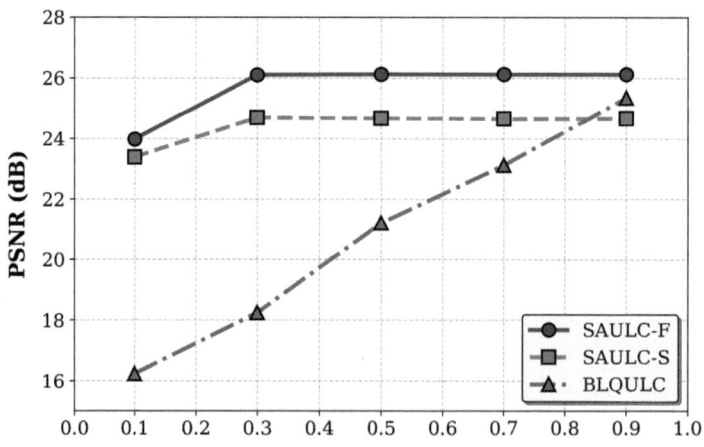

Fig. 5. Comparison of PSNR values of different codecs with different Doppler shift compensation rates.

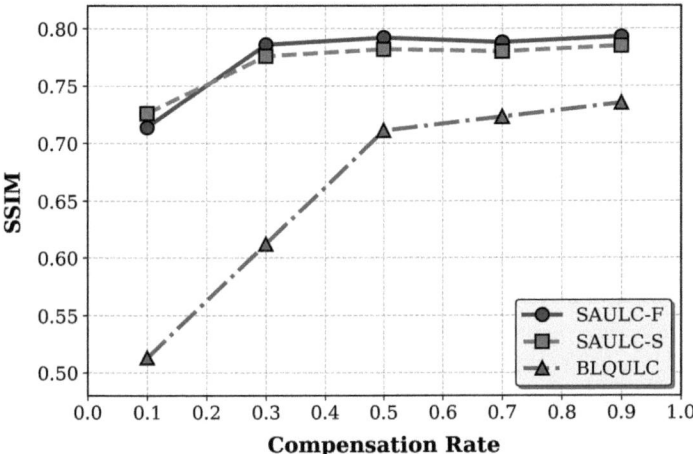

Fig. 6. Comparison of SSIM values of different codecs with different Doppler shift compensation rates.

5 Conclusion

This paper proposes the SAULC method to address the critical challenges of high-bandwidth and low-latency image transmission in dynamic time-varying channels. Experimental validation shows 12–14% PSNR enhancement and 11–40% SSIM improvement over baseline methods under varying Doppler compensation rates.

These advancements demonstrate enhanced robustness in harsh channel conditions and validate the architecture's potential for improving next-generation air-space integrated network reliability.

References

1. Tao, B., Masood, M., Gupta, I., Vasisht, D.: Transmitting, fast and slow: scheduling satellite traffic through space and time. In: ACM MobiCom 2023, New York, NY, USA (2023). https://doi.org/10.1145/3570361.3592521
2. Tang, Q., Fei, Z., Li, B., Han, Z.: Computation offloading in leo satellite networks with hybrid cloud and edge computing. IEEE Internet Things J. **8**(11), 9164–9176 (2021)
3. Lin, X., Liu, A., Han, C., Liang, X., Pan, K., Gao, Z.: Leo satellite and UAVs assisted mobile edge computing for tactical ad-hoc network: a game theory approach. IEEE Internet Things J. **10**(23), 20 560–20 573 (2023)
4. Li, D., Liu, X., Yin, Z., Cheng, N., Liu, J.: CWGAN-based channel modeling of convolutional autoencoder-aided SCMA for satellite-terrestrial communication. IEEE Internet Things J. **11**(22), 36 775–36 785 (2024)

5. Wang, H., Wang, L., Wu, W.: Resource allocation and intelligent trajectory optimization for UAV-assisted semantic communication system. In: 2023 IEEE 23rd International Conference on Communication Technology (ICCT), pp. 1370–1374 (2023)
6. Ren, Y., et al.: Sateriot: high-performance ground-space networking for rural IoT. Association for Computing Machinery, New York (2024). https://doi.org/10.1145/3636534.3690659
7. Li, D., et al.: Near-pareto multiobjective routing optimization for space–air–sea-integrated networks. IEEE Internet Things J. **12**(8), 11 194–11 204 (2025)
8. Zhu, G., et al.: Semantic-based channel state information feedback for AAV-assisted ISAC systems. IEEE Internet Things J. **12**(5), 4981–4991 (2025)
9. Lakew, D.S., Tran, A.-T., Dao, N.-N., Cho, S.: Intelligent self-optimization for task offloading in leo-mec-assisted energy-harvesting-UAV systems. IEEE Trans. Netw. Sci. Eng. **11**(6), 5135–5148 (2024)
10. Zhang, H., Tao, M., Sun, Y., Letaief, K.B.: Improving learning-based semantic coding efficiency for image transmission via shared semantic-aware codebook. IEEE Trans. Commun. **73**(2), 1217–1232 (2025)
11. Liao, Q., Tung, T.-Y.: Adasem: adaptive goal-oriented semantic communications for end-to-end camera relocalization. In: IEEE INFOCOM 2024 - IEEE Conference on Computer Communications, pp. 1111–1120 (2024)
12. Cheng, R., Wu, N., Le, V., Chai, E., Varvello, M., Han, B.: Magicstream: bandwidth-conserving immersive telepresence via semantic communication, pp. 365–379 (2024). https://doi.org/10.1145/3666025.3699344
13. Zhang, Z., Zhao, J., Huang, C., Li, L.: Learning visual semantic map-matching for loosely multi-sensor fusion localization of autonomous vehicles. IEEE Trans. Intell. Veh. **8**(1), 358–367 (2023)
14. Zhang, P., Liu, Y., Song, Y., Zhang, J.: Advances and challenges in semantic communications: a systematic review. Natl. Sci. Open **3**(4), 20230029 (2024). http://www.sciengine.com/publisher/Science%20China%20Press/journal/National%20Science%20Open/3/4/10.1360/nso/20230029
15. Bourtsoulatze, E., Burth Kurka, D., Gündüz, D.: Deep joint source-channel coding for wireless image transmission. IEEE Trans. Cogn. Commun. Netw. **5**(3), 567–579 (2019)
16. Li, Y., et al.: Stable hierarchical routing for operational leo networks. Association for Computing Machinery, New York (2024). https://doi.org/10.1145/3636534.3649362
17. Hu, H., Zhu, X., Zhou, F., Wu, W., Hu, R.Q.: Semantic-oriented resource allocation for multi-modal UAV semantic communication networks. In: GLOBECOM 2023 - 2023 IEEE Global Communications Conference, pp. 7213–7218 (2023)
18. Liao, Q., Tung, T.-Y.: Adasem: adaptive goal-oriented semantic communications for end-to-end camera relocalization. In: IEEE INFOCOM 2024 - IEEE Conference on Computer Communications, pp. 1111–1120 (2024). https://api.semanticscholar.org/CorpusID:267027921
19. Bian, C., Shao, Y., Gündüz, D.: Deepjscc-1++: robust and bandwidth-adaptive wireless image transmission. In: GLOBECOM 2023 - 2023 IEEE Global Communications Conference, pp. 3148–3154 (2023)
20. Zhang, W., Zhang, H., Ma, H., Shao, H., Wang, N., Leung, V.C.M.: Predictive and adaptive deep coding for wireless image transmission in semantic communication. IEEE Trans. Wirel. Commun. **22**(8), 5486–5501 (2023)

21. Li, D., et al.: Dual network computation offloading based on DRL for satellite-terrestrial integrated networks. IEEE Trans. Mob. Comput. **24**(3), 2270–2284 (2025)
22. Han, D., Pan, X., Han, Y., Song, S., Huang, G.: Flatten transformer: Vision transformer using focused linear attention. In: IEEE/CVF International Conference on Computer Vision (ICCV), pp. 5938–5948 (2023)
23. Liu, Z., et al.: Swin transformer: Hierarchical vision transformer using shifted windows. In: 2021 IEEE/CVF International Conference on Computer Vision (ICCV), pp. 9992–10 002 (2021)
24. Pan, M., Hu, J., Yuan, J., Liu, J., Su, Y.: An efficient blind doppler shift estimation and compensation method for leo satellite communications. In: 2020 IEEE 20th International Conference on Communication Technology (ICCT), pp. 643–648 (2020)

A Low Migration and Low Energy Consumption Fog Computing Workflow Scheduling Framework for Multiple Constraints

Tianqi Zhao[1], Wei Duan[2], Li He[2], and Ruihan Hu[1(✉)]

[1] Institute of Corporate and Service Intelligent Computing, Harbin Institute of Technology, Harbin, China
rh.hu@hit.edu.cn
[2] Data Systems Department, ZTE CORPORATION, Nanjing, China
{duanwei,heli}@zte.com.cn

Abstract. Nowadays, mobile device manufacturers employ workflow scheduling. This involves moving data from devices to cloud centers using fog nodes. Such a system facilitates a serverless workflow scheduling scheme. However, existing scheduling schemes have not yet considered the communication problems between cloud and fog nodes, and neglected the heterogeneity of node resource(e.g., CPU, memory, and bandwidth) under limited conditions. To solve these problems, we propose a novel low migration and low energy consumption framework aiming at improving the performance of meta-heuristic algorithms for workflow scheduling in fog computing. Under resource-constrained conditions, we focus on maximizing the task completion efficiency and improving the resource utilisation of the heterogeneous node resource, and consider the communication link latency between the cloud and fog nodes. As a preprocessing part, we propose a "Memory segmentation algorithm" that partitions workflow tasks based on their memory usage patterns. This segmentation strategy reduces the task migration time, and both improves the task allocation accuracy. In addition, we use the "Heterogeneous node resource service score evaluation model" to quantify the service quality and accurately evaluate the scheduling strategy to help the algorithm converge. Comparative and ablation experiments on the publicly available "Bitbrains" dataset suggest that our framework can help the original naive algorithm increase task completion by 50%, reduce task migration times by 27% and reduce energy loss by 12%.

Keywords: Fog Computing · Scheduling Strategy · Multiple Constraints · Quality of Service

1 Introduction

With the rapid increase in the number of IoT devices, a new parallel and distributed computing architecture has emerged, known as mobile edge-cloud computing. This paradigm combines cloud computing and edge devices, utilizing the computing capabilities of fog nodes to provide low-latency services to users. It can also offload computation-intensive tasks to the cloud, enhancing the computational capacity of edge devices[1].

In scientific research and business research, certain computational tasks are not independent; they exhibit temporal dependencies and collectively complete specific service process. We call an ordered and interconnected set of tasks a workflow. In fog computing environments, tasks often require collaborative processing across multiple devices, making task migration an inevitable operation. Poorly designed workflow scheduling algorithms (referred to as "scheduling algorithms") can lead to frequent task migrations, increasing execution delays and network transmission energy consumption for mobile devices, ultimately significantly degrading user Quality of Service (QoS). Therefore, reducing migration count, minimizing energy consumption during scheduling, and improving user QoS are key challenges in fog computing research. Existing workflow scheduling researchs mainly assume homogeneous cloud resources, making it unsuitable for heterogeneous fog computing environments. Moreover, tasks processed in fog environments exhibit high heterogeneity in terms of type, priority, data volume, etc. necessitating consideration of the complexity of heterogeneous computing power distribution[2].

In this paper, we propose "QoS_CEMPF: A Low-Migration, Low-Energy Fog Computing Workflow Scheduling Framework for Multiple Constraints." Our goal is to provide a standardized paradigm, by integrating the update strategies of original algorithms, not only retains their core advantages but also effectively enhances the quality of QoS.

The main contributions of this paper are as follows:

We propose "Memory segmentation algorithm", which divides tasks into small parts, and the corresponding containers can be dynamically adjusted according to the specific memory requirements of each segment, thereby improving the utilization rate of the computing resources of the host machine.

We propose "Heterogeneous node resource service score evaluation model", which extracts and merges workflow features through convolutional layers, thenfurther refines them through linear layers, and finally outputs parameters to evaluate the algorithm's QoS.

The effectiveness of the framework is thoroughly validated through comparative and ablation experiments with multiple baseline algorithms using the public dataset Bitbrains. After applying the proposed framework, the original algorithm's performance showed significant improvement: task completion count increased by 50%, task migration frequency decreased by 27% and energy consumption loss reduced by 12%.

2 Related Work

2.1 Existing Workflow Scheduling Algorithms Are Mainly Categorized as Heuristic and Meta-Heuristic

Heuristic algorithms

Such schemes typically employ goal-driven optimization strategies, with scheduling objectives including but not limited to average response time, energy consumption, execution cost and resource utilization. For example, Abdullatif[3] et al. sent data to edge devices to use edges for central neural network training. The aim was to maximize the selection of user tasks with the highest data weights. Xiao[4] et al. proposed energy -

saving strategies for data - intensive workflows. They used traditional HEFT and CPOP algorithms, adding virtual data access nodes to assess data access energy costs and find the least - energy path. However, the majority of existing work often models and optimizes only a single scheduling objective, which limits the potential of algorithms to achieve high- quality scheduling in heterogeneous environments.

Meta-heuristic algorithms

Zhang[5] et al. innovatively used the number of Pareto individuals and Pareto entropy in the GA algorithm to distinguish the evolutionary state of NSGA-ll and, based on this, proposed an adaptive multi-objective constrained scheduling problem. They also introduced mechanisms such as random perturbations and local search to enhance search diversity and global exploration capabilities.Wang[6] et al. proposed CLOSURE, which uses an attack-defense game theory approach for scheduling. Unlike most existing homogeneous resource assumptions, it has been proven to effectively manage heterogeneous devices through Nash equilibrium, which is crucial in fog computing. Additionally, some researchers have begun exploring the combination of multiple meta-heuristic algorithms to address scheduling problems[7].

Overall, heuristic algorithms utilized the specific information of the problem to conduct efficient searches. Especially when solving complex problems, they could generate high-quality results. However, due to their reliance on specific rules, they are prone to falling into local optima and are sensitive to minor changes in the problem. In contrast, metaheuristic algorithms had a wider range of applicability and can provided good approximate solutions for NP problems. Although it was designed for general problems, if theimplementation lacks consideration of problem characteristics, even originally well-performing algorithms might exhibit significant performance degradation, and the involved parameter tuning is relatively complex.

3 Fog Computing Workflow Scheduling Issues and Solutions

3.1 Computational Framework for Workflow Tasks

QoS-CEMPF is suitable for the typical "Edge-Fog-Cloud" framework[8]. This architecture consists of three main components: IoT devices, fog computing nodes, and cloud computing centers, as shown in Fig. 1. Details are listed as follows:

IoT Devices: located at the edge of the network and are categorized into sensors and performer. They are connected to nearby gateways via Bluetooth, Wi-Fi, responsible for environmental perception and data collection. They usually lack the capacity for data processing.

Fog Nodes: at the edge proximity and primarily handle latency-sensitive tasks, featuring functions such as task scheduling, resource management, and data storage. The three functionally specialized nodes work collaboratively to ensure the reliable performance of applications: **1) General computing nodes:** referred to as "computing nodes," equipped with different computational resources like memory, disk capacity, and CPU processing power. When resources are sufficient, they can execute tasks in parallel to enhance system efficiency. **2) Agent nodes:** responsible for assigning tasks to computing nodes and monitoring their working status such as resource utilization,system stability and so on to ensure system reliability and security. **3) Storage nodes:** enable persistent data storage, facilitate data sharing among nodes, and support system data recovery, thereby improving system security.

Cloud Computing Centers: have high storage capacity and processing speed, handling compute-intensive tasks that fog nodes cannot complete. Although they possess greater computational power, they are usually placed at a relatively far distance from the edge side which brings non-negligible transmission latency.

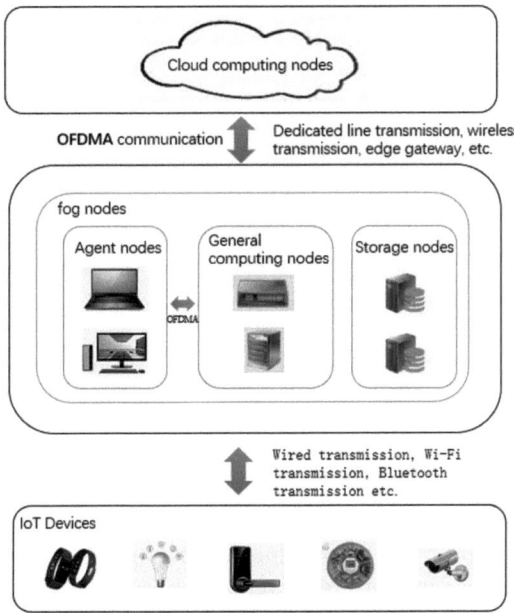

Fig. 1. "Edge-Fog-Cloud" framework

3.2 Cloud-Edge Channel Communication Model

According to Shannon's theorem, the channel transmission rate R (bits per second) is:

$$R = B\log_2(1 + \frac{S}{N}) \#(1)$$

B is the bandwidth, S is the signal's average power, and N is the noise power. On this basis, suppose that at time t_r, the size of p_k that is allocated from P to the computing node h_s is M_k (bits). Then the uplink channel transmission time T_{P,h_s}^{up} is expressed by formula (2), where R_{P,h_s} represents the channel transmission rate between the agent node and h_s.

$$T_{P,h_s}^{up} = \frac{M_k}{R_{P,h_s}} \#(2)$$

When p_k is migrated from h_s to the cloud for computation, it will be allocated a transmission gap τ_k. The node should complete the data transmission of p_k within τ_k. That is to say, it is necessary to satisfy that the uplink channel transmission time $T_{h_s,cloud}^{up}$ of p_k is less than the time slot τ_k allocated to this task by the cloud, as shown in formula (3). Among them, $T_{h_s,cloud}^{up} = \frac{M_k}{R_{h_s,cloud}}$.

$$T_{h_s,cloud}^{up} \leq \tau_k \#(3)$$

Assume task p_k is assigned to the node h_s (distanced $_{h_s,cloud}$ to cloud), the time its migration to the cloud is:

$$T_{h_s,cloud}^{spread} = \frac{d_{h_s,cloud}}{v_{light}} \#(4)$$

v_{light} denotes speed of light. Since the cloud is far away from h_s, this paper assumes that the transmission medium between h_s and the cloud is optical fiber. To avoid an excessively long total response time, the total execution time time$_{total}$ needs to be limited within the deadline deadline$_k$ given by the user. Based on the model assumptions in Sect. 3.1, we ignore the propagation delay from the agent node P to h_s and only consider the channel transmission time T_{P,h_s}^{up} from P to h_s, the execution time T_k^{cmp} of p_k on h_s, the channel transmission time $T_{h_s,cloud}^{up}$ from h_s to the cloud, and the propagation delay $T_{h_s,cloud}^{spread}$. Therefore, the following inequality should hold:

$$2T_{P,h_s}^{up} + T_k^{cmp} + 2T_{h_s,cloud}^{up} + 2T_{h_s,cloud}^{spread} \leq deadline_k \#(5)$$

3.3 Fog Heterogeneous Node Resource Construction Problem

At time t_r, the agent node merges the task sequence $\{p_1, p_2, ..., p_n\}$ with the pending task sequence $p_1^*, p_2^*, ..., p_k^*\}$ for joint scheduling. The task p_k, which awaits fine-grained partitioning, consists of several task segments arranged sequentially in time. Let the available computing nodes be $\{h_1, h_2, ..., h_m\}$. P is responsible for scheduling $\{p_1, p_2, ..., p_n, p_1^*, p_2^*, ..., p_k^*\}$ to be executed on $\{h_1, h_2, ..., h_m\}$. The scheduling algorithm needs to find the best target $h_s \in \{h_1, h_2, ..., h_m\}$ for $p_k \in \{p_1, p_2, ..., p_n, p_1^*, p_2^*, ..., p_k^*\}$. To this end, we construct an independent variable matrix x for the objective optimization function for heterogeneous node resources. The element x_{ks} represents the probability that p_k is scheduled to h_s. For the m available nodes, the probability sequence of p_k being scheduled is $\{x_{k1}, x_{k2}, ..., x_{km}\}$.

We aim to minimize the overall completion time for temporal service scores of tasks. Moreover, it is taken into account that for some tasks p_k, although their execution time is long, their deadlines are also long. Eventually, the optimization objective is defined as minimizing $\frac{time_{total}}{deadline_k}$. In summary, the problem of constructing heterogeneous node resources can be formulated as follows:

$$min_{x_{ks}} \left(\sum_k \frac{T_k^{cmp} + 2\sum_s x_{ks}(T_{P,h_s}^{up} + T_{h_s,cloud}^{up} + T_{h_s,cloud}^{spread})}{deadline_k} \right) \#(6)$$

4 Low-Migration, Low-Energy Fog Computing Workflow Scheduling Framework

However, fog computing faces the following constraints in practical implementation:

(1) Computation and Communication Latency: This paper focuses on the task execution time on computing nodes, the channel transmission time between agent nodes and computing nodes as well as between computing nodes and the cloud, and the propagation delay caused by geographical distance.
(2) The time-varying nature of memory in workflows: This paper conducts a detailed analysis of the memory usage status of workflows in different time periods, covering the memory occupation at different moments, the time threshold of memory occupation, and the maximum number of task segments.

Based on the above considerations, we propose the "Quality of Service-oriented, Low-migration and Low-energy Consumption Fog Computing Workflow Scheduling Framework" (QoS-CEMPF), aiming to improve the performance of the scheduling algorithm when facing these challenges. QoS-CEMPF can effectively integrate metaheuristic algorithms such as "Firefly Optimization", "Cuckoo Optimization", and "Moth Search". As shown in Fig. 2: For problem (1), the framework increases the probability of selecting an effective solution. When the local state reaches a non-convergent state, a better state is selected by comparing the service scores and computing nodes are allocated accordingly; for problem (2), a memory segmentation algorithm is adopted to appropriately partition all tasks of the workflow according to their memory occupation characteristics; for problem (3), in order to solve the problem that it is difficult to quantitatively evaluate the scheduling algorithm, a Heterogeneous node resource service score evaluation model is introduced, which assesses the system's current QoS from multiple perspectives and improves the accuracy of service evaluation.

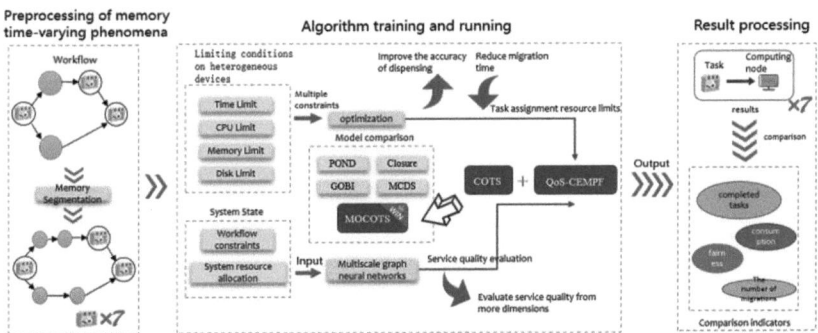

Fig. 2. Quality-of-Service (QoS)-Enhanced Multi-Constrained Task Segmentation and Scheduling Framework

4.1 Container Memory Segmentation Method Under Time-Varying Memory Phenomena

In fog computing, when not involving massive IoT devices, the computational tasks received by nodes from agent nodes at each moment are generally identical or similar. Thus, a complete workflow analysis can be run in advance to observe memory usage. We found that fog computing frequently exhibits "time-varying memory phenomena." In the Bitbrains dataset, for example, roughly 80% of tasks typically maintain stable memory usage. However, at specific moments or intervals, this usage can fluctuate significantly. Such variations may lead to improper resource allocation, thereby affecting overall execution efficiency and resource utilization. Two typical scenarios are illustrated in Figs. 3 and 4. First, Task 20 shows almost zero memory occupancy in [0, 300], but at t = 300, it suddenly spikes sharply. In contrast, Task 36 undergoes substantial memory usage fluctuations over the continuous interval [5300, 7800].

In response to the above phenomena, we propose the "memory segmentation method". When the agent node receives the tasks transmitted from the gateway node, the scheduler will segment the tasks in time according to their memory usage.

As shown in Fig. 3, the tasks are divided into small parts. Containers can dynamically adjust based on each part's memory needs. This enhances the utilization of the host's computing resources. The method operates by carefully managing resource allocation. First, periods of high memory usage are assigned to nodes with ample resources. Concurrently, those with low memory usage can be scheduled on any available node. This targeted allocation ensures that low-memory tasks do not tie up valuable resources. As a result, we maintain the capacity to deploy tasks with high memory requirements whenever needed.

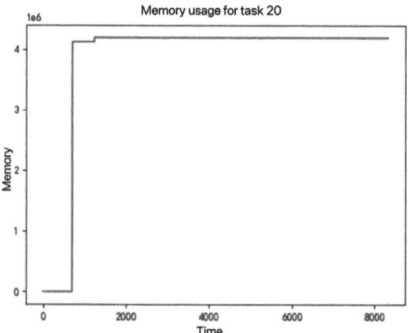

Fig. 3. Memory usage for task 20(left) **Fig. 4.** Memory usage for task 36(right)

First, this paper argues that the k-th task p_k over the time interval $[t_1, t_2]$. Within this interval, any two memory measurements, m_1 and m_2, taken at arbitrary times, must meet a requirement. This requirement is expressed as $|m_1 - m_2| < \varepsilon$, where ε is a hyperparameter. After task segmentation, each moment corresponds to a segmented part. Therefore, the specific implementation idea is to assign a segment label to each moment of p_k. Considering the memory occupation at the current moment and the previous moment, if the absolute difference is within ε, the label is the same as the previous segment. Otherwise, it is different. Here, the dynamic programming method can be used for calculation, as follows:

$$dp[i] = \begin{cases} dp[i-1] + 1, |RAM[i] - RAM[i-1]| \geq \varepsilon \\ dp[i-1], |RAM[i] - RAM[i-1]| < \varepsilon \end{cases} \#(7)$$

However, the above method has limitations. As shown in Fig. 5, there may be "cusps" (at the red pentagrams) in the memory occupation curve of the task. Such mutations are very short-lived. If the duration of the mutation is less than the given threshold $T_{minperiod}$, division should not be performed at this point. Therefore, it is necessary to further process the labels, as detailed in Algorithm 1: First, count the duration of each segment. If the duration of a certain segment is less than $T_{minperiod}$, then the segment divided according to formula 7 is invalid and needs to be merged with the previous segment. Then, after the label update, the point q with different labels before and after is identified as a breakpoint, and thus the breakpoint sequence $[q_1, q_2, ..., q_n]$ is obtained. Finally, each pair of adjacent breakpoints $[q_i, q_{i+1}]$ defines the i-th segment of the task($1 \leq i \leq n$).

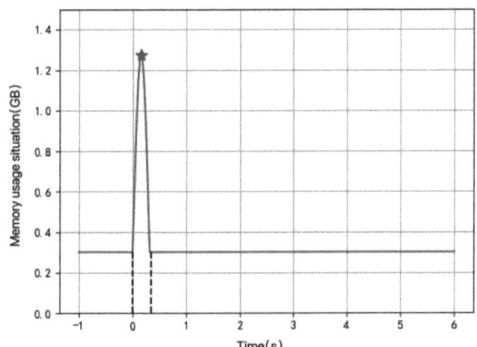

Fig. 5. Schematic diagram of the cusp point of the memory occupancy curve

---

Algorithm 1. Filter Short Segments(*Length* < $T_{minperiod}$)

Input: Segment labels *dp* from Algorithm 1
Output: Updated segment labels *dp*

(continued)

(*continued*)

Algorithm 1. Filter Short Segments(*Length* < $T_{minperiod}$)

1. Len = execution_time
2. **if** (dp[len-1] > *MaxLabel*) **then**
3. **For** i ∈ [0,len-1] **do**
4. dp[i] = 1
5. End for
6. Return
7. end if
8. Count occurrences of each segment label
9. Pos = []
10. **for** i ∈ [1,len-2](reverse iteration) **do**
    if (*dp*[i] != *dp*[i-1]) and (occurrences[*dp*[i]] > $T_{minperiod}$) **then**
12. Cnt += 1
13. **End if**
14. *dp*[i] = cnt
    if *dp*[i] != *dp*[i + 1] **then**
16. Pos.append(i)
17. End if
18. End for

Still taking the tasks in Figs. 3 and 4 as examples, Figs. 6 and 7 show the processing results of the segmentation method. Segments sharing the same color represent the same subtask. Each subtask strictly corresponds to the memory occupation situation in terms of time. As shown in Fig. 6, Task No. 20 is divided into three segments, and the duration of each sub-segment satisfies the condition of being greater than $T_{minperiod}$. As shown in Fig. 7, the occupation curve of Task No. 36 in the interval [5300, 7800] fluctuates significantly and is not divided into sub-segments.

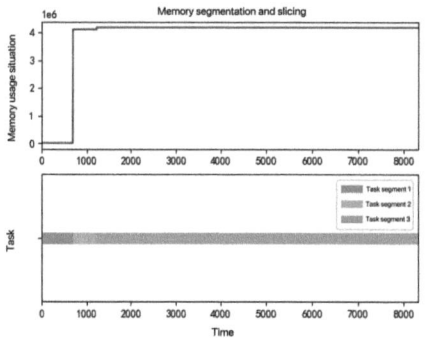
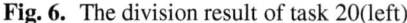
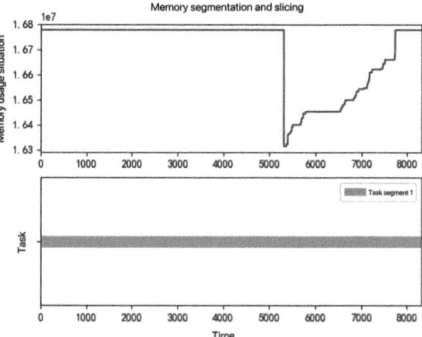

Fig. 6. The division result of task 20(left)

Fig. 7. The division result of task 36(right)

4.2 Heterogeneous Node Resource Service Score Evaluation Model

We utilize the Graph Convolutional Network (GCN) to extract the features of the workflow and adopt a linear layer to extract the workflow and resource consumption. We establish the "Heterogeneous Node Resource Service Score Evaluation Model", which is divided into three stages according to its functions, as shown in Fig. 8: (1) The multi-order feature merging stage: It adopts a three-layer structure. As shown in the blue part, graph convolutional layers with different scales containing 50, 150, and 200 neurons respectively are used to extract features with different connection sparsities and merge them; (2) The feature extraction stage: In the light green part, the merged features are further mined through a linear layer. (3) The service score evaluation stage: It has a dual-branch structure, which outputs the memory parameter θ_E and the time-consuming parameter θ_T respectively to evaluate the QoS of the algorithm.

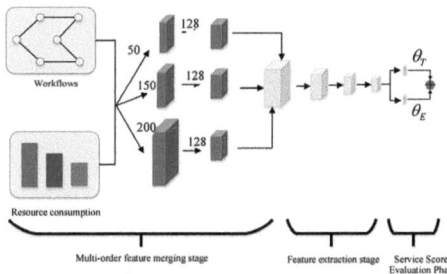

Fig. 8. Heterogeneous Node Resource Service Score Evaluation Model

Finally, the KL divergence is used as the loss function to compare the evaluation score ψ_{score} with the cloud state data distribution ψ_{cloud}, so as to ensure the accuracy and reliability of the evaluation results. The calculation formulas are as follows:

$$\psi_{score} = \theta_E \cdot O_1 + \theta_T \cdot O_2 \#(8)$$

$$loss = KL(\psi_{score}, \psi_{cloud}) \#(9)$$

According to formula (9), the calculated evaluation score is compared with the state of the workflow that has been scheduled in the cloud, and the difference between the evaluation score and the scheduled workflow is obtained.

4.3 Solving Strategy of the Scheduling Framework

According to the basic update strategy of the meta-heuristic algorithm, we record the update from state x^i to state x^j. As shown in Eq. (10), an incremental update method is adopted, where β is the update coefficient.

$$x^{i*} = x^i + \beta(x^j - x^i) \#(10)$$

The updated result x^{i*} should satisfy all the constraint conditions mentioned in Section III.C. If x^{i*} does not satisfy the constraints, then a new state x^{i_new} should be selected between x^i and x^{i*}. At this time, the algorithm used is the target optimization algorithm for constructing resources of heterogeneous fog nodes mentioned in Section III.C. Different from the original algorithm, in order for the new state to be between the two states, we introduce an additional constraint condition. As defined in formula (11), any value in the matrix of x^{i_new} needs to be between x^i and x^{i*}.

$$\forall k, smin(x_{ks}^i, x_{ks}^{i*}) \le x_{ks}^{i\_new} \le max(x_{ks}^i, x_{ks}^{i*}) \#(11)$$

All steps in this subsection can be summarized as Algorithm 2.

Algorithm 2. Decision-Solving Strategy for Scheduling Framework
---
Input: Original state x^i and Target migration state x^j
Ouput: New state x^{i_new}

1. Compute x^{i*} using Eq. (10)
2. **if** x^{i*} satisfies constraints in Sect. 3 **then**
3. **Return** x^{i*}
4. End if
5. Randomly initialize state array $\{x^1, x^2, \ldots, x^n\}$
6. **While** iter < max_iterations **do**
7. **For** i ∈ [1,n] **do**
8. Randomly select j ∈ [1,n] where $j \ne i$
9. **if** $f_i < f_j$ **then**
10. Generate new solution x^{i_new} for x^i
11. **End if**
12. End while
13. **Return** x^{i_new}

After obtaining the optimal solution x, in order to get the computing node h_s assigned to p_k, the general idea is to select the subscript label corresponding to the maximum value in the probability sequence $\{x_{k1}, x_{k2}, \ldots, x_{km}\}$ for p_k as the candidate node. However, since $x_{ks} \in [0,1]$, this may lead to conflicts in task allocation. For example, the allocation probabilities of p_1 and p_2 for the y-th node may both be 0.5, and they are both the maximum values in the sequence. But when actually allocating p_1 and p_2 to node y, the requirements of these two tasks cannot be satisfied simultaneously. For the sake of rigor, as shown in Algorithm 3:

Algorithm 3. Actual Resource Allocation Function

Input: Optimal solution x and tasks to be allocated *tasks*
Output: Array representing resource allocation decisions

(continued)

(continued)

Algorithm 3. Actual Resource Allocation Function
1. Sort the x vector of the task p_k in descending order
2. Sort the tasks in ascending order of their deadlines.
3. **For** each sorted task i **do**
4. **For** each computing node j sorted by x **do**
5. **if** check(i, j)**then**
6. Decisions.append((i, j))
7. **End if**
8. End for
9.end for

The probability sequence needs to be sorted in descending order, and a priority is defined for the tasks, allowing the tasks in the front to preferentially select computing nodes according to the probability sequence. As long as the resource check by the check function passes, node allocation can be carried out. Doing so can not only improve the scheduling efficiency but also meet the limitations of actual resources. According to the characteristics of the tasks, this priority is defined as the deadline, that is, a task with a shorter deadline has a higher priority. It is worth noting that the reason for not considering the limitations of CPU and RAM is that when a task is urgent, it must be deployed on the node preferentially and given sufficient resources.

5 Experimental Design

The experiments in this paper are based on the public dataset Bitbrains[9], provided by an enterprise-level service provider of the same name. The dataset tracks 500 tasks hosted in a cloud computing center over 2–3 months, recording their memory and CPU usage.

We assume that these samples are the tasks that an agent node may receive at a certain moment, and we consider the recorded memory and CPU usage as a complete execution process on a computing node. Meanwhile, three types of virtual nodes are set. The algorithm is executed for 100 rounds, and 50 virtual nodes are set up to simulate the situation where tasks arrive at the agent node.

5.1 Practical Application of the Framework Compared to Baseline Algorithms

We will prove through simulation experiments that: for the workflow scheduling in fog computing, QoS-CEMPF can effectively improve the performance of the metaheuristic algorithm. Taking the "Naive Cuckoo" algorithm after applying the framework (abbreviated as MOCOTS) as an example. The evaluation metrics include: execution throughput per time segment, task migration frequency, energy consumption overhead, etc.

The baseline algorithm is as follows:

COTS[10]: the naive cuckoo algorithm; CLOUSURE: constructs an attack-defense model to determine the optimal mixed defense strategy's probability distribution via

Nash equilibrium; POND[11]: Utilize reinforcement learning and Lyapunov optimization techniques for scheduling, with a primary focus on resource budget and fairness; GOBI: employs gradients to optimize QoS which requires minimal computation and thus fits resource-limited systems; Random Search[12]: substitutes exhaustive enumeration with randomized selection from all available combinations.

Execution Throughput per Time Segment.

We first counted the total number of tasks executed by computing nodes in each time period, as shown in Fig. 9. In the initial stage, MOCOTS executed fewer tasks but still more than the baseline model. Starting from the 40th segment, MOCOTS began to outperform other algorithms in terms of task execution.

From the 80th segment onward, tasks in the system began to decrease, and the number of tasks executed by all algorithms also started to decline.

We also count the average total number of tasks completed by computing nodes during various time periods, as shown in Fig. 10. The average total task completion count of MOCOTS is significantly higher than other algorithms, demonstrating that our proposed "segmentation algorithm" improves resource utilization.

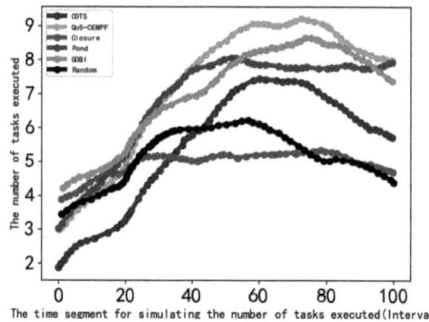

Fig. 9. The curve chart of the number of tasks executed by each algorithm changing over time (left)

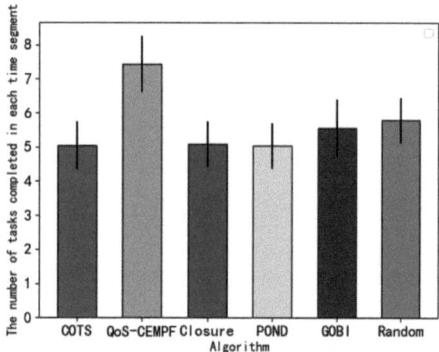

Fig. 10. The average number of tasks completed by each algorithm(right)

Task Migration Frequency.

We also recorded the total number of task migrations between computing nodes during each time period and the total migration time across all nodes in all time intervals, as illustrated in Fig. 11 and Fig. 12.

In Fig. 11, it is easy to observe that the random algorithm has the highest number of migrations, but its total migration time is less than POND. This may be because POND requires a longer average time per migration (affected by transmission distance, channel bandwidth, etc.). In Fig. 12, MOCOTS exhibits the lowest number of migrations and migration time, as we modeled the communication constraints of fog computing and introduced multiple optimization constraints to avoid unnecessary migrations caused by resource and deadline factors.

Energy Consumption Overhead.

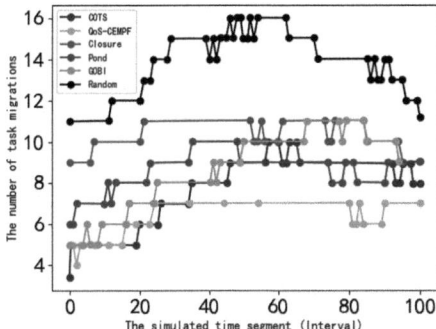

Fig. 11. The curve chart of the number of task migrations changing over time(left)

Fig. 12. The total migration time of tasks in 100 rounds (right)

During experiments, we recorded the CPU and RAM usage of the algorithms. Figures 13 and 14 show the average RAM and CPU utilization across all nodes, clearly indicating that MOCOT has the lowest consumption. Further, assuming the energy calculation method is model power multiplied by unit time.

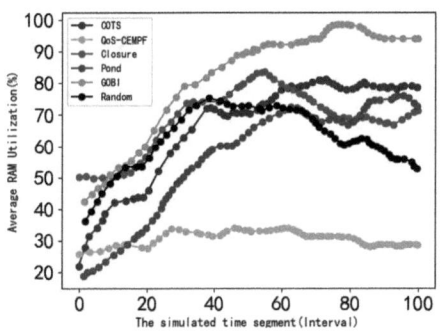 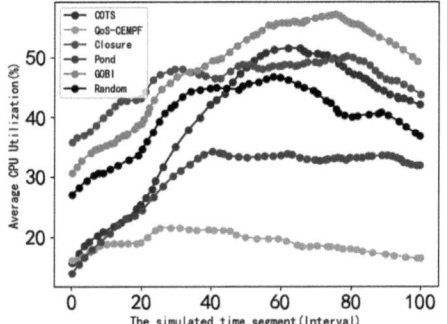

Fig. 13. Average RAM Utilization Rate(left) **Fig. 14.** Average CPU Utilization Rate(right)

Figure 15 demonstrates that MOCOTS also has the lowest energy consumption. This is because we split tasks based on memory requirements, reducing resource occupancy on deployment nodes. Thus, MOCOTS significantly lowers resource consumption compared to baseline algorithms. Lower RAM and CPU usage not only helps reduce system energy consumption but also leaves more resources available for nodes to handle newly arrived tasks. Therefore, our framework ensures high task completion and execution rates while effectively reducing costs for the original algorithm.

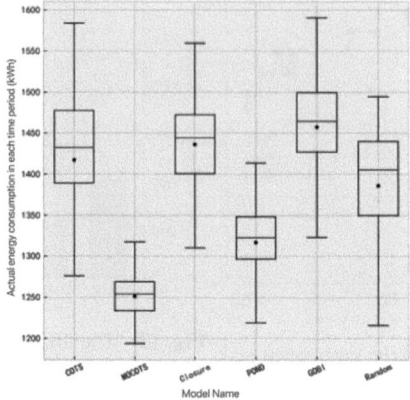

Fig. 15. The average energy consumption in each time period

5.2 Ablation Experiment

Based on MOCOTS, consider the necessity of each component in the framework:

Remove the memory segmentation part and consider the algorithm's performance when there is no task segmentation (in this case, the algorithm is denoted as "w/o task segmentation").

Remove the target optimization part and consider the situation where there are no restrictions when allocating resources (in this case, the algorithm is denoted as "w/o resource restriction").

Remove the part of the "Heterogeneous Node Resource Service Score Evaluation Model" and use the original evaluation function of the algorithm to compare the differences between different evaluation functions (in this case, the algorithm is denoted as "w/o service evaluation").

With other configurations remaining unchanged, the performance of COTS and MOCOTS is also compared. The experimental results are shown in the following table:

TABLE 1 The results of the ablation experiment

algorithm unit	Execution Throughput per Time Segment (pieces/ time)	Task Migrations Frequency (times/unit time)	Energy Consumption Overhead (kilowatt-hours per unit time)
MOCOTS	7.5	6.0	1252.34
w/o task segmentation	5.5	8.2	1453.28
w/o service evaluation	7.0	9.0	1426.26
w/o service evaluation	6.7	8.5	1370.92
COTS	5.0	8.3	1425.85

Analyzing Table 1, the following conclusions can be drawn:

When the task segmentation part is removed from the framework, the energy consumption loss increases significantly. This is because the memory utilization efficiency of tasks is not high, and tasks that require high memory occupancy cannot be immediately allocated the required computing resources. As a result, they wait in the system for too long, leading to an increase in energy consumption.

When the resource restriction part is removed from the framework, the number of task migrations increases significantly, and the fairness decreases slightly. This is because when there are no resource restrictions, the allocated computing nodes may not meet the requirements of the task, resulting in incorrect allocation and ultimately leading to migrations.

By comparing COTS and MOCOTS, it can be found that when the original meta-heuristic algorithm applies QoS-CEMPF, all indicators are significantly improved. Specifically, the number of completed tasks increases by 50%, the number of task migrations decreases by 27%, and the energy consumption loss is reduced by 12%.

In conclusion, the effectiveness and practicality of the framework proposed in this paper can be seen.

6 Conclusion

In this paper, by investigating the meta-heuristic algorithms used in workflow scheduling, we sort out the similar general steps of these methods and summarize as the "Quality-of-Service (QoS)-Enhanced Multi-Constrained Task Segmentation and Scheduling Framework". For the fog computing scheduling problem, this framework can serve as a paradigm for general meta-heuristic algorithms. As long as the update strategy of a given algorithm is plugged into the framework, the QoS can be enhanced on the basis of the original algorithm. Firstly, the tasks are segmented according to the memory occupancy. Then, the target optimization is taken as the constraint for resource allocation. After that, the evaluation model of the resource service score of heterogeneous nodes is used as the QoS evaluation standard. Finally, an efficient scheduling scheme is obtained through solving. A large number of experiments on real-world datasets have all proved the effectiveness of the QoS-CEMPF.

References

1. Tuli, S., Mahmud, R., Tuli, S., et al.: Fogbus: a blockchain-based lightweight framework for edge and fog computing. Journal of Systems and Software 154, 22–36 (2019)Author, F., Author, S.: Title of a proceedings paper. In: Editor, F., Editor, S. (eds.) CONFERENCE 2016, LNCS, vol. 9999, pp. 1–13. Springer, Heidelberg (2016)
2. Peng, Q., Xia, Y., Zheng, W.: A decentralized online edge task scheduling and resource allocation method. Chinese Journal of Computers **45**(7), 1462–1477 (2022)
3. Albaseer, A., Abdallah, M., Al-Fuqaha, A., et al.: Data-driven participant selection and bandwidth allocation for heterogeneous federated edge learning. IEEE Transactions on Systems, Man, and Cybernetics: Systems (2023)
4. Xiao, X., Hu, Z., Qu, X.: An energy-aware scheduling strategy for data-intensive workflows. Journal on Communications **36**(1), 153–162 (2015)

5. Zhang, M., Li, H., Liu, L., et al.: An adaptive multi-objective evolutionary algorithm for constrained workflow scheduling in clouds. Distributed and Parallel Databases **36**, 339–368 (2018)
6. Wang, Y., Guo, Y., Guo, Z., et al.: CLOSURE: a cloud scientific workflow scheduling algorithm based on attack–defense game model. Futur. Gener. Comput. Syst. **111**, 460–474 (2020)
7. Li, H., Wang, D., Zhou, M.C., et al.: Multi-swarm co-evolution based hybrid intelligent optimization for bi-objective multi-workflow scheduling in the cloud. IEEE Trans. Parallel Distrib. Syst. **33**(9), 2183–2197 (2022)
8. Xu, M., Fu, Z., Ma, X., et al.: From cloud to edge: a first look at public edge platforms. pp. 37–53 (2021)
9. Liu, X., et al.: POND: pessimistic-optimistic online dispatch. arXiv preprint arXiv:2010.09995 (2020)
10. Adhikari, M., Amgoth, T., Srirama, S.N.: A survey on scheduling strategies for workflows in cloud environment and emerging trends. ACM Computing Surveys (CSUR) **52**(4), 1–36 (2019)
11. Tuli, S., Poojara, S.R., Srirama, S.N., Casale, G., Jennings, N.R.: COSCO: container orchestration using co-simulation and gradient based optimization for fog computing environments. IEEE Trans. Parallel Distrib. Syst. **33**(1), 101–116 (2022)
12. Ahmad, M.A., Mok, R., Ismail, R.M.T.R., Nasir, A.N.K.: Model-free wind farm control based on random search. In: 2016 IEEE International Conference on Automatic Control and Intelligent Systems (I2CACIS), pp. 131–134. IEEE, Selangor, Malaysia (2016)

A Method for Recognition and Analysis of Industrial Sewing Machine Operating States Based on Edge Computing

Huojin Xie and Yangbo Wu[✉]

Faculty of Electrical Engineering and Computer Science, Ningbo University, Ningbo, China
{2311100261,wuyangbo}@nbu.edu.cn

Abstract. With the continuous advancement of the digital transformation of the manufacturing industry, traditional industrial equipment faces many challenges in data collection and status monitoring. This paper focuses on the intelligent transformation of sewing machines in the clothing manufacturing industry and proposes a sewing machine working state recognition and analysis method based on edge computing. The current mainstream solution usually relies on uploading the collected sewing machine working data to the cloud for processing. However, this method has strong dependence on the network and has problems such as high bandwidth occupancy, large latency and data loss. To this end, this paper designs and implements an edge computing platform based on the embedded Linux system architecture. The Allwinner V3S chip is selected to realize local real-time processing of sewing machine working data. At the software level, this paper builds an algorithm based on one-dimensional convolutional neural network (1D-CNN) to identify the sudden change points in the working current waveform, thereby judging the completion of the process. The algorithm model is trained using a supervised learning method, by introducing a loss function and applying the gradient descent method to optimize the training results, the model's generalization ability is enhanced while ensuring high recognition accuracy. Results show that the proposed model has a good deployment effect on embedded edge computing devices and has high real-time and stability. This study provides a practical solution for the intelligent upgrading of traditional industrial equipment, which has important practical significance for the digital and intelligent transformation of the manufacturing industry.

Keywords: sewing machine · status recognition · embedded system · edge computing · intelligent manufacturing

1 Introduction

In recent years, manufacturing enterprises have encountered many problems during the process of industrial digital transformation. How to count the working information of old industrial equipment is an important issue in the process of industrial digital transformation [1]. Therefore, how to carry out digital transformation for these manufacturing enterprises is particularly important in the era of industrial digitalization [2, 3]. The current statistical scheme for detecting the working status of sewing machines in garment factories and statistically analyzing working information is to upload the collected sewing machine working current waveform data to the cloud, analyze these data in the cloud, count the factory's workload, analyze the degree of coordination between the front and rear stages of the assembly line, and monitor the factory's working conditions [4]. However, this method of uploading all data to the cloud for processing not only greatly increases the consumption of network bandwidth, but also has a high risk of data loss due to network delays. Currently, the intelligentization of smart sewing machines is mostly developing in the direction of automation and intelligence, and the idea of using sewing working current waveform to identify sewing machine process information is rarely mentioned.

Using edge computing to replace the previous cloud computing method has become a reliable solution in many application scenarios and a research hotspot in the field of industrial data processing [5]. Therefore, this paper proposes a solution to process the collected sewing machine working current waveform data in real time at the edge, which can not only reduce the dependence on network bandwidth, but also meet the needs of enterprises to count the workload of the assembly line and monitor the working status of the assembly line [6]. Identifying the working current waveform of an industrial sewing machine can obtain the machine's working quantity information, and at the same time can realize the digital transformation of old industrial equipment, greatly improving the factory's statistical efficiency [8]. Therefore, monitoring and accurately identifying the working current waveform data of sewing machines plays a significant role in ensuring the progress of industrial digitalization.

2 Preliminaries

2.1 Computing Platform Design

In order to process the current waveform data of industrial sewing machines in real time, it is necessary to build an edge computing platform for real-time data processing [6]. The edge computing platform is the core device of this project. The device is based on the structure of the embedded Linux operating system. The selection of embedded System on Chip (SOC) can be roughly measured from the following aspects: Central Processing Unit (CPU) category, which determines the amount and availability of development resources; CPU model and operating frequency, mainly to evaluate whether it meets product requirements, generally the performance under the operating frequency; power consumption, mainly including power consumption in general working mode, supported sleep mode, power consumption in various sleep modes; peripheral components, such as Universal Asynchronous Receiver Transmitter (UART), Inter-Integrated Circuit (I2C), Serial Peripheral Interface (SPI), etc., whether they meet the needs of the application.

The hardware structure of the edge computing platform is mainly composed of the following parts: SOC core board module, power management module, Wireless Fidelity (Wi-Fi) module, and various peripheral interface modules. According to the needs of product applications, the chip Allwinner V3S selected by the edge computing platform is a SOC designed by Allwinner Technology. It integrates a variety of peripherals to support a wide range of Internet of Things (IoT) and consumer electronics applications. The power management module uses EA3036QDR, which is a high-performance bipolar complementary symmetric logic buffer and driver. It is a three-way power management chip used for DC 5V power supply or single-cell lithium battery power supply. It has a built-in three-way synchronous power adjustment module, provides light load high-efficiency mode, and a built-in compensation circuit to simplify customer design. The Wi-Fi module uses RTL8723DS, which is mainly used in embedded devices and mobile devices to provide Wi-Fi connection function. The platform hardware is based on the structure of embedded Linux operating system. The system structure of the platform is shown in Fig. 1.

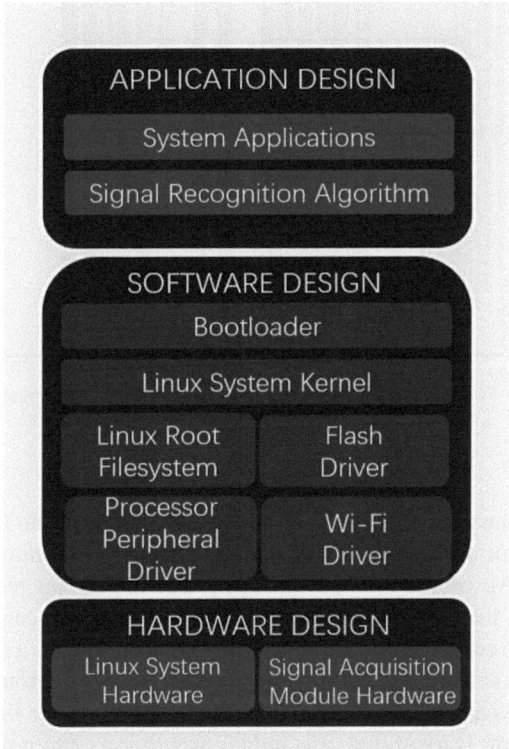

Fig. 1 Edge computing system architecture

The hardware layer structure is mainly composed of the following parts: SOC core board module, power management module, WIFI module and various peripheral interface modules. According to the needs of product applications, the chip Allwinner V3S selected by the edge computing platform is a SOC (system-on-chip) designed by Allwinner Technology. It integrates a variety of peripherals to support a wide range of IoT and consumer electronics applications. The software layer mainly involves the Linux system porting of edge computing devices and the software design of signal acquisition modules, among which the Linux system porting includes kernel porting and the design of various drivers. Finally, the waveform data recognition algorithm is deployed at the application layer to identify the collected waveform data, obtain the information of the corresponding process, and upload the recognized results to the cloud.

2.2 Sewing Machine Operation Time Series Data Analysis

The collected sewing machine working data is plotted into a waveform, and the waveform of part of the collected working data is shown in Fig. 2:

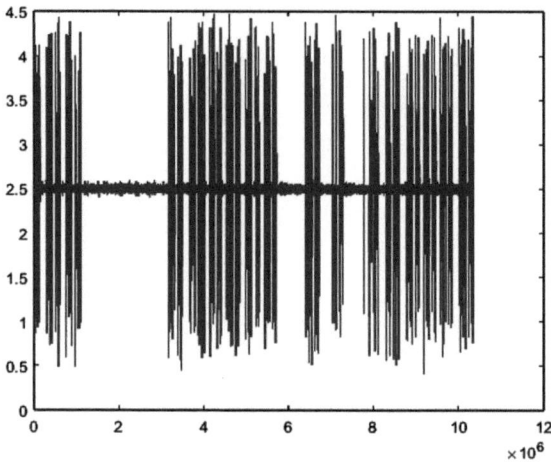

Fig. 2 Working data waveform

The following characteristic information can be obtained from the figure: When the sewing machine is not working, the collected data fluctuates within the range of 2.5 and 0.1, making the curve oscillate more obviously and appear jagged; when the sewing machine is working, the data fluctuates in a larger range, which can clearly reflect their differences. During a certain process, the sewing machine collects a columnar waveform. The waveform of the sewing machine in working state is very obvious, and the waveform of the working state can be identified as a mutation point. After the mutation point of the working waveform is identified, it is judged whether a certain process is completed based on the position information between the mutation points.

The waveform in the above figure is the working waveform of the sewing machine collected by the data acquisition module without any preprocessing. Experiments have found that when the algorithm model directly processes unpreprocessed data, the recognition error is relatively large. However, after data preprocessing, the convergence speed of the model is faster and the recognition accuracy is higher. Therefore, the algorithm model needs to perform preprocessing operations before recognizing the data.

3 Process Identification

3.1 Data Preprocessing

The process identification of sewing machines involves recognizing the collected working current waveforms of sewing machines. The main purpose is to identify the sudden change points in their time series data. Based on the position information between the sudden change points, the completion status of the process can be determined. Since the collected sewing machine operation data is unstable, standardizing the data can make the data differences more stable, making the model training more stable and also making it easier to find the optimal solution. Before identifying time series data, the data is standardized and converted into time series data with a mean of 0 and a standard deviation of 1. The preprocessing part is actually the process of segmenting the traffic data, erasing irrelevant information, and setting the input length. The result of the preprocessing directly affects the accuracy and efficiency of the model [9]. Standardize the input data to optimize the data amplitude into a distribution with a mean of 0 and a standard deviation of 1. Doing so can speed up the convergence process, because after the scale of the features is unified, the algorithm can find the optimal solution faster. First, calculate the mean, traverse the data array, and calculate the mean of all data points; at the same time, traverse the array, calculate the square difference between all data points and the mean, sum and divide by the number of data points, and then square to get the standard deviation; finally, subtract the mean from each data point and divide by the standard deviation to get the standardized data. Data standardization is achieved through the following formula:

$$data[i] = (data[i] - mean)/std \qquad (1)$$

In the above formula, *mean* is the average value calculated from a data stream, and *std* is the standard deviation of that data stream. The following Fig. 3 is a waveform schematic diagram after standardizing the time series data of the sewing machine's operation collected. After standardization, the amplitude of the data waveform is already quite different from that of the unprocessed waveform.

From the waveform in the figure, we can see that after preprocessing, the data is converted to a distribution with a mean of 0 and a standard deviation of 1. This can accelerate the convergence process because the algorithm can find the optimal solution faster after the scale of the features is unified. Compared with the sewing machine working data before preprocessing, the standardized time series data does not change any of the previously held characteristics. However, after the data is standardized, the convergence process can be accelerated and better adapted to the one-dimensional convolutional neural network model.

3.2 Training of Convolutional Neural Network Model

Convolutional neural network convolution kernel training is essentially a parameter optimization process [10]. In convolutional neural networks, convolution kernels are trained in a data-driven way, so that they can automatically learn effective features in

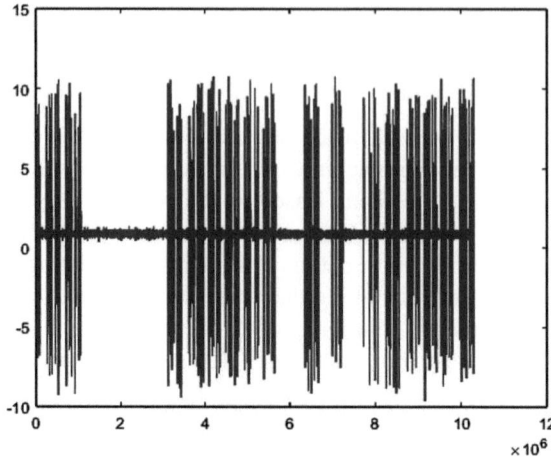

Fig. 3 Normalized waveform

the data [11]. The training process is as follows: Firstly, at the beginning stage, the data is processed by the convolutional layer first. The purpose of the convolution operation is to extract some of the time series data features we need during the time series data analysis process, such as mutations, rises and falls. For a one-dimensional input signal x, the convolution calculation can be expressed as:

$$y(i) = \sum_{k=0}^{K-1} w(k) \times x(i+k) \qquad (2)$$

Among them, $y(i)$ is the convolution output, $w(k)$ is the kth convolution kernel parameter, and K is the length of the convolution kernel. Through convolution, the weighted sum in the local area can highlight important signal features. After the convolution output is processed by a nonlinear activation function, the activation function can enhance the neural network's ability to fit more complex patterns, and can also make the neural network approach nonlinear relationships and alleviate the gradient disappearance problem. After the feature extraction stage of forward propagation, the output of the convolution layer passes through the fully connected layer to generate a prediction result. In order to make the prediction result of the model more accurate, a loss function is introduced to calculate the gap between the predicted value and the true label y. During the training process, binary cross-entropy is used to measure the difference between the true label and the predicted probability, and its mathematical expression is:

$$Loss = -(y \log(\bar{x}) + (1-y) \times \log(1-\bar{y})) \qquad (3)$$

y is the true label, and the other is the predicted probability output by the model. This loss function can effectively measure the gap between the predicted result and the true value. When the model predicts an error, the loss value will be relatively large, thereby guiding the model to optimize the predicted value of error-prone samples. After the loss value is obtained through training, the weight of the convolution kernel will be updated through the back propagation algorithm. The core idea of the algorithm is to pass the loss from the output layer back to each layer layer by layer according to the chain rule, and calculate the gradient of each weight. For each parameter $w(k)$ in the convolution kernel, its gradient can be calculated by the following formula:

$$\frac{\partial Loss}{\partial w(k)} = \sum_i \frac{\partial Loss}{\partial y(i)} \times x(i+k) \tag{4}$$

This means that the change in the convolution kernel parameters is determined by the input data segment $x(i + k)$ and its impact on the overall error. Through the backpropagation algorithm, this model can adjust the parameters obtained from training and reduce the overall prediction error. After obtaining the gradient of each convolution kernel parameter, the next step is to update them. We use the "gradient descent" method to update the parameters, adjusting each weight in the direction that reduces the loss the fastest. The flowchart of training convolutional neural network is shown in Fig. 4:

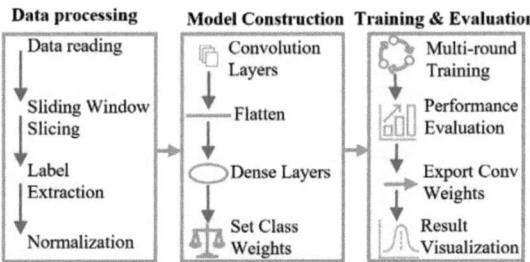

Fig. 4 Training flow chart

In order to achieve the automatic identification of sudden change points in time series, this study constructed an algorithm model based on one-dimensional convolutional neural network (1D-CNN). This model uses a single-layer one-dimensional convolutional neural network, with 3 convolutional kernels, and employs the Leaky ReLU activation function. The training process of the convolution kernel is shown in Fig. 4. The entire training process is divided into the following stages: data preprocessing, model construction, and model training evaluation. First, data preprocessing is carried out in the first stage, the original sensor signal is read and sliced according to a sliding window of fixed length to ensure that the model can learn the local feature changes in the sequence segment by segment. Subsequently, combined with the manually annotated mutation point information, a label is assigned to the center point of each window (1 represents a mutation point, 0 represents normal), and all input data are standardized to improve training stability. In the model construction stage, a lightweight 1D-CNN architecture is adopted. The convolution layer is used to extract the local change pattern of the time series, the Flatten layer flattens the convolution output into a vector, and the Dense layer outputs a binary classification probability, indicating whether the center of the segment is a mutation point. At the same time, in order to solve the problem of the scarcity of mutation point samples, a class weighting mechanism is introduced in the training to improve the model's ability to identify mutation points. Finally, in the training and evaluation phase, the model is optimized through supervised learning on the training set, and its performance is evaluated on the validation set, using indicators including Accuracy, Precision, Recall, and F1 Score. After training, the weights of the convolutional layer are exported and deployed in the embedded system for online detection of mutation points in actual industrial signals. In addition, the loss function and the accuracy change curve were visualized to assist in analyzing the convergence of the algorithm.

3.3 Convolutional Neural Network Architecture

Convolutional neural network is one of the deep learning algorithms, which is capable of large-scale image processing tasks [12]. One-dimensional convolutional neural network (1D-CNN) uses trainable convolution kernels to autonomously learn the characteristic expression of mutation points in time series, and has significant advantages such as few parameters, light calculation, strong noise resistance, and strong generalization ability. Compared with traditional methods, convolutional neural network can achieve more stable and accurate mutation recognition in complex scenarios, and is particularly suitable for deployment and application in embedded and edge computing devices [13]. By analyzing the waveform of sewing machine working current data, it is concluded that when the data value is concentrated around 2.5, the equipment is idling or stable; the data fluctuation shows an obvious block rhythm, which is suspected to be sewing work or feeding process; different time intervals and amplitudes vary greatly, indicating that it may be different working conditions or fabric types. The sewing machine current signal contains a large number of nonlinear mutations with strong local characteristics. The one-dimensional convolutional neural network can efficiently extract these mutation features through sliding convolution kernels, and then identify the process quantity information through the position difference between mutation points. It has strong robustness and real-time performance, and is one of the best solutions for identifying such signals, especially suitable for application scenarios such as industrial sites, embedded deployment and online monitoring [14].

Figure 5 below is the recognition architecture of a one-dimensional convolutional neural network:

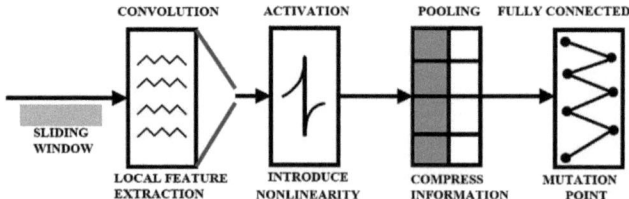

Fig. 5 Recognition architecture of one-dimensional convolutional neural network

4 Experiments and Analysis

This experiment mainly deplores the time series data recognition algorithm model obtained through training at the edge end to identify the key mutation points in the working current signal of the sewing machine [15]. The light-weight network structure is used to extract and judge the characteristics of the university, and finally it is deployed in the embedded system. In order to verify the effectiveness of the proposed solution, the data processing task is carried out on the constructed edge computing platform, and the results are printed to the PC through the serial port after processing. The sewing machine working data to be processed is stored in the memory of the edge computing platform in the form of a data file. The physical edge computing device is shown in Fig. 6:

Fig. 6 Real picture

4.1 Experimental Data Source

The data used in the experiment originated from the output of the current sensor collected during the actual operation of the industrial sewing machine, and the data were from different sewing machines in multiple workshops. The sampling frequency is 10 kHz, and the working waveforms of the sewing process in different steps are recorded, which can clearly reflect the working status of the equipment. The total length of the data is about 10 million points, covering multiple alternating processes between working and non-working states.

4.2 Model Training and Evaluation

This experiment uses supervised learning to train the constructed one-dimensional convolutional neural network model, with the goal of identifying mutation points in the sequence [16]. During the training process, the Adam optimizer and Binary Cross entropy are selected as the loss function, the batch size is set to 64, and the number of training rounds is set to 30. In order to deal with the problem of uneven distribution of positive and negative samples, a category weight mechanism is introduced in the training process. After the model is trained, the performance is evaluated on the validation data using indicators such as Accuracy, Precision, Recall, and F1 Score. The results show that this algorithm can identify the mutation points more accurately and the F1 value is stable. The use of the trained convolution kernel has obvious advantages in capturing mutation features, which provides a reliable foundation for subsequent deployment on embedded systems. Figure 7 below is a fitted image of the convolution kernel trained using real data:

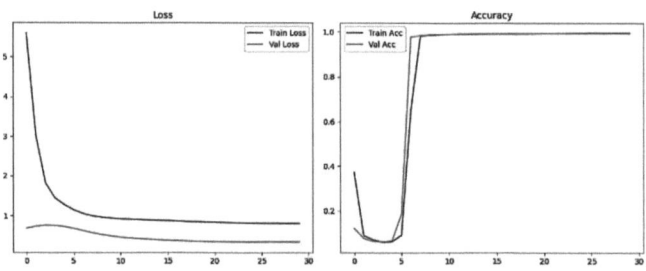

Fig. 7 Fitting image

In the above-mentioned Loss image on the left, the blue curve is the training loss, and the orange curve is the validation loss. It can be seen from the loss function training image that the training loss value keeps decreasing during the training process, and the validation loss is also decreasing. Although there is a slight increase in the later stage of training, it still remains below a low value. In the Accuracy image on the right, the blue curve is the training accuracy, and the orange curve is the validation accuracy. In the early stage of training, the training accuracy and validation accuracy are very low, almost not exceeding 0.5. After Epoch5, the training accuracy is close to 1, and the validation accuracy is almost the same as the training accuracy. The accuracy is close to saturation, indicating that the model can identify the key features required in the data and there is no serious overfitting phenomenon. The accuracy in this figure is based on the overall classification accuracy, which is completely sufficient for the recognition task in this study.

4.3 Embedded Deployment Experiment

Embedded devices are unable to complete the model training process. In this experiment, the required convolution kernel values are trained on the PC side, and the model is embedded into the edge computing device. When running in an embedded environment, the model processes the data stream through a sliding window, uses the pre-trained convolution kernel to perform convolution operations, activation, and pooling, and then calculates the feature gradient to determine whether a mutation occurs. The experiment shows that the deployed model can accurately output the location of the mutation point and has good practicality. The experiment is carried out on the embedded edge device, and the results are printed out through the serial port, as shown in Fig. 8.

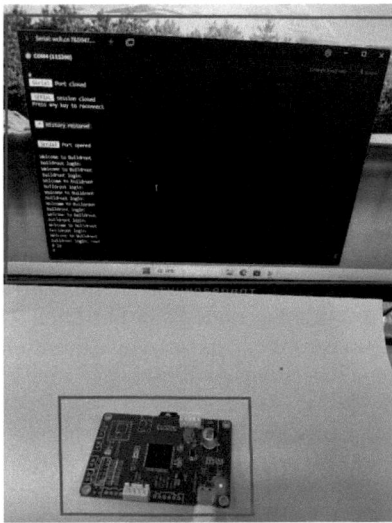

Fig. 8 Experimental scene diagram

Several rounds of experiments with small-scale data were conducted to prove that the recognition scheme proposed in this paper can effectively identify the process status of the sewing machine. Table 1 is an experiment conducted on the collected work data with a long time period and a large order of magnitude. The results are shown in the table.

Table 1 Experimental Data and Results

Number of time series data	Experiments with different data		
	10355360	*6735799*	*1065270*
Recognition accuracy	0.91	0.92	0.92

5 Conclusion

Aiming at the data processing and state recognition problems faced by traditional industrial sewing machine equipment in the process of digital transformation, this paper proposes a one-dimensional convolutional neural network (1D-CNN) recognition method based on edge computing, and builds an embedded edge computing platform to realize local data processing. Through the analysis and modeling of the sewing machine working current waveform, the process status is accurately identified, which effectively reduces the dependence on the cloud and improves the real-time and stability of the system. The experimental results show that the method has good deployment effect on embedded devices, high recognition accuracy, good practicality and promotion prospects. Future research will further optimize the model structure, improve the adaptability to different working conditions, and explore the application potential of this method in more industrial scenarios to provide technical support for the intelligent development of the manufacturing industry.

References

1. Hendrawan S A, Chatra A, Iman N, et al. Digital transformation in MSMEs: Challenges and opportunities in technology management. Jurnal Informasi dan Teknologi, 2024: 141–149
2. Xia, H., Ye, P., Jasimuddin, S.M., et al.: Evolution of digital transformation in traditional enterprises: Evidence from China. Technol. Anal. Strat. Manag. **36**(9), 2014–2034 (2024)
3. Liu Y, Zhang Y, Xie X, et al. Affording digital transformation: The role of industrial Internet platform in traditional manufacturing enterprises digital transformation. Heliyon, 2024, 10(7)
4. Yao Z. Application of cloud computing platform in industrial big data processing[J]. arXiv preprint arXiv:2407.09491, 2024
5. Sodiya, E.O., Umoga, U.J., Obaigbena, A., et al.: Current state and prospects of edge computing within the Internet of Things (IoT) ecosystem. Int. J. Sci. Res. Arc. **11**(1), 1863–1873 (2024)
6. Romero, D.A.V., Laureano, E.V., Betancourt, R.O.J., et al.: An open source IoT edge-computing system for monitoring energy consumption in buildings. Results Eng. **21**, 101875 (2024)
7. Narkhede G, Mahajan S, Narkhede R, et al. Significance of Industry 4.0 technologies in major work functions of manufacturing for sustainable development of small and medium-sized enterprises[J]. Business Strategy & Development, 2024, **7**(1): e325
8. Romero D A V, Laureano E V, Betancourt R O J, et al. An open source IoT edge-computing system for monitoring energy consumption in buildings. Results Eng. **21**: 101875 (2024)
9. Xie Y, Wang S, Zhang G, et al. A review of data-driven whole-life state of health prediction for lithium-ion batteries: Data preprocessing, aging characteristics, algorithms, and future challenges. J. Energy Chem. (2024)
10. HASAN M D A, Bhargav T, SANDEEP V, et al. Image classification using convolutional neural networks[J]. International Journal of Mechanical Engineering Research and Technology, 2024, **16**(2): 173–181
11. Reyes D, Sánchez J. Performance of convolutional neural networks for the classification of brain tumors using magnetic resonance imaging. Heliyon **10**(3) (2024)
12. Zhao, X., Wang, L., Zhang, Y., et al.: A review of convolutional neural networks in computer vision. Artif. Intell. Rev. **57**(4), 99 (2024)
13. Zheng, Y., Wu, C., Cai, P., et al.: Tiny-PPG: A lightweight deep neural network for real-time detection of motion artifacts in photoplethysmogram signals on edge devices[J]. Internet of Things **25**, 101007 (2024)
14. Chen, J., Teo, T.H., Kok, C.L., et al.: A novel single-word speech recognition on embedded systems using a convolution neuron network with improved out-of-distribution detection. Electronics **13**(3), 530 (2024)
15. Tran, V.L., Vo, T.C., Nguyen, T.Q.: One-dimensional convolutional neural network for damage detection of structures using time series data. Asian J. Civil Eng. **25**(1), 827–860 (2024)
16. Nafea, A.A., Alameri, S.A., Majeed, R.R., et al.: A short review on supervised machine learning and deep learning techniques in computer vision. Babylonian J. Mach. Learn. **2024**, 48–55 (2024)

Trustworthy Distributed Decision-Making for Multi-view Sensing Data

Zishuo Song[1,2], Yuzhu Pan[1,2], Mingshu Zhao[1,2], Muhammad Ameen[1,2], Zhenwei Wang[1,2], and Pengfei Wang[1,2,3,4(✉)]

[1] School of Computer Science and Technology, Dalian University of Technology, Dalian 116024, China
wangpf@dlut.edu.cn
[2] Key Laboratory of Social Computing and Cognitive Intelligence, Ministry of Education, Dalian 116024, China
[3] Yantai Yiteng Intelligence Co., Ltd., Yantai 264003, China
[4] Yidatec Co., Ltd., Dalian 116085, China

Abstract. In the context of the rapid development of the Internet of Things (IoT), although a large number of sensing devices have been deployed and vast amounts of data are available, single-view data often suffers from limited perspectives and high uncertainty, leading to issues such as incomplete information and susceptibility to noise, which undermine the reliability of model decisions. To address these challenges, this paper focuses on the problem of distributed trustworthy decision-making using multi-view sensing data, and proposes a distributed and reliable classification method based on multi-view fusion. The proposed method incorporates evidence-based uncertainty modeling and fully accounts for both the consistency and complementarity of multi-view information at the decision-making level. Specifically, in the single-view learning stage, deep neural networks are employed to extract evidence from each view. Based on subjective logic, Dirichlet distributions are used to construct credibility models for each perspective within the Dempster-Shafer evidence theory framework. In the multi-view fusion stage, an improved Dempster combination rule is used to effectively integrate the credibility models from different views. Additionally, a conflict-aware inconsistency measure is designed to assess decision-level disagreements across views. Experimental results on six datasets demonstrate that the proposed method outperforms existing mainstream feature-level and decision-level fusion algorithms in both classification accuracy and uncertainty modeling, under both conventional and challenging scenarios, thereby validating its effectiveness and advantages.

Keywords: Multi-view learning · Distributed trustworthy decision-making · Evidence theory · Dirichlet distribution · Dempster-shafer theory

1 Introduction

With the rapid development of the Internet of Things, the volume and diversity of sensing data have grown explosively (IoT) [16,31]. This data not only forms a rich foundation of information but also provides critical support for intelligent services and automated decision-making [19]. However, the potential value of massive sensing data remains underexploited. As a result, how to efficiently leverage these data resources to ensure that the costs of data acquisition, storage, communication, and processing yield meaningful returns has become one of the central challenges in current research [26,30].

Among the vast array of sensing data, multi-view data is widely present, as illustrated in Fig. 1. Regarding the same object, different sensing devices or data sources have the capacity to furnish information from a wide array of perspectives and in distinct forms. Such multi-view data often contains complementary features, and appropriately designed learning mechanisms can achieve better performance than single-view approaches [33]. Multi-view learning has demonstrated significant advantages in areas such as knowledge graph construction, object detection, and pattern recognition, and is gradually becoming a key technique for improving decision accuracy in IoT scenarios [2].

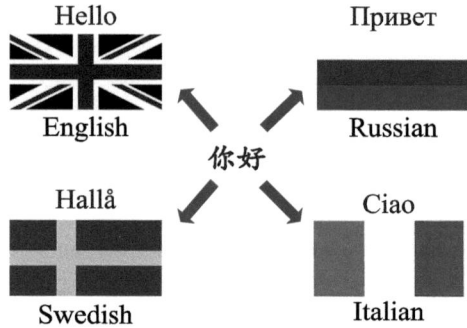

Fig. 1. Different representations of multi-view data.

In recent years, deep neural networks have been widely adopted in multi-view learning due to their powerful capabilities in feature extraction and representation, enabling effective modeling of complex relationships among multi-view data [18]. However, existing multi-view algorithms generally assume that the quality and importance of each view are stable, assigning fixed weights accordingly. This static strategy limits the model's flexibility and robustness in real-world applications [23]. In practice, the quality of sensing devices in IoT environments is influenced by various factors such as environmental conditions, communication interference, and device aging, making the reliability of each view dynamically variable [22,29]. For example, RGB images perform well under good lighting conditions, while infrared images offer advantages in low-light environments [7]. Therefore, models should be capable of dynamically adjusting the weights of different views [11]. Moreover, when decision conflicts arise among different views, how to effectively integrate these conflicting outputs remains a critical challenge [6].

Although deep learning techniques have significantly improved decision-making performance through the construction of complex neural network models, their black-box nature makes it difficult to provide reliable uncertainty estimates [5,25]. This is particularly problematic in distributed IoT decision systems, where high-risk or irreversible scenarios require not only high accuracy, but also trustworthy and interpretable decisions [4]. Therefore, it is essential not only to know what the result is, but also to understand how trustworthy that result is [1]. Current uncertainty modeling methods still suffer from limited expressiveness and overly confident predictions, making them inadequate for reliable deployment in real-world scenarios [21]. To address this, we aim to develop a distributed trustworthy decision-making framework that not only delivers high decision performance but also provides well-calibrated uncertainty estimates by integrating heterogeneous multi-view views, thereby enhancing the intelligence and reliability of IoT systems.

In summary, this paper proposes a distributed trustworthy decision-making method based on multi-view fusion to address the issues of value mining of multi-view sensing data and insufficient decision credibility in the IoT. The method is divided into two stages. In the single-view learning stage, evidence from each perspective is extracted using deep neural networks, and a confidence model is constructed based on Dirichlet distribution within the Dempster-Shafer evidence theory framework. In the multi-view fusion stage, an improved Dempster combination rule is used to effectively integrate the confidence models, while a mechanism considering inconsistency is designed to resolve decision conflicts between different perspectives. This method fully leverages the complementarity and consistency of multi-view data, enhancing the decision-making reliability and accuracy of the model.

2 Problem Analysis and Overall Framework

2.1 Problem Analysis and Definition

For multi-view data describing the same object, it is essential to explore how to effectively fuse such data to not only improve decision accuracy but also quantify uncertainty in order to assess decision reliability [28]. Meanwhile, the diversity of data sources introduces challenges in ensuring consistency and leveraging complementarity among the views for effective decision-making [17]. To address these challenges, this paper proposes a distributed trustworthy decision-making algorithm for multi-view sensing data. The approach performs evidence-based uncertainty estimation on multi-view inputs and fuses them at the decision level, thereby enabling reliable and interpretable decision-making.

The research problem is defined as follows: we consider a multi-view classification task, where the dataset consists of N groups of instances. Each group contains data from V different views, and K denotes the total number of possible classes. Let the feature vector of the n-th instance under the v-th view be denoted as x_n^v, and $x_n^v \in \mathcal{R}^{D_v} (1 \leq v \leq V, 1 \leq n \leq N)$. Here, D_v represents the dimensionality of the feature space for the v-th view, and $\mathcal{R}^{D_v}$ refers to the real-valued space D_v corresponding to the feature vector dimension. The label of the n-th instance is represented by a K-dimensional binary vector y_n, where $y_n \in \{0,1\}^K$. The training dataset is denoted as $\{\{x_n^v\}_{v=1}^V, y_n\}_{n=1}^{N_{train}}$, which contains N_{train} groups of multi-view instances along with their corresponding labels. An evidential deep learning model $\{f^v(\cdot)\}_{v=1}^V$ is constructed, with the training objective of learning how to effectively fuse evidence from different views. The remaining $N - N_{train}$ groups of instances constitute the test set, which is used to evaluate the performance of the model.

This paper aims to develop a distributed trustworthy decision-making algorithm for multi-view sensory data based on multi-view evidence fusion. The primary objective of the algorithm is to perform classification on a set of test instances while simultaneously providing an estimate of the uncertainty associated with each decision. This uncertainty is denoted by u^n, with a value ranging from 0 to 1. A higher value of u^n indicates weaker supporting evidence for the classification result, implying lower confidence in the model's prediction. The proposed algorithm is expected to integrate evidence from multiple views to enable more accurate and reliable decisions, while also quantifying the associated uncertainty.

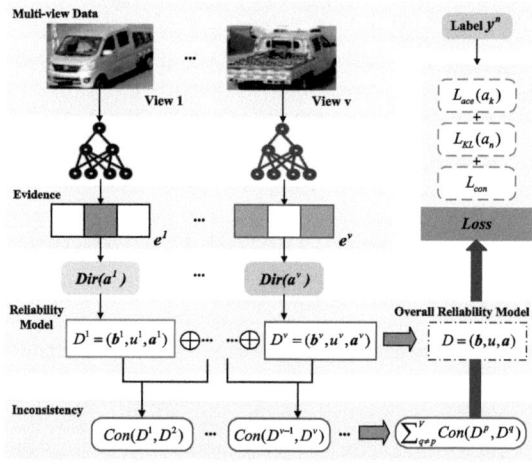

Fig. 2. Different representations of multi-view data.

2.2 Overall Framework

This section introduces the proposed distributed trustworthy decision-making algorithm for multi-view sensory data, referred to as the DTM-DM algorithm

(Distributed Trusted Multi-view Decision-Making). First, we present an overview of the overall framework. Then, we provide a detailed explanation of the key components, including single-view evidential deep learning, multi-view trustworthy correlation fusion, multi-view inconsistency measurement, and the design of the loss function.

As illustrated in Fig. 2, the overall architecture of the proposed algorithm consists of two main stages: single-view evidence learning and multi-view evidence fusion. In the single-view evidence learning stage, deep neural networks are utilized to extract evidence specific to each view, generating supporting evidence for each class. Subjective logic is then incorporated by modeling the view-specific class probability distributions using the Dirichlet distribution, which enables the estimation of classification probabilities and their associated uncertainties.

Based on this, in the multi-view fusion stage, the Dempster combination rule is applied to integrate the confidence and uncertainty from each view, resulting in an overall classification probability and uncertainty estimation. Furthermore, a conflict-aware fusion strategy is proposed, taking into account that inconsistency among views can negatively impact the overall decision and its associated uncertainty. To address this, a measure of uncertainty is introduced to quantify the conflict across different views.

3 Method

3.1 Single-View Evidence Deep Learning

Most traditional multi-view learning methods adopt a softmax layer in neural networks for classification. However, the output of softmax merely represents the probability of class assignment and does not reflect the confidence of the decision. As a result, softmax often leads to overconfident predictions. Fundamentally, this is because softmax scores provide only a point estimate of the predictive distribution, which limits the model's ability to capture uncertainty. Consequently, even when the model makes incorrect decisions, the outputs may still appear overly confident. This inherent limitation hinders the model's performance in scenarios requiring uncertainty handling and risk-aware decision-making.

To address this issue, this paper incorporates Evidential Deep Learning (EDL) in the single-view learning stage. EDL is a deep learning approach based on evidence theory for uncertainty estimation, aiming to enable models to assess the confidence of their own predictions. Unlike traditional deep learning models, EDL not only provides classification outcomes but also quantifies the confidence in its decisions. Specifically, according to Dempster-Shafer Theory (DST) of evidence, EDL assigns a belief mass to each class label, representing the degree of confidence (hereafter referred to as "belief"). Additionally, DST offers a metric for measuring the overall uncertainty within the recognition framework. Subjective Logic (SL) [13] transforms the belief assignment concept in Dempster-Shafer Theory (DST) within the recognition framework into a Dirichlet distribution. This establishes a connection between the Dirichlet distribution and belief distribution, with the parameters of the Dirichlet distribution inferred from the observed evidence.

In this context, evidence refers to the metrics collected from the inputs to support the classification decision. Traditional deep neural networks can be modified and converted into evidential deep neural networks [24]. Specifically, this paper replaces the softmax layer in deep neural networks with a Softplus activation layer, treating its non-negative output as evidence. For an instance $\{x_n^v\}_{n=1}^V$, an evidence vector $e_n^v = f^v(x_n^v)$ is obtained through model training. This evidence is used to assign a degree of belief to an element or a set of elements within the recognition framework.

In addressing a K-class classification problem, the uncertainty information obtained from training a single-view instance is defined as a belief model, denoted by a triplet $\mathcal{D} = (\boldsymbol{b}, \boldsymbol{u}, \boldsymbol{a})$. In this triplet, $\boldsymbol{b} = (b_1, \ldots, b_K)^T$ represents the belief masses calculated for each class, reflecting the likelihood that the instance belongs to each category. The term u denotes the corresponding overall uncertainty. According to the theory of Subjective Logic, the belief mass b and uncertainty u must satisfy the following constraint conditions:

$$u + \sum_{k=1}^{K} b_k = 1 \tag{1}$$

Meanwhile, the conditions $u^v \geq 0$ and $b_k^v \geq 0$ must be satisfied. Let $\boldsymbol{a} = (a_1, \ldots, a_K)^T$ represent the prior probability distribution over all categories. Typically, the prior is manually defined based on prior knowledge or experience. A common approach is to assume a uniform prior distribution, denoted as $a_k = \frac{1}{K}$, which reflects a non-informative assumption that all categories are equally likely at the initial stage—implying that the number of instances in each category is approximately balanced in the dataset. Furthermore, the posterior probability distribution for single-view decision making can be expressed as follows:

$$P_k = b_k + a_k u. \tag{2}$$

where $1 \leq k \leq K$. This study considers a more realistic scenario by employing Subjective Logic to bridge the belief model and the Dirichlet distribution. The Dirichlet distribution is widely recognized as the conjugate prior for categorical distributions, and its probability density function naturally aligns with the random sampling process of statistical events. This enables the transformation of subjective opinions into probabilistic statistical measures. On this basis, the measure of uncertainty can be effectively expressed using the Dirichlet probability density function. A K-dimensional Dirichlet distribution is characterized by K parameters, denoted as $\alpha = [\alpha_1, \alpha_2, \ldots, \alpha_K]$, and its probability density function is given by:

$$Dir(\boldsymbol{p}|\alpha) = \begin{cases} \frac{1}{B(\alpha)} \prod_{i=1}^{K} p_k^{\alpha_k - 1} & \boldsymbol{p} \in \mathcal{S}_K \\ 0 & \text{others} \end{cases} \tag{3}$$

where, $\mathcal{S}_K = \{\boldsymbol{p} | \sum_{i=1}^{K} p_i = 1 \text{ and } 0 \leq p_i \leq 1\}$ denotes the K-dimensional standard simplex, where the classification probability of an instance belonging to class k is represented by $p = (p_1, \ldots, p_K)^T$. The elements of p are non-negative and sum to 1, meaning p lies within the standard $(K-1)$-dimensional simplex. In the Dirichlet probability density function, the denominator term $B(\alpha)$ serves as a normalization factor and corresponds to the multivariate Beta function, which can be reformulated using the Gamma function, denoted as $B(\alpha) = \frac{\prod_{i=1}^{K} \Gamma(\alpha_i)}{\Gamma(\sum_{i=1}^{K} \alpha_i)}$. This ensures that the probability density function of the Dirichlet distribution integrates to 1.

For the v-th view, subjective logic connects the evidence $e^v = [e_1^v, \ldots, e_K^v]$ to the parameters $\alpha^v = [\alpha_1^v, \ldots, \alpha_K^v]$ of a Dirichlet distribution, mapping a four-dimensional Dirichlet distribution onto a standard 3-simplex. Each point on the simplex corresponds to a specific belief (credibility) and uncertainty, thereby providing the model's classification perspective. Specifically, the evidence e_K^v determines the Dirichlet parameters α_K^v, with $\alpha_k^v = e_k^v + 1$ to ensure all parameter values exceed 1, thereby maintaining a non-sparse Dirichlet distribution and avoiding bias due to insufficient prior information. The mapping between the belief model and the Dirichlet distribution is given by the following equation, which is then used to compute the belief b_k^v for each class and the overall confidence u^v:

$$b_k^v = \frac{e_k^v}{S^v} = \frac{\alpha_k^v - 1}{S^v} \tag{4}$$

$$u^v = \frac{K}{S^v} \tag{5}$$

where, S^v denotes the Dirichlet strength, calculated as the sum of $S^v = \sum_{i=1}^{K}(e_i^v + 1) = \sum_{i=1}^{K} \alpha_i^v$. Additionally, the mean of the class probability under the Dirichlet distribution for each category is computed based on the evidence $p_k^v = \frac{\alpha_i^v}{s}$. It can be observed that for the k-th category, the more evidence is observed, the higher the belief b_k^v assigned to that class. Conversely, from a global perspective, fewer total observed evidence leads to lower overall confidence and higher uncertainty u^v. In the absence of any evidence, the belief from each view is zero, and the uncertainty reaches its maximum value of 1.

By performing evidential deep learning at the single-view stage, a belief model can be constructed for each view to model both the class-wise probability and the overall uncertainty of the current decision. This not only enhances the reliability and robustness of the model but also lays a solid foundation for the subsequent multi-view decision fusion stage.

3.2 Trusted Multi-view Correlation Fusion

In the previous section, we established a credibility model for each individual view. Although these single-view models can provide some basis for decision-making, they fail to fully exploit the complementary information and correlations among multiple views. Therefore, this section focuses on how to effectively fuse multiple views to maximize the utilization of inter-view relationships, thereby improving both decision accuracy and reliability.

To achieve this objective, we introduce a multi-view fusion strategy based on Dempster-Shafer Theory (DST) of evidence. DST is capable of integrating evidence from diverse sources to form a decision that comprehensively reflects the combined information. However, when applied to large-scale or high-dimensional data, the computational complexity of DST can become prohibitive. For K-class classification tasks, the original Dempster combination rule may result in an exponential increase in processing complexity, making it impractical in real-world applications. Therefore, to make our model more compatible with the Dempster combination rule, we propose a simplified version of the Dempster rule. This simplification aims to retain the core advantages of evidence theory while reducing the computational burden. It enables the integration of information from multiple sources in a manner that is both efficient and reliable, thus making the fusion process more suitable for complex classification tasks in practical applications.

Suppose that $D^1 = (b^1, u^1, a^1)$ and $D^2 = (b^2, u^2, a^2)$ represent the belief models from two different views of the same instance. The joint belief model is denoted as $D^{1\oplus 2} = D^1 \oplus D^2$, where $\oplus$ is defined by the following Dempster combination rule:

$$b_k^{1\oplus 2} = \frac{b_k^1 b_k^2 + b_k^1 u^2 + b_k^2 u^1}{1 - C} \tag{6}$$

$$u^{1\oplus 2} = \frac{u^1 u^2}{1 - C} \tag{7}$$

$$a_k^{1\oplus 2} = \frac{a_k^1 + a_k^2}{2} \tag{8}$$

where, $C = \sum_{i \neq j} b_i^1 b_j^2$ is a normalization factor that not only captures the underlying divergence between the two belief models but also ensures the normalization of probabilities during the fusion process, thereby maintaining the rationality and consistency of the decision probabilities.

Based on the aforementioned computation method, the combination rule can be recursively applied to construct the final fused belief model from multiple views:

$$D = D^1 \oplus D^2 \oplus ... \oplus D^V \tag{9}$$

The aforementioned fusion strategy follows the following principles: (1) If all views exhibit high uncertainty, the final result will have low confidence; (2) If all views have low uncertainty, the final result will also have low uncertainty; (3) If only one view presents low uncertainty, the final decision will primarily rely on that view; (4) When inconsistencies exist among different views, both the normalization factor C and the overall uncertainty u will increase accordingly.

Based on the aforementioned fusion rule, the final multi-view fused belief model $D = (b, u, a)$ can be obtained, which provides both the final prediction probabilities for each class and an overall uncertainty measurement. Furthermore, by computing the Dirichlet strength $S = \frac{K}{u}$ and integrating the multi-view fused evidence $e_k = b_k * S$, the corresponding Dirichlet parameter distribution $\alpha_k = e_k + 1$ is derived, enabling the generation of final class-wise decision probabilities and overall uncertainty. Compared to traditional neural network classifiers, the proposed trustworthy multi-view association fusion method, by incorporating Subjective Logic, not only enables effective decision-level fusion of multi-view information but also provides an additional uncertainty measure. This enhances both the trustworthiness and interpretability of the model.

3.3 Multi-view Inconsistency Computation

In the previous section, a multi-view trusted association model was constructed based on Dempster-Shafer evidence theory, enabling the estimation of probability distributions and uncertainty measures through belief models. However, the handling of inconsistencies between different views remains insufficient. For instance, when View 1 and View 2 both produce confident predictions (i.e., low uncertainty) but their probability distributions are contradictory, the current fusion method may still yield a result with low uncertainty. This contradicts the real-world expectations and undermines the original intention of reliable decision-making in multi-view fusion. Although the Dempster combination rule introduces a normalization factor to account for conflict and decreases belief reliability as conflict increases, it primarily focuses on belief-level conflicts and does not adequately capture the relationship between belief and uncertainty. Therefore, this section delves into the degree of inconsistency between different views and proposes a method to quantitatively measure this inconsistency.

Specifically, assuming that the belief models of two views for the same instance are denoted as $D^1 = (b^1, u^1, a^1)$ and $D^2 = (b^2, u^2, a^2)$, a metric named inconsistency degree is introduced to quantify the discrepancy between the two view-specific decisions, denoted as $Con(D^1, D^2)$. This metric measures the extent of disagreement between different views and can also be interpreted as the degree of decision conflict. A high inconsistency degree indicates significant divergence in predictions from different views for the same category, which would result in increased uncertainty in the final fused outcome. By capturing this, the method allows for a more comprehensive integration of multi-view information, enabling more accurate and trustworthy decisions. This metric jointly considers the projected distance between probability distributions and the combined uncertainty, and is calculated as follows:

$$Con(D^1, D^2) = d(D^1, D^2) * (1 - u(D^1, D^2)) \tag{10}$$

where, $d(D^1, D^2)$ refers to the projected distance between the probability distributions D^1 and D^2, which measures the divergence between the two distributions. $u(D^1, D^2)$ represents the combined uncertainty derived from D^1 and D^2. These quantities are computed using the following method:

$$d(D^1, D^2) = \frac{\sum_{k=1}^{K} |p_k^1 - p_k^2|}{2} \tag{11}$$

$$u(D^1, D^2) = 1 - (1 - u^1)(1 - u^2) \tag{12}$$

For joint uncertainty u(D^1, D^2), its value approaches 0 when one or both views exhibit maximum uncertainty (i.e., resembling a vacuous opinion). Conversely, when all views provide sufficient evidence and exhibit low uncertainty, u(D^1, D^2) approaches 1. This joint uncertainty reflects the overall uncertainty level across both views. In general, the inconsistency metric satisfies the following properties: when the probability distributions from View 1 and View 2 are completely aligned, the projection distance is 0, indicating no conflict and thus an inconsistency score $Con(D^1, D^2) = 0$. When the views provide confident but conflicting opinions—i.e., their predicted probability distributions differ significantly—the projection distance becomes large and joint uncertainty is low, resulting in an inconsistency score $Con(D^1, D^2)$ close to 1.

By introducing the inconsistency metric, the model can more comprehensively capture the degree of disagreement among different views, thereby allowing more accurate assessment of the uncertainty and reliability of the fusion result. This enhances the accuracy and robustness of multi-view fusion and improves its overall performance and practical applicability.

3.4 Loss Function

To facilitate effective training of the multi-view evidential deep learning model, we propose a loss function that jointly incorporates classification accuracy, uncertainty modeling, and consistency constraints. During the single-view stage, the softmax layer of the neural network is replaced by a Softplus activation function, enabling the outputs to be interpreted as supporting evidence for each class. Under this framework, a modified cross-entropy loss is employed to encourage the model to generate sufficient evidence for the ground-truth labels. To prevent the model from generating excessive evidence for incorrect classes, a Kullback-Leibler divergence term is incorporated to constrain the divergence between the predicted distribution and a non-informative prior. The influence of this term is progressively increased during training to balance exploration and convergence. In addition, a consistency loss is introduced to quantify inter-view prediction conflicts, thereby promoting coordinated decision-making in the fusion stage. The final loss function integrates these three components to jointly enhance classification accuracy and uncertainty estimation, establishing a solid foundation for trustworthy multi-view decision-making.

4 Experimental Validation

4.1 Datasets

In our experiments, we select the following datasets, including classic datasets for classification tasks: (1) HandWritten [3], which contains 2,000 handwritten digit samples ranging from "0" to "9", represented using six different feature sets; (2) HMDB5 [15], which consists of 6,718 human action recognition samples across 51 action categories. HOG and MBH features are extracted to form multi-view representations; (3) Scene15 [9], which includes 4,485 scene classification images from 15 indoor and outdoor scene categories. Three types of features—GIST, PHOG, and LBP—are extracted to serve as multi-view inputs; (4) Caltech101 [8], which comprises 8,677 object images across 101 categories. We select 10 of these categories and extract two deep feature views using DECAF and VGG19 models; (5) PIE [10], which includes 680 facial image samples from 68 different individuals. Brightness, LBP, and Gabor features are extracted as three different views; (6) Stanford Cars dataset [14], which contains 16,185 car images across 196 vehicle categories. HOG and LBP features are extracted to provide multiple views.

4.2 Comparison Methods

The following comparative algorithms are employed. First, for feature-level fusion baselines: (1) DCCAE [27], Deep Canonically Correlated Autoencoders, is a classical method that leverages autoencoders to learn a shared representation. (2) CPM-Nets [32], Cross-Part Multi-view Networks, is a recent and advanced multi-view feature fusion method that focuses on learning a universal representation to capture complex correlations across different views. Additionally, decision-level fusion baselines are considered: (1) TMC [12], Trusted Multi-view Classification, is a pioneering uncertainty-aware method that addresses the challenge of uncertainty estimation and produces reliable classification outcomes. (2) TMDL-OA [20], Trusted Multi-view Deep Learning with Opinion Aggregation, is a state-of-the-art decision fusion method. It is also built upon evidential deep neural networks and introduces a consistency-aware loss to achieve trustworthy learning outcomes.

4.3 Data Processing

To create a test set containing conflicting instances, the following transformations were applied: (1) For noisy views, Gaussian noise with varying standard deviation levels σ was added to a subset of test instances. (2) For misaligned views, we selected a portion of instances and altered the information in a randomly chosen view, resulting in a mismatch between that view's label and the instance's ground-truth label. Each method was run 10 times, and the average results and standard deviations were reported.

4.4 Experimental Results

Tables 1 and 2 present the classification performance on the regular and conflicting test sets, respectively. The following conclusions can be drawn: (1) Even on the regular test set, DTM-DM outperforms all other baseline methods. For example, on the HMDB dataset, DTM-DM achieves an accuracy improvement of approximately 2.64% compared to the second-best model, TMDL. This can be attributed to the incorporation of the consistency loss, which enhances the

Table 1. Conventional Experimental Results

Dataset	DCCAE	CPM	TMC	TMDL	DTM-DM
HandWritten	95.45 ± 0.35	94.55 ± 1.36	98.51 ± 0.13	99.25 ± 0.45	**99.40 ± 0.00**
HMDB5	49.12 ± 1.07	63.32 ± 0.43	65.17 ± 2.42	88.20 ± 0.58	**90.84 ± 1.86**
Scene15	55.03 ± 0.34	67.29 ± 1.01	67.71 ± 0.30	75.57 ± 0.02	**76.19 ± 0.12**
Caltech101	89.56 ± 0.41	90.35 ± 2.12	92.80 ± 0.50	94.63 ± 0.04	**95.36 ± 0.38**
PIE	81.96 ± 1.06	88.53 ± 1.23	91.85 ± 0.23	92.33 ± 0.36	**94.71 ± 0.02**
Standford Cards	78.35 ± 0.86	80.17 ± 1.27	83.18 ± 0.93	82.38 ± 0.15	**85.43 ± 0.33**

Table 2. Conflict Experiment Results

Dataset	DCCAE	CPM	TMC	TMDL	DTM-DM
HandWritten	82.85 ± 0.38	83.34 ± 1.07	92.76 ± 0.15	93.05 ± 0.05	**95.36 ± 0.00**
HMDB5	29.62 ± 1.79	42.62 ± 1.43	47.17 ± 0.15	67.62 ± 0.28	**70.84 ± 1.19**
Scene15	25.97 ± 2.86	29.63 ± 1.12	42.27 ± 1.61	48.42 ± 1.02	**56.97 ± 0.52**
Caltech101	60.90 ± 2.31	66.54 ± 2.89	90.16 ± 2.50	90.33 ± 0.24	**92.36 ± 1.38**
PIE	26.89 ± 1.10	53.19 ± 1.22	61.65 ± 1.03	68.16 ± 0.34	**82.01 ± 0.12**
Standford Cards	37.18 ± 0.76	42.87 ± 1.89	47.58 ± 0.23	57.28 ± 0.55	**78.13 ± 0.63**

model's learning capability. (2) When evaluated on the conflicting test set, all comparison methods experience a significant drop in accuracy. However, thanks to the incorporation of inconsistency-aware opinion aggregation, DTM-DM demonstrates a strong ability to perceive potential inconsistency among different views, thus achieving impressive results across all datasets. This highlights the effectiveness of DTM-DM in handling both regular and conflicting multi-view data.

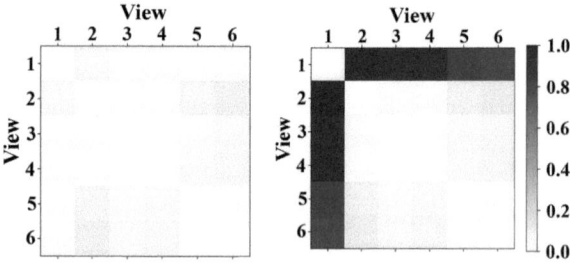

Fig. 3. Visualization of the degree of conflict.

4.5 Performance Analysis

Conflict intensity was visualized using the HandWritten dataset. Figure 3 illustrates the conflict levels across six views, with the left and right parts displaying the conflict intensity for normal and conflicting instances, respectively. To induce conflict, we modified the content of the first view, causing inconsistencies in the attributes of the other views. The results clearly show that DTM-DM can effectively capture and quantify the degree of inconsistency between perspectives, further validating the reliability of DTM-DM.

To further evaluate the effectiveness of the algorithm in uncertainty estimation, the distribution of normal and conflicting test sets was visualized on the PIE dataset. To construct the conflicting test set, Gaussian noise with a standard deviation of $\sigma = \{0.1, 1, 5, 10\}$ was introduced to 50% of the test instances. The experimental results are shown in Fig. 4. The results indicate that when the noise intensity is low ($\sigma = 0.1$), the distribution curve of conflicting instances is closely related to that of normal instances. However, as the noise intensity increases, the uncertainty of the conflicting instances also increases. This finding suggests that the estimated uncertainty is associated with the quality of the instances, thus validating the capability of our method in uncertainty estimation.

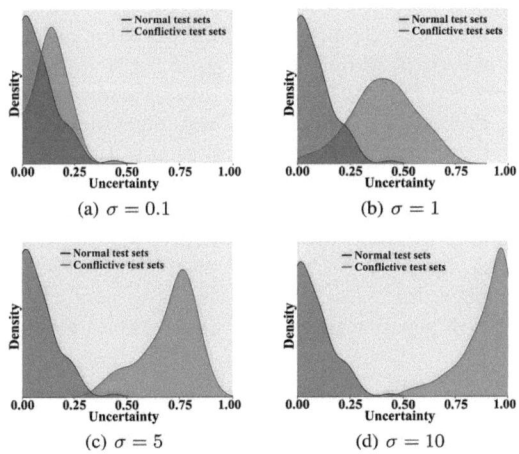

Fig. 4. Uncertainty density map.

5 Conclusion

In this paper, we propose a distributed trustworthy decision-making method for multi-view sensory data, named DTM-DM. This method centers on two key components—trustworthy modeling and multi-view fusion—and is structured into two stages: single-view learning and multi-view fusion. In the first stage, deep neural networks are employed to extract evidence from each view. Leveraging subjective logic, a belief model under the Dempster-Shafer framework is constructed using the Dirichlet distribution. In the second stage, the belief models from different views are integrated using an improved Dempster combination rule, and an inconsistency-aware mechanism is introduced to effectively mitigate conflicts across views. Experimental results on several benchmark classification datasets demonstrate that the DTM-DM algorithm outperforms state-of-the-art feature-level and decision-level fusion methods in both classification accuracy and uncertainty modeling, validating its effectiveness and robustness in complex IoT scenarios.

Acknowledgments. This work was supported by the National Natural Science Foundation of China under grant 62202080, the Science and Technology Project of Liaoning Province under grant 2023JH1/10400083, the Dalian Science and Technology Talent Innovation Support Plan for Outstanding Young Scholars under grant 2023RY023, the Yantai Smart City Innovation Lab under grant 202310-06-SJYS-03, and Xiaomi Young Talents Program.

References

1. Adler, R.H.: Trustworthiness in qualitative research. J. Hum. Lact. **38**(4), 598–602 (2022)
2. Ahmad, H.M., Rahimi, A.: Deep learning methods for object detection in smart manufacturing: a survey. J. Manuf. Syst. **64**, 181–196 (2022)
3. Asuncion, A., Newman, D., et al.: UCI machine learning repository (2007)
4. Bampatsikos, M., Politis, I., Ioannidis, T., Xenakis, C.: Trust score prediction and management in IoT ecosystems using Markov chains and MADM techniques. IEEE Trans. Consum. Electron. (2025)
5. Bohanec, M., Robnik-Šikonja, M., Kljajić Borštnar, M.: Decision-making framework with double-loop learning through interpretable black-box machine learning models. Ind. Manag. Data Syst. **117**(7), 1389–1406 (2017)
6. Carroll, M.B., Sanchez, P.L.: Decision making with conflicting information: influencing factors and best practice guidelines. Theor. Issues Ergon. Sci. **22**(3), 296–316 (2021)
7. Deng, Q., Tian, W., Huang, Y., Xiong, L., Bi, X.: Pedestrian detection by fusion of RGB and infrared images in low-light environment. In: 2021 IEEE 24th International Conference on Information Fusion (FUSION), pp. 1–8. IEEE (2021)
8. Fei-Fei, L.: Learning generative visual models from few training examples. In: Workshop on Generative-Model Based Vision, IEEE Proceedings CVPR, 2004 (2004)
9. Fei-Fei, L., Perona, P.: A Bayesian hierarchical model for learning natural scene categories. In: 2005 IEEE Computer Society Conference on Computer Vision and Pattern Recognition (CVPR 2005), vol. 2, pp. 524–531. IEEE (2005)
10. Gross, R., Matthews, I., Cohn, J., Kanade, T., Baker, S.: Multi-pie. In: 2008 8th IEEE International Conference on Automatic Face & Gesture Recognition (2009)
11. Han, Y., Huang, G., Song, S., Yang, L., Wang, H., Wang, Y.: Dynamic neural networks: a survey. IEEE Trans. Pattern Anal. Mach. Intell. **44**(11), 7436–7456 (2021)
12. Han, Z., Zhang, C., Fu, H., Zhou, J.T.: Trusted multi-view classification with dynamic evidential fusion. IEEE Trans. Pattern Anal. Mach. Intell. **45**(2), 2551–2566 (2022)

13. Jsang, A.: Subjective Logic: A Formalism for Reasoning Under Uncertainty. Springer (2018)
14. Krause, J., Stark, M., Deng, J., Fei-Fei, L.: 3D object representations for fine-grained categorization. In: Proceedings of the IEEE International Conference on Computer Vision Workshops, pp. 554–561 (2013)
15. Kuehne, H., Jhuang, H., Garrote, E., Poggio, T., Serre, T.: HMDB: a large video database for human motion recognition. In: 2011 International Conference on Computer Vision, pp. 2556–2563. IEEE (2011)
16. Leiyang, X., Xiaolong, Z., Liang, L.: Vortex EM wave-based rotation speed monitoring on commodity wifi. Chin. J. Electron. (2024)
17. Li, X., Wang, Y., Yao, J., Li, M., Gao, Z.: Multi-sensor fusion fault diagnosis method of wind turbine bearing based on adaptive convergent viewable neural networks. Reliab. Eng. Syst. Saf. **245**, 109980 (2024)
18. Li, Y., Yang, M., Zhang, Z.: A survey of multi-view representation learning. IEEE Trans. Knowl. Data Eng. **31**(10), 1863–1883 (2018)
19. Li, Z., Yu, Y., Wang, S.: Practical evaluation of intelligent algorithms in ESG management of manufacturing enterprises. Sci. Rep. **14**(1), 19394 (2024)
20. Liu, W., Yue, X., Chen, Y., Denoeux, T.: Trusted multi-view deep learning with opinion aggregation. In: Proceedings of the AAAI Conference on Artificial Intelligence, vol. 36, pp. 7585–7593 (2022)
21. Ma, F., et al.: Monocular 3D lane detection for autonomous driving: Recent achievements, challenges, and outlooks. IEEE Trans. Intell. Transp. Syst. (2025)
22. Mois, G., Folea, S., Sanislav, T.: Analysis of three IoT-based wireless sensors for environmental monitoring. IEEE Trans. Instrum. Meas. **66**(8), 2056–2064 (2017)
23. Roy, P.: Enhancing real-world robustness in AI: challenges and solutions. J. Recent Trends Comput. Sci. Eng. (JRTCSE) **12**(1), 34–49 (2024)
24. Sensoy, M., Kaplan, L., Kandemir, M.: Evidential deep learning to quantify classification uncertainty. In: Advances in Neural Information Processing Systems, vol. 31 (2018)
25. Thuy, A., Benoit, D.F.: Explainability through uncertainty: trustworthy decision-making with neural networks. Eur. J. Oper. Res. **317**(2), 330–340 (2024)
26. Tien, J.M.: Big data: unleashing information. J. Syst. Sci. Syst. Eng. **22**, 127–151 (2013)
27. Wang, W., Arora, R., Livescu, K., Bilmes, J.: On deep multi-view representation learning. In: International Conference on Machine Learning, pp. 1083–1092. PMLR (2015)
28. Xie, Z., Yang, Y., Zhang, Y., Wang, J., Du, S.: Deep learning on multi-view sequential data: a survey. Artif. Intell. Rev. **56**(7), 6661–6704 (2023)
29. Xing, L.: Reliability in internet of things: current status and future perspectives. IEEE Internet Things J. **7**(8), 6704–6721 (2020)
30. Xu, L., Zheng, X., Du, X., Liu, L., Ma, H.: Wicamera: vortex electromagnetic wave-based wifi imaging. IEEE Trans. Mob. Comput. (2024)
31. Yalli, J.S., Hasan, M.H., Badawi, A.: Internet of things (IoT): origin, embedded technologies, smart applications and its growth in the last decade. IEEE Access (2024)
32. Zhang, C., Cui, Y., Han, Z., Zhou, J.T., Fu, H., Hu, Q.: Deep partial multi-view learning. IEEE Trans. Pattern Anal. Mach. Intell. **44**(5), 2402–2415 (2020)
33. Zhang, W., Deng, Z., Choi, K.S., Wang, J., Wang, S.: Dual representation learning for one-step clustering of multi-view data. Artif. Intell. Rev. **58**(7), 201 (2025)

From Sparse Labels to Accurate Models: Active Semi-Supervised Learning for mmWave Radar Target Classification

Liyang Zheng[1,2], Miao Zhang[1], Siyuan Fang[2], Lishen Guan[2], and Xing Chen[2(✉)]

[1] Harbin Institute of Technology, Shenzhen, China
zhangmiao@hit.edu.cn
[2] HeyiSpace, Shenzhen, China
{fangsiyuan,guanlishen,chenxing}@szh1.com.cn

Abstract. Millimeter-wave (mmWave) radar provides robust and privacy preserving sensing for indoor object detection on edge devices, which is critical for smart environments and healthcare applications. However, acquiring numerous labeled data for training neural networks is often prohibitively expensive in practice. Furthermore, conventional self-supervised and semi-supervised methods offer limited effectiveness in leveraging unlabeled data when applied to lightweight models. To address these challenges, we propose an Active Semi-Supervised Learning (ASSL) framework for mmWave radar target classification in label-scarce scenarios. ASSL actively selects a small subset of informative samples for annotation to further guide model effectively learning from unlabeled data, tailored for lightweight networks. Central to this framework is a novel Hybrid Uncertainty-Diversity Query (HUDQ) strategy, which combines model uncertainty (identifying ambiguous samples) with data diversity (ensuring broad coverage) to enhance model generalization. By balancing these complementary criteria, ASSL effectively identifies representative samples for labeling, following with a robust semi-supervised learning scheme to train a lightweight CNN classifier. Experimental results on a real-world mmWave radar dataset demonstrate that ASSL, using only 20% labeled data, surpasses the performance of fully supervised models on a lightweight architecture. This highlights its practical value for AIoT applications on edge devices with lightweight models and limited annotated data.

Keywords: Active Semi-Supervised Learning · Data efficiency · Millimeter-wave radar · Target Classification · IoT Data

1 Introduction

The integration of Artificial Intelligence (AI) within the Internet of Things (IoT), often termed AIoT, represents a significant technological trend with broad applicability across various domains [7]. Millimeter-wave (mmWave) radar sensors are increasingly being considered for AIoT systems due to their operational advantages, including robustness in diverse environmental conditions, inherent privacy-preserving characteristics, and the capability to detect fine-grained motion via micro-Doppler analysis [32]. These characteristics render mmWave radar suitable for applications like Human Activity Recognition (HAR) [2,21] and target classification [31] (e.g., human vs. non-human detection) in contexts like smart environments and healthcare (Fig. 1).

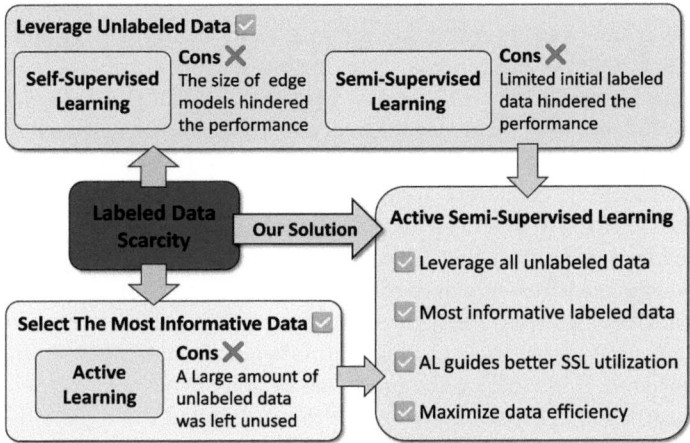

Fig. 1. Comparison of approaches for addressing label scarcity on edge device.

However, the practical deployment of deep learning models for radar-based target classification within typical IoT constraints encounters substantial challenges. Firstly, the acquisition of large-scale, accurately labeled radar datasets is often labor-intensive [9,18]. Manual annotation is not only time-consuming but also often demands specialized expertise for interpreting radar signatures, thereby posing a significant barrier to dataset collection [19,28]. Secondly, deployment targets such as edge devices frequently impose strict limitations on model complexity due to constraints on computational power, memory, and energy consumption [1]. Consequently, lightweight models are often employed for AIoT applications, which usually offer limited capacity particularly when labeled data is scarce.

Addressing these challenges with standard machine learning paradigms proves difficult. The performance of standard supervised learning approaches, for instance, degrades significantly when faced with insufficient labeled data [19,22,28]. Although self-supervised learning (SelfSL)and semi-supervised learning (SSL) paradigm [4,6,25,27] show their capacity in leveraging unlabeled data, more recent works found lightweight models could hardly take advantage of unlabeled data [33]. Active Learning (AL) optimizes labeling by selecting informative samples [13,14], but it typically ignores the remaining unlabeled data during model training.

The limitations of semi-supervised learning (SSL) with small models are often exacerbated by the quality and representativeness of the initial labeled set [10], which hinder the learning from unlabeled data for AIoT model. Strategically selected labeled samples, identified by active learning (AL), could provide a substantially more robust foundation for SSL. To enhance the capacity of AIoT model in a label-scarce scenario, where solely relying on self-supervised or semi-supervised learning fails, this paper introduce the active learning paradigm to identify informative and generalized samples to improve the model learning from unlabeled data. Specifically, we propose an **Active Semi-Supervised Learning (ASSL)** framework tailored for data-efficient mmWave radar target classification using lightweight models. Our active learning approach integrates a novel **Hybrid Uncertainty-Diversity Query (HUDQ)** strategy, which balances model **U**ncertainty and data **D**iversity metrics, to actively identify the most informative unlabeled samples. We determined that a hybrid approach, which simultaneously considers model uncertainty (**exploitation of known weaknesses**) and data diversity (**exploration of new data space**), is crucial for efficient learning. This is particularly true for complex data such as mmWave radar signatures, which exhibit subtle variations and potential ambiguities. These actively selected samples, after annotation, are incorporated into the training process alongside the remaining unlabeled data using a semi-supervised learning algorithm. This allows the model to benefit from both targeted labeling of critical samples and broad exposure to the unlabeled data distribution.

Experiments on our collected mmWave radar dataset demonstrate remarkable data efficiency: our proposed ASSL framework, starting with only 500 labeled samples and actively acquiring 1500 samples, **surpasses the performance of a fully supervised model trained on the entire 10,565 labeled samples (10,000+ labels) on lightweight architectures**. This highlights the practical value of our approach for AIoT deployments facing limited labels and computational budgets.

The main contributions of this work are as follows:

- An ASSL framework tailored for data-efficient mmWave radar target classification on lightweight AIoT model.
- A novel Hybrid Query Strategy (HUDQ) combining uncertainty and diversity for effective active selection in the context of mmWave radar data.
- A demonstration of the synergistic effect between HUDQ active selection and a robust semi-supervised training methodology, leading to significant performance gains with limited labels.
- Experimental results verify the effectiveness of the proposed ASSL framework, which surpassed a fully supervised model while achieving approximately an 80% reduction in labeling requirements, offering a practical AIoT solution highly suitable for deployments facing limited labeled data and constrained computational resources.

The remainder of this paper is organized as follows: Sect. 2 reviews related work. Section 3 details the proposed ASSL framework and its components.

Section 4 presents the experimental setup and results. Section 5 discusses the findings, and Sect. 6 concludes the paper.

2 Related Work

This section reviews methods addressing data scarcity, focusing on radar-based applications, categorizing them into Self-Supervised Learning (SelfSL), Semi-Supervised Learning (SSL), Active Learning (AL), and their combination, Active Semi-Supervised Learning (ASSL).

2.1 Self-Supervised Learning

Self-supervised methods learn representations from unlabeled data via pretext tasks like contrastive learning (e.g., SimCLR [6]) or reconstruction [12]. While effective for large models and datasets [6], the benefits of self-supervised methods often diminish when applied to lightweight models commonly used in edge devices [11,16]. For instance, while self-supervision has shown benefits for larger models on micro-Doppler data [19], our preliminary experiments with a shallow CNN yielded limited performance gains. This was likely due to the model's restricted representational capacity, which motivated our exploration of alternative data efficiency techniques.

2.2 Semi-Supervised Learning (SSL)

Semi-supervised learning (SSL) leverages both labeled (L) and unlabeled (U) data for the target task. Key paradigms include consistency regularization (e.g., Mean Teacher [26], UDA [29]) and pseudo-labeling [5]. One such widely used baseline, FixMatch [25], combines confident pseudo-labels generated from weakly augmented unlabeled data with a consistency loss against predictions on strongly augmented versions of the same data.

SSL techniques have also been explored in the context of radar-based HAR. For example, Li et al. [15] employed a method based on domain adaptation and knowledge distillation (JDS-TL), which requires a labeled source domain. Similarly, Shi et al. [24] utilized a Mean Teacher framework with consistency loss for radar-based hand gesture recognition, incorporating a pre-training phase with an autoencoder. While these studies demonstrate the potential of SSL, they often rely on specific assumptions (e.g., the availability of source domain data or adherence to a Mean Teacher framework) that differ from our approach. Our work, in contrast, focuses on integrating active selection without the prerequisite of auxiliary domains.

2.3 Active Learning (AL)

Active Learning (AL) aims to minimize labeling effort by selecting the most informative unlabeled samples, often based on uncertainty or diversity metrics [13,14]. This enhances learning efficiency, especially when labels are scarce.

Although the principles of AL are broadly applicable, their direct application to enhance label efficiency in radar classification tasks has been less explored in the literature. For instance, Coutino et al. [8] applied AL to radar system performance verification, specifically for function approximation. This objective differs from our focus, which is to select data efficiently for classifier training.

3 Methodology: The Proposed ASSL Framework

To address the challenges of label scarcity and model constraints in mmWave radar target classification for IoT applications, we propose an Active Semi-Supervised Learning (ASSL) framework. This framework integrates an intelligent active sample selection strategy with a robust semi-supervised learning algorithm to efficiently leverage both limited labeled data and abundant unlabeled data. An overview of the iterative process is depicted in Fig. 2.

ASSL synergizes AL's sample selection with SSL's use of unlabeled data. In ASSL, AL guides the selection of informative samples for labeling. These labeled samples then serve as a crucial component or foundation for the subsequent SSL process, which utilizes the entire data pool (L and U). Existing ASSL approaches vary, ranging from sequential application of AL and SSL [10] to tighter integrations. These tighter integrations may involve using consistency metrics for both the SSL loss calculation and the AL query strategy [33], or employing teacher-student frameworks [20]. Benz et al. [3] proposed an ASSL method for automotive radar object classification. Their approach leverages track consistency to generate pseudo-labels for SSL and employs relevance/quality metrics for AL, making it particularly suited for tracking scenarios.

Our work differs significantly by targeting **data-efficient mmWave target classification with lightweight models**. Unlike approaches such as [15], our method does not assume the availability of source domains. Instead, we focus on integrating AL with a robust SSL framework (distinct from Mean Teacher-based approaches [24] or tracking-specific methods [3]). Furthermore, we introduce a novel **Hybrid Query Strategy (HUDQ)** that combines uncertainty and diversity, specifically tailored for this constrained setting. A key aspect of our work is the systematic evaluation of the **synergy between AL and SSL**, through which we demonstrate substantial improvements in data efficiency.

3.1 Framework Overview

Our ASSL framework operates iteratively. Starting with a small initial labeled dataset L_0 and a large pool of unlabeled data U_0, each iteration t involves two main phases:

1. **Active Sample Selection (Sect. 3.2):** The current model M_t evaluates the unlabeled pool U_t. Our Hybrid Query Strategy (HUDQ) then selects the K most informative samples denoted as B_K based on uncertainty and diversity metrics.

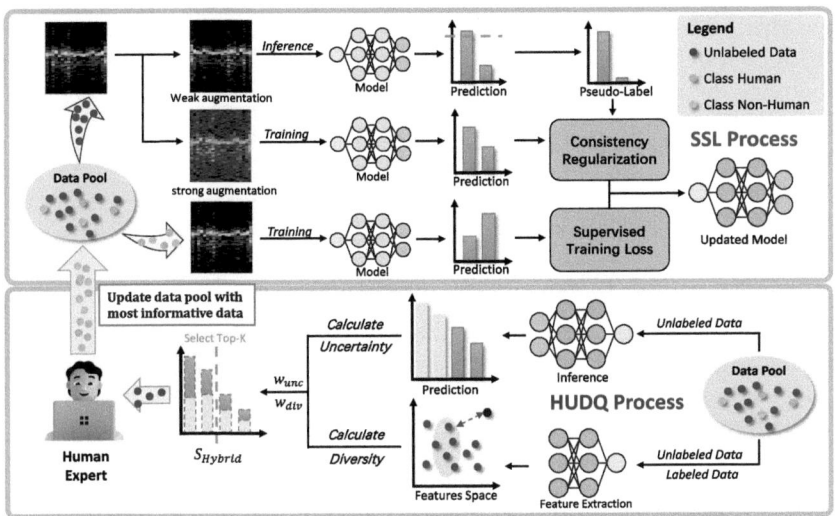

Fig. 2. Overview of the proposed Active Semi-Supervised Learning (ASSL) framework

2. **Semi-Supervised Model Training (Sect. 3.3):** The selected samples B_K are annotated (simulated using ground-truth labels in our experiments) to obtain their corresponding labels Y_K. The labeled set is updated to $L_{t+1} = L_t \cup (B_K, Y_K)$, and the unlabeled pool becomes $U_{t+1} = U_t \setminus B_K$. Subsequently, a new model M_{t+1} is trained using an SSL algorithm on the updated labeled set L_{t+1} and unlabeled pool U_{t+1}.

This loop continues until a predefined labeling budget or stopping criterion is met. The final model M_T is used for evaluation.

Central to our framework is a base classification model, which is iteratively trained and utilized for sample selection. Given the typical resource constraints of edge IoT devices, we employ a lightweight 2D Convolutional Neural Network (CNN) as this base classifier. The architecture of this CNN, designed to balance classification performance with computational efficiency. It consists of two convolutional blocks (16 and 32 filters of size 3×3, respectively, each followed by Batch Normalization, ReLU, and Max Pooling) and a final fully connected layer with Softmax activation to output probabilities for the C target classes (here, $C = 2$).

3.2 Active Sample Selection via HUDQ

The core of our active learning component is the Hybrid Query Strategy (HUDQ), designed to select samples that offer a balance between informing the model about uncertain regions and ensuring coverage of the data distribution. For each unlabeled sample $x_u \in U_t$, we calculate a hybrid score $S_{\text{Hybrid}}(x_u)$.

Uncertainty Measurement. We quantify model uncertainty using the **Least Confidence (LC)** metric. Given the model M_t with parameters θ_t, it predicts a probability distribution $P(y|x_u; \theta_t)$ for sample x_u. The LC score is:

$$U_{\text{LC}}(x_u) = 1 - \max_y P(y|x_u; \theta_t) \tag{1}$$

A higher U_{LC} indicates greater uncertainty.

Diversity Measurement. To promote diversity, we measure how dissimilar an unlabeled sample is from the current labeled set L_t in the feature space learned by the model's encoder $E(x; \theta_{e_t})$. We use the **Minimum Euclidean Distance** to the labeled set as the diversity metric. Let $f(x) = E(x; \theta_{e_t})$ denote the feature vector extracted by the encoder (e.g., the output before the final fully connected layer or after the GAP layer). The diversity score is:

$$D_{\text{MD}}(x_u) = \min_{x_l \in L_t} \|f(x_u) - f(x_l)\|_2 \tag{2}$$

A higher D_{MD} suggests that x_u is located far from any currently labeled sample in the feature space, thus representing a potentially novel region.

Score Normalization and Hybridization. As uncertainty and diversity scores may inherently possess different scales, we normalize them to a common $[0, 1]$ range across the current unlabeled pool U_t using min-max scaling:

$$\tilde{U}_{\text{LC}}(x_u) = \frac{U_{\text{LC}}(x_u) - \min_{z \in U_t} U_{\text{LC}}(z)}{\max_{z \in U_t} U_{\text{LC}}(z) - \min_{z \in U_t} U_{\text{LC}}(z) + \epsilon} \tag{3}$$

$$\tilde{D}_{\text{MD}}(x_u) = \frac{D_{\text{MD}}(x_u) - \min_{z \in U_t} D_{\text{MD}}(z)}{\max_{z \in U_t} D_{\text{MD}}(z) - \min_{z \in U_t} D_{\text{MD}}(z) + \epsilon} \tag{4}$$

where ϵ is a small constant to prevent division by zero.

Relying solely on uncertainty might select redundant samples near the decision boundary, neglecting broader feature space coverage essential for generalization, especially for mmWave Doppler data with subtle intra-class variations (e.g., different gaits). Conversely, prioritizing only diversity might select easily classifiable outliers, offering little immediate value for refining decision boundaries. Therefore, our HUDQ strategy (Eq. 5) employs a weighted combination of these complementary aspects. The uncertainty term ($\tilde{U}_{\text{LC}}$) targets samples challenging the model's current understanding, aiding accuracy. The diversity term ($\tilde{D}_{\text{MD}}$) promotes sampling from underrepresented feature regions, fostering generalization and robustness against radar signal variability. The weights w_{unc} and w_{div} allow controlling this trade-off.

The final hybrid score for an unlabeled sample x_u is computed as a weighted sum:

$$S_{\text{Hybrid}}(x_u) = w_{\text{unc}} \cdot \tilde{U}_{\text{LC}}(x_u) + w_{\text{div}} \cdot \tilde{D}_{\text{MD}}(x_u) \tag{5}$$

where w_{unc} and w_{div} are hyperparameters that control the balance between uncertainty and diversity, satisfying $w_{\text{unc}} + w_{\text{div}} = 1$. In our experiments, we typically set $w_{\text{unc}} > w_{\text{div}}$ (e.g., 0.7 and 0.3) to prioritize uncertainty while still incorporating diversity.

Selection. At each iteration t, we select the K samples from U_t with the highest S_{Hybrid} scores to form the batch B_K for simulated annotation.

3.3 Semi-Supervised Training Module

After updating the labeled set L_{t+1} and unlabeled pool U_{t+1}, the model M_{t+1} is trained using a chosen SSL method (FixMatch [25] in our case), leveraging both data types. This SSL approach heavily relies on data augmentation to create different views of the input data, which is crucial for enforcing consistency in the learning process and improving model generalization.

Data Augmentation. FixMatch uses two augmentation levels:

- **Weak Augmentation** (A_{weak}): Standard transformations preserving semantics (e.g., horizontal flips used here).
- **Strong Augmentation** (A_{strong}): Aggressive transformations designed for consistency regularization (e.g., jittering used here, suitable for micro-Doppler [23]).

Loss Function. The training objective combines supervised and unsupervised losses:

$$\mathcal{L} = \mathcal{L}_s + \lambda_u \mathcal{L}_u \tag{6}$$

where $\mathcal{L}_s$ is the standard cross-entropy loss on weakly augmented labeled samples $(x_i, y_i) \in L_{t+1}$:

$$\mathcal{L}_s = \frac{1}{B_L} \sum_{i=1}^{B_L} H(y_i, P(M_{t+1}(A_{\text{weak}}(x_i)))) \tag{7}$$

and $\mathcal{L}_u$ is the consistency loss on unlabeled samples $u_j \in U_{t+1}$. It uses confident pseudo-labels $\hat{y}_j$ (derived from $P(M_{t+1}(A_{\text{weak}}(u_j)))$ if confidence $\max(q_j) \geq \tau$) as targets for predictions on strongly augmented versions $P(M_{t+1}(A_{\text{strong}}(u_j)))$:

$$\mathcal{L}_u = \frac{1}{B_U} \sum_{j=1}^{B_U} \mathbb{1}(\max(q_j) \geq \tau) \cdot H(\hat{y}_j, P(M_{t+1}(A_{\text{strong}}(u_j)))) \tag{8}$$

Here, B_L, B_U are batch sizes, H is cross-entropy, $P(\cdot)$ denotes model predictions, λ_u weights the unsupervised loss (typically 1), and τ is the confidence threshold.

Model Training Iteration. Within each ASSL iteration, the model M_{t+1} is trained by minimizing $\mathcal{L}$. We typically re-initialize the model weights at the beginning of the training phase in each ASSL iteration to mitigate error accumulation from previous iterations, although continuous training (fine-tuning) presents an alternative strategy. Specific optimizers, learning rates, schedulers, and epoch counts are detailed in Sect. 4.2.

4 Experiments

This section details the experimental setup, evaluation metrics, baseline methods, and corresponding results used to validate the effectiveness of our proposed Active Semi-Supervised Learning (ASSL) framework. The framework integrates the HUDQ strategy with a well-established SSL technique (FixMatch) for mmWave radar target classification using a lightweight model.

4.1 Dataset and Preprocessing

We collected a dataset using a Texas Instruments IWR6432 low-power mmWave radar (57–64 GHz band) for human vs. non-human target classification. Data were captured at 10 Hz across 10 diverse indoor office environments, featuring various human subjects and non-human objects (e.g., fans, plants) positioned typically 0.5–5 m from the sensor. The dataset comprises 15,092 instances, each a 2-second sequence (20 frames). Following real-time processing on the IWRL6432 (as described in [30]), the input micro-Doppler features for our classifier are represented as matrices with dimensions of (64, 20).

Data Split: We reserved 30% (4,527 samples) for testing. From the remaining 10,565 samples, we created an initial balanced labeled set L_0 of 500 samples (250 human, 250 non-human) and used the rest (10,065 samples) as the initial unlabeled pool U_0. Table 1 details the split.

Table 1. Dataset Split Summary

Set Type	Human	Non-Human	Total
Initial Labeled L_0	250	250	500
Unlabeled Pool U_0	4932	5133	10065
Test Set T	2240	2287	4527
Total	7432	7660	15092

4.2 Implementation Details

Base Model: We used the lightweight 2D CNN described in Sect. 3.

Active Learning Setup: Starting with L_0 (500 samples), we ran $T = 15$ iterations, querying $K = 100$ samples per iteration using HUDQ ($w_{\text{unc}} = 0.7, w_{\text{div}} = 0.3$ default) up to 2000 total labeled samples.

Semi-supervised Learning Setup: For the SSL component (FixMatch), we set the confidence threshold $\tau = 0.95$ and the unsupervised loss weight $\lambda_u = 1$. A_{weak} was random horizontal flips; A_{strong} was jittering (Gaussian noise, std=0.15).

Training Details: Models were trained from scratch for $E_{\text{iter}} = 70$ epochs within each ASSL iteration using AdamW [17] (initial LR 10^{-4}, weight decay 10^{-5}) with a Cosine Annealing scheduler (min LR 10^{-6}). We used a batch size of 32 for both labeled and unlabeled data ($B_L = B_U = 32$). All experiments were conducted using PyTorch on an NVIDIA A100 GPU.

Evaluation Metrics: Model performance was evaluated on the test set using Accuracy and Micro F1-Score.

4.3 Compared Methods

We compare our proposed ASSL method against several relevant baselines and variations:

- **Supervised-Small (Sup-500):** The baseline model trained only on the initial 500 labeled samples (L_0). This establishes the reference performance achievable with the minimal set of initial labeled data, highlighting the challenge our method aims to address.
- **Supervised-Full (Sup-10k+):** The baseline model trained on the entire available training data (10,565 samples: $L_0 \cup U_0$) with their ground-truth labels. This serves as an approximate upper bound, indicating the potential performance if all data were labeled.
- **Supervised-Small + Random Sample (Sup-2k):** The baseline model trained on the initial 500 labeled samples (L_0) augmented with an additional 1,500 randomly selected and labeled samples from the unlabeled pool (U_0). This baseline helps quantify the benefit of intelligent sample selection by comparing active learning against simple random sampling for data acquisition.
- **SSL-Only (using FixMatch):** A standard SSL method (FixMatch [25]) trained using the initial labeled set L_0 and the full unlabeled pool U_0 for E_{iter} epochs. This comparison evaluates the effectiveness of a pure SSL approach (without active selection) in leveraging unlabeled data.
- **ASSL (Random Sampling + SSL):** Our ASSL framework employing the chosen SSL method (FixMatch), but using random sampling instead of HUDQ to select K samples at each iteration. This allows for a direct assessment of the contribution of AL by comparing it against a non-intelligent active selection within the same ASSL framework.

- **ASSL (Uncertainty Sampling + SSL):** Our ASSL framework with the SSL method (FixMatch), using only the uncertainty component (Least Confidence) of HUDQ ($w_{\text{unc}} = 1, w_{\text{div}} = 0$) for sample selection. This is included to analyze the specific impact of the uncertainty criterion in our hybrid query strategy.
- **ASSL (Diversity Sampling + SSL):** Our ASSL framework with the SSL method (FixMatch), using only the diversity component (Minimum Euclidean Distance) of HUDQ ($w_{\text{unc}} = 0, w_{\text{div}} = 1$) for sample selection. This helps to understand the specific contribution of the diversity criterion in our hybrid query strategy.
- **AL (HUDQ Sampling) Only:** The HUDQ active learning strategy is used to select 1,500 samples, which are added (with true labels) to L_0. The model is then trained only on this final labeled set (2,000 samples) using standard supervised learning, without any SSL component. This isolates and quantifies the benefit derived purely from the HUDQ active selection strategy.
- **SimCLR + Supervised (Self-Sup Pretrain):** The base model is first pre-trained using SimCLR [6], a representative contrastive self-supervised learning method, on all 10,565 available training samples (treated as unlabeled). Subsequently, it is fine-tuned on the initial 500 labeled samples (L_0). This comparison positions our ASSL approach against another prominent paradigm for leveraging unlabeled data.
- **Proposed ASSL (HUDQ + SSL) (Ours):** Our proposed full framework, which synergistically combines the HUDQ active selection strategy with the chosen SSL method (FixMatch) for semi-supervised training. This is the main approach whose performance and data efficiency are evaluated against the aforementioned baselines.

4.4 Main Results

Table 2 summarizes the performance of all methods, demonstrating the effectiveness of our proposed approach. As observed in Fig. 3, our proposed ASSL method (integrating HUDQ and SSL) consistently outperforms all other strategies that utilize the same number of labeled samples. Starting from the initial 500 labeled samples, our method shows a steeper performance increase with each active learning iteration compared to random sampling combined with SSL, highlighting the effectiveness of the HUDQ selection strategy. Notably, compared to the HUDQ + Supervised approach (which reaches 96.02% accuracy with 2000 labels), our full ASSL framework leverages the remaining unlabeled data via SSL to achieve a significantly higher accuracy of **97.31%** with the same 2000 actively selected labels. This demonstrates the crucial synergistic effect between intelligent sample selection and robust semi-supervised learning.

Furthermore, standard FixMatch-Only (pure SSL starting with 500 random labels) achieves 92.35% accuracy, indicating the benefit of SSL over the Supervised-Small baseline (90.13%). However, our ASSL approach surpasses this performance significantly, showcasing the advantage of actively guiding the learning process. Most strikingly, with only 2000 labeled samples (500 initial + 1500

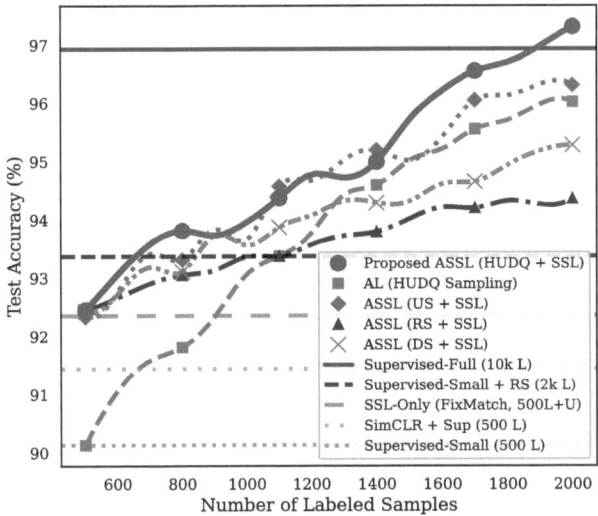

Fig. 3. Test accuracy comparison of different methods as the number of labeled samples increases through active learning iterations.

actively selected), **our proposed ASSL method (HUDQ + SSL) (97.31%) exceeds the performance of the fully supervised model trained on the entire 10,565 labeled samples (96.93%)**. This result underscores the remarkable data efficiency of our proposed framework, achieving superior performance with approximately **80% fewer labels** compared to full supervision in this resource-constrained setting.

4.5 Ablation Studies

To further understand the contribution of different components, we perform ablation studies.

Query Strategy Components. We compare our full HUDQ strategy against using only uncertainty (US-SSL) or only diversity (DS-SSL) within the ASSL framework. As shown in Fig. 3 and Table 2, the hybrid HUDQ strategy consistently outperforms using either uncertainty or diversity alone, confirming the benefit of balancing both aspects for selecting informative samples.

Synergy Between AL and SSL. The powerful synergistic effect of integrating active learning (AL) with semi-supervised learning (SSL) is clearly demonstrated by comparing our **Proposed ASSL (HUDQ + SSL)** framework (achieving 97.31% accuracy) with the **AL (HUDQ Sampling) Only** baseline (96.02% accuracy), as detailed in Fig. 3 and Table 2. The notable performance improvement of approximately 1.3% is not merely an additive contribution from the

Table 2. All Methods Performance Comparison

Method	Accuracy (%)	F1-Score (Micro)
Supervised-Small (500)	90.13	89.75
FixMatch (500 L + U)	92.35	91.98
SimCLR + Sup (500 L)	91.44	90.96
Supervised-Small + RS (2000 L)	93.37	93.02
ASSL (RS + SSL)	94.36	94.12
ASSL (US + SSL)	96.31	96.05
ASSL (DS + SSL)	95.28	94.93
HUDQ (2000 L)	96.02	95.78
ASSL (HUDQ + SSL) (Ours)	**97.31**	**97.05**
Supervised-Full (10k L)	96.93	96.72

SSL mechanism (specifically, FixMatch); rather, it highlights a strong interplay between AL and SSL. HUDQ first pinpoints the most informative samples for annotation. The SSL component then effectively leverages these high-value annotated samples along with the vast unlabeled data pool. This synergy—combining AL's precise selection with SSL's enhanced data utilization—yields a more significant performance gain than AL could achieve alone with the same labeled set.

Table 3. Impact of w_{unc} and w_{div} on Final Accuracy

Weight Combination	Final Accuracy (%)
$w_{unc} = 1.0, w_{div} = 0.0$	96.31
$w_{unc} = 0.9, w_{div} = 0.1$	96.64
$\boldsymbol{w_{unc} = 0.7, w_{div} = 0.3}$	**97.31**
$w_{unc} = 0.5, w_{div} = 0.5$	96.06
$w_{unc} = 0.3, w_{div} = 0.7$	95.80
$w_{unc} = 0.1, w_{div} = 0.9$	95.55
$w_{unc} = 0.0, w_{div} = 1.0$	95.28

Hybrid Query Weights (w_{unc}, w_{div}). To validate the effectiveness of the hybrid nature of our HUDQ strategy, we systematically evaluated the impact of the weighting factors w_{unc} and w_{div}, ensuring $w_{unc} + w_{div} = 1$. Table 3 illustrates the final test accuracy achieved after 15 active learning iterations for various weight combinations.

Pure uncertainty sampling ($w_{unc} = 1.0$; "exploitation") yielded 96.31% accuracy, targeting difficult boundary samples but risking redundancy and neglecting coverage, especially for varied mmWave signatures. Pure diversity sampling

$w_{\text{div}} = 1.0$; "exploration") achieved 95.28%, ensuring feature space exploration but potentially selecting easily classifiable, less critical samples.

Our experiments show that several hybrid strategies outperform both pure approaches. Notably, the configuration with $w_{\text{unc}} = 0.7$ and $w_{\text{div}} = 0.3$ yielded the best performance, reaching **97.31%** accuracy. Other combinations like $w_{\text{unc}} = 0.9, w_{\text{div}} = 0.1$ (96.64%) also performed strongly, generally showing that a blend prioritizing uncertainty while incorporating diversity is beneficial. Over-emphasizing diversity (e.g., $w_{\text{unc}} = 0.1, w_{\text{div}} = 0.9$ resulted in 95.55%) led to suboptimal results compared to the best hybrid mix.

This confirms the synergistic value of HUDQ. By balancing uncertainty (addressing model weaknesses, improving accuracy) and diversity (ensuring data coverage, improving generalization), HUDQ achieves a more effective and efficient learning trajectory. This balanced approach is particularly crucial for navigating the complexities of mmWave radar data, which features both fine-grained distinctions and broad variations.

5 Discussion

Our experiments validate the integration of SSL substantially boosted AL performance, confirming the value of Our ASSL framework. Ablation studies confirmed the hybrid nature of HUDQ, balancing uncertainty and diversity, is crucial for optimal performance, outperforming strategies based solely on either criterion. This suggests that selecting both "difficult" (uncertain) and "representative" (diverse) samples is key for efficiently training lightweight models. Limitations include the diversity metric's dependence on learned feature quality and the need for hyperparameter tuning. Although data efficiency has been significantly improved, a portion of the data still requires manual annotation.

6 Conclusion

We presented an Active Semi-Supervised Learning (ASSL) framework tailored for mmWave radar target classification, addressing label scarcity and model constraints prevalent in AIoT. Our approach integrates a novel Hybrid Query Strategy (HUDQ) with an SSL algorithm, a synergistic combination of intelligent sample selection and semi-supervised learning that results in significantly enhanced model performance and data efficiency. Experiments demonstrated significant data efficiency, with our proposed ASSL method requiring substantially fewer labeled samples (approx. 2000) to outperform a fully supervised model trained on a much larger dataset (approx. 10,000). This work offers a practical and effective solution for building accurate, resource-efficient radar-based AIoT systems. Future work may explore adaptive query strategies and integration with online learning settings.

Acknowledgments. This work was conducted while the first author Liyang Zheng was an intern at HeyiSpace. The authors would like to thank HeyiSpace for providing the necessary resources and support for this research.

Disclosure of Interests. Liyang Zheng and Miao Zhang have received research grants from HeyiSpace.

References

1. Abadade, Y., Temouden, A., Bamoumen, H., Benamar, N., Chtouki, Y., Hafid, A.S.: A comprehensive survey on tinyml. IEEE Access **11**, 96892–96922 (2023). https://doi.org/10.1109/ACCESS.2023.3294111
2. An, S., Ogras, U.Y.: Mars: mmwave-based assistive rehabilitation system for smart healthcare. ACM Trans. Embed. Comput. Syst. **20**(5s) (2021). https://doi.org/10.1145/3477003
3. Benz, J., Weiss, C., Aponte, A.A., Hakobyan, G.: Semi-supervised active learning for radar based object classification using track consistency. In: 2023 IEEE Radar Conference (RadarConf23), pp. 1–6 (2023). https://doi.org/10.1109/RadarConf2351548.2023.10149705
4. Bonab, M.N., Tanha, J., Masdari, M.: A semi-supervised learning approach to quality-based web service classification. IEEE Access **12**, 50489–50503 (2024). https://doi.org/10.1109/ACCESS.2024.3385341
5. Cascante-Bonilla, P., Tan, F., Qi, Y., Ordonez, V.: Curriculum labeling: revisiting pseudo-labeling for semi-supervised learning. In: Proceedings of the AAAI Conference on Artificial Intelligence, vol. 35, pp. 6912–6920 (2021)
6. Chen, T., Kornblith, S., Norouzi, M., Hinton, G.: A simple framework for contrastive learning of visual representations. In: International Conference on Machine Learning, pp. 1597–1607. PmLR (2020)
7. Conteh, A., Al-Turjman, F.: A review of artificial intelligence of things. In: The International Conference on Forthcoming Networks and Sustainability (FoNeS 2022), vol. 2022, pp. 832–837 (2022). https://doi.org/10.1049/icp.2022.2559
8. Coutino, M., Cox, P., Lascaris, Z.: Active learning for radar system performance verification. In: 2024 IEEE Radar Conference (RadarConf24), pp. 1–6 (2024). https://doi.org/10.1109/RadarConf2458775.2024.10548721
9. Cui, H., Zhong, S., Wu, J., Shen, Z., Dahnoun, N., Zhao, Y.: Milipoint: a point cloud dataset for mmwave radar. In: Advances in Neural Information Processing Systems, vol. 36, pp. 62713–62726 (2023)
10. Gao, M., Zhang, Z., Yu, G., Arık, S.Ö., Davis, L.S., Pfister, T.: Consistency-based semi-supervised active learning: towards minimizing labeling cost. In: Vedaldi, A., Bischof, H., Brox, T., Frahm, J.-M. (eds.) ECCV 2020. LNCS, vol. 12355, pp. 510–526. Springer, Cham (2020). https://doi.org/10.1007/978-3-030-58607-2_30
11. Gao, Y., et al.: Disco: remedying self-supervised learning on lightweight models with distilled contrastive learning. In: European Conference on Computer Vision, pp. 237–253. Springer (2022)
12. Hinton, G.E., Salakhutdinov, R.R.: Reducing the dimensionality of data with neural networks. Science **313**(5786), 504–507 (2006)
13. Kumar, P., Gupta, A.: Active learning query strategies for classification, regression, and clustering: a survey. J. Comput. Sci. Technol. **35**, 913–945 (2020)
14. Li, D., Wang, Z., Chen, Y., Jiang, R., Ding, W., Okumura, M.: A survey on deep active learning: recent advances and new frontiers. IEEE Trans. Neural Netw. Learn. Syst. (2024)
15. Li, X., He, Y., Fioranelli, F., Jing, X.: Semisupervised human activity recognition with radar micro-doppler signatures. IEEE Trans. Geosci. Remote Sens. **60**, 1–12 (2021)

16. Liu, H., Ye, M.: Improving self-supervised lightweight model learning via hard-aware metric distillation. In: European Conference on Computer Vision, pp. 295–311. Springer (2022)
17. Loshchilov, I., Hutter, F.: Decoupled weight decay regularization. arXiv preprint arXiv:1711.05101 (2017)
18. Rahman, M.M., et al.: MMVR: millimeter-wave multi-view radar dataset and benchmark for indoor perception (2024). https://arxiv.org/abs/2406.10708
19. Rahman, M.M., Gurbuz, S.Z.: Self-supervised contrastive learning for radar-based human activity recognition. In: 2023 IEEE Radar Conference (RadarConf23), pp. 1–6 (2023). https://doi.org/10.1109/RadarConf2351548.2023.10149770
20. Rangnekar, A., Kanan, C., Hoffman, M.: Semantic segmentation with active semi-supervised learning. In: Proceedings of the IEEE/CVF Winter Conference on Applications of Computer Vision, pp. 5966–5977 (2023)
21. Sen, A., Das, A., Pradhan, S., Chakraborty, S.: A dataset for multi-intensity continuous human activity recognition through passive sensing. arXiv preprint arXiv:2407.21125 (2024)
22. Seyfioğlu, M.S., Gürbüz, S.Z.: Deep neural network initialization methods for micro-doppler classification with low training sample support. IEEE Geosci. Remote Sens. Lett. **14**(12), 2462–2466 (2017). https://doi.org/10.1109/LGRS.2017.2771405
23. She, D., Lou, X., Ye, W.: Radarspecaugment: a simple data augmentation method for radar-based human activity recognition. IEEE Sensors Letters **5**(4), 1–4 (2021)
24. Shi, Y., et al.: Semi-supervised FMCW radar hand gesture recognition via pseudo-label consistency learning. Remote Sens. **16**(13), 2267 (2024)
25. Sohn, K., et al.: Fixmatch: simplifying semi-supervised learning with consistency and confidence. Adv. Neural. Inf. Process. Syst. **33**, 596–608 (2020)
26. Tarvainen, A., Valpola, H.: Mean teachers are better role models: weight-averaged consistency targets improve semi-supervised deep learning results. In: Advances in Neural Information Processing Systems, vol. 30 (2017)
27. Upretee, P., Khanal, B.: Fixmatchseg: fixing fixmatch for semi-supervised semantic segmentation. arXiv preprint arXiv:2208.00400 (2022)
28. Wu, C., Ye, W.: Generative adversarial network for radar-based human activities classification with low training data support. In: 2021 IEEE 4th International Conference on Electronic Information and Communication Technology (ICEICT), pp. 415–419 (2021). https://doi.org/10.1109/ICEICT53123.2021.9531147
29. Xie, Q., Dai, Z., Hovy, E., Luong, T., Le, Q.: Unsupervised data augmentation for consistency training. Adv. Neural. Inf. Process. Syst. **33**, 6256–6268 (2020)
30. Yanik, M., Chandrasekaran, A.K., Rao, S.: Machine learning on the edge with the mmWave radar device IWRL6432 (2023). https://www.ti.com/lit/wp/swra774/swra774.pdf
31. Yanik, M.E., Rao, S.: Radar-based multiple target classification in complex environments using 1D-CNN models. In: 2023 IEEE Radar Conference (RadarConf23), pp. 1–6. IEEE (2023)
32. Zhang, J., et al.: A survey of mmwave-based human sensing: technology, platforms and applications. IEEE Commun. Surv. Tutor. **25**(4), 2052–2087 (2023). https://doi.org/10.1109/COMST.2023.3298300
33. Zhang, W., et al.: Boostmis: boosting medical image semi-supervised learning with adaptive pseudo labeling and informative active annotation. In: Proceedings of the IEEE/CVF Conference on Computer Vision and Pattern Recognition, pp. 20666–20676 (2022)

Runtime Heterogeneous Sensor Selection with Data Reuse in Multi-device Environments

Chun Li, Yu Zhang(✉)[ID], Hira Khyzer, Yu Yan, and Xingshe Zhou

Northwestern Polytechnical University, Xi'an 710129, China
{lichun2993,hirakhyzer,yyan}@mail.nwpu.edu.cn,
{zhangyu,zhouxs}@nwpu.edu.cn

Abstract. With the increasing deployment of diverse sensory devices on or near a human body, perception resource usage across various applications has been significantly enhanced. However, frequent sensor switching during concurrent multi-sensor operation on mobile devices for different tasks leads to excessive energy consumption. To this end, we propose an energy-efficient data-reuse-based runtime framework called DataReuse, which dynamically selects the best sensor for task execution. It utilizes a single sensor to handle different tasks by sharing data, thereby reducing sensor switching when new tasks arrive. DataReuse includes an adaptive sensor selection mechanism that selects the best sensor based on the current sensor state and the capability values of candidate sensors, and operates under both fixed-period and threshold-triggered policies to ensure optimal performance and timely response. The framework provides a Jacobson/Karels algorithm to evaluate sensor capability and combines it with LFU (Least Frequently Used) to update sensor capability, in order to improve estimation accuracy and reduce unnecessary updates. Our evaluation using the Opportunity dataset shows that DataReuse provides 24% energy savings with comparable accuracy across locomotion and activity recognition tasks when compared to state-of-the-art methods.

Keywords: Sensor selection · Runtime selection · Data reuse · Capability update · Multi-device environments

1 Introduction

The rapid proliferation of sensor-equipped devices, such as smartphones, wearables, and IoT systems, has revolutionized data collection and utilization for applications like autonomous driving [1], object tracking [6], and emotion recognition [12]. In multi-device environments, heterogeneous sensors are deployed on or around the human body to capture physiological and environmental data, enabling real-time inference tasks such as posture recognition, activity classification, and emotion detection. While sensor-fusion techniques [5,11] that leverage multiple devices can achieve high accuracy, they suffer from significant energy consumption in dynamic environments. Consequently, selecting an optimal sensor from a candidate set to execute tasks independently has emerged as a promising approach to balance accuracy and energy efficiency.

Although existing research has contributed to energy-efficient and accurate perception applications through sensor selection, our analysis of prior work reveals the following unresolved issues: 1) To achieve higher recognition accuracy, existing selection methods [3,9,10] only consider one sensor supporting one specific task, neglecting data sharing. In terms of selection timing, they increase execution frequency by shortening the selection duration or increasing the triggering threshold. This leads to frequent sensor switching, introducing significant energy overhead, as shown in Fig. 1.

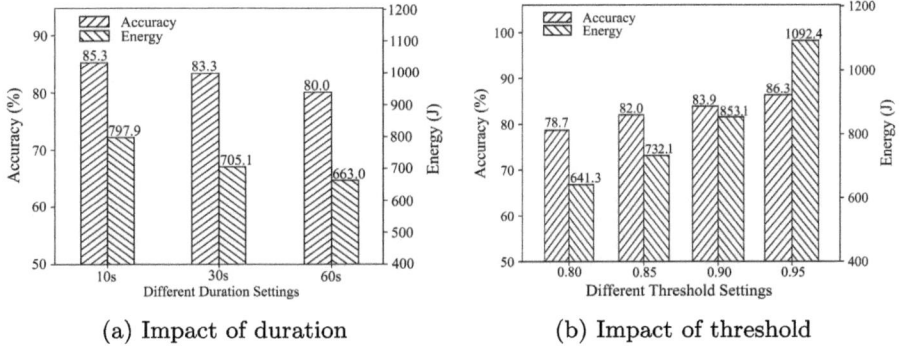

(a) Impact of duration (b) Impact of threshold

Fig. 1. Impact of duration and threshold on task execution. A smaller duration results in higher accuracy but increased energy consumption; a larger threshold similarly improves accuracy at the cost of higher energy consumption.

2) Capability values, which serve as the basis for runtime sensor selection, vary over time, as shown in Fig. 2. However, the update strategies of existing methods [8,13] either determine the capability value based solely on the current moment, leading to a decrease in recognition accuracy, or they are integrated with the sensor selection module, resulting in inefficient resource utilization. Moreover, updating efficiency values for all candidate sensors incurs energy consumption.

Given these two considerations, determining an appropriate strategy and timing for both sensor selection and capability updates while trading off energy consumption and accuracy constitutes the core of our research problem. To this end, we design and implement DataReuse, a runtime sensor selection scheme for multi-device environments, integrating sensor selection with adaptive updates. Through experiments, we demonstrate that single-sensor data can independently support diverse perception tasks, providing the basis for data reuse. Leveraging this, we explore opportunities for data reuse to enhance task efficiency, particularly in reducing energy consumption. For capability assessment, we adopt a

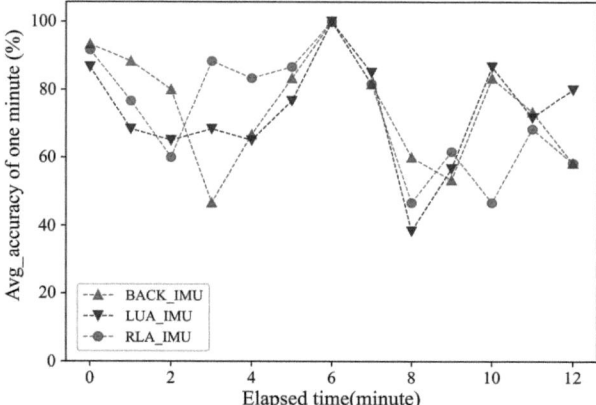

Fig. 2. Runtime characteristics of perception resources. The recognition average accuracy of the same perception resource varies over time when executing tasks; the perception resource with the highest recognition average accuracy differs across different time points.

Jacobson/Karels-based algorithm that integrates historical efficiency values from multiple past timestamps to determine updated values, thereby mitigating the impact of estimation lag. To manage energy during updates, we introduce an LFU (Least Frequently Used) policy that restricts the number of candidate sensors updated per cycle, effectively reducing energy consumption caused by data transmission. Finally, we develop a prototype system for locomotion and activity recognition, validating the significance of data reuse in improving accuracy and reducing energy consumption in multi-device environments.

In summary, the key contributions of the work are as follows:

- A preliminary experiment is conducted to analyze the performance of existing sensor selection methods in recognition tasks. During this experiment, we identify shortcomings in selection mechanisms and update strategies. Current approaches employ shorter sensor usage intervals and higher trigger thresholds, leading to increased energy consumption. Furthermore, their reliance on capability updates based only on the current moment compromises recognition accuracy.
- A data-reuse-based framework that includes a runtime sensor selection algorithm and a capability update algorithm to balance energy consumption and accuracy for task execution is proposed. The selection algorithm enables a single sensor to concurrently execute multiple tasks under shorter selection cycles and higher triggering thresholds, thereby minimizing redundant data transmission. The update algorithm combines the Jacobson/Karels algorithm with the LFU algorithm.
- A prototype system for implementation and evaluation is developed. Evaluations on locomotion recognition and activity recognition tasks show that

DataReuse achieves a 24% reduction in average energy consumption with only a 3% accuracy drop compared to state-of-the-art methods.

The rest of this paper is organized as follows. Section 2 summarizes related work. Section 3 presents the proposed DataReuse framework and the details of the runtime sensor selection and capability update algorithms. Sections 4 and 5 present the prototype implementation and performance evaluation, and Sect. 6 concludes the paper.

2 Related Work

We summarize two key research areas related to our method: sensor selection mechanisms and adaptive capability update strategies in multi-device environments.

Kang et al. [2] introduced Orchestrator, a rule-based framework that selects sensors statically using predefined contextual parameters. Min et al. [4] proposed a dynamic runtime method leveraging heuristic functions and a siamese neural network to assess sensor quality in real time, aiming to optimize both accuracy and energy efficiency. Zhong et al. [13] developed a sparsity-based selection method for distributed systems, employing convex optimization to reduce computational overhead while addressing correlated noise. Yasuo et al. [8] further advanced data-driven techniques with a greedy algorithm that minimizes a regularized ridge-regression cost function, enabling rapid computation of sensor sets to enhance estimation accuracy while mitigating overfitting. However, these methods exhibit critical limitations: rule-based or static approaches lack adaptability to runtime sensor performance variations, while dynamic and data-driven strategies often fail to holistically balance accuracy gains against the energy costs of frequent computations, or redundant data processing.

Capability updates are critical to sustaining selection accuracy in dynamic sensing environments. Min et al. [3] proposed SensiX, a runtime component that refreshes sensor capabilities at fixed intervals. In contrast, Taleb et al. [9,10] introduced EGO, an ontology-based framework using threshold-triggered updates. However, fixed-cycle updates inefficiently balance accuracy and energy: shorter intervals improve capability freshness but incur high update costs, whereas longer intervals risk outdated evaluations. Similarly, threshold-triggered updates depend on threshold configuration: a small threshold causes delayed adaptation to performance fluctuations, while a large threshold triggers unnecessary updates, leading to unnecessary energy consumption. Furthermore, existing methods often perform full updates across all candidate sensors, neglecting the energy overhead of redundant data collection, processing, and transmission. Our method decouples the selection and update modules, employing distinct strategies and triggering mechanisms, thereby dynamically balances accuracy-energy trade-offs.

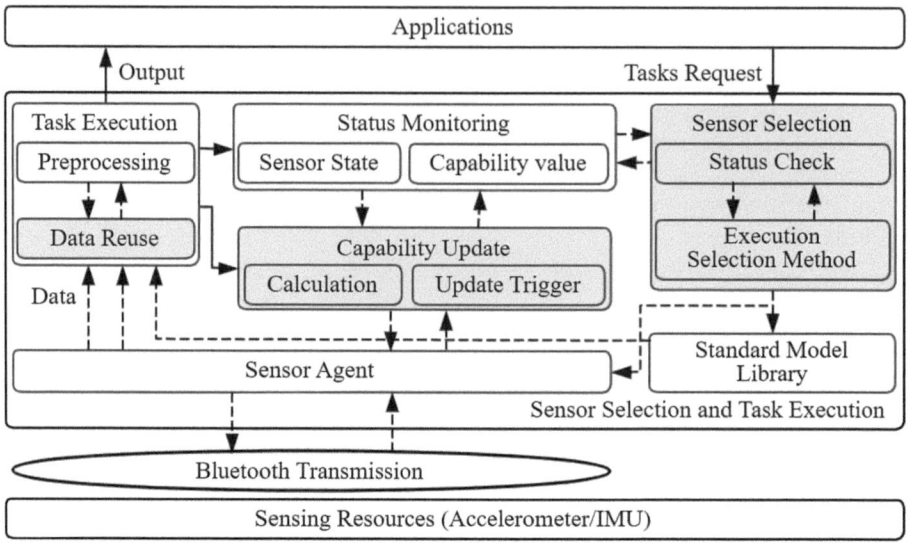

Fig. 3. The general framework of DataReuse. The colored blocks denote the core components of our framework: the data reuse minimizes redundant sensor activations through contextual reuse strategies; the sensor selection optimally balances accuracy and energy consumption; and the capability update dynamically maintains sensor capabilities for optimal selection.

3 System Design

In this section, we present the detailed design of our DataReuse framework, which aims to optimize runtime sensor selection in multi-device environments. The framework combines two main components: a data-reuse-based runtime sensor selection algorithm, and a Jacobson/Karels and LFU-based runtime capability update algorithm. They work in coordination to minimize energy consumption during sensing device switching and enhance the accuracy of perception task execution.

3.1 DataReuse Framework Overview

The DataReuse framework addresses the challenges associated with dynamic sensor management in multi-device environments. These challenges include the need for efficient sensor utilization, energy conservation, and accurate sensing task execution. The system architecture is divided into three primary layers: application layer, sensor selection and task execution layer, and perception resource layer, as illustrated in Fig. 3. At the application layer, task requests (e.g., locomotion recognition and activity recognition) are processed and forwarded to the sensor selection layer. The sensor selection layer employs a runtime selection mechanism that queries real-time sensor efficiency values from the state monitoring module, executes a selection algorithm to identify an optimal sensor, and deploys a sensor proxy to drive data collection. Simultaneously, the efficiency update module dynamically adjusts sensor performance metrics using a history-based assessment algorithm, reducing assessment errors. The perception resource layer abstracts physical devices for Bluetooth communication. During task execution, the inference module loads pretrained models from the model library to process collected data, with results returned to the application layer.

The runtime sensor selection module and capability update module are the core components of DataReuse, providing essential support for high-accuracy and low-energy execution of sensing tasks. These two modules are designed to be highly decoupled in architecture while functionally interconnected through a shared capability matrix, working collaboratively to optimize sensing task performance. The following subsections will separately elaborate on the runtime sensor selection and capability update modules.

3.2 Runtime Sensor Selection Algorithm

In multi-device environments, the dynamic selection of heterogeneous sensors during runtime is critical for balancing task execution accuracy and energy efficiency. Existing methods often prioritize higher accuracy by frequently switching sensors, leading to increased energy consumption. To this end, we propose a data-reuse-based runtime sensor selection mechanism that leverages the capability of a single sensor to concurrently execute multiple tasks, thereby reducing redundant data transmission and energy overhead.

Problem Formulation. The sensor selection aims to dynamically choose the optimal sensor for each task request while minimizing energy consumption and maintaining high accuracy. Let M denote concurrent task requests $L = \{l_1, l_2, ..., l_m\}$, and N represent available sensors $S = \{s_1, s_2, ..., s_n\}$. Each task l_m can be executed by a subset of sensors $S_{l_m} \subseteq S$.

Define a binary decision variable $Y_{l_m}^{s_n} \in \{0, 1\}$, where $Y_{l_m}^{s_n} = 1$ if sensor s_n is assigned to task l_m. The total energy consumption E_{total} comprises data transmission energy E_{s_n} and inference energy $E_{s_n}^{\text{inference}}$. Inference energy is determined by the computational load of executing the task-specific inference model on s_n. For data reuse, E_{s_n} is counted only once if a sensor executes multiple tasks. Therefore, the total energy consumption is formulated as:

$$E_{\text{total}} = \sum_{i=1}^{T} \sum_{n=1}^{N} \left(E_{S_n} \cdot Y_{l_m}^{S_n} + \sum_{m=1}^{M} E_{S_n}^{\text{inference}} \cdot Y_{l_m}^{S_n} \right) \quad (1)$$

The execution efficiency C_{total} is defined as the sum of the energy consumption of each task during the entire execution:

$$C_{\text{total}} = \sum_{i=1}^{T} \sum_{m=1}^{M} \left(s_{l_m}^{t} \left(\sum_{n=1}^{N} C_{l_m}^{s_n} \cdot Y_{l_m}^{s_n} \right) \right) \quad (2)$$

where $s_{l_m}^t$ indicates whether task l_m needs to be executed at time t. Specifically, $s_{l_m}^t = 1$ denotes that task l_m is requested, which depends on the application layer's task requirements. $C_{l_m}^{s_n}$ represents the capability value of sensor s_n for task l_m. If s_n does not support task l_m, $C_{l_m}^{s_n}$ is set to $-\infty$.

The optimization objective of the sensor selection problem is to minimize the total energy consumption E_{total} while maximizing the task execution efficiency C_{total}. This dual objective is formalized as a ratio-based optimization problem:

$$\text{minimize} \frac{\sum_{i=1}^{T} \sum_{n=1}^{N} \left(E_{s_n} \cdot Y_{l_m}^{s_n} + \sum_{m=1}^{M} E_{s_n}^{\text{inference}} \cdot Y_{l_m}^{s_n} \right)}{\sum_{i=1}^{T} \sum_{m=1}^{M} \left(s_{l_m}^t \left(\sum_{n=1}^{N} C_{l_m}^{s_n} \cdot Y_{l_m}^{s_n} \right) \right)} \quad (3)$$

Data Reuse Factor for Accuracy-Energy Trade-Off. The motivation for data reuse stems from the fact that transmitting sensing data constitutes the primary energy consumption source in perception task execution. Additionally, since a single sensor can support multiple tasks simultaneously, it becomes feasible to utilize the same sensor to respond to different tasks during execution. This approach effectively reduces energy consumption by minimizing redundant data transmission. For instance, as illustrated in Fig. 4, a sensor S_2 can execute both tasks T_1 and T_2 using the same data collected at a given timestep. Here, T_1 and T_2 correspond to distinct inference models (M_2^1 and M_2^i) but share the input data from S_2. Similarly, tasks T_j and T_n are assigned to another sensor S_j, which processes both requests through models M_j^i and M_j^n without additional data transmission.

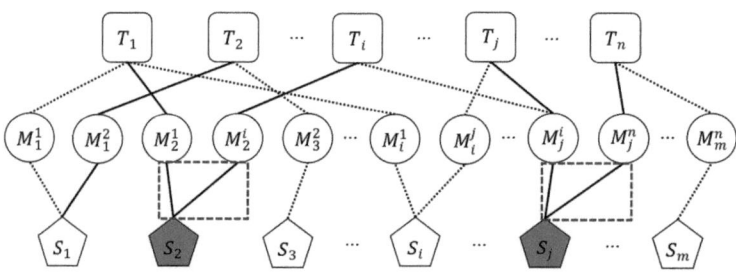

Fig. 4. Example of data reuse. The highlighted red boxes demonstrate reuse mechanism: although two data paths exist from S_2 to M_2^1 and M_2^i, only one instance of data transmission is required. (Color figure online)

Data reuse reduces energy consumption by allowing a single sensor to serve multiple tasks using the same data. However, this may compromise accuracy if the selected sensor is not optimal for some tasks. To balance this trade-off, we introduce a data reuse factor σ, which dynamically adjusts the preference between accuracy and energy efficiency.

As shown in Algorithm 1, the selection process first checks if any active sensor $s_n \in S_{\text{running}}$ (already executing tasks) can support the new task l_m. The sensor with the highest capability value $C_{l_m}^{s_n}$ in S_{running} is compared against a threshold derived from σ. If $\sigma' < \sigma$, the sensor is reused; otherwise, the algorithm selects the best available sensor from $S_{\text{not\_running}}$. Here, σ is configurable: a smaller σ prioritizes accuracy, while a larger σ favors energy savings.

Algorithm 1: Runtime Sensor Selection Based on Data Reuse Factor

---
Input: Task request L_m
Output: Selected sensor S_n
1 Initialize parameters: σ, S_n, $C_{\max}$;
2 **for** each S_i in active sensor set S_{running} **do**
3   **if** S_i supports L_m **then**
4   Retrieve $C_{l_m}^{s_i}$;
5   Update $C_{\max}$ and S_n if $C_{l_m}^{s_i} > C_{\max}$;
6   **end**
7 **end**
8 Compute $\sigma' = 1/(\sigma \cdot C_{\max} + 1)$;
9 **if** $\sigma' < \sigma$ **then**
10   Reuse S_n, update $\sigma \leftarrow \sigma'$;
11   **return** S_n;
12 **end**
13 **for** each S_i in inactive sensor set $S_{\text{not\_running}}$ **do**
14   **if** S_i supports L_m **then**
15   Retrieve $C_{l_m}^{s_i}$;
16   Update $C_{\max}$ and S_n if $C_{l_m}^{s_i} > C_{\max}$;
17   **end**
18 **end**
19 Set $Y_{l_m}^{s_n} = 1$;
20 **return** S_n

The time complexity of sensor selection is $O(N)$, where N is the number of candidate sensors, and the execution result can be obtained through a single traversal. Compared with the two-stage selection method proposed in the existing research EGO, our approach has a lower time complexity. This is because our study focuses on selecting a single optimal sensor to execute the task, rather than selecting a group of sensors. Therefore, we do not need to consider the permutations and combinations among various resources. As a result, the time complexity is reduced from $O(V)$ to $O(N)$, where V is the number of permutations and combinations when the number of resources is N, and $V > N$.

Hybrid Selection Strategy with Fixed-Period and Threshold-Triggered. The data-reuse-based sensor selection method identifies the optimal sensor from a candidate set for task execution. However, the efficiency values of these sensors exhibit temporal fluctuations, causing the most efficient sensor for specific sensing tasks to vary over time. Given the runtime characteristics of sensors, the selection method must be executed dynamically during task execution cycles to maintain task accuracy requirements.

To avoid excessive sensor switching, we combine fixed-period with threshold-triggered selection. The selection is invoked either at fixed intervals d or when the margin sampling P_{t_i} of inference results falls below a threshold th_r. The threshold th_r is adaptively adjusted using a control-theoretic feedback mechanism:

$$th_r = 0.7 \cdot V_{t'} + 0.5 \cdot P \cdot (r - V_{t'}) - \omega + P \cdot (r - V_t) \qquad (4)$$

where V_t and V'_t represent the historical and current threshold states, respectively. Here, t denotes the previous calculation timestep, while t' corresponds to the current timestep. Notably, t' does not strictly align with real-world execution time, as threshold updates and selection methods are not invoked at every timestep. $r = 0.84$ is the target convergence value for the threshold, while $P = 3.25$ and $\omega = -0.2$ are control coefficients, typically set as constants.

Specifically, the system periodically evaluates sensor efficacy at intervals d while simultaneously monitoring real-time task performance. The inference model's output, represented by the margin sampling P_{t_i} (the difference between the highest and second-highest probabilities), serves as an indicator of current sensor performance. If P_{t_i} falls below a dynamically adjusted threshold th_r, the system triggers an immediate sensor reselection to replace underperforming sensors. This dual-trigger approach ensures timely adaptation to runtime variations in sensor efficacy while mitigating excessive energy overhead from overly frequent selections.

3.3 Runtime Capability Update Algorithm

In multi-device environments, the dynamic nature of heterogeneous sensors necessitates an adaptive capability update strategy to maintain task accuracy while minimizing energy consumption. Traditional methods often rely on static or periodic updates, which either fail to capture real-time performance variations or incur excessive energy costs due to redundant data transmission. To this end, we propose a hybrid capability update strategy that integrates the Jacobson/Karels algorithm with an LFU (Least Frequently Used) policy, enabling efficient and adaptive updates of sensor capabilities during runtime.

Problem Formulation. The capability value of a sensor serves as the fundamental basis for selection decisions, reflecting its performance in executing perception tasks. However, due to the runtime characteristics of sensors, their actual capabilities can fluctuate over time, leading to discrepancies between the accessed capability values used during selection and the real-time capabilities. This phenomenon, known as capability lag, can significantly degrade the accuracy of task execution if not properly addressed. The primary objective of the capability update mechanism is to minimize this lag by dynamically adjusting the evaluated capability values to align with the current state of the sensors.

Formally, consider a candidate sensor set denoted as $S_{l_m} = \{s_{l_n}^1, s_{l_n}^2, ..., s_{l_n}^p\}$, where each element $s_{l_n}^i$ represents a sensor available for task l_m. The corresponding capability assessment set is $C_{l_m} = \{C_{l_m}^{s_1}, C_{l_m}^{s_2}, ..., C_{l_m}^{s_p}\}$, with $C_{l_m}^{s_i}$ representing the accessed capability value of sensor s_i for task l_m. The goal of capability update is to optimize the alignment between the accessed capability values and the real-time capabilities, which can be mathematically formulated as minimizing the sum of squared differences between the ranks of accessed and real-time capabilities:

$$\text{minimize} \frac{\sum_{i=1}^{P}(g(s_i) - f(s_i))^2}{P} \tag{5}$$

Here, P denotes the size of the candidate set S_{l_m}, while $g(s_i)$ and $f(s_i)$ represent the ranks of sensor s_i in the real-time capability sequence and the assessed capability sequence, respectively.

Jacobson/Karels-Based Efficiency Assessment. To mitigate the impact of capability lag, we propose a capability evaluation method inspired by the Jacobson/Karels algorithm, which is renowned for its effectiveness in network congestion control. This method leverages historical capability data and their temporal variations to compute updated capability values, thereby capturing the dynamic nature of sensors more accurately.

The initial SR_{t_0} and DR_{t_0} are calculated using the margin sampling P_{t_0} (the difference between the maximum and second-highest probabilities in the model's output distribution) and the energy consumption D_s in sensing tasks:

$$SR_{t_0} = \frac{P_{t_0}}{1 + D_s} \tag{6}$$

$$DR_{t_0} = \frac{SR_{t_{i-1}}}{2} = \frac{P_{t_0}}{2 \cdot (1 + D_s)} \tag{7}$$

For subsequent updates, the SR_{t_i} and DR_{t_i} are refined using adaptive factors α and β:

$$SR_{t_i} = SR_{t_{i-1}} + \alpha \cdot \left(\frac{P_{t_i}}{1 + D_s} - SR_{t_{i-1}}\right) \tag{8}$$

$$DR_{t_i} = (1 - \beta) \cdot \frac{SR_{t_{i-1}}}{2} + \beta \cdot (|SR_{t_i} - SR_{t_{i-1}}|) \tag{9}$$

The final updated capability value $C_{l_m}^{t_i}$ is derived as a weighted sum of the SR_{t_i} and DR_{t_i}:

$$C_{l_m}^{t_i} = \mu \cdot SR_{t_i} + \rho \cdot DR_{t_i} \tag{10}$$

In these equations, P_{t_i} represents the margin sampling at time t_i, and $\alpha = 0.125$, $\beta = 0.25$, $\mu = 1$ and $\rho = 4$ are tuning parameters that balance the influence of historical data and current observations. As can be seen from (8) and (10), the capability value $C_{l_m}^{t_i}$ is inversely proportional to D_s and directly proportional to P_{t_i}. This is because D_s measures the energy consumption during task execution: a larger D_s indicates higher energy consumption, resulting in a lower efficiency value. Conversely, P_{t_i} measures accuracy: a larger P_{t_i} signifies a higher efficiency value and better task execution performance.

As shown in Algorithm 2, capability update process dynamically refines sensor performance metrics to balance historical trends and real-time observations. For each timestep t_i, the final capability value $C_{l_m}^{s_i}$ is computed as a weighted sum of SR_{t_i} and DR_{t_i}. This design ensures that capability values regulate energy consumption and recognition accuracy.

Algorithm 2: Runtime Capability Update

Input: Task request L_m, S_n
Output: void
1 Initialize parameters: D_s, P_{t_0}, SR_{t_0}, DR_{t_0};
2 **for** time t_i from 1 to T **do**
3   Set margin sampling P_{t_i} at time t_i;
4   **if** sensor S_n is triggered at time t_i **then**
5        $SR_{t_i} \leftarrow SR_{t_{i-1}} + \alpha \cdot \left(\frac{P_{t_i}}{1+D_s} - SR_{t_{i-1}}\right)$;
6        $DR_{t_i} \leftarrow (1-\beta) \cdot \frac{SR_{t_{i-1}}}{2} + \beta \cdot \left(|SR_{t_i} - SR_{t_{i-1}}|\right)$;
7        $C_{l_m}^{s_i} \leftarrow \mu \cdot SR_{t_i} + \rho \cdot DR_{t_i}$;
8   Update capability $C_{l_m}^{s_i}$ of S_n on task L_m;
9   **end**
10 **end**

LFU-Based Runtime Capability Update. To minimize the energy overhead associated with frequent capability updates, we adopt an LFU (Least Frequently Used) strategy that selectively updates the capabilities of sensors. By prioritizing updates for sensors with the lowest update frequency, this strategy effectively reduces redundant data collection and transmission, thereby conserving energy.

The LFU-based capability update strategy dynamically manages sensor refresh cycles by maintaining an array to track the update frequency of each sensor, which is updated in real-time during capability refreshes. During each update cycle, the algorithm selects a sensor with the lowest recorded frequency for capability enhancement, ensuring timely updates for less frequently accessed sensors. Upon completing the update, the corresponding entry in the frequency array is incremented to reflect the recent activity, thereby dynamically adjusting the update priority for subsequent cycles.

4 System Implementation

This section elaborates on the system implementation details, adhering to the principles of modularity to ensure seamless interaction between hardware and software components.

The proposed DataReuse framework is implemented as a Client-Server (C/S) architecture-based system. As illustrated in Fig. 5, the client primarily handles interactions with physical sensing devices, controlling data collection/transmission and maintaining real-time communication with the server. In practical deployments, the software code for the client can be installed on edge devices equipped with sensors, such as Raspberry Pi, ESP32, smartphones and smartwatches. The server, typically deployed on more powerful central devices (e.g., cloud servers or high-performance smartphones), executes modules including sensor selection, capability updates, and task inference.

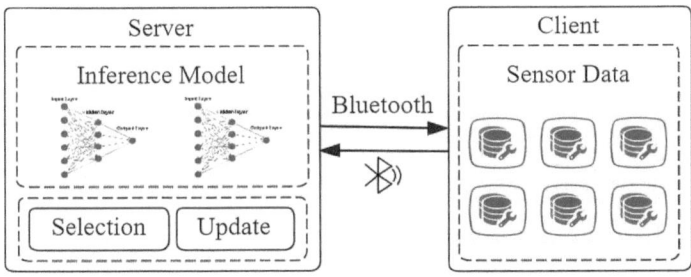

Fig. 5. Client-Server architecture based on Bluetooth communication.

Fig. 6. Smartphones: Honor 30S and Honor X10 Max.

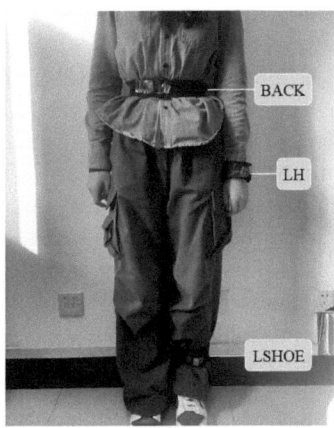

Fig. 7. Sensor deployment locations: back, left hand, left shoe.

Upon receiving sensor data from the client, the server loads the corresponding model to execute task inference and generate results. Meanwhile, it performs runtime sensor selection and capability update algorithms, then notifies the client of selection decisions. Based on these decisions, the client determines whether to switch to new sensors for continued task execution. This architectural separation streamlines sensor management workflows, enabling efficient adaptation to dynamic sensing environments.

The DataReuse system is implemented in Java using the Android Studio. It adopts a client-server architecture, where both the client and server are deployed on smartphones equipped with Kirin 820 and MediaTek Dimensity 800 processors to execute locomotion recognition and activity recognition tasks, as shown in Fig. 6. Pretrained deep learning models, built using PyTorch, are stored in a standardized model library for real-time inference. Bluetooth communication employs a Reactor pattern with multiplexed I/O to efficiently handle concurrent connections.

5 Performance Evaluation

To validate the effectiveness of the proposed DataReuse scheme, we conduct extensive experiments on the Opportunity dataset [7] under different task scenarios and system configurations. The evaluation focuses on balancing energy efficiency and task accuracy in dynamic multi-device environments.

5.1 Experimental Setup

Dataset and Tasks. We use the Opportunity dataset, which provides six hours of data from four participants to simulate real-world multi-device scenarios. 7 heterogeneous sensors are selected as candidate sensors, comprising 4 Inertial Measurement Units (IMUs) positioned at the back (BACK), left upper arm (LUA), right lower arm (RLA), and right shoe (RSHOE), along with 3 accelerometers located at the left hand (LH), hip (HIP), and right knee (RKN). We also selected three critical measurement points for the physical deployment of IMU sensors, as illustrated in Fig. 7. The information about the candidate sensors is shown in Table 1. All devices operate at a sampling frequency of 30 Hz, and there are variations in the dimensionality of the collected sensing data. We evaluate two tasks: Locomotion recognition (stand, walk, sit, lie) and HIActivity recognition (relaxing, coffee time, early morning, cleanup, sandwich time). Three task modes (each lasting 35 min) are designed to simulate real-world scenarios, as shown in Fig. 8:

- Task mode I: Both Locomotion recognition and HIActivity recognition tasks execute simultaneously.
- Task mode II: The Locomotion recognition task runs continuously, while the HIActivity recognition task executes intermittently.
- Task mode III: Locomotion recognition and HIActivity recognition tasks operate at separate intervals.

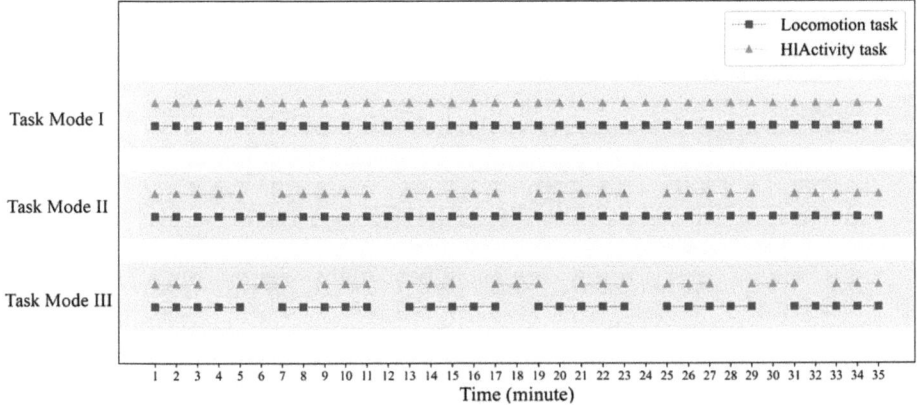

Fig. 8. Sensing task mode.

Baselines. We test four baseline methods in comparison with DataReuse. Although the authors of SensiX and EGO have conducted extensive research beyond sensor selection methods and achieved significant results in their studies, this study specifically focuses on comparing the sensor selection components addressed in these papers.

- SensiX [3]: A runtime sensor selection framework with periodic updates.
- EGO [10]: A threshold-triggered ontology-based selection method.
- BA (Best Accuracy): Greedy selection prioritizing accuracy.
- BE (Best Energy): Greedy selection minimizing energy consumption.

Performance Metrics. We consider three metrics in the evaluation:

- Accuracy: The percentage of samples where the inference results match the ground truth.
- Energy: The energy consumption of the specified application is measured using Google's ADB tool during a 35-min task execution period.
- V_b value: The discrepancy between the runtime-evaluated efficiency values of sensing tasks under the update strategy and their ground-truth efficiency values. A smaller V_b indicates that the evaluation results more accurately reflect the true efficiency hierarchy among candidate sensors, enabling selection methods to prioritize sensors with higher actual efficiency.

5.2 Overall Performance

Figure 9 shows the performance of our method compared to baselines across three task modes. It can be observed that the BA achieves the highest task accuracy but higher energy consumption than other methods. Conversely, the

Table 1. Information of candidate sensors.

Number	Type	Sensor	Sampling Frequency	Data Dimension
1	IMU	BACK	30 Hz	13
2	IMU	LUA	30 Hz	13
3	IMU	RLA	30 Hz	13
4	IMU	RSHOE	30 Hz	15
5	Accelerometer	LH	30 Hz	3
6	Accelerometer	HIP	30 Hz	3
7	Accelerometer	RKN	30 Hz	3

BE, while exhibiting lower accuracy, outperforms other approaches in energy efficiency. This discrepancy arises because the BA prioritizes accuracy during sensor selection, favoring IMU sensors with higher data dimensions. In contrast, the BE selects sensors solely based on minimizing energy consumption, opting for low-dimensional accelerometer sensors, thereby reducing energy usage at the cost of lower accuracy. On the other hand, DataReuse achieves accuracy rates of 83.3%, 83.3%, and 82.0% for Locomotion recognition tasks and 69.8%, 68.1%, and 68.7% for HIActivity recognition tasks across the three task modes. These results are about 3% lower than SensiX and EGO. In terms of energy consumption, DataReuse consumes only 1058.6 J, 876.4 J, and 682.3 J respectively in the three task modes, significantly outperforming all methods except BE.

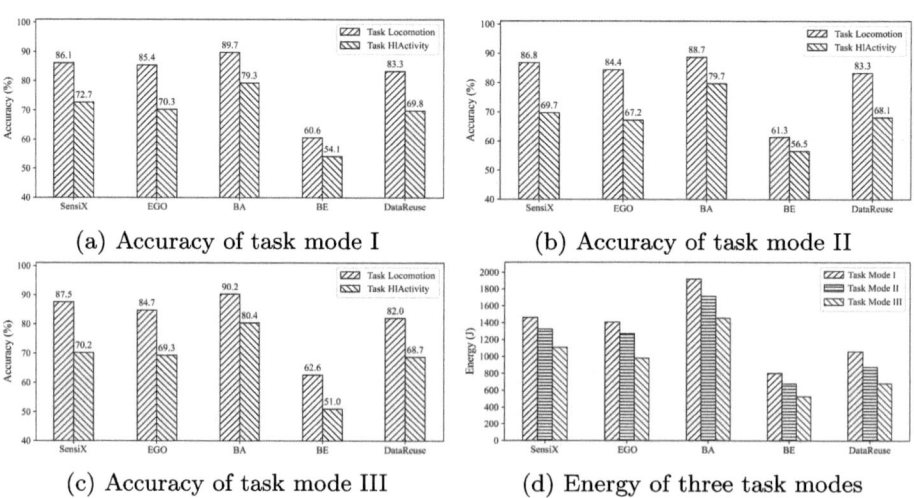

(a) Accuracy of task mode I   (b) Accuracy of task mode II

(c) Accuracy of task mode III   (d) Energy of three task modes

Fig. 9. Comparison of accuracy and energy consumption.

Overall, DataReuse outperforms BA and BE in task execution. Compared to SensiX and EGO, it reduces energy use by 24% with just a 3% accuracy

loss, thanks to data reuse and efficiency-focused updates. Its unique traits are:
1) enabling one sensor to handle two tasks at once, unlike others that assign one task per sensor and only pick idle ones; 2) using past performance data in updates to optimize sensor allocation, updating just one sensor each time to save energy.

We further analyze the usage of sensors for Locomotion and HIActivity recognition tasks in task mode I. As shown in Fig. 10, SensiX exhibits frequent sensor switching under small selection durations, achieving the highest accuracy but incurring the highest energy consumption. EGO, with smaller thresholds, triggers sensor selection irregularly and infrequently, resulting in the lowest energy consumption but also the lowest accuracy. In contrast, our method combines duration-based and threshold-based selection mechanisms, balancing accuracy and energy consumption under identical parameter settings. Crucially, by enabling data reuse, sensors do not need to be switched during task transitions, effectively reducing energy consumption without significant accuracy degradation.

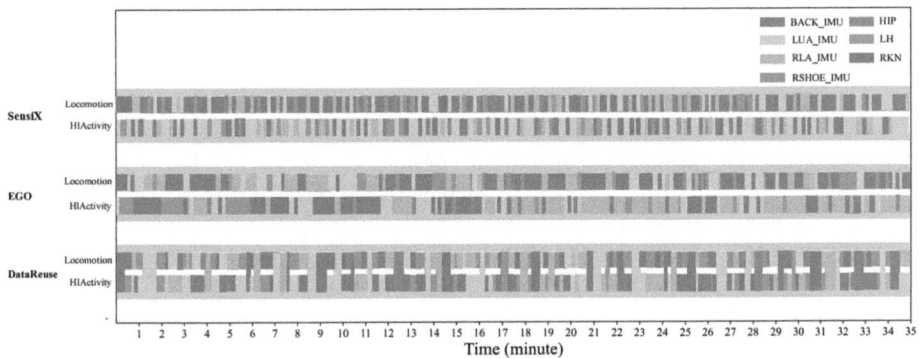

Fig. 10. Detailed utilization of sensors during the execution of task mode I.

5.3 Performance Based on the Number of Candidate Sensors

In real-world multi-device sensing scenarios, the number of available sensors dynamically changes over time. We evaluate the performance of our proposed method under varying numbers of available sensors (ranging from 1 to 7, as listed in Table 1) in task mode I.

As shown in Fig. 11, the task execution accuracy of all three methods gradually improves as the number of candidate sensors increases. When only the BACK sensor is available, SensiX and EGO require executing tasks twice, whereas DataReuse reuses BACK's sensing data to simultaneously perform both Locomotion and HIActivity recognition tasks, consuming only 58% and 56% of the energy of SensiX and EGO, respectively. However, the accuracy achieved by BACK is relatively low, with 73.0% for Locomotion and 64.1% for HIActivity recognition.

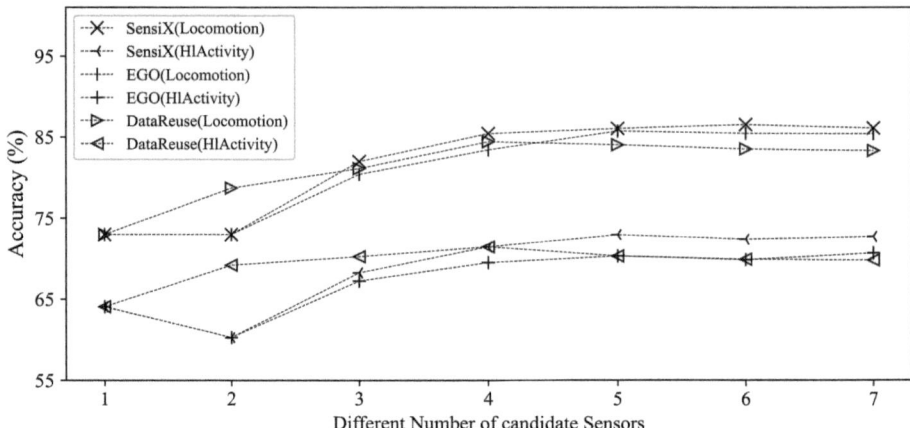

Fig. 11. Accuracy of increasing the numbers of candidate sensors.

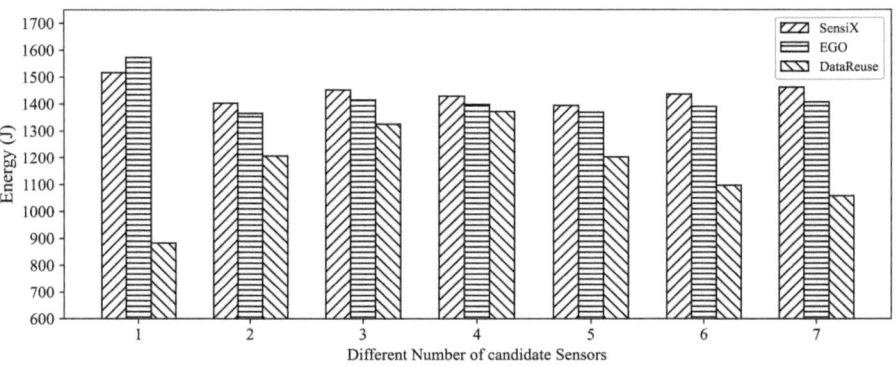

Fig. 12. Energy consumption of increasing the numbers of candidate sensors.

As more sensors become available, the selection methods gain opportunities to replace BACK with higher-efficiency sensors at specific moments, thereby improving accuracy. When the number of candidate sensors is small, DataReuse outperforms others. For example, with 2 candidate sensors, DataReuse achieves 78.7% (Locomotion) and 69.2% (HIActivity) accuracy, compared to SensiX's 73.0% and EGO's 60.3%. This is because, with limited sensors, SensiX and EGO suffer from resource contention: once one task claims a resource, the other must use the remaining one. In contrast, DataReuse's data reuse mechanism provides more candidate sensors under the same conditions. Notably, the marginal improvement in accuracy diminishes as the number of sensors increases.

As shown in Fig. 12, increasing the number of candidate sensors reduces the proportion of data reuse moments, leading to higher energy consumption as tasks increasingly rely on separate sensors. SensiX and EGO's energy consumption remains less affected by the number of sensors. When the number reaches 5, newly added low-dimensional sensors (LH, HIP, RKN) with 3-dimensions data reduce energy consumption when selected, causing DataReuse's overall energy usage to decline. However, these sensors' lower data dimensions also reduce task accuracy when utilized.

5.4 Effect of Selection and Update Timing

Effect of Selection Duration and Threshold. The selection duration (d) and threshold (th_r) create a trade-off between task execution accuracy and energy consumption. To address this, we evaluate DataReuse's performance under varying d and th_r values in task mode I. As shown in Fig. 13, compared to SensiX, DataReuse exhibits smaller fluctuations in accuracy and energy consumption as d changes. When d increases from 10 s to 60 s, the accuracy of Locomotion and HIActivity recognition drops are 3.4% and 3.8%, lower than SensiX's reductions of 4.9% and 5.5%. In energy consumption, DataReuse reduces energy usage by 285.8 J as d increases, outperforming SensiX's reduction of 538.9 J.

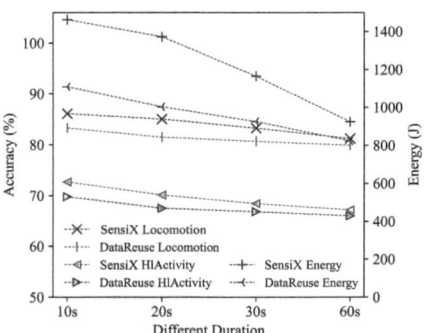
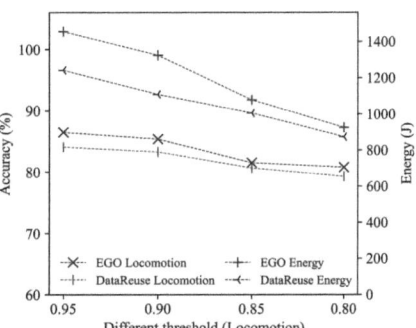

Fig. 13. Effect of selection duration.   **Fig. 14.** Effect of selection threshold.

For threshold analysis, we fix the threshold of HIActivity recognition and evaluate the impact of th_r on Locomotion tasks. From Fig. 14, compared to EGO, DataReuse demonstrates greater stability in accuracy and energy consumption as th_r decreases. Reducing th_r from 0.95 to 0.80 causes DataReuse's Locomotion accuracy to decline by 4.8%, outperforming EGO's 5.7% drop. Energy consumption decreases by 368.1 J for DataReuse, significantly better than EGO's 529.7 J reduction. Overall, DataReuse's performance is less sensitive to d and th_r variations than existing methods, achieving substantial energy savings with only minor accuracy loss.

Effect of Update Duration. We analyze DataReuse's runtime capability update performance. Figure 15 illustrates the V_b metric (discrepancy between runtime-evaluated and ground-truth efficiency values) for different methods in task mode I. As the update duration increases, V_b values also exhibit an upward trend, indicating larger deviations between estimated and actual efficiency. Although DataReuse's V_b values are higher than SensiX's under the same update durations, its update counts (317, 154, 78 for $d = 10$ s, 30 s, 60 s) are far lower than SensiX's (1470, 490, 245). This is because SensiX updates all candidate sensors each time, whereas DataReuse selectively updates the least frequently used one. While smaller V_b values improve accuracy, frequent updates incur higher energy costs. DataReuse strategically sacrifices slight V_b accuracy to reduce update frequency, fitting its energy-efficient design and consistent task performance.

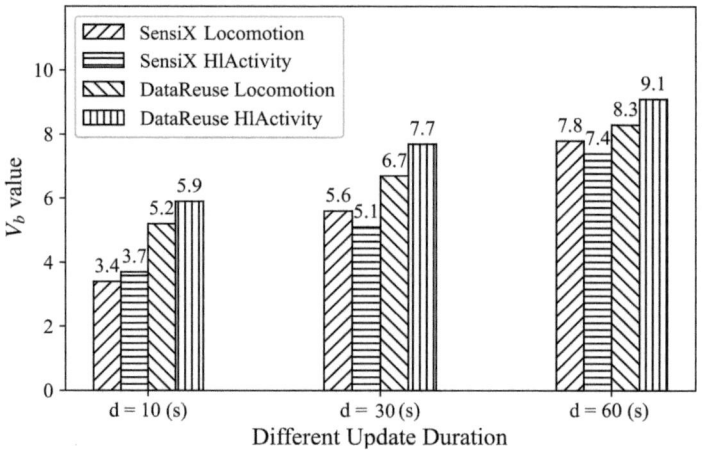

Fig. 15. Effect of update duration.

6 Conclusion

In this paper, we presented DataReuse: a runtime framework addressing the dynamic selection and capability update challenges of heterogeneous sensors in multi-device environments. DataReuse integrates data reuse strategies with adaptive capability update mechanisms, achieving a balance between energy efficiency and task accuracy. By enabling a single sensor to serve multiple tasks through data reuse, redundant data transmission is minimized, which reduces energy consumption. The Jacobson/Karels-based capability assessment leverages historical performance trends to mitigate estimation lag, while the LFU-driven update strategy optimizes energy overhead by prioritizing infrequently updated sensors. Furthermore, we develop a DataReuse prototype system, validating its effectiveness. Through experimental evaluations in human posture and activity recognition scenarios, DataReuse demonstrates a 24% average energy reduction with only a 3% accuracy trade-off compared to SensiX and EGO.

Acknowledgments. This research was supported by the National Natural Science Foundation of China (Nos. 62172336 and 62032018).

Disclosure of Interests. The authors have no competing interests to declare that are relevant to the content of this article.

References

1. Hu, S., Chen, L., Wu, P., Li, H., Yan, J., Tao, D.: ST-P3: end-to-end vision-based autonomous driving via spatial-temporal feature learning. In: European Conference on Computer Vision, pp. 533–549. Springer (2022)
2. Kang, S., et al.: Orchestrator: an active resource orchestration framework for mobile context monitoring in sensor-rich mobile environments. In: 2010 IEEE International Conference on Pervasive Computing and Communications (PERCOM), pp. 135–144. IEEE (2010)
3. Min, C., Mathur, A., Montanari, A., Kawsar, F.: Sensix: a system for best-effort inference of machine learning models in multi-device environments. IEEE Trans. Mob. Comput. **22**(9), 5525–5538 (2023). https://doi.org/10.1109/TMC.2022.3173914
4. Min, C., Montanari, A., Mathur, A., Kawsar, F.: A closer look at quality-aware runtime assessment of sensing models in multi-device environments. In: Proceedings of the 17th Conference on Embedded Networked Sensor Systems, pp. 271–284 (2019)
5. Ordóñez, F.J., Roggen, D.: Deep convolutional and LSTM recurrent neural networks for multimodal wearable activity recognition. Sensors **16**(1), 115 (2016)
6. Ren, H., Han, S., Ding, H., Zhang, Z., Wang, H., Wang, F.: Focus on details: online multi-object tracking with diverse fine-grained representation. In: 2023 IEEE/CVF Conference on Computer Vision and Pattern Recognition (CVPR), pp. 11289–11298 (2023). https://doi.org/10.1109/CVPR52729.2023.01086
7. Roggen, D., et al.: Collecting complex activity datasets in highly rich networked sensor environments. In: 2010 Seventh International Conference on Networked Sensing Systems (INSS), pp. 233–240. IEEE (2010)
8. Sasaki, Y., Yamada, K., Nagata, T., Saito, Y., Nonomura, T.: Fast data-driven greedy sensor selection for ridge regression. IEEE Sens. J. **25**(6), 10030–10045 (2025). https://doi.org/10.1109/JSEN.2025.3537702
9. Taleb, S., Hajj, H., Dawy, Z.: VCAMS: viterbi-based context aware mobile sensing to trade-off energy and delay. IEEE Trans. Mob. Comput. **17**(1), 225–242 (2018). https://doi.org/10.1109/TMC.2017.2706687
10. Taleb, S., Hajj, H., Dawy, Z.: EGO: optimized sensor selection for multi-context aware applications with an ontology for recognition models. IEEE Trans. Mob. Comput. **18**(11), 2518–2535 (2019). https://doi.org/10.1109/TMC.2018.2879864
11. Yao, S., Hu, S., Zhao, Y., Zhang, A., Abdelzaher, T.: Deepsense: a unified deep learning framework for time-series mobile sensing data processing. In: Proceedings of the 26th International Conference on World Wide Web, pp. 351–360 (2017)
12. Yi, X., et al.: Egolocate: real-time motion capture, localization, and mapping with sparse body-mounted sensors. ACM Trans. Graph. (TOG) **42**(4), 1–17 (2023)
13. Zhong, Y., Yang, N., Huang, L., Shi, G., Shi, L.: Sparse sensor selection for distributed systems: an L1-relaxation approach. Automatica **165**, 111670 (2024)

Can Time-Series Foundation Models Enhance Wireless Sensing Data Analytics? An Empirical Study

Shuangping Li[1], Ruifeng Wang[1], Ke Xu[2,3](✉), and Jiangtao Wang[1,2,3](✉)

[1] School of AI and Data Science, USTC, Hefei, China
{lishuangping,wrf3210}@mail.ustc.edu.cn, wangjiangtao@ustc.edu.cn
[2] Suzhou Institute for Advanced Research of USTC, Suzhou, China
nick_xuke@ustc.edu.cn
[3] Suzhou Big Data & AI Research and Engineering Center, Suzhou, China

Abstract. WiFi Channel State Information (CSI) has emerged as a promising modality for device-free human sensing tasks, such as human activity recognition (HAR) and identity recognition (Human ID), due to its ubiquity, low cost, and non-intrusive nature. Traditional model-driven approaches, based on physical propagation theories (e.g., Fresnel zone models), offer interpretable features but lack the capacity to capture complex semantic patterns. Learning-based methods improve performance by integrating domain knowledge with deep neural architectures, yet they remain limited under task shifts, scarce data, and unseen classes. In this work, we explore the use of large pre-trained time-series foundation models (TSFMs)—originally developed for domains such as weather forecasting and electric signal analysis—for CSI-based human sensing. These models exhibit strong cross-task generalization, jointly model frequency–temporal dependencies, and are inherently suited for few-shot learning and multimodal fusion via a modular pretraining–finetuning paradigm. We further investigate parameter-efficient fine-tuning (PEFT) strategies, including LoRA and prompting, to adapt state-of-the-art TSFMs to CSI data. Experimental results on HAR and Human ID tasks demonstrate that our approach consistently outperforms existing baselines across various settings, providing a scalable, data-efficient, and generalizable framework for wireless sensing in AIoT applications.

Keywords: WiFi Sensing · Time-Series Foundation Models · Large Model

1 Introduction

The rapid development of the Artificial Intelligence of Things (AIoT) has enabled pervasive sensing, real-time interaction, and intelligent decision-making across diverse domains such as smart homes, healthcare, industrial automation, and security. As AIoT systems advance, there is an increasing demand for human sensing technologies that are accurate, low-cost, privacy-preserving, and easily deployable without relying on invasive infrastructure. Device-free sensing has emerged as a promising approach, offering non-intrusive and scalable alternatives to traditional wearable and camera-based systems.

Among various sensing modalities, WiFi-based Channel State Information (CSI) has attracted growing attention due to its unique advantages. CSI captures fine-grained physical-layer characteristics of wireless signal propagation, enabling the detection of subtle human movements and body-related signatures. Unlike wearable sensors, CSI-based sensing requires no user instrumentation, supporting natural interaction. Compared to vision-based methods, it offers enhanced privacy and robustness in low-light or occluded environments. Moreover, CSI can be extracted from commodity WiFi hardware, making it a cost-effective and scalable solution for AIoT applications.

In the field of CSI-based human sensing, two main paradigms have emerged: model-based and learning-based methods. Model-based approaches utilize physical signal propagation models—such as Fresnel zone theory or Doppler shift analysis—to derive handcrafted features that are invariant to environmental changes. These methods are interpretable and robust but are often constrained in their representational capacity and lack the flexibility to capture complex, high-level human behaviors or identity-related patterns.

Learning-based approaches, on the other hand, leverage deep learning techniques to learn task-relevant representations from raw CSI data. When integrated with domain knowledge (e.g., via Doppler analysis or antenna response modeling), they demonstrate strong performance on tasks such as Human Activity Recognition (HAR) and human identification (Human ID). However, these models typically require extensive labeled data for each deployment scenario, generalize poorly to unseen classes (i.e., limited few-shot and zero-shot capabilities), and often overfit to specific devices or environments.

In contrast to task-specific deep learning models, the emergence of Time Series Foundation Models (TSFMs) offers a promising alternative for general-purpose time series representation learning. Inspired by the success of large-scale pretraining in natural language processing, TSFMs typically adopt transformer-based architectures and are trained with self-supervised objectives (e.g., masked prediction, contrastive learning) on diverse, unlabeled time series corpora. This paradigm introduces two key advantages over traditional task-specific approaches:

- **Data Efficiency:** By leveraging rich, pre-learned representations, TSFMs support parameter-efficient fine-tuning (PEFT), significantly reducing the dependence on large labeled datasets—one of the main bottlenecks in CSI-based human sensing.
- **Strong Generalization:** Pretraining on heterogeneous time series data across domains (e.g., healthcare, energy, sensor telemetry) enables TSFMs to capture broad temporal dynamics and inductive biases, facilitating transfer to new tasks and domains with minimal supervision.

These properties make TSFMs particularly appealing for wireless sensing applications, where labeled CSI data is often scarce, and deployed models must generalize across diverse users, devices, and environments. However, despite their growing success in conventional time series domains, the potential of TSFMs for CSI-based human sensing tasks remains largely unexplored. Several important questions remain unanswered: **1)** *Can the knowledge learned by TSFMs from non-CSI data effectively benefit CSI-based human sensing tasks such as Human Activity Recognition (HAR) and Human Identification (HumanID)?* **2)** *Can Parameter-Efficient Fine-Tuning (PEFT) achieve comparable or even superior performance compared to full parameter fine-tuning?* **3)** *How well do TSFMs perform when transferred to new domains or tasks?*

To address these questions, we conduct the first systematic investigation into the applicability of TSFMs for CSI-based human sensing. Specifically, we evaluate the state-of-the-art TSFM, MOMENT, which was originally pretrained on heterogeneous non-CSI time series data, on two representative CSI-based tasks: Human Activity Recognition (HAR) and Human Identification (HumanID). Our contributions are three-fold:

- We present the first comprehensive study on applying large-scale TSFMs to CSI-based human sensing tasks, covering both Human Activity Recognition (HAR) and Human Identification (HumanID).
- We conduct extensive empirical evaluations of state-of-the-art models (MOMENT) under cross-domain and few-shot scenarios using real-world CSI datasets.
- We provide in-depth analyses on the transferability and limitations of TSFMs in CSI-based applications, offering practical insights for deploying foundation models in real-world AIoT sensing systems.

2 Related Work

2.1 CSI Based Human Sensing

CSI-based human sensing methods can be broadly categorized into model-based and learning-based approaches.

Model-based methods rely on physical models to interpret signal variations caused by human motion. The Fresnel zone model divides the space between transmitter and receiver into concentric regions to explain how motion affects CSI. It has been applied to walking direction inference [26], and multi-person respiration monitoring [30]. The Doppler-Velocity model uses Doppler Frequency Shifts (DFS) to capture motion-induced path changes. DFS has been utilized for gesture recognition [19], velocity estimation [20], and multi-parameter extraction via EM algorithms [21].

Learning-based methods use data-driven models to map CSI to target tasks. Early works adopted MLPs for people counting [31], followed by CNNs and RNNs for activity recognition [3,27]. Recent studies have explored transformer architectures for long-range temporal modeling [16,28]. To enhance generalization, hybrid approaches integrate physical priors into deep models. For instance, Zhang et al. [32] proposed the Body-Coordinate Velocity Profile (BVP) for cross-domain gesture recognition.

2.2 Time-Series Foundation Models

Foundation models, inspired by pretraining advances in language and vision, have emerged as a unified solution for diverse time series tasks. Unlike few-shot methods based on semantic priors or meta-learning, Time Series Foundation Models (TSFMs) leverage self-supervised pretraining on unlabeled data to learn task- and modality-agnostic representations, enabling robust generalization across tasks, sensors, and domains. Current research follows two main directions: training TSFMs from scratch or adapting large language models (LLMs) to time series.

The first line of work designs large Transformer-based models tailored for time series, using objectives such as forecasting or reconstruction. These models adopt various architectures (e.g., encoder-only, decoder-only) and strategies like patching to capture temporal dependencies. Representative examples include MOMENT [10], Moirai [25], Lag-Llama [23], TimesFM [5], and TimeGPT [9].

The second adapts LLMs to time series through input transformation. Text-visible adaptation converts numeric values into text tokens via quantization or statistics (e.g., PromptCast [29]), while embedding-visible adaptation maps time series into the LLM embedding space via encoders, often requiring fine-tuning (e.g., TimeLLM [14], TEMPO [2]).

Recent work [24] applies vision transformers to image-like RF representations for activity recognition. In contrast, our approach directly models raw or lightly processed CSI sequences, preserving temporal structure for fine-grained physiological sensing.

2.3 PEFT

Parameter-efficient fine-tuning (PEFT) adapts large pre-trained models by updating only a small subset of parameters, reducing resource requirements and mitigating overfitting and catastrophic forgetting. As model sizes grow, full-model tuning becomes increasingly impractical. Methods such as adapters [12], prefix tuning [17], and LoRA [13] introduce lightweight, task-specific modules to achieve competitive performance with minimal parameter updates.

3 Method

3.1 Problem Definition

We address two human sensing tasks based on CSI signals: **HAR** and **Human ID**. Both are formulated as supervised classification problems over time series extracted from CSI measurements.

Given an input sequence $X \in \mathbb{R}^{T \times F}$, where T is the number of time steps and F is the number of features after preprocessing, the goal for HAR is to predict an activity label $y_{\text{act}} \in \{1, 2, \ldots, C_{\text{act}}\}$ and for Human ID, to predict an identity label $y_{\text{id}} \in \{1, 2, \ldots, C_{\text{id}}\}$, where C_{act} and C_{id} are the numbers of activity classes and subjects, respectively. A model $f(\cdot; \theta)$ with parameters θ is trained to minimize the classification loss:

$$\mathcal{L} = \mathcal{L}_{\text{cls}}(f(X; \theta), y)$$

where y refers to either y_{act} or y_{id}.

We assume CSI sequences are collected under controlled conditions with synchronized labels, without assuming temporal stationarity, enabling a unified evaluation of large time series models across both tasks.

3.2 CSI Data Preprocessing

CSI data comprises complex-valued measurements containing both amplitude and phase information. Prior studies have investigated three primary processing strategies:

First, amplitude-only processing is the most widely adopted due to the instability of raw phase caused by hardware-induced offsets. In contrast, amplitude remains relatively stable and suitable for Wi-Fi sensing. Basic denoising techniques, such as wavelet filtering, are commonly applied to suppress high-frequency noise.

Second, antenna pairwise phase difference processing exploits the relative stability of inter-antenna phase differences to capture fine-grained human motion. Methods like CSI ratio further enhance robustness by attenuating noise through division-based normalization. These techniques are typically used in model-based systems requiring clean and calibrated CSI.

Third, Doppler-based processing extracts motion-sensitive features, such as the Body-coordinate Velocity Profile (BVP), which simulates Doppler shifts to highlight dynamic motion and suppress static components in the environment.

In this work, we adopt the amplitude-only strategy. The CSI amplitude is normalized and downsampled to align with the input requirements of large time-series models, facilitating stable feature extraction and improved computational efficiency.

3.3 MOMENT

Current time series base models vary in terms of model architecture, support for multivariate time series, and out-of-the-box support for time series tasks. This article surveys eight different time series base models that are currently at the forefront. Their information is in Table 1. Given that the wireless sensing task explored in this study, along with the selected dataset, falls under the category of multivariate time series classification, we adopt the MOMENT model family as the foundational time series model. MOMENT is designed to be readily adaptable to a wide range of downstream tasks in multivariate time series analysis. By simply replacing the linear classification head, MOMENT can be repurposed for different time series tasks. For these reasons, the MOMENT model family is selected as the baseline architecture for analyzing wireless sensing data in this experiment.

MOMENT is a family of open-source, pre-trained transformer-based foundation models specifically designed for general-purpose time series analysis. The models are trained using self-supervised learning over a broad collection of public time series datasets called the Time Series Pile.

Table 1. Implementation details of foundational time series models. Task abbreviations: F = Forecasting, I = Imputation, C = Classification, AD = Anomaly Detection.

Model	#Params (Small to Large)	Tasks
TinyTimeMixers	1M	F
TimeGPT	Unknown	F
Chronos	8M, 46M, 201M, 710M	F
Lag-LLaMa	200M	F
Timer	29M, 50M, 67M	F, I, AD
TimesFM	17M, 70M, 200M	F
Moment	40M, 125M, 385M	F, C, I, AD
Moirai	14M, 91M, 311M	F

To enable large-scale time series pretraining, we adopt the **Time Series Pile**, a unified corpus combining datasets from multiple public repositories. It covers forecasting (e.g., ETT [33], Traffic [1]), classification (UCR [6]), and anomaly detection (TSB-UAD [18]), spanning diverse domains such as healthcare, engineering, and finance. The datasets exhibit varied temporal properties, including sampling rates, sequence lengths, and multivariate structures.

3.4 Model Architecture

MOMENT is a transformer-based model designed for general-purpose time series representation learning. A univariate time series is divided into fixed-length, non-overlapping patches, each projected into a D-dimensional embedding space. During pretraining, a subset of these patches is masked using a learnable token, and the model is tasked with reconstructing the original inputs from corrupted sequences.

The encoder is based on the T5 [22] architecture but modified for time series. It employs reversible instance normalization to normalize inputs and uses relative positional embeddings along with optional sinusoidal absolute embeddings.

The design supports varying sequence lengths by padding shorter series and sub-sampling longer ones. Multivariate data is handled channel-wise, treating each univariate stream independently.

To retain flexibility in downstream applications, a lightweight reconstruction head is used instead of a heavy decoder. This separation allows task-specific heads (e.g., for forecasting) to be plugged in during fine-tuning.

3.5 Pretraining Objective

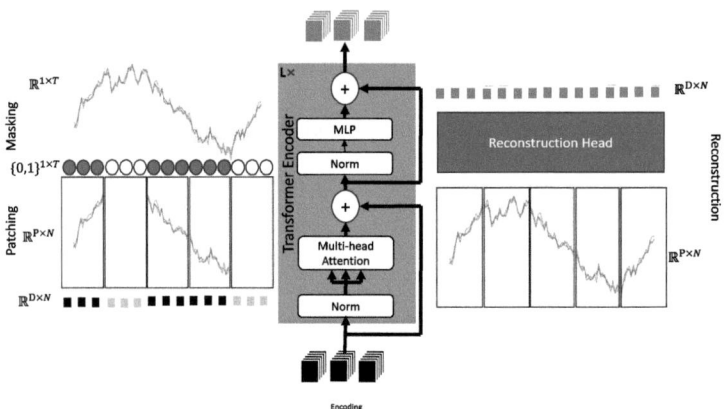

Fig. 1. Overview of the MOMENT architecture.

As shown in Fig. 1, the weights used for MOMENT are obtained through pretraining with a masked time series modeling objective. Specifically, during pretraining, 30% of the patches extracted from the input time series are randomly masked, and the model is trained to reconstruct these masked regions using a lightweight prediction head. The training objective is to minimize the mean squared error (MSE) between the predicted and original values within the masked regions. In our experiments, three variant of MOMENT are used, corresponding in capacity to T5-Small (40M parameters), Base (125M), and Large (385M). All models are pretrained for two epochs on 64 patches per sequence, each of length 8, totaling 512 time steps per input.

3.6 Fine-Tuning and Task Adaptation

It is important to note that MOMENT is pretrained exclusively on general-purpose time series corpora and does not leverage any CSI-based data or human sensing supervision during pretraining. To evaluate its applicability to CSI-based human sensing, we fine-tune the pretrained MOMENT models on downstream tasks including HAR and Human ID using labeled WiFi CSI datasets.

We investigate three fine-tuning strategies: (i) Full-parameter fine-tuning, which updates all model parameters; (ii) Linear probing, which freezes the backbone and only trains a linear classifier to assess representation quality; (iii) LoRA [13], which inserts trainable low-rank matrices into transformer layers, reducing trainable parameters while maintaining competitive performance. Specifically, LoRA decomposes the weight updates of the self-attention and feedforward layers into two low-rank matrices $A \in \mathbb{R}^{r \times d}$ and $B \in \mathbb{R}^{d \times r}$, where $r \ll d$. During training, only A and B are updated, while the original weight $W_0 \in \mathbb{R}^{d \times d}$ remains frozen. The modified forward pass becomes:

$$W = W_0 + \alpha BA,$$

where α is a scaling factor. This technique enables efficient adaptation to domain-specific tasks such as CSI-based human sensing, particularly when labeled data is limited.

By comparing full fine-tuning and LoRA, we aim to assess not only the performance of foundation models on wireless sensing tasks but also the trade-off between adaptability and computational efficiency under different tuning regimes.

Table 2. Details of the NTU-Fi Dataset

Dataset	NTU-Fi HAR	NTU-Fi HumanID
Collection Platform	Atheros CSI Tool	Atheros CSI Tool
Number of Classes	6	14
Class Names	Box, Circle, Clean, Fall, Run, Walk	Fourteen gait IDs
Data Dims.	3, 114, 2000	3, 114, 2000
Dims. Description	Antennas, Subcarriers, Time Steps	Antennas, Subcarriers, Time Steps
Training Samples	936	546
Testing Samples	264	294

4 Experiment

4.1 Dataset

The NTU-Fi dataset is the selected wireless sensing CSI dataset used in this study. It is a publicly available benchmark dataset widely adopted in the field of wireless human sensing with CSI signals. NTU-Fi supports two core time series classification tasks: HAR and HumanID.

The dataset was collected using the Atheros CSI tool, which provides high-resolution CSI with 114 subcarriers per antenna pair. Each CSI sample is pre-segmented and labeled. For the HAR task, data were collected under three distinct environmental layouts. For the HumanID task, gait data were recorded under three different clothing conditions: wearing a T-shirt, wearing a coat, and carrying a backpack. The data acquisition process is thoroughly documented in the original literature.

Table 2 summarizes the detailed specifications of both subsets in the NTU-Fi dataset.

4.2 Baseline

To evaluate the performance of the MOMENT foundation models on NTU-Fi classification tasks, we compare them with several classical neural network baselines: (1) Multilayer Perceptron (MLP) [8]: A three-layer feedforward network for modeling nonlinear relationships. (2) Convolutional Neural Network (CNN) [15]: Extracts spatial and temporal features through convolutional operations. (3) Recurrent Neural Network (RNN) [7]: Captures temporal dependencies in sequential data via recurrent connections. (4) Gated Recurrent Unit (GRU) [4]: An RNN variant with gating mechanisms to model long-term dependencies efficiently. (5) Long Short-Term Memory (LSTM) [11]: Enhances RNNs with memory cells and gates to better preserve long-range information.

Table 3. Cross-user evaluation results on the NTU-Fi-HumanID dataset. Each user group (Group 1: IDs 0–4, Group 2: 5–8, Group 3: 10–13) is used as the source domain in turn; the remaining two serve as target domains. Half of the target data is used for fine-tuning, and the rest for testing. MOMENT-S, MOMENT-B, and MOMENT-L denote the small, base, and large model variants, respectively. **Bold** indicates the best, and underlined the second-best performance.

User Group	MOMENT-S	MOMENT-B	MOMENT-L	MLP	CNN	RNN	GRU	LSTM
Group 1	**0.9704**	0.9593	0.9667	0.9037	0.9667	0.8370	0.9259	0.8630
Group 2	**0.9333**	0.9037	0.8815	0.8222	0.8741	0.5593	0.6667	0.6815
Group 3	**0.9500**	**0.9500**	0.9000	0.7600	0.9300	0.6433	0.7400	0.7800
Average	**0.9512**	0.9377	0.9161	0.8286	0.9236	0.6799	0.7775	0.7748

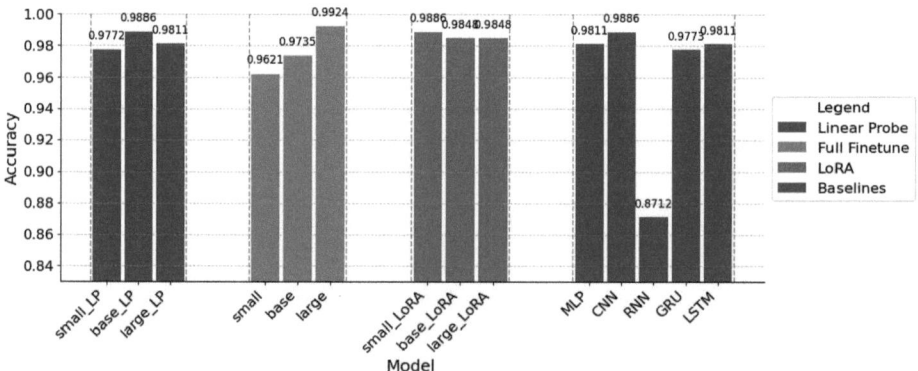

Fig. 2. Performance of the pretrained MOMENT model with different fine-tuning methods on the HAR task.

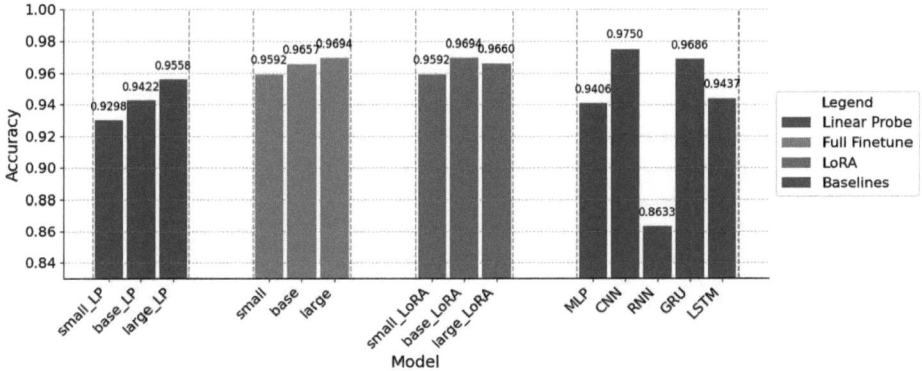

Fig. 3. Performance of the pretrained MOMENT model with different fine-tuning methods on the Human ID task.

4.3 Direct Adaptation via Fine-Tuning

To assess the transferability of the pretrained MOMENT model, we evaluate its small, base, and large configurations on the NTU-Fi HAR and HumanID datasets using three fine-tuning strategies: linear probing, full fine-tuning, and LoRA-based fine-tuning. Results are compared against baseline models.

As shown in Fig. 2, all models except RNN perform comparably on HAR tasks. Linear probing achieves competitive results, indicating that MOMENT effectively captures critical features even without extensive finetuning. MOMENT-large with full fine-tuning yields the highest accuracy, while LoRA achieves performance close to full fine-tuning with significantly fewer trainable parameters.

In HumanID tasks (Fig. 3), CNN slightly outperforms MOMENT by 0.54.

Overall, these results validate MOMENT as an effective time-series foundation model for wireless sensing. While full fine-tuning maximizes performance, LoRA provides a practical trade-off, reducing both data and computational demands.

4.4 Cross Domain Result

Table 3 reports the cross-user evaluation results of MOMENT and baseline models on the NTU-Fi-HumanID dataset. MOMENT consistently outperforms all baselines across different user groups, demonstrating strong generalization to unseen users. The MOMENT-small variant achieves the highest average accuracy of 95.12%, while MOMENT-base and MOMENT-large also perform competitively with 93.77% and 91.61%, respectively. This indicates that even smaller MOMENT models effectively capture discriminative user features, benefiting from self-supervised pretraining and domain adaptation.

The largest performance gap appears in Group 2, suggesting MOMENT's robustness under challenging domain shifts with limited source data. Overall,

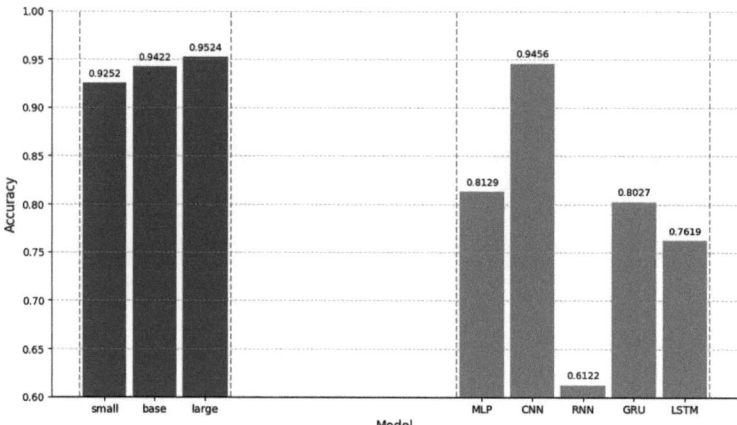

Fig. 4. Performance that model trained on HAR tasks and evaluated on the Human ID tasks.

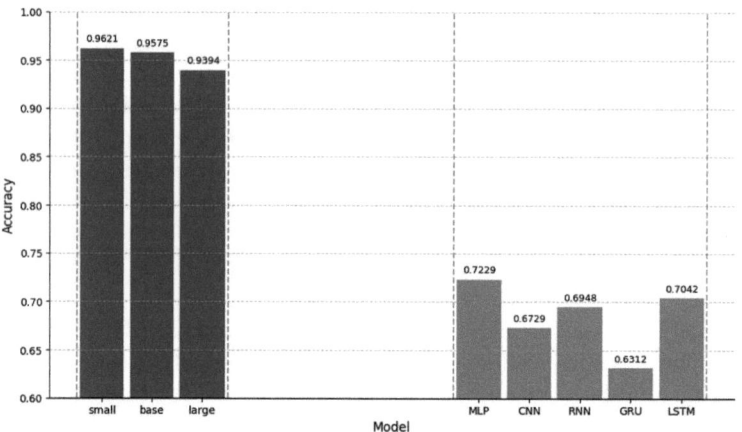

Fig. 5. Performance that model trained on Human ID tasks and evaluated on the HAR tasks.

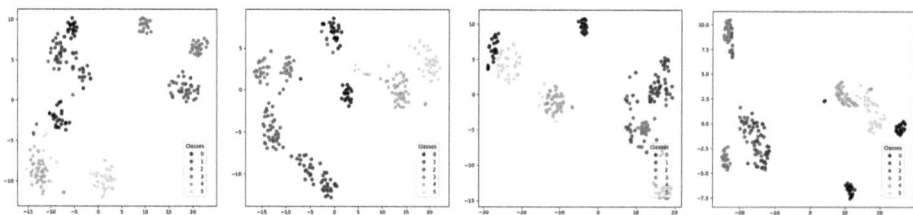

Fig. 6. Visualization results from left to right: MOMENT-small, MOMENT-base, CNN, and MLP on cross-task experiments with the NTU-Fi HumanID dataset.

these results validate the scalability and effectiveness of MOMENT in cross-user scenarios.

4.5 Task Transferring

To evaluate the cross-task generalization of time-series foundation models in wireless sensing, we conduct two transfer learning experiments.

In the first, models are trained on the NTU-Fi HAR dataset and tested on the HumanID dataset. As shown in Fig. 4, MOMENT and CNN outperform other baselines by over 10%, with MOMENT-large achieving the highest accuracy (95.24%). This demonstrates that temporal features learned from activity recognition transfer well to identity recognition due to shared motion patterns.

In the second experiment, models are trained on HumanID and tested on HAR. As shown in Fig. 5, MOMENT again outperforms all baselines, with a larger performance gap. This indicates that transferring from identity to activity recognition is easier, benefiting from the HumanID dataset's motion diversity. Interestingly, MOMENT-small achieves the best results here, suggesting smaller models generalize better under diverse motion patterns.

In summary, MOMENT models exhibit strong cross-task generalization. They effectively transfer learned temporal features across tasks, and smaller models show superior generalization when trained on diverse datasets, highlighting their efficiency and adaptability for wireless sensing.

4.6 Visualization

Figure 6 shows visualization results on the NTU-Fi HumanID dataset. MOMENT models produce more compact and separable features than MLP and CNN, indicating stronger discriminative power. MOMENT-base exhibits the clearest class boundaries, highlighting its superior feature extraction and generalization.

5 Limitation

The current framework has several limitations requiring further investigation: 1) Limited PEFT Evaluation. Only partial exploration of PEFT methods is conducted, without systematic comparison of other methods such as Adapter. 2) Insufficient Dataset Diversity. Experiments rely solely on NTU-Fi, lacking validation on broader datasets, which limits generalizability. 3) Incomplete Model Benchmarking. MOMENT is not thoroughly compared with other temporal models, preventing a comprehensive evaluation of its effectiveness. 4) Narrow Task Focus. The study only considers HAR and Human ID, overlooking applications such as fall detection and gesture recognition. 5) Device Heterogeneity

Overlooked. Lack of experiments across different hardware configurations limits insights into deployment robustness. Future work should address these gaps to improve model generalization and practical applicability.

6 Conclusion

This paper presents the first systematic exploration of applying TSFMs to CSI-based human sensing tasks. Experimental results demonstrate that TSFMs, particularly the MOMENT family, significantly outperform conventional models in terms of data efficiency, cross-domain generalization, and adaptability under limited supervision. Additionally, parameter-efficient fine-tuning methods such as LoRA achieve competitive performance with reduced computational overhead. These findings highlight the potential of TSFMs as a unified backbone for robust and scalable human sensing in AIoT systems, paving the way for future research on multimodal fusion and efficient deployment in real-world environments.

References

1. California department of transportation. performance measurement system (PEMS). http://pems.dot.ca.gov/. Accessed 30 Apr 2025
2. Cao, D., et al.: TEMPO: prompt-based generative pre-trained transformer for time series forecasting. In: The Twelfth International Conference on Learning Representations (2024)
3. Chen, Z., Zhang, L., Jiang, C., Cao, Z., Cui, W.: Wifi CSI based passive human activity recognition using attention based BLSTM. IEEE Trans. Mob. Comput. **18**(11), 2714–2724 (2018)
4. Chung, J., Gulcehre, C., Cho, K., Bengio, Y.: Empirical evaluation of gated recurrent neural networks on sequence modeling. arXiv preprint arXiv:1412.3555 (2014)
5. Das, A., Kong, W., Sen, R., Zhou, Y.: A decoder-only foundation model for time-series forecasting (2024). https://arxiv.org/abs/2310.10688
6. Dau, H.A., et al.: The UCR time series archive. IEEE/CAA J. Automatica Sinica **6**(6), 1293–1305 (2019)
7. Elman, J.L.: Finding structure in time. Cogn. Sci. **14**(2), 179–211 (1990)
8. Gardner, M.W., Dorling, S.R.: Artificial neural networks (the multilayer perceptron)–a review of applications in the atmospheric sciences. Atmos. Environ. **32**(14–15), 2627–2636 (1998)
9. Garza, A., Challu, C., Mergenthaler-Canseco, M.: Timegpt-1 (2024). https://arxiv.org/abs/2310.03589
10. Goswami, M., Szafer, K., Choudhry, A., Cai, Y., Li, S., Dubrawski, A.: Moment: a family of open time-series foundation models. In: Forty-First International Conference on Machine Learning (2024)
11. Hochreiter, S., Schmidhuber, J.: Long short-term memory. Neural Comput. **9**(8), 1735–1780 (1997). https://doi.org/10.1162/neco.1997.9.8.1735
12. Houlsby, N., et al.: Parameter-efficient transfer learning for NLP. In: International Conference on Machine Learning, pp. 2790–2799. PMLR (2019)
13. Hu, E.J., et al.: Lora: low-rank adaptation of large language models. ICLR **1**(2), 3 (2022)

14. Jin, M., et al.: Time-LLM: time series forecasting by reprogramming large language models. In: The Twelfth International Conference on Learning Representations (2024)
15. Lecun, Y., Bottou, L., Bengio, Y., Haffner, P.: Gradient-based learning applied to document recognition. Proc. IEEE **86**(11), 2278–2324 (1998)
16. Li, B., Cui, W., Wang, W., Zhang, L., Chen, Z., Wu, M.: Two-stream convolution augmented transformer for human activity recognition. In: Proceedings of the AAAI Conference on Artificial Intelligence, vol. 35, pp. 286–293 (2021)
17. Li, X.L., Liang, P.: Prefix-tuning: optimizing continuous prompts for generation. arXiv preprint arXiv:2101.00190 (2021)
18. Paparrizos, J., Kang, Y., Boniol, P., Tsay, R.S., Palpanas, T., Franklin, M.J.: TSB-UAD: an end-to-end benchmark suite for univariate time-series anomaly detection. Proc. VLDB Endow. **15**(8), 1697–1711 (2022)
19. Pu, Q., Gupta, S., Gollakota, S., Patel, S.: Whole-home gesture recognition using wireless signals. In: Proceedings of the 19th Annual International Conference on Mobile Computing & Networking, pp. 27–38 (2013)
20. Qian, K., Wu, C., Yang, Z., Liu, Y., Jamieson, K.: Widar: decimeter-level passive tracking via velocity monitoring with commodity wi-fi. In: Proceedings of the 18th ACM International Symposium on Mobile Ad Hoc Networking and Computing, pp. 1–10 (2017)
21. Qian, K., Wu, C., Zhang, Y., Zhang, G., Yang, Z., Liu, Y.: Widar2.0: passive human tracking with a single wi-fi link. In: Proceedings of the 16th Annual International Conference on Mobile Systems, Applications, and Services, pp. 350–361 (2018)
22. Raffel, C., et al.: Exploring the limits of transfer learning with a unified text-to-text transformer. J. Mach. Learn. Res. **21**(140), 1–67 (2020)
23. Rasul, K., et al.: Lag-llama: towards foundation models for time series forecasting. In: R0-FoMo: Robustness of Few-Shot and Zero-Shot Learning in Large Foundation Models (2023)
24. Weng, Y., Wu, G., Zheng, T., Yang, Y., Luo, J.: Large model for small data: foundation model for cross-modal RF human activity recognition. In: Proceedings of the 22nd ACM Conference on Embedded Networked Sensor Systems, pp. 436–449 (2024)
25. Woo, G., Liu, C., Kumar, A., Xiong, C., Savarese, S., Sahoo, D.: Unified training of universal time series forecasting transformers. In: Forty-First International Conference on Machine Learning (2024)
26. Wu, D., Zhang, D., Xu, C., Wang, Y., Wang, H.: Widir: walking direction estimation using wireless signals. In: Proceedings of the 2016 ACM International Joint Conference on Pervasive and Ubiquitous Computing, pp. 351–362 (2016)
27. Xiao, C., Lei, Y., Ma, Y., Zhou, F., Qin, Z.: Deepseg: deep-learning-based activity segmentation framework for activity recognition using wifi. IEEE Internet Things J. **8**(7), 5669–5681 (2020)
28. Xu, K., Wang, J., Zhang, L., Zhu, H., Zheng, D.: Dual-stream contrastive learning for channel state information based human activity recognition. IEEE J. Biomed. Health Inform. **27**(1), 329–338 (2022)
29. Xue, H., Salim, F.D.: Promptcast: a new prompt-based learning paradigm for time series forecasting. IEEE Trans. Knowl. Data Eng. **36**(11), 6851–6864 (2024)
30. Zeng, Y., Wu, D., Xiong, J., Liu, J., Liu, Z., Zhang, D.: Multisense: enabling multi-person respiration sensing with commodity wifi. Proc. ACM Interact. Mob. Wearable Ubiquitous Technol. **4**(3), 1–29 (2020)

31. Zhang, J., Tang, Z., Li, M., Fang, D., Nurmi, P., Wang, Z.: Crosssense: towards cross-site and large-scale wifi sensing. In: Proceedings of the 24th Annual International Conference on Mobile Computing and Networking, pp. 305–320 (2018)
32. Zhang, Y., et al.: Widar3.0: zero-effort cross-domain gesture recognition with wi-fi. IEEE Trans. Pattern Anal. Mach. Intell. **44**(11), 8671–8688 (2021)
33. Zhou, H., et al.: Informer: beyond efficient transformer for long sequence time-series forecasting. In: Proceedings of the AAAI Conference on Artificial Intelligence, vol. 35, pp. 11106–11115 (2021)

DGA-Based Power Transformer Fault Diagnosis via Knowledge Distillation of Large Language Model

Xinhai Li[1], Lingcheng Zeng[1], Qingzhu Zeng[1], Yunan Lu[2], Yi Guo[3], and Lei Yang[2(✉)]

[1] Guangdong Zhongshan Power Supply Bureau, China Southern Power Grid Co., Ltd., Zhongshan, China
[2] Department of Computing, The Hong Kong Polytechnic University, Hong Kong, China
ray.yang@polyu.edu.hk
[3] Department of Automation, Shanghai Jiao Tong University, Shanghai, China

Abstract. Transformer fault diagnosis (TFD) plays a critical role in the maintenance and management of power systems. Dissolved gases in insulating oil have emerged as critical indicators of transformer faults, as they effectively reveal the internal failures of power transformers. The primary methods for dissolved gas analysis (DGA) encompass knowledge-driven algorithms and data-driven algorithms. Despite their significant advancements, they are still subject to certain limitations. Specifically, knowledge-driven algorithms are constrained by their excessive reliance on predetermined rules, and the uncertainties inherent in industry situations inevitably diminish diagnostic accuracy. On the other hand, data-driven algorithms face challenges such as overfitting with limited samples and the difficulty in interpreting complex machine learning models. Therefore, we propose a transformer fault diagnosis algorithm based on knowledge distillation of large language models, aiming to mitigate the drawbacks of knowledge-driven and data-driven approaches. The main idea is to encode prior knowledge of large language models as prior fault distributions and employ label distribution learning techniques to integrate this prior knowledge with the fault classifier. Finally, extensive experimental results validate the effectiveness of our proposed method.

Keywords: Transformer fault diagnosis · Large language models · Label distribution learning

1 Introduction

Power transformers serve as critical equipment in power systems, playing an important role in both transmission and distribution networks. In the power transmission stage, transformers can significantly reduce power loss over long-distance and high-capacity transmission through voltage transformation. In the power distribution stage, transformers ensure safe voltage levels that directly determine end-user power quality and supply reliability. Therefore, the operational condition of transformers fundamentally impacts the stability of the power grid and national economy [30]. However, during prolonged service, the internal materials of transformers undergo progressive degradation due to the combined stresses of electrical, thermal, and mechanical. This deterioration mechanism underscores the significant engineering importance of developing accurate fault diagnosis methodologies to maintain power system security and operational reliability [1].

Dissolved Gas Analysis (DGA) is an effective technique for detecting internal faults in power transformers through the gases dissolved in insulating oil. When abnormalities such as partial discharge, overheating, or arcing occur within a transformer, the decomposition of insulating oil and solid insulation materials generates gases such as hydrogen (H_2), methane (CH_4), acetylene (C_2H_2), ethylene (C_2H_4), and ethane (C_2H_6). By evaluating the types, concentrations, and evolving trends of these gases, DGA enables effective identification of potential fault types and their severity levels, which has been established as a standardized fault detection methodology in the power industry.

Existing works of the DGA-based transformer fault diagnosis algorithms can be roughly divided into two groups: knowledge-driven algorithms and data-driven algorithms. In terms of the knowledge-driven diagnostic algorithms, they identify and diagnose faults through domain knowledge, fault rules, or logical reasoning. The typical examples include the IEC ratio method, Rogers ratio method, and Duval triangle method. In terms of the data-driven diagnostic algorithms, they collect historical data of dissolved gas and the corresponding transformer faults, and utilize machine learning techniques to learn the quantitative relationship between the dissolved gas and the transformer faults. Representative examples include the diagnosis algorithms based on support vector machine [34,39], the diagnosis algorithms based on boosting models (such as adaptive boosting algorithm [26], gradient boosting decision tree [28]), and the diagnosis algorithms based on deep neural networks (such as convolutional neural networks [11,35], and gated recurrent networks [42]). Notwithstanding the success of these two kinds of algorithms, they remain subject to certain deficiencies. On the one hand, the knowledge-driven algorithms are hindered by an over-reliance on predetermined rules and the empirical knowledge of maintenance personnel, as well as by uncertainties in gas indicator values stemming from unpredictable operating environment factors, all of which unavoidably degrade diagnostic accuracy. On the other hand, the data-driven algorithms, when confronted with a paucity of samples, are inherently at risk of overfitting, thereby diminishing the accuracy of predictions. Furthermore, the utilization of complex models typically engenders a heightened risk of opacity in their interpretability.

Therefore, in this paper, we aim to simultaneously mitigate the drawbacks of knowledge-driven and data-driven approaches by embedding expert knowledge within the machine learning model. To achieve this goal, we propose an algorithm called TFDLM (Power Transformer Fault Diagnosis via Knowledge Distillation of Large Language Model). The TFDLM algorithm can be broadly divided into three stages: Knowledge extraction, knowledge calibration, and knowledge integration. In the stage of knowledge extraction, we input the components of dissolved gas into the large language model (LLM), and require LLM to evaluate the probabilities of various fault types. This strategy Comparing with the prior knowledge handcrafted by domain experts, this strategy is more efficient and takes into account the characteristics of individual instances. In the stage of knowledge calibration, we design a calibration algorithm to avoid the mismatch between the prior knowledge output by the LLM and the ground-truth in the dataset. In the stage of knowledge integration, we utilize the label distribution learning technique to enable the transformer fault classifier to more effectively absorb the prior knowledge output by LLM. Finally, in order to demonstrate the effectiveness of our proposed algorithms, we evaluate the performance of TFDLM algorithm in a real-world TFD dataset by various evaluation metrics, including precision, recall, ROC curve, and confusion matrix.

2 Related Work

The AI-driven transformer fault diagnosis (TFD) methods aim to learn a mapping from the transformer feature to the fault type. The feature of power transformer typically can be obtained from sensors via a variety of communication techniques such as WiFi [4,13,43–45,48,51,55] and cross-technology communication [40,41,54]. The existing transformer fault diagnosis methods can be generally divided into knowledge-driven algorithms and data-driven algorithms. Knowledge-driven algorithms rely on predefined rules to diagnose transformer fault [14,27,38]. These rules have been widely used in transformer fault diagnosis due to their simplicity and ease of implementation [3,36]. However, their accuracy is limited by the quality of the rules and their ability to handle the complexity and uncertainty of unseen fault patterns. Data-driven algorithms can be further divided into traditional machine learning methods and modern deep learning methods. Traditional machine learning algorithms, including support vector machines, decision trees, and clustering algorithms, have been increasingly applied in transformer fault diagnosis because they can handle large and complex datasets [10,17,31,33] For example, [2] applies SVM to establish the power transformers faults classification and to choose the most appropriate gas signature between SVM and the DGA traditional methods. [50] proposes a novel transformer insulation fault diagnosis method using an entropy-based ID3 decision tree, including pruning for noise reduction and rule extraction, which not only exhibits rapid induction learning and classification but also effectively compresses data, thereby conserving memory and enhancing its effectiveness. In recent years, deep learning has made significant advancements. Numerous deep learning algorithms such as convolutional neural networks, recurrent neural networks, and graph convoluational networks have shown superior performance in TFD accuracy compared to traditional machine learning algorithms [15,49]. For example, [16] introduces graph convolutional networks for transformer fault diagnosis. They first construct the adjacency matrix to fully represent the similarity metrics between unknown samples and labeled samples, and then use the graph convolutional layers with strong feature extraction ability to find the complex nonlinear relationship between dissolved gas and fault type. In addition, transfer learning has also found applications in the field of TFD, which is a technique that allows knowledge gained from one dataset to be transferred to another. This approach has been used to develop models that can classify transformer faults with high accuracy even when trained on small amounts of data [25,29]. Furthermore, researchers have developed hybrid AI systems that combine multiple AI techniques to improve fault diagnosis accuracy. For example, a hybrid system combining bidirectional recurrent neural networks and support vector machines has been developed to accurately classify transformer faults based on DGA data [52].

The machine learning paradigm employed for the analysis of dissolved gas data in this paper is Label Distribution Learning (LDL) [6], which is highly effective in learning the mapping from feature vector to label distribution (a vector formally equivalent to a probability distribution, where each element represents the degree to which a corresponding label describes the instance). The studies in label distribution learning primarily focus on how to learn generalizable patterns from data with label distributions. Initially, traditional algorithms such as support vector machine, k-nearest neighbors, and backpropagation algorithm were adapted for LDL problems through problem transformation and algorithm modification [6]. However, these methods overlook the complexity of the LDL problem, which has spurred extensive research into leveraging the intrinsic knowledge of label distribution data. Typical examples include the algorithms based on label correlation or sample correlation [32,37,47,53], the algorithms based on label ranking [7–9,18,19,22,23]. Furthermore, considering the difficulty in obtaining precise label distribution supervision information, certain label distribution learning algorithms also focused on the label distribution learning problem under weakly supervised scenarios [20,21,24,46].

3 Methodology

Initially, we provide a brief explanation of the mathematical notations commonly used in this paper. Let $\mathcal{X}^D = \mathbb{R}^D$ and $\mathcal{Y} = \{1, 2, \ldots, M\}$ denote the D-dimensional feature space and the M-dimensional class space, respectively. Let $\boldsymbol{x}_n \in \mathcal{X}^D$ and $y_n \in \mathcal{Y}$ denote the feature vector and the fault type of the n-th instance, respectively. The goal of transformer fault diagnosis is to learn a fault classifier according to the dataset $\{(\boldsymbol{x}_n, y_n)\}_{n=1}^N$.

In the next subsections, we will illustrate our proposed TFDLM algorithm. Firstly, in Sects. 3.1, 3.2 and 3.3, we illustrate the three stages of the algorithm, respectively, namely knowledge extraction, knowledge calibration, and knowledge integration. Subsequently, we summarize these three stages as the ultimate optimization objective in Sect. 3.4.

3.1 Knowledge Extraction

In conventional approaches, prior knowledge is typically handcrafted by domain experts, which is not only labor-intensive but also tends to result in a set of general rules that fail to account for the characteristics of individual instances.

Large language models, built upon extensive text corpora, can effectively simulate expert decision-making and efficiently process each instance individually at scale. Therefore, we propose to leverage large language models (LLM) to rapidly acquire the prior knowledge for each instance. Specifically, we have devised the following prompt:

"*In terms of the power transformer fault diagnosis, if there are x_1 microliters of hydrogen (H_2), x_2 microliters of methane (CH_4), x_3 microliters of ethane (C_2H_6), x_4 microliters of ethylene (C_2H_4), and x_5 microliters of acetylene (C_2H_2) in one liter of transformer oil, then what are the probabilities of the transformer being in the following states: normal, low-temperature overheating, medium-temperature overheating, high-temperature overheating, local discharge, low-energy discharge, and high-energy discharge?*"

In the above prompt, $x_1, x_2, \ldots, x_5$, i.e. the feature values, vary from one instance to another within the dataset. We denote the output fault probabilities for the instance $\boldsymbol{x}_n$ as the prior fault distribution $\tilde{\boldsymbol{z}}_n$.

3.2 Knowledge Calibration

During the practical experimental process, we observed a misalignment between the prior fault distributions generated by the LLM and the ground-truth faults, i.e., the fault with the maximum probability in the prior fault distribution does not always correspond to the fault in the dataset. To address this discrepancy, we propose a knowledge calibration strategy. Specifically, if the most probable fault based on the prior fault distribution does not match the ground-truth fault, we linearly combine the prior fault distribution with the one-hot encoded representation of the ground-truth fault: $\boldsymbol{z}_n^\star = \alpha \tilde{\boldsymbol{z}}_n + (1-\alpha)\boldsymbol{e}_n$, where α is the weight of prior fault distribution, and $\boldsymbol{e}_n$ is the one-hot encoded representation of the ground-truth fault of the n-th instance. To make sure that $\boldsymbol{z}_n^\star$ aligns with the ground-truth fault, it is required that α falls within the range $(0, 0.5)$. In this paper, α is set to 0.3.

3.3 Knowledge Integration

Based on the results in Sect. 3.2, the prior knowledge of LLM is represented as a fault distribution, i.e. a discrete probability distribution. Therefore, we aim to effectively integrate this prior knowledge into the classifier during the learning process of the fault classifier by enabling the classifier to simultaneously fit the fault probability distribution. To efficiently mine the potential patterns within the fault distribution, we introduce a well-established ranking-aware label distribution learning algorithm called LDL-LRR [9]. The loss function of LDL-LRR comprises two components: Kullback-Leibler divergence term and logistic ranking error term. The Kullback-Leibler divergence term aims to make the class distribution output by the model as close as possible to the prior fault distribution, and its formal definition is as follows:

$$\mathrm{KL}(\boldsymbol{z}_n^\star \| f(\boldsymbol{x}_n; \boldsymbol{\Theta})) = -\sum_{m=1}^{M} z_{nm}^\star \log f_m(\boldsymbol{x}_n; \boldsymbol{\Theta}) + \mathrm{const}, \quad (1)$$

where $f(\boldsymbol{x}_n; \boldsymbol{\Theta})$, a function with $\boldsymbol{x}_n$ as input and $\boldsymbol{\Theta}$ as learnable parameters, output the probability of each type of fault. The logistic ranking error term aims to capture and preserve the ranking structure of faults underlying the prior fault distributions, and its formal definition is as follows:

$$\mathcal{R}_{nkt} = (r_{nkt} \cdot \log \hat{r}_{nkt} + (1 - r_{nk}) \cdot \log(1 - \hat{r}_{nkt})) \cdot w_{nkt}, \tag{2}$$

where $\hat{r}_{nkt} = \text{sigmoid}(\sigma \cdot (f_k(\boldsymbol{x}_n) - f_t(\boldsymbol{x}_n)))$ models the probability that the k-th fault type precedes the t-th fault type, $\text{sigmoid}(v) = (1 + \exp(-v))^{-1}$, $w_{nkt} = (z^\star_{nk} - z^\star_{nt})^2$ denotes the strength of the corresponding ranking relation, and the ground-truth of ranking relation is defined as follows:

$$r_{nkt} = \begin{cases} 1 & \text{if } z^\star_{nk} > z^\star_{nt} \\ 0 & \text{if } z^\star_{nk} < z^\star_{nt} \\ \frac{1}{2} & \text{if } z^\star_{nk} = z^\star_{nt} \end{cases}. \tag{3}$$

3.4 Optimization Objective

According to the discussions in Sects. 3.1, 3.2 and 3.3, we derive the final optimization objective as follows:

$$\arg\min_{\boldsymbol{\Theta}} \sum_{n=1}^{N} \sum_{m=1}^{M} -e_{nm} \log f_m(\boldsymbol{x}_n; \boldsymbol{\Theta}) - \lambda \sum_{k=1}^{M} \mathcal{R}_{nkm} \\ + \beta \cdot \text{KL}(\boldsymbol{z}^\star_n \| f(\boldsymbol{x}_n; \boldsymbol{\Theta})) + \mu \cdot \|\boldsymbol{\Theta}\|_2, \tag{4}$$

where $e_{nm} = \mathbb{I}(y_n = m)$, $\mathcal{R}_{nkm}$ is defined by Eq. (2), $\text{KL}(\boldsymbol{z}^\star_n \| f(\boldsymbol{x}_n; \boldsymbol{\Theta}))$ is defined by Eq. (1). The loss term $-e_{nm} \log f_m(\boldsymbol{x}_n; \boldsymbol{\Theta})$ aims to minimize the classification error of fault type. $\|\boldsymbol{\Theta}\|_2$ is the L_2 regularization term. λ, β, μ are the trade-off parameters. Since Eq. (4) is differentiable with respect to the parameter $\boldsymbol{\Theta}$, we can leverage gradient-based optimization algorithms to learn the model.

4 Experiments

4.1 Dataset and Evaluation Metrics

The dataset [12] used in this paper comprises 2,910 instances, each containing a feature vector of dissolved gases along with the corresponding transformer fault type. The feature vector of dissolved gases consists of 5 elements, corresponding to the concentrations of hydrogen (H_2), methane (CH_4), ethane (C_2H_6), ethylene (C_2H_4), and acetylene (C_2H_2), respectively. The transformer fault types include normal state (NS), low-temperature overheating (LTO), middle-temperature overheating (MTO), high-temperature overheating (MTO), local discharge (LD), low-energy discharge (LED), and high-energy discharge (HED). The evaluation metrics utilized in this paper include precision, recall, F_1 score, and AUC (i.e. the area under the ROC curve). For the m-th fault type, precision measures the

Table 1. Average Classification Performance of Various Algorithms at 10% Training Set Ratio.

	Precision	Recall	F_1 score	Precision	Recall	F_1 score
	Macro-Average			Micro-Average		
Ours	<u>0.662</u>±0.06	<u>0.658</u>±0.05	<u>0.656</u>±0.05	<u>0.734</u>±0.03	<u>0.734</u>±0.03	<u>0.734</u>±0.03
LR	0.573±0.09	0.557±0.02	0.545±0.03	0.646±0.04	0.646±0.04	0.646±0.04
SVM	0.390±0.07	0.201±0.02	0.154±0.02	0.329±0.03	0.329±0.03	0.329±0.03
KNN	0.603±0.02	0.585±0.09	0.588±0.03	0.672±0.02	0.672±0.02	0.672±0.02
NB	0.458±0.09	0.393±0.03	0.346±0.03	0.440±0.03	0.440±0.03	0.440±0.03
DT	0.592±0.09	0.579±0.09	0.581±0.09	0.655±0.01	0.655±0.01	0.655±0.01
MLP	0.601±0.07	0.574±0.06	0.570±0.08	0.657±0.08	0.657±0.08	0.657±0.08

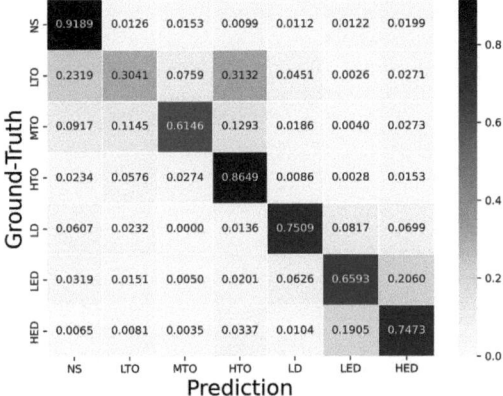

Fig. 1. Confusion matrix of our proposed algorithm at 30% training set ratio. The value in (i,j) entry represents the probability that the model misclassifies the i-th fault type as the j-th fault type.

Table 2. Average Classification Performance of Various Algorithms at 30% Training Set Ratio.

	Precision	Recall	F_1 score	Precision	Recall	F_1 score
	Macro-Average			Micro-Average		
Ours	<u>0.712</u>±0.02	<u>0.694</u>±0.02	<u>0.699</u>±0.02	<u>0.763</u>±0.02	<u>0.763</u>±0.02	<u>0.763</u>±0.02
LR	0.490±0.08	0.478±0.07	0.452±0.09	0.532±0.11	0.532±0.11	0.532±0.11
SVM	0.439±0.06	0.239±0.01	0.208±0.02	0.376±0.02	0.376±0.02	0.376±0.02
KNN	0.679±0.01	0.668±0.02	0.671±0.02	0.738±0.01	0.738±0.01	0.738±0.01
NB	0.450±0.09	0.341±0.02	0.288±0.02	0.401±0.02	0.401±0.02	0.401±0.02
DT	0.670±0.02	0.665±0.02	0.666±0.02	0.726±0.02	0.726±0.02	0.726±0.02
MLP	0.599±0.04	0.565±0.04	0.552±0.05	0.657±0.07	0.657±0.07	0.657±0.07

proportion of true positive instances among all instances predicted as positive; recall measures the proportion of true positive instances that are correctly pre-

Table 3. Classification Performance of Various Algorithms for Different Types of Transformer Fault at 10% Training Set Ratio.

	NS	LTO	MTO	HTO	LD	LED	HED
				Precision			
Ours	$0.843_{\pm0.04}$	$0.354_{\pm0.07}$	$\underline{0.566}_{\pm0.21}$	$\underline{0.780}_{\pm0.05}$	$0.672_{\pm0.25}$	$\underline{0.704}_{\pm0.06}$	$\underline{0.716}_{\pm0.05}$
LR	$0.894_{\pm0.07}$	$0.244_{\pm0.10}$	$0.261_{\pm0.16}$	$0.619_{\pm0.05}$	$0.671_{\pm0.13}$	$0.618_{\pm0.08}$	$0.704_{\pm0.09}$
SVM	$0.296_{\pm0.02}$	$0.000_{\pm0.00}$	$0.275_{\pm0.45}$	$0.610_{\pm0.05}$	$\underline{0.769}_{\pm0.32}$	$0.162_{\pm0.39}$	$0.616_{\pm0.20}$
KNN	$\underline{0.908}_{\pm0.02}$	$0.352_{\pm0.05}$	$0.491_{\pm0.10}$	$0.697_{\pm0.03}$	$0.658_{\pm0.08}$	$0.490_{\pm0.04}$	$0.624_{\pm0.06}$
NB	$0.719_{\pm0.10}$	$0.189_{\pm0.05}$	$0.154_{\pm0.06}$	$0.633_{\pm0.15}$	$0.631_{\pm0.26}$	$0.311_{\pm0.10}$	$0.571_{\pm0.08}$
DT	$0.894_{\pm0.03}$	$\underline{0.387}_{\pm0.08}$	$0.363_{\pm0.06}$	$0.660_{\pm0.04}$	$0.673_{\pm0.12}$	$0.566_{\pm0.05}$	$0.600_{\pm0.04}$
MLP	$0.825_{\pm0.07}$	$0.317_{\pm0.10}$	$0.391_{\pm0.24}$	$0.686_{\pm0.09}$	$0.719_{\pm0.12}$	$0.596_{\pm0.07}$	$0.670_{\pm0.08}$
				Recall			
Ours	$0.924_{\pm0.03}$	$0.327_{\pm0.15}$	$\underline{0.559}_{\pm0.20}$	$0.805_{\pm0.08}$	$0.623_{\pm0.24}$	$\underline{0.633}_{\pm0.08}$	$\underline{0.737}_{\pm0.06}$
LR	$0.722_{\pm0.16}$	$0.147_{\pm0.07}$	$0.184_{\pm0.12}$	$\underline{0.845}_{\pm0.07}$	$\underline{0.712}_{\pm0.09}$	$0.604_{\pm0.19}$	$0.682_{\pm0.10}$
SVM	$1.000_{\pm0.00}$	$0.000_{\pm0.00}$	$0.005_{\pm0.01}$	$0.201_{\pm0.12}$	$0.055_{\pm0.04}$	$0.006_{\pm0.03}$	$0.129_{\pm0.05}$
KNN	$0.918_{\pm0.02}$	$0.332_{\pm0.09}$	$0.394_{\pm0.15}$	$0.784_{\pm0.04}$	$0.573_{\pm0.12}$	$0.484_{\pm0.07}$	$0.609_{\pm0.05}$
NB	$0.964_{\pm0.02}$	$\underline{0.662}_{\pm0.13}$	$0.135_{\pm0.12}$	$0.144_{\pm0.07}$	$0.355_{\pm0.14}$	$0.250_{\pm0.15}$	$0.242_{\pm0.21}$
DT	$0.905_{\pm0.03}$	$0.415_{\pm0.05}$	$0.355_{\pm0.12}$	$0.678_{\pm0.06}$	$0.559_{\pm0.13}$	$0.526_{\pm0.07}$	$0.615_{\pm0.07}$
MLP	$0.831_{\pm0.28}$	$0.208_{\pm0.10}$	$0.373_{\pm0.19}$	$0.756_{\pm0.13}$	$0.610_{\pm0.16}$	$0.558_{\pm0.14}$	$0.679_{\pm0.11}$
				F_1-score			
Ours	$0.881_{\pm0.02}$	$0.327_{\pm0.10}$	$\underline{0.560}_{\pm0.20}$	$\underline{0.792}_{\pm0.03}$	$0.645_{\pm0.24}$	$\underline{0.665}_{\pm0.06}$	$\underline{0.724}_{\pm0.03}$
LR	$0.786_{\pm0.12}$	$0.165_{\pm0.06}$	$0.204_{\pm0.11}$	$0.712_{\pm0.03}$	$\underline{0.677}_{\pm0.06}$	$0.592_{\pm0.11}$	$0.682_{\pm0.05}$
SVM	$0.456_{\pm0.03}$	$0.000_{\pm0.00}$	$0.010_{\pm0.02}$	$0.286_{\pm0.11}$	$0.100_{\pm0.07}$	$0.025_{\pm0.05}$	$0.203_{\pm0.06}$
KNN	$\underline{0.913}_{\pm0.01}$	$0.339_{\pm0.07}$	$0.423_{\pm0.11}$	$0.737_{\pm0.02}$	$0.605_{\pm0.09}$	$0.485_{\pm0.05}$	$0.614_{\pm0.04}$
NB	$0.819_{\pm0.06}$	$0.288_{\pm0.06}$	$0.128_{\pm0.09}$	$0.220_{\pm0.09}$	$0.425_{\pm0.13}$	$0.242_{\pm0.10}$	$0.299_{\pm0.17}$
DT	$0.899_{\pm0.01}$	$\underline{0.396}_{\pm0.04}$	$0.349_{\pm0.08}$	$0.666_{\pm0.03}$	$0.607_{\pm0.12}$	$0.542_{\pm0.04}$	$0.605_{\pm0.04}$
MLP	$0.795_{\pm0.26}$	$0.244_{\pm0.10}$	$0.373_{\pm0.21}$	$0.710_{\pm0.07}$	$0.636_{\pm0.10}$	$0.570_{\pm0.10}$	$0.664_{\pm0.05}$

dicted among all actual positive instances; F_1 score is the harmonic mean of precision and recall. These metrics are defined as follows.

$$\text{Precision}^{(m)} = \frac{TP_m}{TP_m + FP_m}, \text{Recall}^{(m)} = \frac{TP_m}{TP_m + FN_m}, \quad (5)$$

$$F_1 \text{ score}^{(m)} = \frac{\text{Precision}^{(m)} \times \text{Recall}^{(m)}}{\text{Precision}^{(m)} + \text{Recall}^{(m)}} \times 2,$$

where TP_m, FP_m, and FN_m denotes the number of true positives, false positives, and false negatives for the m-th fault type, respectively:

- TP_m (True Positive): The number of samples that belong to m-th fault type and are correctly predicted.
- FP_m (False Positive): The number of samples that do not belong to the m-th fault type but are predicted as the m-th type of fault.
- FN_m (False Negative): The number of samples that belong to m-th fault type but are incorrectly predicted.

To obtain the performance over all types of fault, we average the performance on each type of fault by micro-average and macro-average. The macro-average

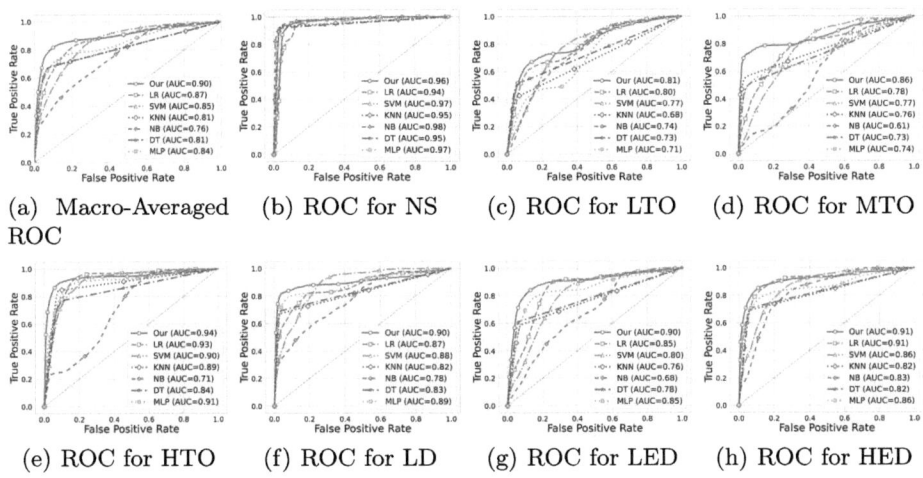

(a) Macro-Averaged ROC (b) ROC for NS (c) ROC for LTO (d) ROC for MTO
(e) ROC for HTO (f) ROC for LD (g) ROC for LED (h) ROC for HED

Fig. 2. ROC curves and the corresponding AUC of comparison algorithms on each type of transformer fault (30% training set ratio).

of the performance is defined as follows. Macro-averaged precision, recall, and F_1 score is $\frac{1}{M}\sum_{m=1}^{M} \text{Precision}^{(m)}$, $\frac{1}{M}\sum_{m=1}^{M} \text{Recall}^{(m)}$, and $\frac{1}{M}\sum_{m=1}^{M} F_1 \text{ score}^{(m)}$ respectively. The micro-average for precision, recall, and F_1 score is the same, defined as $(\sum_{m=1}^{M}(TP_m + FP_m))^{-1}\sum_{m=1}^{M} TP_m$.

4.2 Training Configuration

The training configuration of our proposed algorithm is illustrated as follows.

- LLM for prior knowledge extraction: DeepSeek-R1 [5].
- Model $f_m(x; \Theta)$: A neural network with one hidden layer. There are 128 neurons in the hidden layer. The activation function for the hidden layer is leaky ReLU function, and the activation function for the output layer is the softmax function. The leaky ReLU and softmax functions are defined as follows:

$$\text{LeakyReLU}(v) = \begin{cases} v & \text{if } x \geq 0 \\ 10^{-2}v & \text{if } x < 0 \end{cases}, \qquad (6)$$

$$\text{softmax}(v) = \exp(v) \cdot \|\exp(v)\|_1^{-1}.$$

- Optimization algorithm: The optimization algorithm applied for our model is L-BFGS (Limited-memory Broyden-Fletcher-Goldfarb-Shanno algorithm). The maximum number of iterations is 1,000. If the mean gradient value is smaller than 10^{-5}, or the change of loss value is smaller than 1.5×10^{-8}, the iteration will be stopped. The line search algorithm satisfies the strong Wolfe's conditions.
- Experimental procedure: The dataset is first divided into training and test sets. A portion of the training set is further set aside as a validation set. The

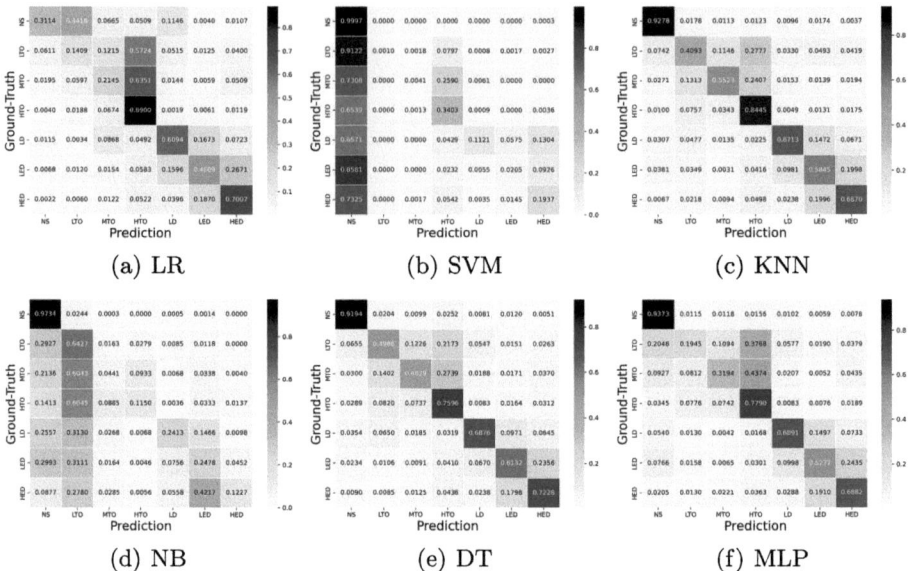

Fig. 3. Confusion matrix of comparison algorithms at 30% training set ratio. The value in (i, j) entry represents the probability that the model misclassifies the i-th fault type as the j-th fault type.

validation set is utilized to identify the best-performing hyperparameters. Based on the optimal hyperparameters, the model is trained on the training data. The trained model is subsequently used to predict fault types for the instances in the test set, and the performance is assessed. The entire procedure is repeated ten times independently.

4.3 Results and Discussions

The overall classification performance of each comparison algorithm is shown in Tables 1 and 2. The baseline algorithms adopted in this paper include logistic regression (LR), support vector machine (SVM), K-nearest neighbor classifier (KNN), naive Bayes classifier (NB), decision tree (DT), and multilayer perceptron (MLP). The experimental result in each cell of Tables 1 and 2 is formatted as "$a_{\pm b}$", where a denotes the mean value of the performance, and b denotes the standard deviation of the performance. The best performance is highlighted by underline. As illustrated in Fig. 2, we present a systematic performance comparison through ROC (Receiver Operating Characteristic) curve analysis across all types of transformer fault. The visualization includes: (1) ROC curves for each comparison algorithm stratified by fault type (NS, LTO, MTO, HTO, LD, LED, and HED), and (2) their corresponding AUC (Area Under the Curve) values. This dual representation enables both qualitative shape analysis of the curves and quantitative comparison through AUC metrics. f

Table 4. Classification Performance of Various Algorithms for Different Types of Transformer Fault at 30% Training Set Ratio.

	NS	LTO	MTO	HTO	LD	LED	HED
				Precision			
Ours	$0.859_{\pm0.03}$	$0.443_{\pm0.05}$	$\underline{0.708}_{\pm0.06}$	$\underline{0.814}_{\pm0.02}$	$0.751_{\pm0.05}$	$\underline{0.689}_{\pm0.05}$	$\underline{0.723}_{\pm0.05}$
LR	$0.624_{\pm0.08}$	$0.121_{\pm0.07}$	$0.247_{\pm0.13}$	$0.628_{\pm0.04}$	$0.527_{\pm0.17}$	$0.595_{\pm0.07}$	$0.687_{\pm0.07}$
SVM	$0.318_{\pm0.01}$	$0.100_{\pm0.02}$	$0.270_{\pm0.42}$	$0.664_{\pm0.03}$	$\underline{0.839}_{\pm0.13}$	$0.294_{\pm0.22}$	$0.588_{\pm0.12}$
KNN	$\underline{0.928}_{\pm0.02}$	$0.427_{\pm0.04}$	$0.629_{\pm0.07}$	$0.771_{\pm0.02}$	$0.684_{\pm0.04}$	$0.593_{\pm0.04}$	$0.721_{\pm0.02}$
NB	$0.642_{\pm0.05}$	$0.149_{\pm0.04}$	$0.093_{\pm0.06}$	$0.699_{\pm0.09}$	$0.644_{\pm0.21}$	$0.267_{\pm0.06}$	$0.659_{\pm0.08}$
DT	$0.919_{\pm0.02}$	$\underline{0.497}_{\pm0.06}$	$0.498_{\pm0.08}$	$0.753_{\pm0.03}$	$0.713_{\pm0.05}$	$0.633_{\pm0.02}$	$0.681_{\pm0.04}$
MLP	$0.817_{\pm0.04}$	$0.367_{\pm0.14}$	$0.472_{\pm0.17}$	$0.687_{\pm0.11}$	$0.643_{\pm0.11}$	$0.579_{\pm0.08}$	$0.627_{\pm0.09}$
				Recall			
Ours	$0.919_{\pm0.04}$	$0.304_{\pm0.09}$	$\underline{0.615}_{\pm0.05}$	$0.865_{\pm0.02}$	$\underline{0.751}_{\pm0.03}$	$\underline{0.659}_{\pm0.05}$	$\underline{0.747}_{\pm0.03}$
LR	$0.311_{\pm0.98}$	$0.141_{\pm0.05}$	$0.215_{\pm0.11}$	$\underline{0.890}_{\pm0.04}$	$0.609_{\pm0.14}$	$0.481_{\pm0.18}$	$0.701_{\pm0.07}$
SVM	$\mathbf{1.000}_{\pm0.00}$	$0.001_{\pm0.00}$	$0.004_{\pm0.01}$	$0.201_{\pm0.12}$	$0.112_{\pm0.04}$	$0.021_{\pm0.02}$	$0.194_{\pm0.05}$
KNN	$0.928_{\pm0.02}$	$0.409_{\pm0.03}$	$0.552_{\pm0.09}$	$0.845_{\pm0.02}$	$0.671_{\pm0.04}$	$0.585_{\pm0.04}$	$0.687_{\pm0.03}$
NB	$0.973_{\pm0.01}$	$\underline{0.643}_{\pm0.08}$	$0.044_{\pm0.07}$	$0.115_{\pm0.05}$	$0.241_{\pm0.13}$	$0.248_{\pm0.19}$	$0.123_{\pm0.03}$
DT	$0.920_{\pm0.02}$	$0.494_{\pm0.05}$	$0.488_{\pm0.05}$	$0.759_{\pm0.04}$	$0.684_{\pm0.04}$	$0.612_{\pm0.06}$	$0.701_{\pm0.04}$
MLP	$0.844_{\pm0.27}$	$0.152_{\pm0.12}$	$0.356_{\pm0.15}$	$0.824_{\pm0.06}$	$0.652_{\pm0.20}$	$0.463_{\pm0.17}$	$0.665_{\pm0.09}$
				F_1 score			
Ours	$0.887_{\pm0.03}$	$0.355_{\pm0.07}$	$\underline{0.656}_{\pm0.04}$	$\underline{0.800}_{\pm0.03}$	$\underline{0.750}_{\pm0.04}$	$\underline{0.673}_{\pm0.04}$	$\underline{0.734}_{\pm0.03}$
LR	$0.350_{\pm0.41}$	$0.112_{\pm0.04}$	$0.218_{\pm0.10}$	$0.735_{\pm0.05}$	$0.544_{\pm0.14}$	$0.514_{\pm0.11}$	$0.689_{\pm0.03}$
SVM	$0.483_{\pm0.02}$	$0.002_{\pm0.01}$	$0.008_{\pm0.01}$	$0.448_{\pm0.05}$	$0.194_{\pm0.07}$	$0.037_{\pm0.03}$	$0.284_{\pm0.05}$
KNN	$\underline{0.927}_{\pm0.01}$	$0.417_{\pm0.02}$	$0.582_{\pm0.06}$	$0.806_{\pm0.01}$	$0.675_{\pm0.05}$	$0.587_{\pm0.03}$	$0.704_{\pm0.02}$
NB	$0.772_{\pm0.04}$	$0.239_{\pm0.05}$	$0.044_{\pm0.04}$	$0.195_{\pm0.08}$	$0.311_{\pm0.10}$	$0.246_{\pm0.09}$	$0.205_{\pm0.05}$
DT	$0.919_{\pm0.01}$	$\underline{0.492}_{\pm0.04}$	$0.487_{\pm0.07}$	$0.755_{\pm0.02}$	$0.697_{\pm0.04}$	$0.621_{\pm0.03}$	$0.690_{\pm0.03}$
MLP	$0.799_{\pm0.22}$	$0.186_{\pm0.11}$	$0.389_{\pm0.12}$	$0.745_{\pm0.08}$	$0.619_{\pm0.14}$	$0.491_{\pm0.13}$	$0.637_{\pm0.05}$

It can be seen from the experimental results in Tables 1 and 2 that our algorithm achieves the best overall performance, as evidenced by its superior macro-average and micro-average for precision, recall, and F_1 score. The detailed performance for each fault type is presented in Tables 3 and 4, which reveals the following key observations:

- Our proposed algorithm achieves best performance on the fault types of MTO, HTO, LED, HED. These faults are critical failure modes where misclassification could lead to significant system damage. Compared to other comparison algorithms, the best detection performance of our algorithm ensures minimal operational risk.
- Our proposed algorithm achieves moderate performance on the fault types of LTO and LD. These are mild or early-stage faults, where misclassification carries lower risk since they rarely escalate into immediate damage.
- Our algorithm shows slightly lower accuracy in detecting normal operating conditions. While this may seem like a drawback, it is actually a safety-first approach that makes perfect sense for industrial applications. In real-world settings, it is safer to occasionally mistake normal operation for a potential fault than to miss an actual fault that could lead to equipment damage.

The observed performance distribution aligns with risk-minimization principles. false positives lead only to additional diagnostic labor, a negligible cost compared to missed faults (i.e. false negatives). This aligns with predictive maintenance strategies, where over-detection is preferable to under-detection. For critical faults (MTO, HTO, LED, and HED), our algorithm is more likely to prevent catastrophic failures compared to other baseline algorithms. For mild faults (LTO and LD), the slightly higher miss rate is tolerable since these rarely cause immediate damage. In summary, our algorithm not only surpasses existing machine learning methods in statistical performance but also delivers a risk-optimized solution tailored for industrial deployment. By maximizing detection of critical faults while tolerating conservative false alarms, it ensures high reliability, which is crucial for real-world predictive maintenance systems.

Furthermore, in order to gain a deeper understanding of our proposed algorithm, we present the confusion matrix of our algorithm at 30% training set ratio in Fig. 1. In this matrix, the value located at the i-th row and j-th column represents the probability that the model misclassifies the fault type corresponding to the i-th row as the fault type corresponding to the j-th column. It is evident from the matrix that our algorithm has a low probability (not exceeding 2%) of misclassifying transformer faults as the normal state. This further underscores the practical safety and reliability of our algorithm in real-world applications. For comparison purposes, we also present the confusion matrices of other comparison algorithms in Fig. 3. It is evident that our algorithm significantly outperforms the other comparison algorithms in terms of practical reliability.

5 Conclusion

This paper reveals the limitations of existing DGA-based transformer fault diagnosis algorithms, i.e. the inflexibility of knowledge-driven methods and the overfitting risks associated with data-driven methods. To address these issues, we propose a transformer fault diagnosis algorithm based on knowledge distillation of large language models. The core idea is to represent the prior knowledge of large language models as prior fault distributions and integrate the fault classifier with the prior knowledge by label distribution learning techniques. Ultimately, experimental results across multiple evaluation methods demonstrate that our algorithm, compared to most commonly used machine learning models, achieves superior classification performance, along with enhanced reliability and safety in practical applications.

Acknowledgements. The research presented in this paper was primarily supported by Science and Technology Project of China Southern Power Grid Limited Liability Company (Grant No. 032000KC23120050/GDKJXM20231537) and supported in part by the NSF of China under Grant 62473251.

References

1. Ali, M.S., Omar, A., Jaafar, A.S.A., Mohamed, S.H.: Conventional methods of dissolved gas analysis using oil-immersed power transformer for fault diagnosis: a review. Electric Power Syst. Res. **216**, 109064 (2023)
2. Bacha, K., Souahlia, S., Gossa, M.: Power transformer fault diagnosis based on dissolved gas analysis by support vector machine. Electric Power Syst. Res. **83**(1), 73–79 (2012)
3. Castro, A.R.G., Miranda, V.: An interpretation of neural networks as inference engines with application to transformer failure diagnosis. Int. J. Electrical Power Energy Syst. **27**(9-10), 620–626 (2005)
4. Dai, D., An, Z., Gong, Z., Pan, Q., Yang, L.: RFID+: spatially controllable identification of UHF RFIDs via controlled magnetic fields. In: USENIX Symposium on Networked Systems Design and Implementation, pp. 1351–1367 (2024)
5. DeepSeek-AI. Deepseek-R1: Incentivizing reasoning capability in LLMs via reinforcement learning (2025)
6. Geng, X.: Label distribution learning. IEEE Trans. Knowl. Data Eng. **28**(7), 1734–1748 (2016)
7. Jia, X., Yunan, L., Zhang, F.: Label enhancement by maintaining positive and negative label relation. IEEE Trans. Knowl. Data Eng. **35**(2), 1708–1720 (2023)
8. Jia, X., Qin, T., Yunan, L., Li, W.: Adaptive weighted ranking-oriented label distribution learning. IEEE Trans. Neural Networks Learn. Syst. **35**(8), 11302–11316 (2024)
9. Jia, X., Shen, X., Li, W., Yunan, L., Zhu, J.: Label distribution learning by maintaining label ranking relation. IEEE Trans. Knowl. Data Eng. **35**(2), 1695–1707 (2023)
10. Kari, T., et al.: Hybrid feature selection approach for power transformer fault diagnosis based on support vector machine and genetic algorithm. IET Gener. Transm. Distrib. **12**(21), 5672–5680 (2018)
11. Li, C., Chen, J., Yang, C., Yang, J., Liu, Z., Davari, P.: Convolutional neural network-based transformer fault diagnosis using vibration signals. Sensors **23**(10), 4781 (2023)
12. Li, E., Wang, L., Song, B.: Fault diagnosis of power transformers with membership degree. IEEE Access **7**, 28791–28798 (2019)
13. Li, R., Zheng, X., Liu, L., Ma, H.: Plug-and-play indoor GPS positioning system with the assistance of optically transparent metasurfaces. In: International Conference on Mobile Computing and Networking, pp. 875–889 (2024)
14. Li, Y., Li, G., Wang, Z.: Rule extraction based on extreme learning machine and an improved ant-miner algorithm for transient stability assessment. PLoS ONE **10**(6), e0130814 (2015)
15. Li, Y., He, S., Li, Y., Ge, L., Lou, S., Zeng, Z.: Probabilistic charging power forecast of EVCS: Reinforcement learning assisted deep learning approach. IEEE Trans. Intell. Veh. **8**(1), 344–357 (2022)
16. Liao, W., Yang, D., Wang, Y., Ren, X.: Fault diagnosis of power transformers using graph convolutional network. CSEE J. Power Energy Syst. **7**(2), 241–249 (2020)
17. Liu, C., Cui, H., Li, G.: Fault diagnosis for the power transformer based on multi-feature fusion algorithm. In: International Conference on Mechatronics, Materials, Chemistry and Computer Engineering, pp. 647–651 (2017)
18. Lu, Y., He, L., Min, F., Li, W., Jia, X.: Generative label enhancement with Gaussian mixture and partial ranking. In: AAAI Conference on Artificial Intelligence, pp. 8975–8983 (2023)

19. Lu, Y., Jia, X.: Predicting label distribution from multi-label ranking. In: Advances in Neural Information Processing Systems, pp. 36931–36943 (2022)
20. Lu, Y., Jia, X.: Predicting label distribution from ternary labels. In: Advances in Neural Information Processing Systems, pp. 70431–70452 (2024)
21. Lu, Y., Li, W., Jia, X.: Label enhancement via joint implicit representation clustering. In: International Joint Conference on Artificial Intelligence, pp. 4019–4027 (2023)
22. Yunan, L., Li, W., Li, H., Jia, X.: Predicting label distribution from tie-allowed multi-label ranking. IEEE Trans. Pattern Anal. Mach. Intell. **45**(12), 15364–15379 (2023)
23. Yunan, L., Li, W., Li, H., Jia, X.: Ranking-preserved generative label enhancement. Mach. Learn. **112**, 4693–4721 (2023)
24. Lu, Y., Li, W., Liu, D., Li, H., Jia, X.: Adaptive-grained label distribution learning. In: AAAI Conference on Artificial Intelligence, pp. 19161–19169 (2025)
25. Mao, W., et al.: Fault diagnosis for power transformers through semi-supervised transfer learning. Sensors **22**(12), 4470 (2022)
26. Mian, Z., et al.: A literature review of fault diagnosis based on ensemble learning. Eng. Appl. Artif. Intell. **127**, 107357 (2024)
27. Miranda, V., Castro, A.R.G.: Improving the IEC table for transformer failure diagnosis with knowledge extraction from neural networks. IEEE Trans. Power Delivery **20**(4), 2509–2516 (2005)
28. Pan, R., Liu, T., Huang, W., Wang, Y., Yang, D., Chen, J.: State of health estimation for lithium-ion batteries based on two-stage features extraction and gradient boosting decision tree. Energy **285**, 129460 (2023)
29. Pei, X., Zheng, X., Jinliang, W.: Rotating machinery fault diagnosis through a transformer convolution network subjected to transfer learning. IEEE Trans. Instrum. Meas. **70**, 1–11 (2021)
30. Raja, B., Venkatakrishnan, G.R., Rengaraj, R.: Power transformer fault diagnosis and condition monitoring using hybrid TDO-SNN technique. Int. J. Hydrogen Energy **68**, 1370–1381 (2024)
31. Rao, U.M., Fofana, I., Rajesh, K.N.V.P.S., Picher, P.: Identification and application of machine learning algorithms for transformer dissolved gas analysis. IEEE Trans. Dielectr. Electr. Insul. **28**(5), 1828–1835 (2021)
32. Ren, T., Jia, X., Li, W., Zhao, S.: Label distribution learning with label correlations via low-rank approximation. In: International Joint Conference on Artificial Intelligence, pp. 3325–3331 (2019)
33. El Amine, M., Senoussaoui, M.B., Fofana, I.: Combining and comparing various machine-learning algorithms to improve dissolved gas analysis interpretation. IET Gener. Transm. Distrib. **12**(15), 3673–3679 (2018)
34. Tan, X., Guo, C., Wang, K., Wan, F.: A novel two-stage dissolved gas analysis fault diagnosis system based semi-supervised learning. High Voltage **7**(4), 676–691 (2022)
35. Thomas, J.B., Chaudhari, S.G., Verma, N.K., et al.: CNN-based transformer model for fault detection in power system networks. IEEE Trans. Instrum. Measur. **72**, 1–10 (2023)
36. LTightiz, L., Nasab, M.A., Yang, H., Addeh, A.: An intelligent system based on optimized ANFIS and association rules for power transformer fault diagnosis. tISA Trans. **103**, 63–74 (2020)
37. Wang, J., Geng, X.: Label distribution learning by exploiting label distribution manifold. IEEE Trans. Neural Networks Learn. Syst. **34**(2), 839–852 (2023)

38. Wang, L., et al.: Method for extracting patterns of coordinated network attacks on electric power CPS based on temporal-topological correlation. IEEE Access **8**, 57260–57272 (2020)
39. Yuhan, W., Sun, X., Dai, B., Yang, P., Wang, Z.: A transformer fault diagnosis method based on hybrid improved grey wolf optimization and least squares-support vector machine. IET Gener. Transm. Distrib. **16**(10), 1950–1963 (2022)
40. Xia, D., Zheng, X., Liu, L., Huang, S., Ma, H.: WiCast: parallel cross-technology transmission for connecting heterogeneous IoT devices. IEEE Trans. Mob. Comput. 1–18 (2025)
41. Xia, D., Zheng, X., Liu, L., Wang, C., Ma, H.: c-Chirp: towards symmetric cross-technology communication over asymmetric channels. IEEE/ACM Trans. Networking **29**(3), 1169–1182 (2021)
42. Xing, Z., He, Y.: Multi-modal information analysis for fault diagnosis with time-series data from power transformer. Int. J. Electr. Power Energy Syst. **144**, 108567 (2023)
43. Xu, C., Zheng, X., Ren, Z., Liu, L., Ma, H.: UHead: driver attention monitoring system using UWB radar. In: Proceedings of the ACM on Interactive, Mobile, Wearable and Ubiquitous Technologies, vol. 8, no. 1 (2024)
44. Leiyang, X., Zheng, X., Xinrun, D., Liu, L., Ma, H.: WiCamera: vortex electromagnetic wave-based WiFi imaging. IEEE Trans. Mob. Comput. **24**(5), 3633–3649 (2025)
45. Leiyang, X., Zheng, X., Liu, L.: Vortex EM wave-based rotation speed monitoring on commodity WiFi. Chin. J. Electron. **34**, 1–13 (2024)
46. Xu, N., Liu, Y.P., Geng, X.: Label enhancement for label distribution learning. In: International Joint Conference on Artificial Intelligence, pp. 1632–1643 (2018)
47. Suping, X., Hengrong, J., Shang, L., Pedrycz, W., Yang, X., Li, C.: Label distribution learning: a local collaborative mechanism. Int. J. Approximate Reasoning **121**, 59–84 (2020)
48. Yang, L., Chen, Y., Li, X.Y., Xiao, C., Li, M., Liu, Y.: Tagoram: real-time tracking of mobile RFID tags to high precision using COTS devices. In: International Conference on Mobile Computing and Networking, pp. 237–248 (2014)
49. Zhang, M., Li, J., Li, Y., Runnan, X.: Deep learning for short-term voltage stability assessment of power systems. IEEE Access **9**, 29711–29718 (2021)
50. Zhao, F., Su, H.: A decision tree approach for power transformer insulation fault diagnosis. In: World Congress on Intelligent Control and Automation, pp. 6882–6886 (2008)
51. Zhao, X., An, Z., Pan, Q., Yang, L.: Nerf2: neural radio-frequency radiance fields. In: International Conference on Mobile Computing and Networking, pp. 1–15 (2023)
52. Zhao, X., Chen, S., Gao, K., Luo, L.: Bidirectional recurrent neural network based on multi-kernel learning support vector machine for transformer fault diagnosis. Int. J. Adv. Comput. Sci. Appl. **14**(1), (2023)
53. Zheng, X., Jia, X., Li, W.: Label distribution learning by exploiting sample correlations locally. In: AAAI Conference on Artificial Intelligence, pp. 4556–4563 (2018)
54. Zheng, X., Xia, D., Yu, F., Liu, L., Ma, H.: Enabling cross-technology communication from WiFi to LoRa with IEEE 802.11ax. IEEE/ACM Trans. Networking **32**(3), 1936–1950 (2024)
55. Zheng, X., Yang, K., Xiong, J., Liu, L., Ma, H.: Pushing the limits of WiFi sensing with low transmission rates. IEEE Trans. Mob. Comput. **23**(11), 10265–10279 (2024)

GraphSAGE-Enhanced Reinforcement Learning for Optimizing Load-Aware Microservice Deployment

Keli Liu[1], Jing Yang[1(✉)], Xiaoli Ruan[1], Qing Hou[2(✉)], Xianghong Tang[1], and Jianhong Cheng[3]

[1] State Key Laboratory of Public Big Data, Guizhou University, Guiyang, China
{gs.klliu23,jyang23,xlruan,xhtang}@gzu.edu.cn
[2] Guizhou Communication Industry Service Co., Ltd., Guiyang, China
houqing.gz@chinaccs.cn
[3] Institute of Guizhou Aerospace Measuring and Testing Technology, Guiyang, China
jianhong_cheng@csu.edu.cn

Abstract. Edge microservice deployment encounters three primary challenges: inter-service dependency constraints that impede optimal resource provisioning, mobility-driven service quality degradation, and geographically heterogeneous load distribution patterns. To address these issues, we propose a GraphSAGE-enhanced reinforcement learning framework for microservice deployment optimization, aimed at reducing deployment response time and balancing edge server loads. The framework leverages GraphSAGE's inductive feature aggregation capabilities to efficiently extract and encode topological characteristics of microservice dependencies and edge server interconnections. Subsequently, deployment decisions are optimized through a Twin Delayed Deep Deterministic Policy Gradient (TD3) algorithm incorporating dual Critic networks to mitigate Q-value overestimation. Experimental results demonstrate that, compared to existing methods such as DQN, SAC, and TRPO, our proposed approach achieves significant improvements in average response time and load balancing, reducing average response time by up to 28.10% and improving load balancing by 68.66%. Notably, the proposed method exhibits excellent performance stability under resource-constrained conditions and large-scale application scenarios.

Keywords: Microservices · Edge Computing · Load Balancing · Reinforcement Learning · Graph Neural Networks

1 Introduction

With the proliferation of 5G networks, microservice architecture has demonstrated significant advantages in latency-sensitive application domains such as vehicle networks, telemedicine, and autonomous driving [30]. Due to its flexibility, scalability, and maintainability, microservice architecture is gradually becoming the mainstream approach for modern application development [2]. Currently, global technology giants including Netflix, Amazon, and Apple have widely adopted microservice architecture to support the operational requirements of their large-scale complex systems.

However, deploying microservices in edge-cloud environments faces multidimensional challenges stemming from both the complexity of microservice architecture and the dynamic characteristics of edge-cloud environments [13, 14, 19]. First, existing research often overlooks the complex dependencies among microservices, which include functional call chains, data flows, and resource sharing [12, 16, 21]. As system scale increases, the complexity of microservice dependencies grows exponentially. Second, frequent changes in user geographical locations may lead to severely uneven distribution of service requests across different regions [1, 22]. For instance, during e-commerce shopping festival peaks, specific regions may experience dramatic increases in service requests due to large numbers of concurrent online shoppers, resulting in server overload. Therefore, how to extract microservice dependencies and design efficient resource allocation strategies has become one of the key challenges in microservice deployment research within edge computing environments.

In this work, we propose a microservice deployment framework for edge-cloud environments aimed at reducing microservice deployment response time and balancing edge server load. To effectively address the complexity and dynamicity of microservice deployment in edge-cloud environments, we employ graph neural networks to extract feature information from edge servers and microservices, further enhancing the optimization capabilities of reinforcement learning algorithms. The main contributions of this work are summarized as follows:

- We construct a heterogeneous graph that models microservices and edge servers as nodes, and employ GraphSAGE's neighbor sampling and aggregation mechanisms to capture complex service dependencies. Node embeddings are transformed into graph-level representations, providing topology-aware state inputs for reinforcement learning deployment optimization.
- We formalize the microservice deployment problem as a Markov Decision Process (MDP) with multi-objective reward functions targeting response time minimization and load balancing. A TD3-based reinforcement learning framework is implemented with dual Critic networks and delayed policy updates to mitigate Q-value overestimation, achieving stable policy learning in dynamic edge environments.
- Comparative experiments with DQN, SAC, TRPO, and Random baselines demonstrate that the proposed method exhibits excellent performance under various resource capacity conditions and application scales, reducing average response time by up to 28.10% and improving load balancing efficiency by 68.66%.

2 Related Work

In recent years, microservice deployment in edge-cloud environments has become a research hotspot. Numerous scholars have focused on deployment methods based on heuristic algorithms [4,6,7]. For instance, Guo et al. [7] proposed a heuristic algorithm based on random rounding for coordinated deployment of microservices in edge computing environments, considering both intra-server and inter-server layer sharing to maximize edge throughput. However, traditional heuristic methods still face limitations when dealing with resource-constrained and highly dynamic environments, particularly performing poorly under high-load conditions. To address these issues, recent research has gradually employed Deep Reinforcement Learning (DRL) methods to optimize microservice deployment [3,5,23,26,28,29]. For example, Gasmi et al. [5] adopted a Q-Learning-based approach to learn the dynamic characteristics of the environment and optimize microservice deployment decisions. Some studies have also proposed deployment methods that incorporate graph neural networks [17]. The aforementioned research primarily focuses on reducing deployment response latency for microservices but overlooks the issue of load balancing among system servers.

Some studies have been dedicated to orchestration and load balancing in containerized environments [9,15,25]. For example, Liu et al. [15] jointly considered service latency and cluster load in designing heuristic request scheduling and service placement algorithms. Shafiq et al. [25] proposed a load balancing algorithm (LB) that improves load distribution in cloud environments through task scheduling; however, this algorithm assumes user requests arrive in random order and may lack sufficient adaptability when facing complex and highly dynamic load situations, making it difficult to effectively respond to load fluctuations in real environments. Additionally, some research utilizes DRL methods to address the load allocation problem for microservices. For instance, Maia et al. [18] proposed a DRL scheme based on load allocation probability and request load. Nevertheless, the above studies primarily rely on single metrics to optimize request load, making it challenging to provide more fine-grained resource management and load balancing strategies.

Inspired by the aforementioned research, we will jointly consider fine-grained microservice deployment and load balancing in our study, aiming to achieve efficient load balancing while satisfying the QoS requirements of microservices.

3 System Model and Problem Formulation

This section introduces the system model and problem formulation for microservice deployment.

3.1 System Model

As illustrated in Fig. 1, the system model comprises three layers: the user layer, the edge layer, and the central cloud layer.

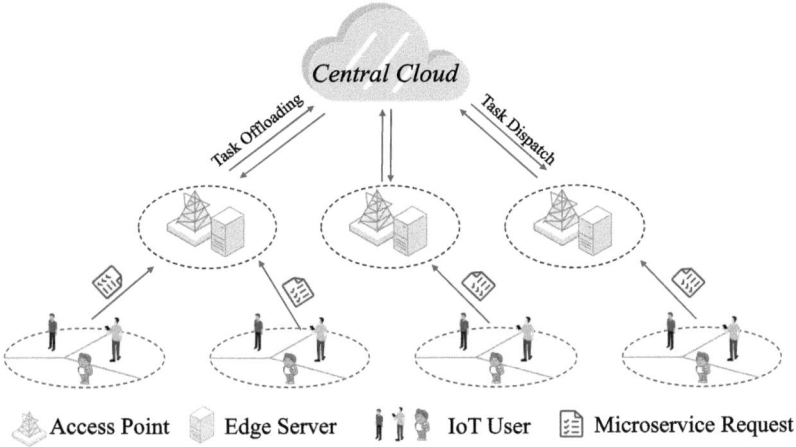

Fig. 1. System architecture

User Layer. The user layer consists of terminal devices distributed across different geographical locations, connected to edge computing servers via wireless networks. Considering user mobility and the time-varying characteristics of request patterns, we model the system operation time in 24-hour cycles. User-requested applications can be decomposed into multiple interdependent microservices, which are dynamically allocated to either the edge layer or central cloud layer for execution based on their computational requirements and resource allocation strategies.

Edge Layer. Composed of N edge servers, each edge server possesses limited CPU computing capability and memory resources. Network resources are represented by uplink and downlink bandwidth, while geographical location is expressed through longitude and latitude coordinates. The coverage range of an edge server is determined by its radius, with users located within the circular service area able to connect directly to that server.

Central Cloud Layer. Equipped with powerful computing and storage resources, the central cloud layer can process complex tasks that cannot be completed by the edge layer, ensuring all tasks are efficiently processed. However, communication latency between users and the central cloud is typically higher.

3.2 Problem Formulation

In an edge-cloud environment, the microservice deployment problem can be formulated as determining the optimal deployment location (edge server or central cloud) for each microservice, subject to resource constraints and QoS requirements, to minimize the average response latency L_{avg} while ensuring load balancing V_e among edge servers.

Based on the deployment location of microservices, we consider the following three scenarios. When all microservices are processed by the same server, the response time is calculated as follows:

$$L = L_e^{up} + L_e^{down} + L_e^{ex} \tag{1}$$

where L_e^{up} and L_e^{down} represent the latency for uploading data to the edge server and returning computation results to the user, respectively. L_e^{ex} denotes the execution time of microservices. When some microservices are processed on the current edge server and subsequent microservices are transferred to other edge servers due to high load, considering the communication latency between microservices, the response time is calculated as:

$$L = L_e^{up} + L_e^{ex} + L_{com} + L_{e'}^{ex} + L_{e'}^{down} \tag{2}$$

where L_{com} represents the communication latency between edge servers. If the server cannot process subsequent microservices and they are instead processed by the cloud, the response time becomes:

$$L = L_e^{up} + L_e^{ex} + L_{com}^c + L_c^{ex} + L_c^{down} \tag{3}$$

where L_{com}^c represents the communication latency between the edge server and the cloud, L_c^{down} denotes the download latency from the cloud to the user, and L_c^{ex} indicates the execution latency of microservices in the cloud.

To quantify the degree of load balancing among edge servers, we use the variance of comprehensive resource utilization V_e as an evaluation metric:

$$V_e = \frac{1}{N} \sum_{i=1}^{N} (u_i - \mu)^2 \tag{4}$$

where u_i is the comprehensive resource utilization of the i-th edge node, and μ is the average comprehensive resource utilization of all edge nodes. A smaller variance indicates more balanced load among edge nodes.

4 Methodology

This section presents an optimization framework for microservice deployment in edge computing environments. To address system dynamics and resource heterogeneity, we design a reinforcement learning method based on TD3, incorporating graph neural networks for topology-aware state representation to achieve multi-objective optimization of response time minimization and resource balancing.

4.1 MDP Modeling for Microservice Deployment

In edge computing environments, the microservice deployment problem can be formalized as a Markov Decision Process (MDP). The MDP is formally defined as $M = (S, A, P, R, \gamma)$, with the following components:

State Space. S encompasses microservice request queues, CPU and memory requirements of the current scheduling request, resource status of edge servers, and user geographical location.

Action Space. A represents all possible actions $a_t \in A$ that an agent can take in state $s_t \in S$. In the microservice deployment problem, actions correspond to assigning microservices to specific computing nodes, which can be one of N edge nodes or a cloud node, resulting in $N + 1$ discrete action choices.

State Transition Probability. $P(s_{t+1}|s_t, a_t)$ describes the probability of transitioning from the current state s_t to the next state s_{t+1} after executing a specific action a_t, which is influenced by microservice deployment adjustments and resource utilization changes.

Reward Function. $R(s_t, a_t, s_{t+1})$ quantifies the immediate reward for executing action a_t in state s_t and transitioning to the next state s_{t+1}. To achieve multi-objective optimization of response time minimization and load balancing, we design a composite reward function that includes: (i) response time optimization reward, providing positive feedback based on the degree of reduction in average response time; (ii) load balancing reward, evaluating the degree of load balancing across edge nodes through resource utilization variance, encouraging microservice deployment to nodes with lighter loads. The agent's objective is to find the optimal deployment policy that maximizes the expected discounted cumulative reward:

$$\pi^* = \arg\max_{\pi} \mathbb{E}\left[\sum_{t=0}^{\infty} \gamma^t R(s_t, a_t, s_{t+1}) \Big| a_t = \pi(s_t)\right] \tag{5}$$

where $\gamma \in [0, 1)$ is the discount factor, balancing immediate rewards and long-term gains.

Algorithm 1: GraphSAGE-Enhanced Reinforcement Learning for Microservice Deployment

Input: Graph $G = (V, E)$, learning rates η_a, η_c, discount factor γ
Output: Deployment policy π_θ
1: Initialize networks π_θ, Q_{ϕ_1}, Q_{ϕ_2} and target networks
2: Initialize replay buffer $\mathcal{D}$.
3: **for each** episode **do**
4:    **GraphSAGE Representation:**
5:       Sample node neighbors using Eq.8
6:       Update node embeddings using Eq.7
7:       state representation $s_t = [z_G, t]^T$
8:    **TD3 Training:**
9:       Execute action $a_t = \pi_\theta(s_t) + \epsilon$ and observe (R_t, s_{t+1})
10:      Store transition (s_t, a_t, R_t, s_{t+1}) in $\mathcal{D}$
11:      Compute target using Eq.10
12:      Update Critics using Eq.12
13:      Update Actor using Eq.13
14:      Soft update target networks
15: **return** Policy π_θ

4.2 GraphSAGE-Based State Representation

A key challenge in microservice deployment is effectively capturing the complex topological structure and node dependencies in edge computing environments. Traditional vector-based state representations struggle to express this structured information, making it difficult for reinforcement learning algorithms to learn effective deployment strategies. To address this challenge, we propose an inductive graph representation learning method based on GraphSAGE [8], which explicitly models system topology through message passing mechanisms, enhancing the expressiveness of state representations. The main procedure of the proposed method is detailed in Algorithm 1.

The edge computing system is formalized as a graph $G = (V, E)$, where the node set V contains microservice nodes V_S and edge server nodes V_E. Each microservice features include computational requirements, storage needs, and dependencies with other microservices. Each edge node features encompass computational capability, storage capacity, and network bandwidth. E represents the edge set, indicating relationships between nodes.

Unlike traditional GCN, GraphSAGE adopts an inductive learning approach capable of handling dynamically changing graph structures, making it more suitable for microservice deployment scenarios in edge computing environments. For each node $v \in V$, neighbor information $N(v)$ is integrated through an aggregation function:

$$h_{N(v)}^{(k)} = \text{AGGREGATE}_k(\{h_u^{(k-1)}\}) \tag{6}$$

Node representation is updated by combining self-information with aggregated neighbor information:

$$h_v^{(k)} = \sigma(W^{(k)} \cdot \text{CONCAT}(h_v^{(k-1)}, h_{N(v)}^{(k)}) + b^{(k)}) \tag{7}$$

where $h_v^{(k)}$ is the embedding vector of node v at layer k, AGGREGATE$_k$ is the aggregation function at layer k, $W^{(k)}$ and $b^{(k)}$ are the weight matrix and bias vector at layer k, and σ is the RELU activation function.

We apply L2 normalization to the newly generated embeddings $h_v^{(k)}$ to enhance numerical stability and generalization capability. To avoid high computational complexity associated with full-graph training, we employ a local subgraph training strategy based on neighbor sampling. For each target node v, a fixed number of neighbors are sampled at layer k:

$$N_k(v) = \text{AGGREGATE}(N(v), S_k) \tag{8}$$

where S_k is the sampling quantity at layer k. The node-level embeddings generated by GraphSAGE need to be transformed into state representations suitable for reinforcement learning. We adopt mean pooling operations to aggregate node-level embeddings into graph-level embeddings:

$$z_G = f_{\text{pool}}(\{h_v^{(K)}\}) \tag{9}$$

The final state representation $s_t = [z_G, t]^T$ provides the reinforcement learning algorithm with state input that integrates topological structure information and temporal dynamic characteristics.

4.3 Training Process

To optimize microservice deployment, we design a reinforcement learning algorithm based on GraphSAGE. As illustrated in Fig. 2, we leverage node feature representations extracted from GraphSAGE along with task state information to learn optimal deployment policies through an Actor-Critic architecture. The details are as follows:

In the microservice deployment optimization problem, the Actor network maps system states to service placement adjustment decisions. The Actor network employs a three-layer feedforward neural network structure with hidden layer dimensions of 512, mapping state space S to action space A. The output layer utilizes a tanh activation function to ensure action values satisfy system constraints: $\pi_\theta(s_t) = \tanh(f_\theta(s_t))$, where f_θ represents the nonlinear mapping function of the feedforward neural network. The Critic network is responsible for estimating the value function of state-action pairs (s_t, a_t), evaluating the long-term returns of executing specific actions in given states. TD3 maintains two structurally identical but parametrically independent Critic networks Q_{ϕ_1} and Q_{ϕ_2}, along with corresponding target networks $Q_{\phi_1'}$ and $Q_{\phi_2'}$. Target Q-value calculation adopts the minimum estimated value from two Critic networks, formally expressed as:

$$y_t = R_t + \gamma \min_{i=1,2} Q_{\phi_i'}(s_{t+1}, a_{t+1}) \tag{10}$$

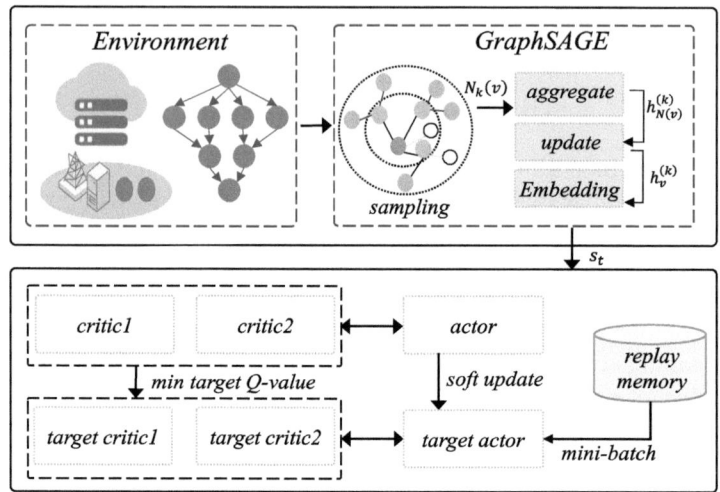

Fig. 2. Microservice deployment framework.

where a_{t+1} is the action output of the target policy in the next state s_{t+1}. Critic networks are optimized by minimizing temporal difference error, with the loss function defined as:

$$L(\phi_i) = \mathbb{E}_{(s_t, a_t, R_t, s_{t+1}) \sim \mathcal{D}} \left[(Q_{\phi_i}(s_t, a_t) - y_t)^2 \right] \tag{11}$$

where $\mathcal{D}$ represents the experience replay buffer. The Actor network is optimized by maximizing the expected Q-value of the first Critic network, with objective function $J(\theta) = \mathbb{E}_{s_t \sim \mathcal{D}}\left[Q_{\phi_1}(s_t, \pi_\theta(s_t))\right]$ and corresponding loss function $L(\theta) = -J(\theta)$. Network parameter update formulas are as follows:

$$\phi_i \leftarrow \phi_i - \eta_c \nabla_{\phi_i} L(\phi_i) \tag{12}$$

$$\theta \leftarrow \theta - \eta_a \nabla_\theta L(\theta) \tag{13}$$

where η_c and η_a represent the learning rates of Critic and Actor networks, respectively. Additionally, the TD3 algorithm employs delayed policy updates and target network soft updates to further enhance training stability.

5 Performance Evaluation

This section presents the performance evaluation results of the microservice deployment framework.

5.1 Experimental Setup

Experiments utilize the real-world EUA dataset from Australia [11], which contains geographical location information of edge servers and user devices in major

Australian cities. We implement the microservice deployment framework using PyTorch, with all experiments conducted on a Windows 11 desktop computer equipped with 128 GB RAM, 12th Gen i9-12900K CPU, and RTX4090 GPU.

Edge-Cloud Environment. Based on the EUA dataset, we construct a cross-regional scenario consisting of 4 heterogeneous edge servers and 47 user devices, with unlimited central cloud resources. IoT devices are deployed in each region to simulate users sending service requests.

Microservice Requests. Each microservice has different resource requirements. Typical application scales comprise approximately 20 applications, each invoking 10–20 microservices. Microservice dependencies are simulated through an affinity matrix.

Baseline Methods. To comprehensively evaluate the performance advantages of our proposed method, we select the following four representative algorithms as benchmarks for comparison:

- Trust Region Policy Optimization (TRPO) [24]: A policy optimization algorithm based on trust regions that ensures monotonic improvement by constraining policy update step sizes, being one of the classic algorithms in the reinforcement learning domain.
- Deep Q-Network (DQN) [27]: A classic value function method combining deep learning with Q-learning, implemented using a double DQN variant.
- Soft Actor-Critic (SAC) [10]: An algorithm based on maximum entropy reinforcement learning that balances exploration and exploitation by incorporating an entropy term in the objective function.
- Random [20]: A random deployment strategy that randomly selects a feasible edge server or cloud server for each deployment decision.

5.2 Performance Evaluation

We evaluate the performance of various methods using two metrics: average response time and variance of comprehensive utilization across edge servers.

Ablation Study. We conduct a detailed analysis of OURS versus OURS w/o GraphSAGE over 100 training epochs, with results demonstrating significant advantages of our complete method in microservice deployment optimization. As illustrated in Figs. 3 and 4, our proposed method rapidly converges to a stable state after initial training, maintaining an average response time of approximately 580ms with extremely low system load variance of around 0.2. In contrast, the method without GraphSAGE integration exhibits notable volatility throughout the training process, with response times frequently fluctuating between

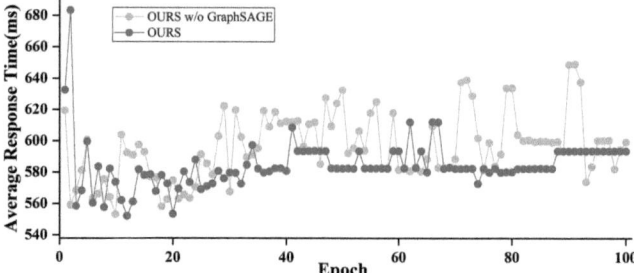

Fig. 3. Average Response Time under Different Ablation Strategies across Iterations

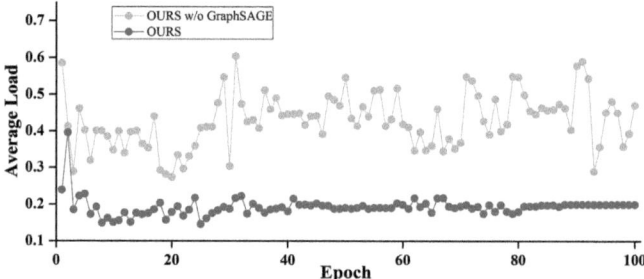

Fig. 4. Average Load under Different Ablation Strategies across Iterations

580–650ms and system load consistently remaining at higher levels (averaging approximately 0.4–0.5). Particularly noteworthy is that by the 40th epoch, our proposed method achieves stable operational status, while the version without GraphSAGE continues to display significant performance fluctuations. This comparative result conclusively demonstrates the critical role of GraphSAGE in perceiving microservice dependencies and optimizing resource allocation, providing superior performance for deployment.

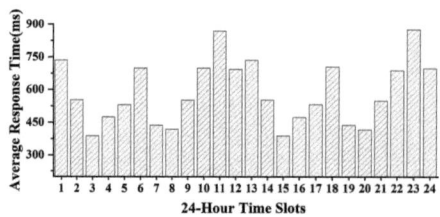

Fig. 5. Average Response Time Across 24-Hour Time Slots.

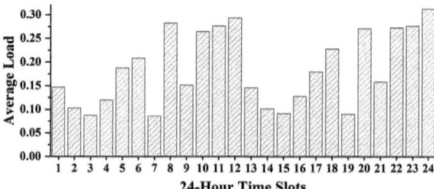

Fig. 6. Average Load Across 24-Hour Time Slots

Performance Analysis Over 24-H Cycle. We conduct a comprehensive analysis of the proposed method's average response time and system load over a 24-h operational cycle. As illustrated in Fig. 5, the method exhibits distinct bimodal fluctuation characteristics throughout the day, primarily reflecting the time-varying nature of user requests in real network environments. Specifically, during peak business hours (11th and 23rd hours), the average response time reaches peak values of approximately 900ms; while during off-peak periods (3rd and 15th hours), response times significantly decrease to below 400ms. This fluctuation pattern clearly presents two complete cycles, corresponding to user behavior characteristics during daytime working hours and nighttime activity periods.

Figure 6 reveals the dynamic patterns of system load variation over time. Experimental results demonstrate a high positive correlation between load changes and response times, with system load reaching maximum values of 0.29–0.31 during request-intensive peak periods, while dropping below 0.1 during off-peak periods. Notably, even under maximum load conditions, system resource utilization remains at relatively low levels (maximum of only 0.31), which conclusively demonstrates the superior performance of the proposed method in resource scheduling and load balancing, achieving efficient resource utilization while ensuring service quality.

Impact of Different Resource Capacities on Performance Metrics. We comprehensively evaluate the performance of our proposed method against four baseline approaches (DQN, SAC, TRPO, Random) under varying resource capacity conditions, as illustrated in Fig. 7. Across all resource capacity levels ranging from 50% to 400%, our proposed method significantly outperforms other methods in both average response time and load balancing. Particularly under extreme resource-constrained conditions (50%), our method achieves an average response time of only approximately 700ms, reducing response time by up to 30% compared to other methods; simultaneously, its system load remains at around 0.1, while other algorithms generally exhibit high loads of 0.7–0.8. As resource capacity gradually increases, response times for all methods show a declining trend, yet our proposed method consistently maintains a significant performance advantage. Notably, although our method exhibits slight load fluctuations with increasing resources, it consistently maintains the lowest level among all methods. Even under extremely abundant resource conditions (400%), other methods still show loads of approximately 0.4, indicating significantly inefficient resource utilization.

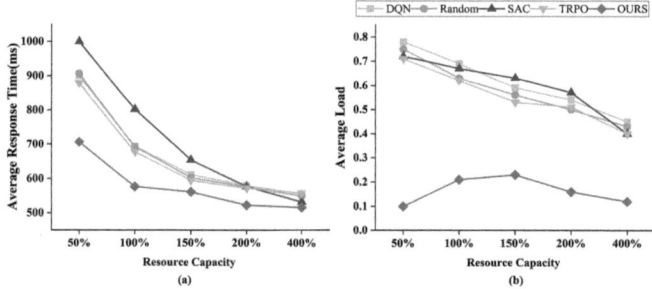

Fig. 7. Impact of Resource Capacity on Average Response Time and Load

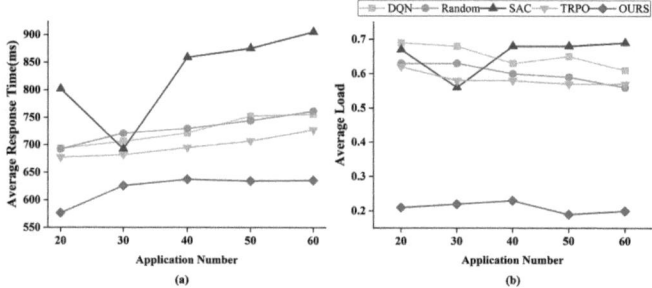

Fig. 8. Impact of Application Numbers on Average Response Time and Load

Impact of Different Application Numbers on Performance Metrics. We conduct a systematic study of how varying application quantities (20–60) affect microservice deployment performance, as illustrated in Fig. 8. As the number of applications increases, our proposed method demonstrates superior performance across all test scenarios, with response times showing only moderate growth from approximately 580ms to around 640ms, an increase of about 10%, while maintaining consistently low load levels between 0.19–0.23 with minimal fluctuation. In contrast, SAC exhibits the most dramatic response time fluctuations, initially decreasing from approximately 800 ms at 20 applications to 695 ms at 30 applications, then sharply climbing to 910 ms at 60 applications. DQN, Random, and TRPO display relatively gradual upward trends, yet with response times significantly higher than our method. Regarding system load, all baseline algorithms generally maintain loads between 0.56–0.69, approximately 65% higher than our proposed method. These results indicate that our proposed method possesses excellent scalability in complex multi-task edge computing environments, with its performance advantages becoming particularly pronounced in high-load scenarios.

6 Conclusion

This work presented an optimization framework for microservice deployment in edge-cloud environments, combining GraphSAGE neural networks with reinforcement learning algorithm to achieve multi-objective optimization of response time minimization and load balancing. By modeling the topological structure of edge servers and microservices, our proposed method effectively captured the complex dependencies within edge-cloud systems, significantly enhancing the expressiveness of state representations. Ablation studies confirmed that GraphSAGE improved deployment stability. In comprehensive comparisons with DQN, SAC, TRPO, and Random baseline methods, our approach demonstrated clear performance advantages across various load conditions and resource configurations. Particularly in challenging environments with limited resource capacity and large-scale application deployments, our method consistently maintained stable low response times and system loads, exhibiting exceptional robustness and scalability.

Future work will introduce federated learning framework to enable collaborative optimization and privacy protection between edge nodes, further improving the efficiency and security of microservice deployment in edge computing environments.

Acknowledgments. This work was supported by the national natural science foundation of China (62441608), the science and technology project of Guizhou Province (QKHZC[2023]368, QKHCG[2025]007, QKHZC[2025]023), the high level innovative talent project of Guizhou Province (QKHGCC[2023]101), the developing objects and projects of scientific and technological talents in Guiyang city (No. ZKHT[2023]48-8), Guizhou University Science and Technology Innovation Team ([2024]No.07), Guizhou University Basic Research Fund ([2024]08), the National Natural Science Foundation of China under Grant (No. 62302119), and Guiyang City Science and Technology Plan Project (No. [2024]2-18).

References

1. Bhattacharya, R., Gao, Y., Wood, T.: Dynamically balancing load with overload control for microservices. ACM Trans. Autonom. Adapt. Syst. **19**(4), 1–23 (2024)
2. Du, F., Shi, J., Chen, Q., Li, L., Guo, M.: A microservice graph generator with production characteristics. arXiv preprint arXiv:2412.19083 (2024)
3. Fan, W., Liu, X., Yuan, H., Li, N., Liu, Y.: Time-slotted task offloading and resource allocation for cloud-edge-end cooperative computing networks. IEEE Trans. Mob. Comput. (2024)
4. Fan, W., et al.: Collaborative service placement, task scheduling, and resource allocation for task offloading with edge-cloud cooperation. IEEE Trans. Mob. Comput. **23**(1), 238–256 (2022)
5. Gasmi, K., Abassi, K., Romdhani, L., Debauche, O.: Reinforcement learning-based approach for microservices-based application placement in edge environment. In: 2023 IEEE Symposium on Computers and Communications (ISCC), pp. 1098–1103 (2023)

6. Gu, L., Chen, Z., Xu, H., Zeng, D., Li, B., Jin, H.: Layer-aware collaborative microservice deployment toward maximal edge throughput. In: IEEE INFOCOM 2022-IEEE Conference on Computer Communications, pp. 71–79 (2022)
7. Guo, F., Tang, B., Tang, M.: Joint optimization of delay and cost for microservice composition in mobile edge computing. World Wide Web **25**(5), 2019–2047 (2022)
8. Hamilton, W., Ying, Z., Leskovec, J.: Inductive representation learning on large graphs. In: Advances in Neural Information Processing Systems, vol. 30 (2017)
9. Hardikar, S., Ahirwar, P., Rajan, S.: Containerization: cloud computing based inspiration technology for adoption through docker and kubernetes. In: 2021 Second International Conference on Electronics and Sustainable Communication Systems (ICESC), pp. 1996–2003 (2021)
10. Hazarika, B., Singh, K., Biswas, S., Mumtaz, S., Li, C.P.: Sac-based resource allocation for computation offloading in iov networks. In: 2022 Joint European Conference on Networks and Communications & 6G Summit (EuCNC/6G Summit), pp. 314–319 (2022)
11. He, Q., et al.: A game-theoretical approach for user allocation in edge computing environment. IEEE Trans. Parallel Distrib. Syst. **31**(3), 515–529 (2019)
12. Hu, B., Cao, Z.: Minimizing resource consumption cost of dag applications with reliability requirement on heterogeneous processor systems. IEEE Trans. Industr. Inf. **16**(12), 7437–7447 (2019)
13. Hua, W., Liu, P., Huang, L.: Energy-efficient resource allocation for heterogeneous edge-cloud computing. IEEE Internet Things J. (2023)
14. Jamshidi, P., Pahl, C., Mendonça, N.C., Lewis, J., Tilkov, S.: Microservices: the journey so far and challenges ahead. IEEE Softw. **35**(3), 24–35 (2018)
15. Liu, H., Li, Y., Wang, S.: Request scheduling combined with load balancing in mobile-edge computing. IEEE Internet Things J. **9**(21), 20841–20852 (2022)
16. Liu, M., Tu, Z., Xu, H., Xu, X., Wang, Z.: Dysr: a dynamic graph neural network based service bundle recommendation model for mashup creation. IEEE Trans. Serv. Comput. **16**(4), 2592–2605 (2023)
17. Lv, W., et al.: Graph-reinforcement-learning-based dependency-aware microservice deployment in edge computing. IEEE Internet Things J. **11**(1), 1604–1615 (2023)
18. Maia, A.M., Ghamri-Doudane, Y.: A deep reinforcement learning approach for the placement of scalable microservices in the edge-to-cloud continuum. In: GLOBECOM 2023-2023 IEEE Global Communications Conference, pp. 479–485 (2023)
19. Nain, A., Sheikh, S., Shahid, M., Malik, R.: Resource optimization in edge and sdn-based edge computing: a comprehensive study. Cluster Comput. 1–29 (2024)
20. Psychas, K., Ghaderi, J.: Randomized algorithms for scheduling multi-resource jobs in the cloud. IEEE/ACM Trans. Networking **26**(5), 2202–2215 (2018)
21. Qian, L., Li, J., He, X., Gu, R., Shao, J., Lu, Y.: Microservice extraction using graph deep clustering based on dual view fusion. Inf. Softw. Technol. **158**, 107171 (2023)
22. Santos, J., Wauters, T., De Turck, F., Steenkiste, P.: Towards optimal load balancing in multi-zone kubernetes clusters via reinforcement learning. In: 2024 33rd International Conference on Computer Communications and Networks (ICCCN), pp. 1–9. IEEE (2024)
23. Santos, J., et al.: Efficient microservice deployment in kubernetes multi-clusters through reinforcement learning. In: NOMS 2024-2024 IEEE Network Operations and Management Symposium, pp. 1–9 (2024)
24. Schulman, J., Levine, S., Abbeel, P., Jordan, M., Moritz, P.: Trust region policy optimization. In: Proceedings of the 32nd International Conference on Machine Learning, vol. 37, pp. 1889–1897 (2015)

25. Singh, N., et al.: Load balancing and service discovery using docker swarm for microservice based big data applications. J. Cloud Comput. **12**(1), 4 (2023)
26. Yang, J., Lu, J., Zhou, X., Li, S., Xiong, C., Hu, J.: Ha-a2c: hard attention and advantage actor-critic for addressing latency optimization in edge computing. IEEE Trans. Green Commun. Networking (2024)
27. Zeng, L., Liu, Q., Shen, S., Liu, X.: Improved double deep q network-based task scheduling algorithm in edge computing for makespan optimization. Tsinghua Sci. Technol. **29**(3), 806–817 (2023)
28. Zhou, X., Yang, J., Li, Y., Li, S., Su, Z.: Deep reinforcement learning-based resource scheduling for energy optimization and load balancing in sdn-driven edge computing. Comput. Commun. **226**, 107925 (2024)
29. Zhou, X., Yang, J., Li, Y., Li, S., Su, Z., Lu, J.: Ec-trl: evolutionary-weighted clustering and transformer-augmented reinforcement learning for dynamic resource scheduling in edge cloud environments. IEEE Internet Things J. (2024)
30. Zuo, X., Wang, M., Xiao, T., Wang, X.: Low-latency networking: architecture, techniques, and opportunities. IEEE Internet Comput. **22**(5), 56–63 (2018)

STCL-Dynamic Sparsity-Driven Transition Feature Replay Continual Learning for Edge Devices

Peng Zhang[1], Jing Yang[1,2](✉), Xiaoli Ruan[2], Qing Hou[3](✉), Xianghong Tang[2], and Jianhong Cheng[4]

[1] College of Computer Science and Technology, Guizhou University, Guiyang, China
{gs.zhangp24,gs.jyang23}@gzu.edu.cn
[2] State Key Laboratory of Public Big Data, Guizhou University, Guiyang, China
{xlruan,xhtang}@gzu.edu.cn
[3] Guizhou Communication Industry Service Co., Ltd., Guiyang, China
houqing.gz@chinaccs.cn
[4] Institute of Guizhou Aerospace Measuring and Testing Technology, Guiyang, China
jianhong_cheng@csu.edu.cn

Abstract. With the growing adoption of AI models and the increasing scale and complexity of inference tasks, continual learning (CL) has gained prominence for edge devices. However, traditional CL methods rely on large replay buffers to mitigate forgetting, which conflicts with the limited resources of edge systems and struggles with task ambiguity and sudden data shifts. To address these issues, we proposes Dynamic Sparsity-driven Transition Feature Replay Continual Learning (STCL), a lightweight CL framework tailored for edge devices. STCL tackles three key challenges: *(i)* limited on-chip memory for CNN parameters, *(ii)* insufficient compute power for multi-epoch training, and *(iii)* the need for real-time adaptation to new environments. It achieves this through weight sparsity, transition feature activation, and adaptive dual-memory replay. Extensive experiments on large-scale CL benchmarks show that STCL achieves 95.6% adaptive accuracy on CORe50, outperforming state-of-the-art methods while significantly reducing memory usage.

Keywords: Dynamic sparsification compression · Continual learning · Dual-memory replay · User adaptation

1 Introduction

Continual Learning [1,2], also known as incremental learning or lifelong learning, is a machine learning method that dynamically updates neural network parameters to enable the model to continuously acquire new knowledge from non-stationary data streams, and its core goal is to adapt to the data distribution changes of new tasks while effectively retaining the knowledge memory of historical tasks during the evolution of task sequences. However, existing methods face dual challenges: *i)* Traditional continual learning relies on intensive parameter updating and large-scale replay buffer mechanisms, whose memory and computation overheads far exceed the resource upper limit of edge devices (e.g., FPGA, EdgeTPU), making it difficult to realize efficient deployment under limited arithmetic; and *ii)* in scenarios of fuzzy task boundaries (e.g., incremental task switching) and dynamic data drifting, replay-based mainstream approaches usually adopt a uniform sampling strategy that fails to differentially evaluate the importance of tasks and samples, resulting in the loss of key knowledge due to storage redundancy or inefficient replay. Therefore, in order to continuously learn new classes without forgetting previously learned old classes, a flexible framework that can incrementally store new instances and classes on edge devices is required. Current continual learning approaches can be summarized into three main technical challenges: *i)* regularization-based approaches (e.g., online EWC++ [1], LWF [2]) inhibit catastrophic forgetting by constraining weight updates, but their reliance on global parameter optimization and high-precision floating-point computation makes it difficult to adapt to the low-computing-power requirements of edge devices; *ii)* architecture-scaling-based approaches reduce task disruption by parameter isolation or dynamically expanding network capacity to reduce task interference, but the storage overhead far exceeds the resource limit of edge devices due to the linear growth of parameter scale; *iii)* Replay-based approaches (e.g., ER [3], SLDA [4]) mitigate forgetting by mixing historical and new data for training, but are limited by storage and computation bottlenecks.

The Chameleon framework proposed by Shivam et al. [27] innovatively employs the dual memory replay technique with a tiered storage architecture to address catastrophic forgetting and ephemeral cycling of data sets in non-independent and identified distributed (Non-IID) data streams on edge devices. Non-IID data streams in edge devices address catastrophic forgetting and ephemeral cycling of data sets. The framework deploys Short Memory in On-Chip Memory to cache user-favored incremental data in real time through user-centered training, and Real-Time Memory in Off-Chip Memory to cache user-preferred incremental data in real time. Off-Chip Memory is used to build Long-Term Memory to preserve global knowledge representations through a class-balanced sampling strategy. However, when deploying continuous learning models in end devices, the limited on-chip memory struggles to carry dense CNN parameters, still retains a large number of zero-valued weights and potential activations, and the static sparse strategy cannot dynamically adapt to task conflicts and hardware acceleration requirements.

To overcome the resource limitations and catastrophic forgetting problems in edge devices, we propose a sparsity - dual Memory cooperative continuous learning framework (STCL), as shown in Fig. 1. Compared to Latent Replay [5], which stores intermediate layer features via uniform random sampling and faces high latency and energy costs when accessing off-chip memory for training, STCL achieves efficient anti-forgotten continuous learning through joint optimization of weight sparsity compression, transition feature activation, and behavioral adaptive dual-memory replay. The WSCM performs dynamic sparsification using double-threshold pruning to balance model compression rate and accuracy loss. The TFAM stores activation quantities in the feature replay layer, reducing model computation and storage requirements. The BARM dynamically balances mitigating catastrophic forgetting and addressing resource limitations through a three-dimensional sampling strategy and revisiting old tasks. Experimental results on the CORE50 dataset show that STCL achieves state-of-the-art performance. The main contributions of this work are summarized as follows:

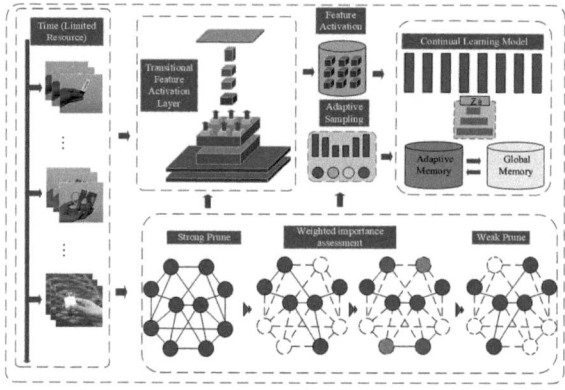

Fig. 1. Dynamic Sparsity-driven Transition Feature Replay Continual Learning for Edge Devices.

- **Weighted sparsity compression:** to address the deployment bottleneck of CL models on edge devices, a dual-threshold dynamic pruning mechanism is proposed to adaptively regulate the sparsity rate through hardware-aware sparsity rate as a means to reduce computational resource demand and load.
- **Transitional Feature Activation:** to break through the dependence of traditional replay methods on raw data storage, a lightweight Feature Replay Layer (FRL) is designed. Only fixed layer activation values are cached, and the underlying general-purpose feature extractor is stabilized by reducing the learning rate of the network below the Feature Replay Layer.
- **Behavioral Adaptive Dual Memory Replay:** For dynamic coupling of task boundary blurring and data drift, a synergistic architecture of Adaptive Memory (AM) and Global Memory (GM) is constructed.

2 Related Work

The edge device-oriented continuous learning scenario focuses on the knowledge playback optimization problem facing continuous learning models with resource constraints during incremental adaptation to new categories. To address this problem, this section will systematically sort out the research progress of sparsification-oriented compression technique continuous learning with continuous learning models on edge end.

2.1 Continual Learning for Sparsification Compression

Current research utilizes neural network pruning to reduce the computational complexity of continuous learning models, primarily through weight pruning and sparse training to reduce computational load, shorten iteration time, and lower hardware resource consumption. Unstructured weight pruning employs heuristic rules or regularization-driven methods to randomly remove redundant parameters. Han et al. [6] used an iterative algorithm to eliminate small-magnitude weights and then retrained the model to restore accuracy; however, the irregular sparse pattern resulted in limited actual acceleration efficiency. Sparse training in continuous learning models involves constraining model parameter updates using fixed masks. While Lee et al. [7] and Wang et al. [8] improved the execution efficiency of pruning models, their offline pruning characteristics limit their adaptability under dynamic task distributions.

Dynamic sparse training paradigms adjust sparse topologies via dynamic masks to maintain low memory consumption, such as Li et al. [9]'s factorized convolutional filters, FreezeNets' [10] sparse gradient computation. [12] end-to-end sparse training algorithm. However, most current research mainly looks at single-task situations and does not effectively model how knowledge can be shared between tasks in continuous learning, which makes the problem of forgetting previous tasks worse. The continuous learning methods proposed in [13–16] incorporate weight pruning concepts, assigning each task a dedicated sparse subnetwork to reduce interference between tasks, but this gradually reduces the model's sparsity. This study proposes a dual-threshold neural network weight sparsity compression method integrated with a dual memory replay mechanism to efficiently compress the resource-intensive CNN parameters on-chip while maintaining prediction accuracy.

2.2 Continual Learning on Edge Devices

Edge devices have become the core platform for AI deployment due to their low latency and data privacy advantages. However, continuous learning (CL) models face deployment challenges due to resource constraints such as energy efficiency and computational power. Replay-based mechanisms offer an effective approach for continuous learning on edge devices. Early studies, such as GSS [17], proposed sample selection strategies based on gradient directions. Some research (e.g., parallelization algorithms for MNIST [18] and CIFAR [19]) was limited to simple image classification, while OCS [20] and Coreset [21] maintained data summaries of input streams by constructing weighted subsets of data. Other studies (such as DER [22] and PRE-DFKD [23]) combine knowledge distillation [24] with replay and distillation loss fusion, achieving performance superior to multiple baseline methods. More advanced continuous learning methods (such as C-FSCIL [25]) propose solutions based on hyper-dimensional computation, suitable for class incremental learning in small-sample scenarios. Potential Replay [5] replaces the original input images with uniformly randomly sampled intermediate layer feature maps from the network, storing more samples within the same memory budget. It leverages cross-domain high transferability to enhance the early layers of the network while simplifying the training of the remaining half of the network to reduce costs. SparCL [26] proposes a task-aware dynamic path mechanism for sparse continuous learning on edge devices, learning sparse networks throughout the CL process.

Chameleon [27] proposed a hardware-friendly continuous learning scheme for dual-memory replay on edge devices, leveraging their hierarchical storage to build a user-centric dual-replay buffer. This boosts acceleration and energy efficiency. However, edge devices' limited on-chip storage and computational power cannot handle dense CNN parameters and traditional continuous learning's multi-round training. To solve this, we propose a new streaming data continuous learning framework. By combining incremental dynamic sampling with dual-threshold pruning in a dynamic sparse coordination strategy, we balance model compression and accuracy loss while controlling sparsity in stages.

3 Framework

For Chameleon [27], the goal is to overcome the CF problem within the limited on-chip available memory of embedded devices by introducing a dual memory replay mechanism with short-term (ST) and long-term (LT) buffers through a user preference-aware CL method. Based on this foundation, we further proposes a novel user-adaptive dynamic sparsification method, STCL, to improve the model's adaptability and storage efficiency in constrained environments. The method mainly constructs a weight sparsity compression module, a transition feature activation module, and a behavior-adaptive memory replay module. STCL firstly performs a weak pruning of 0.1 of the weights in the initialization phase of the model, and initializes sparsity for connections with weights less than a threshold value. Then, the incoming data are subjected to transition feature extraction, which extracts the input high-dimensional feature representations (transition feature activations) for model training and maps the transition activations to their classification. Finally, the behavior-driven STCL dynamically maps and caches the transition feature activations in adaptive memory and global memory to achieve adaptive playback of historical knowledge.

3.1 Formulation of the Problem

We focuses on a **Resource-Constrained Continual Learning** problem for edge devices, where a continual learning model needs to receive multiple tasks consecutively and accomplish incremental learning in a supervised image classification task. We define a set of data flow sources $\mathcal{D} = \{D_1, D_2, D_3, \ldots, D_t\}$, where each D_t comes from a different domain, i.e., satisfies $D_i \cap D_j = \varnothing, \forall i \neq j$.

At each time step, the model accepts the current batch sample $B_t(X_t, Y_t)$, where X_t is the input image and Y_t is its corresponding supervised label. We define a non-smooth task sequence $\mathcal{T} = \{T_1, T_2, \ldots, T_t\}$, where each task T_k is associated with an independent data distribution $\mathcal{D}_k \sim P_k(X, Y)$, and satisfies the domain exclusivity constraints: $P_i(X, Y) \neq P_j(X, Y)$ and $y_i \cap y_j = \varnothing$, $\forall i \neq j$. The input space is $\mathcal{X} \subseteq \mathbb{R}^{H \times W \times C}$, where $H \times W$ is the image resolution and C is the number of channels. The labeling space $y_k = \{1, 2, \ldots, K_k\}$ is dynamically expanded with the task. The model architecture consists of a transition feature extractor f_ψ and a task-adaptive module $\{g_{v_k}\}$: the parameter ψ is constrained to cross-task feature stability by elastic weight curing, and the task-specific parameter v_k is activated with dynamic sparsity to activate local features. The overall representation is defined as

$$h_\theta(x) = f_\psi(x) \cdot g_{v_k} \tag{1}$$

where $\theta = \psi \cup \{v_1, v_2, \cdots, v_T\}$, f_ψ is a shared feature extractor which extracts cross-task generalized features via a convolutional encoder, and g_{v_k} is a task-adaptive module that obeys hardware-aware sparsity constraints to activate only the feature subspaces relevant to the current task.

3.2 Overview

Through multi-module coordination, the STCL framework can achieve efficient continuous learning on edge devices. The framework diagram is shown in Fig. 2. The TFAM converts raw images into high-dimensional transition features, reducing memory usage by eliminating raw data storage. The WSCM uses dynamic pruning with gradient-based evaluation to compress model parameters to 1/5 of a dense model's size without accuracy loss. The BARM enhances feature replay via a 3D dynamic sampling strategy: incremental sampling retains uncertain samples, KL divergence-based selection captures user preferences, high-reconstruction-error samples prioritize information density, and diversity-weighted global memory samples cover low-frequency categories. STCL also employs dual-threshold dynamic pruning for model optimization, balancing efficiency and accuracy. A lightweight feature replay layer caches intermediate activations to reduce computational and memory overhead. In complex, dynamic environments, its collaborative adaptive-global memory adjusts architecture sampling flexibly for task ambiguity and data drift, effectively mitigating catastrophic forgetting and ensuring stable knowledge accumulation and adaptation.

3.3 Weight Sparsity Compression Module(WSCM)

WSCM achieves hardware-aware compression of model parameters through a dual-threshold dynamic pruning mechanism, which is specifically divided into weak pruning in the initialization phase and dynamic strong pruning in the training phase, combined with a gradient redirection compensation strategy to balance the compression rate and classification accuracy.

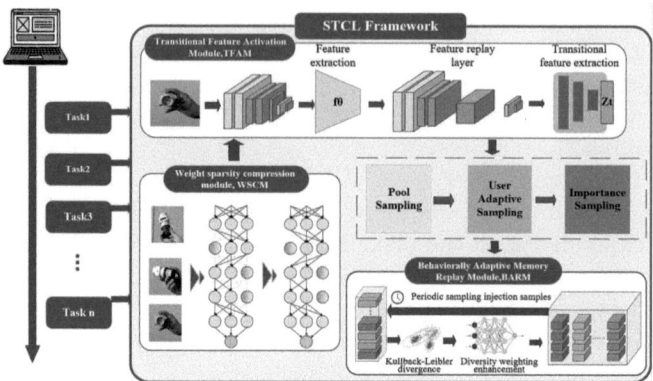

Fig. 2. STCL Framework

3.3.1 Weak Pruning in the Initialization Phase

Before training, a conservative pruning strategy removes redundant parameters while preserving important connections to avoid performance degradation. During initialization, untrained weights often contain many near-zero values with little impact on output. Based on this, the safe pruning criterion uses weight absolute values to assess importance and prunes only the least significant 30%, retaining critical structures and ensuring model stability. This establishes a solid sparsity foundation for dynamic pruning. The entire pruning process—threshold calculation, mask generation, and weight update—is unified into a joint expression:

$$W^l_{\text{sparse}} = W^l \odot I(|W^l| \geq \text{quantile}(|W^l|, 1 - s_{\text{init}})) \tag{2}$$

where quantile($\cdot$) is used to calculate the quantile threshold. Firstly, based on the set of absolute values of the l-th layer weight matrix W^l, its $(1 - s_{\text{init}})$ quantile is calculated as the pruning threshold T_{weak}. Subsequently, a binary mask matrix is generated by the indicator function, where the weight positions with absolute values greater than or equal to T_{weak} are labeled as 1, and the rest are labeled as 0. Eventually, the original weight W^l is multiplied element-wise with this mask matrix (i.e., the Hadamard product), removing the weights below the threshold and directly outputting the sparsified weights W^l_{sparse}.

3.3.2 Dynamic Strong Pruning in the Training Phase

During the model training process, the pruning strategy is dynamically adjusted according to the importance of the parameters (joint evaluation of the weights and gradients) to gradually increase the sparsity rate to the target value to achieve a hardware-friendly high compression rate while maintaining model accuracy. The actual contribution of the weights is determined by their absolute value and the training gradient magnitude to avoid deviation of a single indicator, and the sparse rate is smoothly adjusted to gradually approach the target value from the initial value to ensure that the pruning intensity is adapted to the training stage. Progressive sparse scheduling, dynamic importance assessment, mask pruning and gradient compensation are integrated into an end-to-end expression:

$$W_{t+1}^l = \left(W_t^l - \eta \cdot \frac{\mathbb{I}\left(I_{\text{joint}}(W_t^l) \geq T_{\text{strong}}^t\right)}{1 - s_t} \right) \quad (3)$$

where W_t^l represents the weight matrix of layer l at training step t; $I_{\text{joint}}(W_t^l)$ denotes the joint importance score of weights and gradients; T_{strong}^t is the dynamic threshold; $\mathbb{I}(\cdot)$ is the indicator function; and η is the learning rate. During the training process, the layer-l weight matrix W_t^l is evaluated by jointly assessing the dynamic importance based on the absolute value of the weights and the gradient magnitude. The quantile threshold T_{strong}^t is dynamically calculated. In the gradient update phase, the gradients of the retained parameters are amplified to compensate for pruning, and strong pruning of the weight matrix is executed simultaneously.

3.4 Adaptation and Global Memory Updating

3.4.1 Adaptive Memory

To handle the dynamics of streaming data and the non-uniformity of information density, dealing with the fuzziness of task boundaries and the limitations of traditional static storage are the core challenges. Combining three strategies, an adaptive storage design scheme with a three-dimensional dynamic balance mechanism is proposed. Reservoir sampling uses FIFO buffers with fixed capacity for uniform exploration, ensuring global diversity and coverage, and retaining long-tail or low-frequency data. Adaptive frequency sampling monitors the activation frequency change rate of characteristic sensitivity, dynamically enhances the storage of transient events, and responds to changes in local data distribution. Importance sampling quantifies information density by lightweight autoencoder reconstruction errors, prioritizing the processing of high-error samples, filtering redundantly, and improving the information entropy efficiency of storage resources. The specific formula is as follows:

$$P_{\text{adp}}(x_i) = \alpha \cdot \frac{1}{N_{\text{reservoir}}} + \beta \cdot \frac{\Delta f(x_i)}{\sum_j \Delta f(x_j)} + \gamma \cdot \frac{\varepsilon_{\text{rec}}(x_i)}{\sum_j \varepsilon_{\text{rec}}(x_j)} \quad (4)$$

where $N_{\text{reservoir}}$ is the fixed capacity of the first-in-first-out (FIFO) buffer, which guarantees the global coverage of the data distribution; $\Delta f(x_i)$ is the temporal rate of change of the feature frequency; and $\varepsilon_{\text{rec}}(x_i)$ is the transition feature reconstruction error, where a larger error indicates that the sample is more difficult to be captured by the current model. The class strategy is co-evolved by dynamic weighting coefficients α, β and γ. The constraints are $\alpha + \beta + \gamma = 1$ to ensure normalization. Reservoir sampling dominates at the initial stage when the data distribution is unknown, making α tend to 1; frequency sensitivity is enhanced at the mid-term when bursty patterns emerge, making β rise, and information condensation is strengthened at the end when the distribution stabilizes, making γ rise, and combining with hardware-aware optimization to reduce the storage and computation overheads of the edge devices.

3.4.2 Global Memory

To address distribution bias and catastrophic forgetting in edge device continuous learning, the key is to eliminate scale bias and identify key samples that deviate from the existing distribution. The global memory uses a combination of KL scatter, diversity weight, and uncertainty sampling strategies. First, middle layer transition activation features are extracted from the current training batch and L2 normalized to remove scale bias. Next, the KL scatter measures the distributional difference between new and historical features, helping to filter out key samples that deviate significantly. Additionally, a diversity weight mechanism enhances the storage priority of low-frequency samples through inverse logarithmic weighting of category frequencies. This mechanism also screens high uncertainty samples to cover the decision boundary region, forming a candidate set that broadly represents the data distribution. Assume that the set of normalized features $\mathcal{H}_{\text{global}} = \{\hat{h}_1, \hat{h}_2, \cdots, \hat{h}_M\}$ stored in the global memory obeys a Gaussian distribution $\mathcal{N}(\mu_{\text{global}}, \Sigma_{\text{global}})$. The current batch feature set $\mathcal{H}_{\text{current}} = \{\hat{h}_1, \hat{h}_2, \cdots, \hat{h}_N\}$ obeys a Gaussian distribution $\mathcal{N}(\mu_{\text{current}}, \Sigma_{\text{current}})$. KL divergence is computed as:

$$D_{\text{KL}}(\mathcal{N}_{\text{current}} \| \mathcal{N}_{\text{global}}) = \frac{1}{2} \Big[\text{tr}(\Sigma_{\text{global}}^{-1} \Sigma_{\text{current}}) \\ + (\mu_{\text{global}} - \mu_{\text{current}})^\top \\ + \ln\left(\frac{|\Sigma_{\text{global}}|}{|\Sigma_{\text{current}}|}\right) - d \Big] \quad (5)$$

where $\text{tr}(\Sigma_{\text{global}}^{-1} \Sigma_{\text{current}})$ calculates the trace of two covariance matrices (sum of diagonals), which reacts to the similarity of the covariance structure; if the difference between Σ_{current} of the current data and the historical distribution Σ_{global} is big, the value of this item will increase. The distribution width adjustment term $\ln\left(\frac{|\Sigma_{\text{global}}|}{|\Sigma_{\text{current}}|}\right)$ is the logarithm of the determinant of the covariance matrix, which reacts to the difference of the distribution "volume". If the current data distribution is more dispersed than the historical distribution ($|\Sigma_{\text{current}}| > |\Sigma_{\text{global}}|$), the value of this item is negative, suppressing the KL value; conversely, it amplifies the difference. The KL scatter formula quantifies the degree of deviation of the old and new feature distributions through two core dimensions: covariance structure difference and distribution width adjustment. Its design is closely integrated with edge device characteristics (e.g., low computation volume, online update demand), and efficient computation is realized through Gaussian assumption and matrix simplification strategy.

3.5 Training

In the training phase, to ensure efficient learning of new task knowledge, the task classification loss is defined to use a cross-entropy loss function with the mathematical expression:

$$L_{\text{cls}} = -\frac{1}{N} \sum_{i=1}^{N} \sum_{c=1}^{C_t} y_{i,c} \log(p_{i,c}) \qquad (6)$$

where C_t is the number of categories of the t-th task, which scales dynamically with the task flow; $p_{i,c}$ is the predicted probability of sample x_i in the c-th category, which is outputted by Softmax; and N is the batch size, which is limited by the memory constraints of the edge device. Cross-entropy loss drives the model to quickly fit new task data features by minimizing the KL scatter between the predicted distribution and the true label distribution.

In the continuous learning scenario for edge devices, the testing phase requires a comprehensive assessment of the model's performance in terms of historical task knowledge retention and resource efficiency. The average classification performance and stability of the model on the learned tasks are first evaluated:

$$L_{\text{task}} = \sum_{t=1}^{T} \left[\frac{1}{\sqrt{f_t} N_t} \sum_{i=1}^{N_t} \sum_{c=1}^{C_t} y_{i,c} \log(p_{i,c}) \right] \qquad (7)$$

where $\frac{1}{\sqrt{f_t} N_t}$ is the task weight coefficient, f_t is the sample frequency of task t in the test set, which suppresses high-frequency tasks to prevent overfitting. In this way, the performance of adapting to long-tailed tasks is improved and the performance fluctuation between tasks is suppressed.

4 Evaluation

4.1 Settings

Our experiments were realized in the following environment: an Intel(R) Core(TM) i9-10900X CPU; 64 GB RAM; NVIDIA GeForce RTX 3090 GPU and Win10; python 3.9.18 and torch 2.5.1+cu118. We evaluated our method on the CORe50 dataset in an online continual learning task to compare its classification effectiveness with previous methods. The experimental results of our proposed method are also analyzed by ablation experiments, where the model code is implemented based on pytorch framework, and the pre-trained MobileNetV1 is used for the basic experiments, with the playback memory size rm_sz set to 10, and the initial learning rate and initial update rate set to 0.01.

4.2 Datasets

Designed for online continuous learning, CORe50 contains 164,866 video frames across 50 categories and 11 domains, focusing on assessing object diversity and environmental adaptation. Unlike other benchmark tests, it focuses on time-dependent video frames, each containing different objects, providing diverse training and evaluation data.

4.3 Performance Metric

We evaluated the performance of all the methods using the classification accuracy avg_acc. To avoid the impact of randomness in neural network training on the overall performance, we ran all the experiments five times randomly and reported its average performance; we also evaluated the final model accuracy avg_k_acc, which was averaged over the behavioral adaptive classes to obtain the final classification accuracy. In addition, to better represent the contribution of our proposed method to the overall performance, we also performed ablation experiments.

4.4 Comparison Results

4.4.1 Overall Classification Accuracy

Our experiments employ the Core50 image-classification benchmark. We will compare the performance of our method with eight other approaches: EWC++ [1], LWF [2], SLDA [4], GSS [17], ER [3], DER [22], Latent Replay [5], and Chameleon [28]. For each method, we report the avg_acc along with its standard deviation computed over multiple runs. Table 1 provides a comprehensive comparison of all methods.

The STCL framework effectively balances resource efficiency and accuracy for edge device continuous learning. It surpasses existing methods in memory efficiency, classification accuracy, and stability. At a replay buffer of 1500, STCL uses 48.3MB memory—33.1%, 34.2%, and 76.3% less than ER, DER, and GSS—and achieves 0.30% higher accuracy than Chameleon with the same memory. Under low-resource conditions (buffer 100, 3.5MB memory), STCL attains 79.55% accuracy, outperforming ER and Latent Replay by 46.94% and 7.66%. In high-resource settings, it reaches 80.22%, a 57.00% improvement over EWC++. STCL also shows better stability with a standard deviation of ±0.16 to ±1.45 during buffer expansion, outperforming ER and DER. These results underscore the effectiveness of STCL's behavioral adaptive replay strategy with dynamic sparse compression in addressing task conflict and data drift.

4.4.2 User-Adaptive Classification Accuracy

To verify the optimization effect of the stability versus generalization trade-off of the continuous learning model in the dynamic task flow. We do so by comparing the dynamic changes of adaptive class accuracy (avg_K_acc) with average global accuracy (avg_acc) over 8 training batches, as shown in Fig. 3 and Fig. 4.

The average k-accuracy of five classes rises nonlinearly from 0.32 in Batch 1 to 0.95 in Batch 8. Class 3 fluctuates between Batches 4–6 due to task complexity but converges via the dynamic sparse strategy. The global average accuracy stabilizes between 0.65–0.75 across eight batches. Notably, in Batch 5 during a sudden data drift, the double memory replay mechanism boosts accuracy from 0.66 to 0.72. These metrics show the dynamic parameter isolation strategy works. At Batch 8, the model achieves 0.75 accuracy with a 58% sparse rate, offering an efficient edge-based continuous learning method.

Table 1. Memory overhead and replay buffer size versus avg_acc for continuous-learning methods on the Core50 dataset.

Method	Replay Buffer Size	Memory Overhead	CoRe50
EWC++	–	13.0	23.22
LWF	–	12.5	27.91
SLDA	–	1.2	77.20
GSS	100	48.8	43.51
	200	53.6	47.47
	500	68.0	48.57
	1500	204.0	53.19
ER	100	4.8	32.61
	200	9.6	36.07
	500	24.0	62.31
	1500	72.0	63.33
DER	100	4.9	58.72
	200	9.8	62.15
	500	24.5	67.35
	1500	73.5	68.73
Latent Replay	100	3.2	71.89
	200	6.4	72.87
	500	16.0	75.43
	1500	48.0	79.07
Chameleon	$M_s = 10, M_l = 100$	$M_s = 0.3, M_l = 3.2$	79.48
	$M_s = 10, M_l = 200$	$M_s = 0.3, M_l = 6.4$	79.56
	$M_s = 10, M_l = 500$	$M_s = 0.3, M_l = 16$	79.86
	$M_s = 10, M_l = 1500$	$M_s = 0.3, M_l = 48.0$	79.92
DSTF-RCL (ours)	$M_b = 10, M_g = 100$	$M_b = 0.3, M_g = 3.2$	79.55
	$M_b = 10, M_g = 200$	$M_b = 0.3, M_g = 6.4$	79.62
	$M_b = 10, M_g = 500$	$M_b = 0.3, M_g = 16$	79.91
	$M_b = 10, M_g = 1500$	$M_b = 0.3, M_g = 48.0$	80.22

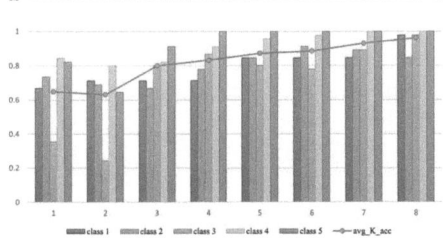

Fig. 3. Adaptive class accuracy with respective adaptive class performance evolution.

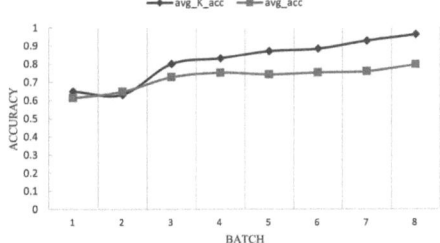

Fig. 4. Dynamics of adaptive class accuracy and average accuracy.

Table 2. Ablation research of the STCL method on the CORE 50 dataset

Method	Memory Overhead	CoRe50
WSCM	3.2	21.67
	6.4	23.79
	16.0	30.16
	48.0	47.28
TFAM	3.2	29.65
	6.4	32.16
	16.0	36.63
	48.0	53.68
BARM	$M_b = 0.3, M_g = 3.2$	42.52
	$M_b = 0.3, M_g = 6.4$	48.76
	$M_b = 0.3, M_g = 16$	54.15
	$M_b = 0.3, M_g = 48.0$	61.38
WSCM+TFAM	3.2	32.54
	6.4	34.37
	16.0	41.96
	48.0	53.57
WSCM+BARM	$M_b = 0.3, M_g = 3.2$	48.68
	$M_b = 0.3, M_g = 6.4$	51.73
	$M_b = 0.3, M_g = 16$	59.85
	$M_b = 0.3, M_g = 48.0$	66.92
TFAM+BARM	$M_b = 0.3, M_g = 3.2$	47.55
	$M_b = 0.3, M_g = 6.4$	53.35
	$M_b = 0.3, M_g = 16$	58.73
	$M_b = 0.3, M_g = 48.0$	71.62
WSCM+TFAM+BARM	$M_b = 0.3, M_g = 3.2$	78.17
	$M_b = 0.3, M_g = 6.4$	79.21
	$M_b = 0.3, M_g = 16$	79.63
	$M_b = 0.3, M_g = 48.0$	80.03

4.5 Ablation Study

To validate the effectiveness of each component of STCL, we conduct ablation experiments in closed-world scenarios using the Core50 dataset. As shown in Table 2, we systematically validate the WSCM (Weight Sparsity Compression Module) and TFAM (Transitional Feature Activation Module) of STCL with BARM (Behaviorally Adaptive Memory Replay Module) and their synergistic effects.

Ablation experiments on Core50 show STCL achieves a resource-accuracy balance breakthrough through module synergy. Single modules have limited performance. Dual-module combinations improve by 5.54%–10.24% but still lag behind full STCL by 8.41%–26.46%. The complete STCL reaches 78.17% accuracy at low memory, 15× more efficient than ER. With memory expanded to 48.0MB, accuracy optimizes to 80.03%, surpassing the dual-module 23.6% boost. This highlights the three-module synergy's effectiveness and sets a new benchmark for edge continuous learning.

5 Conclusion and Future Work

We proposes STCL, a dual-memory replay continual learning framework with sparsity compression for edge devices. STCL integrates weight sparsity compression, transition feature activation, and behavior adaptive dual-memory replay to balance accuracy and resource constraints. It employs a dual-threshold dynamic pruning mechanism to compress model parameters while maintaining classification accuracy. The Transition Feature Activation Module (TFAM) stores fixed-layer activations and stabilizes feature extraction by reducing the learning rate below the replay layer. The Behavior Adaptive Memory Replay Module (BARM) combines Adaptive Memory and Global Memory with a three-dimensional sampling strategy to better manage catastrophic forgetting and maintain knowledge from past tasks.

Experiments on the CORe50 dataset validate STCL's effectiveness, achieving strong performance with significantly reduced model size. While promising, STCL can be further improved in handling new instances and class additions. Future work will explore heterogeneous feature distillation, multimodal feature decoupling, and selective fusion to enhance performance in complex scenarios, along with energy-aware evaluations and hardware-software co-optimization for real-world deployment.

Acknowledgments. This work was supported by the national natural science foundation of China (62441608), the science and technology project of Guizhou Province (QKHZC[2023]368, QKHCG[2025]007, QKHZC[2025]023), the high level innovative talent project of Guizhou Province(QKHGCC[2023]101), the developing objects and projects of scientific and technological talents in Guiyang city (No. ZKHT[2023]48-8), Guizhou University Science and Technology Innovation Team ([2024]No. 07), Guizhou University Basic Research Fund ([2024]08), the National Natural Science Foundation of China under Grant (No. 62302119), and Guiyang City Science and Technology Plan Project (No. [2024]2-18).

References

1. An, H., et al.: A class-incremental learning approach for learning feature-compatible embeddings. Neural Netw. **180**, 106685 (2024). https://doi.org/10.1016/j.neunet.2024.106685
2. Li, Q., Yang, J., Ruan, X., Li, S., Hu, J., Hu, B.: SPIRF-CTA: selection of parameter importance levels for reasonable forgetting in continuous task adaptation. Knowl.-Based Syst. **305**, 112575 (2024)

3. Chaudhry, A., Dokania, P.K., Ajanthan, T., Torr, P.H.S.: Riemannian walk for incremental learning: understanding forgetting and intransigence. In: Ferrari, V., Hebert, M., Sminchisescu, C., Weiss, Y. (eds.) ECCV 2018. LNCS, vol. 11215, pp. 556–572. Springer, Cham (2018). https://doi.org/10.1007/978-3-030-01252-6_33
4. Li, Z., Hoiem, D.: Learning without forgetting. IEEE Trans. Pattern Anal. **40**(12), 2935–2947 (2017)
5. Chaudhry, A., et al.: Continual learning with tiny episodic memories. In: Workshop on Multi-Task and Lifelong Reinforcement Learning (2019)
6. Liu, B.: Lifelong machine learning: a paradigm for continuous learning. Front. Comput. Sci.-CHI **11**(3), 359–361 (2017)
7. Pellegrini, L., Graffieti, G., Lomonaco, V., Maltoni, D.: Latent replay for real-time continual learning. In: 2020 IEEE/RSJ International Conference on Intelligent Robots and Systems (IROS), pp. 10203–10209. IEEE (2020)
8. Han, S., Pool, J., Tran, J., Dally, W.J.: Learning both weights and connections for efficient neural networks. In: Proceedings of the 29th International Conference on Neural Information Processing Systems, Montreal, Canada, vol. 1, pp. 1135–1143. MIT Press (2015)
9. Lee, N., Ajanthan, T., Torr, P.H.: SNIP: single-shot network pruning based on connection sensitivity. arXiv preprint arXiv:1810.02340 (2018)
10. Wang, C., Zhang, G., Grosse, R.: Picking winning tickets before training by preserving gradient flow. arXiv preprint arXiv:2002.07376 (2020)
11. Li, T., Wu, B., Yang, Y., Fan, Y., Zhang, Y., Liu, W.: Compressing convolutional neural networks via factorized convolutional filters. In: 2019 Proceedings of the IEEE/CVF Conference on Computer Vision and Pattern Recognition, pp. 3977–3986 (2019)
12. Wimmer, P., Mehnert, J., Condurache, A.: FreezeNet: full performance by reduced storage costs. In: Proceedings of the Asian Conference on Computer Vision (ACCV), pp. 685–701 (2020)
13. Ding, X., Ding, G., Zhou, X., Guo, Y., Han, J., Liu, J.: Global sparse momentum SGD for pruning very deep neural networks. In: Proceedings of the 33rd International Conference on Neural Information Processing Systems, pp. 6382–6394. Curran Associates Inc. (2019)
14. Liu, J., Xu, Z., Shi, R., Cheung, R.C., So, H.K.: Dynamic sparse training: find efficient sparse network from scratch with trainable masked layers. arXiv preprint arXiv:2005.06870 (2020)
15. Wang, Z., Jian, T., Chowdhury, K., Wang, Y., Dy, J., Ioannidis, S.: Learn-prune-share for lifelong learning. In: 2020 IEEE International Conference on Data Mining (ICDM), pp. 641–650. IEEE (2020)
16. Mallya, A., Lazebnik, S.: PackNet: adding multiple tasks to a single network by iterative pruning. In: Proceedings of the IEEE conference on Computer Vision and Pattern Recognition, pp. 7765–7773 (2018)
17. Mallya, A., Davis, D., Lazebnik, S.: Piggyback: adapting a single network to multiple tasks by learning to mask weights. In: Ferrari, V., Hebert, M., Sminchisescu, C., Weiss, Y. (eds.) ECCV 2018. LNCS, vol. 11208, pp. 72–88. Springer, Cham (2018). https://doi.org/10.1007/978-3-030-01225-0_5
18. Sokar, G., Mocanu, D.C., Pechenizkiy, M.: SpaceNet: make free space for continual learning. Neurocomputing **439**, 1–11 (2021)
19. Aljundi, R., Lin, M., Goujaud, B., Bengio, Y.: Gradient based sample selection for online continual learning. In: Proceedings of the 33rd International Conference on Neural Information Processing Systems, pp. 11817–11826. Curran Associates Inc. (2019)

20. Deng, L.: The MNIST database of handwritten digit images for machine learning research [best of the web]. IEEE Signal Process. Mag. **29**(6), 141–142 (2012)
21. Krizhevsky, A.: Learning multiple layers of features from tiny images, pp. 1–16 (2009)
22. Yoon, J., Madaan, D., Yang, E., Hwang, S.J.: Online coreset selection for rehearsal-based continual learning. arXiv preprint arXiv:2106.01085 (2021)
23. Borsos, Z., Mutny, M., Krause, A.: Coresets via bilevel optimization for continual learning and streaming. Adv. Neural. Inf. Process. Syst. **33**, 14879–14890 (2020)
24. Buzzega, P., Boschini, M., Porrello, A., Abati, D., Calderara, S.: Dark experience for general continual learning: a strong, simple baseline. Adv. Neural. Inf. Process. Syst. **33**, 15920–15930 (2020)
25. Binici, K., Aggarwal, S., Pham, N.T., Leman, K., Mitra, T.: Robust and resource-efficient data-free knowledge distillation by generative pseudo replay. In: 2022 Proceedings of the AAAI Conference on Artificial Intelligence, pp. 6089–6096 (2022)
26. Hinton, G., Vinyals, O., Dean, J.: Distilling the knowledge in a neural network. arXiv preprint arXiv:1503.02531 (2015)
27. Hersche, M., Karunaratne, G., Cherubini, G., Benini, L., Sebastian, A., Rahimi, A.: Constrained few-shot class-incremental learning. In: 2022 Proceedings of the IEEE/CVF Conference on Computer Vision and Pattern Recognition, pp. 9057–9067 (2022)
28. Wang, Z., et al.: SparCL: sparse continual learning on the edge. Adv. Neural. Inf. Process. Syst. **35**, 20366–20380 (2022)
29. Aggarwal, S., Binici, K., Mitra, T.: Chameleon: dual memory replay for online continual learning on edge devices. IEEE Trans. Comput.-Aided Design Integr. Circuits Syst. (TCAD) **43**(6), 1663–1676 (2023)

Towards On-Device NPU-Friendly Neural Network Operator Optimization

Wei Ye[1], Jinrui Zhang[2], Deyu Zhang[1(✉)], Huan Yang[1], and Yin Tang[1]

[1] School of Computer Science and Engineering, Central South University, Changsha, China
{234711040,zdy876,yanghuan9812,tangyin0512}@csu.edu.cn
[2] Department of Computer Science and Technology, Tsinghua University, Beijing, China
jinruizhang@tsinghua.edu.cn

Abstract. In mobile deep learning tasks, particularly for deploying multimodal large language models (mLLMs), the incompatibility between hardware operator support and model complexity poses a key challenge. To address operator parameter constraints (e.g., convolution/pooling kernel size limitations) and operator absence issues in NPU-accelerated multimodal model inference, we propose a dual optimization strategy: 1) An operator composition substitution method based on equivalent computational graphs, achieving NPU-compatible transformation of restricted operators through tensor remapping and weight migration; 2) MLP surrogate networks for unsupported operators, leveraging multilayer perceptrons' universal approximation capability to reconstruct irregular computational paths. By establishing an operator substitution cost model that comprehensively evaluates the Pareto frontier of computational complexity versus accuracy degradation, we ultimately select the optimal deployment scheme. Experimental results demonstrate that the optimized deep learning model exhibits only 0.5% accuracy degradation on NPU platforms, while significantly reducing inference energy consumption by 85.7–91.6%, 83.3–96.6%, and 95–97.5% compared to CPU, GPU, and NPU-CPU heterogeneous architectures, respectively, validating the method's effectiveness in balancing computational efficiency and model performance.

Keywords: Mobile Computing · Deep Learning Model · Mobile NPU · Neural Network Operator Optimization

1 Introduction

In recent years, deep learning models have witnessed rapid growth in edge computing deployments [17], with applications spanning core domains including natural language processing [12], computer vision [16]. The rise of Large Language

---
W. Ye and J. Zhang—Equal contribution.

Models (LLMs) has made on-device AI agents a key research focus, with localized deployment of multimodal LLMs emerging as a critical trend. Models like GPT Mobile [10] and Gemini Nano [1], along with mobile-optimized vision-language architectures, leverage on-device inference to enhance privacy, reduce latency, and enable personalization. Meanwhile, advances in mobile SoC design have led to widespread integration of Neural Processing Units (NPUs), which offer significantly higher energy efficiency than general-purpose processors for core operations like matrix multiplication and convolution. By leveraging customized data paths and compute units, NPUs provide hardware-level acceleration for tensor operations, making them well-suited for efficient multimodal LLM deployment on mobile devices.

However, NPU scalability remains insufficient to resolve key technical challenges in mobile AI deployment. Research reveals persistent incompatibilities between many deep learning operators and NPU instruction sets, attributable to their specialized design: (1) NPUs prioritize cost and complexity control by supporting limited mathematical functions, leaving complex operations (e.g., high-order nonlinear functions) unaccelerated; (2) Hardware constraints—including computational unit scale, on-chip memory, and data throughput—preclude execution of oversized operators. Table 1 categorizes unsupported operators into device-restricted (e.g., high-dimensional convolutions) constrained by NPU resources, and hardware-unsupported types (e.g., *Log* functions, CELU activations) absent from NPU instruction sets.

Table 1. Operator Classification and NPU Platform Support

Operator Name	RKNN	MTK
AvgPool_2d	Spec Limitation	Spec Limitation
Conv_2d	Spec Limitation	Spec Limitation
Abs	Not Supported	Supported
CELU	Not Supported	Supported
Log	Not Supported	Supported

The coexistence of NPU-restricted and unsupported operators poses critical challenges for DNN deployment: (1) **Resource Preemption**: When deploying DNNs on CPUs/GPUs, mobile operating systems' fair scheduling policies enable high-priority system tasks to preempt DNN-allocated resources during concurrent execution. This contention induces intermittent resource starvation for DNN inference, exacerbating latency and energy consumption due to frequent context switching and unstable resource allocation. Such degradation is particularly severe in resource-constrained mobile devices under multitasking workloads. (2) **Operator Fallback Overhead**: NPU-based DNN deployments incur performance penalties when encountering unsupported operators, necessitating data fallback to CPUs for partial execution before resuming NPU computation. This fallback triggers inter-device data transfers/synchronization, introducing communication latency while leaving NPU cores idle. As shown in Table 2, with three fallbacks (e.g., three logarithmic operators in a model containing two 3 × 3 2D convolutional layers), NPU-based inference latency matches or exceeds CPU-only execution while consuming 1.6× more power—nullifying NPU's low-latency, low-power advantages for mobile deployment.

Table 2. Operator Fallback to CPU Performance Test

Fallback Count	Latency/ms		Power/A	
	CPU	NPU-CPU	CPU	NPU-CPU
1	30.16	15.94	0.307	0.539
2	32.21	25.93	0.292	0.518
3	37.26	36.82	0.301	0.524

To tackle these challenges, we propose a neural operator optimization and substitution framework tailored for on-device NPU deployment. For device-restricted operators (e.g., convolutions, pooling), we design NPU-compatible substitution schemes that are computationally equivalent and analytically proven to preserve functionality, enabling direct weight transfer without retraining. A FLOPs-based evaluation selects the lowest-complexity variant. For unsupported operators (e.g., log, CELU), we employ MLP-based functional approximators, guided by an automated neuron search to balance accuracy and efficiency. Experiments show our framework eliminates NPU-CPU fallback overhead while maintaining accuracy (within 0.5% degradation), and reduces energy consumption by 85.7–91.6%, 83.3–96.6%, and 95–97.5% compared to CPU, GPU, and NNAPI NPU-CPU co-execution, respectively.

In summary, the main contributions are as follows:

- For NPU-constrained operators (e.g., convolutional, pooling layers): Establish substitution framework using mathematically equivalent operator sequences. Enables lossless weight transfer without retraining, optimized via minimal-FLOP selection from computationally validated candidates.
- For hardware-unsupported operators (e.g., Log, CELU): Design MLP-based function surrogates coupled with NAS. NAS optimizes MLP configurations for computational efficiency to ensure NPU-compatible implementation.
- Comprehensive benchmarks: Exhibit <0.5% accuracy degradation with energy efficiency improvements of 7–12×, 6–30×, and 20–40× over CPU, GPU, and NPU-CPU baselines.

2 Related Work

On-Device DNN Inference Acceleration. Model optimization frameworks primarily accelerate inference through operator fusion to reduce memory access and computational overhead. DNNFusion [9] employs high-level abstractions (e.g., mapping type analysis and Execution Context Graphs) and graph rewriting based on mathematical properties to optimize operator fusion. GPU-based acceleration leverages mobile GPUs: Sumin et al. [5] reduce GPU inference overhead via kernel launch cycle prediction. NPU-CPU co-execution addresses NPU's operator incompatibility through collaborative computation: Band [4] artitions DNNs into subgraphs with dynamic scheduling, and Mandheling [15] reduces CPU-DSP context switches via execution overlap. However, frequent CPU-NPU synchronization introduces communication overhead, and CPU fallback execution incurs high power costs. Our operator substitution framework replaces unsupported operators with NPU-compatible equivalents, enabling full NPU deployment to eliminate co-execution overhead and CPU-induced energy penalties.

Operator Equivalence Characterization. Operator equivalence characterization refers to structural transformations, parameter reconfigurations, or topological substitutions of computational units under mathematical equivalence or functional similarity principles, optimizing model performance, efficiency, and deployment flexibility. Core techniques include operator fusion, decomposition, and substitution. Edge-oriented Convolution Block [18] merges multi-branch topologies (e.g., spatial derivatives, 3×3 convolutions) into unified operators for efficient inference. Diverse Branch Block [3] trains multi-scale convolution branches during training and converts them to vanilla structures for inference. Neural architecture search (NAS) dynamically selects operators (e.g., convolutions, attention) via differentiable search or reinforcement learning to meet hardware constraints.

3 Operator Optimized Replacement Scheme

In this section, we present a detailed description of the proposed neural network operator optimization and replacement scheme designed for edge-side NPU constraints. As illustrated in Fig. 1, the proposed method consists of two main steps:

Operator Classification: Operators in DNN models intended for mobile NPUs are first analyzed; any operator not supported by the target NPU is then classified—per the platform's requirements—as either a "restricted" operator or an "unsupported" operator. Building on this classification, we substitute non-compliant operators with optimized counterparts, ensuring full compatibility with the NPU platform.

Operator Replacement: Restricted operators—chiefly convolutions—are subject to NPU limits on kernel size and stride. We tackle kernel constraints by parallelizing a large convolution into multiple smaller ones, and address stride limits by replacing a large-stride convolution with a sequence of small-stride convolutions plus filtering. Unsupported operators, primarily mathematical functions, are approximated using a multi-layer perceptron (MLP) to preserve their functionality.

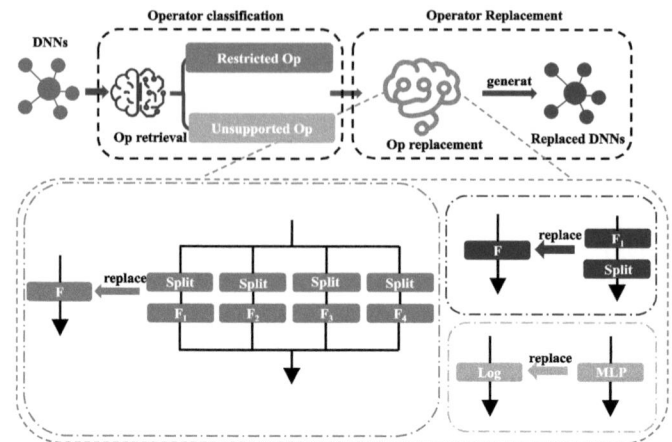

Fig. 1. System Architecture Diagram.

3.1 Convolutional Parallel Combination

A convolution layer has C_{in} input channels, C_{out} output channels, and a convolution kernel of size $K_1 \times K_2$. The weights of the convolutional layer are mainly contained within the convolution kernel and the bias term. The convolution kernel is a fourth-order tensor: $F \in R^{C_{\text{out}} \times C_{\text{in}} \times K_1 \times K_2}$, and the bias is a first-order tensor: $b \in R^D$. The input feature map of the convolutional layer is: $I \in R^{C \times H \times W}$, and the output feature map is: $O \in R^{D \times H' \times W'}$, where H' and W' are determined by the convolution kernel size $K_1 \times K_2$, the stride, and the padding. The convolution operation is defined as shown in Eq. (1):

$$O = I \otimes F + \text{REP}(b) \tag{1}$$

where $\otimes$ denotes the convolution operator, and $\text{REP}(b) \in R^{D \times H' \times W'}$ denotes the bias tensor expanded to match the output feature map dimensions.

The value at position (h, w) in the j-th output channel is given by:

$$O_{j,h,w} = \sum_{c=1}^{C_{\text{in}}} \sum_{u=1}^{K_1} \sum_{v=1}^{K_2} F_{j,c,u,v} \, X_{c,(h \cdot s+u),(w \cdot s+v)} + b_j \tag{2}$$

where $X(c, h, w) \in R^{K_1 \times K_2}$ is the sliding window extracted from the c-th channel of the input I corresponding to the output location, and s is the stride of the convolution. The correspondence between I and O is determined by the stride and padding.

Due to the linearity of convolution, the following distributive property holds:

$$I \otimes F_1 + I \otimes F_2 = I \otimes (F_1 + F_2), \tag{3}$$

where F_1 and F_2 must have the same size along a specific dimension to be concatenated into a larger convolution kernel. Furthermore, the other convolution parameters, such as stride and padding, must also be identical. Based on the linearity of convolution, we can derive the following theorem.

Theorem 1. *Additivity of Parallel Convolution Kernels*

If the convolution kernel F can be decomposed into multiple sub-kernels $F_1, F_2, \ldots, F_n$, and only the bias of F_1 is set to the original bias b, that is:

$$F = \mathrm{ConCat}(F_1, F_2, \ldots, F_n), \quad \mathrm{REP}(b) = \mathrm{REP}(b_1) \tag{4}$$

where ConCat denotes concatenation of the sub-kernels $F_1, F_2, \ldots, F_n$ along a certain dimension to form the original kernel F, then the convolution operation satisfies the additivity property:

$$I \otimes F + \mathrm{REP}(b) = \sum_{i=1}^{n}(I \otimes F_i) + \mathrm{REP}(b_1) \tag{5}$$

Proof. According to the linearity of convolution, when the stride, number of channels, and other parameters are consistent and no padding is used, the convolution operation satisfies both distributive and associative properties.

First, given that the convolution kernel F is decomposed into sub-kernels $F_1, F_2, \ldots, F_n$, and only F_1 retains the original bias b, we have:

$$I \otimes F + \mathrm{REP}(b) = I \otimes \left(\sum_{i=1}^{n} F_i\right) + \mathrm{REP}(b_1) \tag{6}$$

By the distributive property of convolution, it follows that:

$$I \otimes F + \mathrm{REP}(b) = \sum_{i=1}^{n}(I \otimes F_i) + \mathrm{REP}(b_1) \tag{7}$$

Therefore, the convolution operation can be equivalently expressed as:

$$I \otimes F + \mathrm{REP}(b) = \sum_{i=1}^{n}(I \otimes F_i) + \mathrm{REP}(b) \tag{8}$$

As illustrated in Fig. 2, a 9×9 input feature map I is convolved with a 5×5 convolution kernel F with bias b to produce a 5×5 output feature map O. Suppose the convolution kernel F exceeds the kernel size limitations of the NPU. According to Theorem 1, F can be decomposed into four sub-kernels F_1, F_2, F_3, F_4, with kernel sizes 3×3, 3×2, 2×3, and 2×2 respectively. Each sub-kernel is responsible for migrating the corresponding weights from different regions of the original kernel F.

The input feature map I is convolved with F_1, F_2, F_3, F_4, and their outputs are summed to obtain O', which matches the original output O. To ensure consistency in both shape and values, each sub-convolution must first apply a filtering step (i.e., tensor slicing via a Split operator). If the original convolution

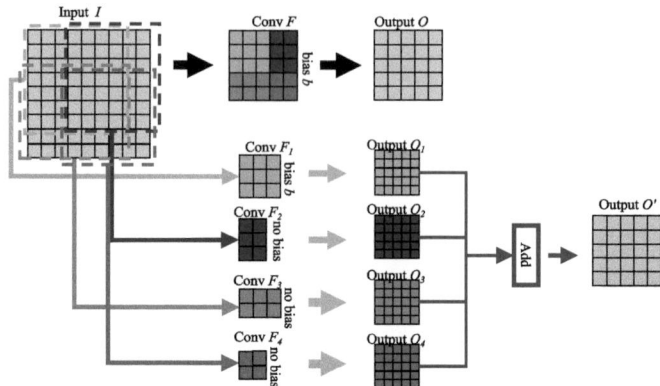

Fig. 2. Illustration of the equivalent additive property for a 9×9 tensor passing through a 5×5 convolutional kernel.

includes padding, it must be applied to I prior to splitting, in accordance with the conditions of the theorem.

Therefore, we can implement restricted operators by summing the outputs of several compliant operators in parallel. Although there are many ways to decompose a convolution, our complexity analysis shows that the total FLOPs remain unchanged. Empirically, deploying parallel convolutions with uniform kernel shapes incurs no NPU performance penalty. Based on these findings, we adopt a fixed parallel decomposition scheme to replace the original convolution at deployment.

3.2 Stride Separation and Combination

In convolution calculations, stride is one of the core hyperparameters, and by controlling the sliding interval of the convolution kernel over the input feature map, it directly determines the spatial resolution of the output feature map. The stride determines the resolution of the output feature map without changing the data within the output.

The formula for calculating the convolution output shape is given by:

$$O = \left\lfloor \frac{I - K + 2P}{S} \right\rfloor + 1 \qquad (9)$$

where O is the output shape, I is the input shape, K is the kernel size, S is the stride size, P is the padding size, and $\lfloor \cdot \rfloor$ denotes the floor operation.

From the convolution calculation and shape formula, the following theorem is derived:

Theorem 2. *Stride Separation Equivalence*

If the stride S of a convolution F can be decomposed into two values S_1 and S_2 such that $S = S_1 \times S_2$, then the stride of the convolution can be separated into a convolution operation with stride S_1 and a filtering operation with stride S_2 using the Split tensor operation:

$$I \otimes F + REP(b) = Split(I \otimes F_1 + REP(b_1))_{S_2} \tag{10}$$

where $Split(\cdot)_{S_2}$ represents the tensor splitting operation with stride S_2.

Proof. According to Eq. (9), the output shape of the convolution with stride S can be decomposed into a stride S_1 convolution and a filtering operation with stride S_2, where $S = S_1 \times S_2$. The output shape after the convolution with stride S_1 is given by:

$$O_1 = \left\lfloor \frac{I - K + 2P}{S_1} \right\rfloor + 1 \tag{11}$$

For the filtering operation with stride S, the output shape is given by:

$$O' = \left\lceil \frac{I}{S} \right\rceil \tag{12}$$

where $\lceil \cdot \rceil$ denotes the ceiling operation. After applying the filtering operation with stride S_2, the final output is:

$$O'' = \left\lceil \frac{\left\lfloor \frac{I-K+2P}{S_1} \right\rfloor + 1}{S_2} \right\rceil = \left\lfloor \frac{I - K + 2P}{S_1 S_2} \right\rfloor + 1 = O \tag{13}$$

Therefore, the convolution with stride S can be separated into a convolution with stride S_1 and a filtering operation with stride S_2, i.e.:

$$I \otimes F + \text{REP}(b) = \text{Split}\left(I \otimes F_1 + \text{REP}(b_1)\right)_{S_2} \tag{14}$$

When the convolution stride exceeds NPU hardware constraints, we leverage Theorem 2 to decompose the operation into a sequence of small-stride convolutions followed by a filtering module. The small-stride convolutions perform dense feature extraction but may introduce redundancy due to overlapping receptive fields. To mitigate this, a filtering operator is applied post-convolution to suppress irrelevant activations, ensuring the composite operation retains the functional equivalence of the original large-stride convolution.

As shown in Fig. 3, a 9×9 input tensor is convolved with a 3×3 kernel using stride 3, producing a 3×3 output feature map. When such a stride violates NPU architectural constraints, we replace the original convolution F with a compatible operation F_1 that shares weights and biases but uses a supported stride.

The output O_1 from F_1 consists of two regions: the blue region matches the original output O, while the yellow region contains redundant computations. A stride-3 filtering operation is then applied to O_1 to remove these redundancies, yielding O_2, which is equivalent to the original output O.

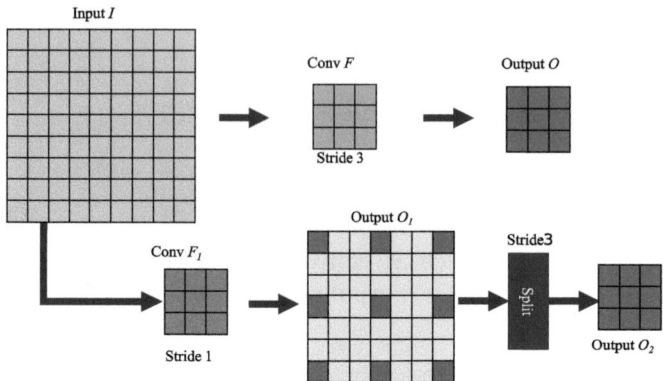

Fig. 3. Diagram shows stride decomposition of a 9 × 9 input with a 3 × 3 kernel.

Therefore, by decomposing large-stride convolutions into combinations of small-stride convolutions and filtering operators, we can overcome NPU constraints on convolution stride sizes. However, there are multiple decomposition strategies for such a stride separation. Since weight migration from the original convolution preserves output equivalence in these substitution schemes, we evaluated different stride separation approaches based on their computational complexity and selected the optimal decomposition strategy.

Notably, the filtering operators introduce negligible computational overhead, as the primary computation cost lies in the small-stride convolution operations. The FLOPs for a standard 2D convolution without bias are given by:

$$\begin{aligned} \text{FLOPs} &= HWC_{\text{out}}(2C_{\text{in}}K_1k_1 - 1) \\ &= 2HWC_{\text{in}}K_1k_1C_{\text{out}} - HWC_{\text{out}} \end{aligned} \tag{15}$$

while the computation with bias becomes:

$$\text{FLOPs} = 2HWC_{\text{in}}K_1k_1C_{\text{out}} \tag{16}$$

From Eq. (9), we know that the output feature size depends on the direction. Substituting Eq. (9) into Eq. (16) yields:

$$\text{FLOPs} = 2C_{\text{in}}K_1k_1C_{\text{out}} \left(\frac{H_{\text{in}} - K_2 + 2P}{S_1}\right)\left(\frac{W_{\text{in}} - k_2 + 2P}{S_1}\right) \tag{17}$$

Equation (17) shows that the computational cost of the stride separation scheme is inversely proportional to the small stride S_1. Thus, our evaluation framework selects the decomposition with the largest permissible S_1, minimizing overall computation.

3.3 MLP Simulation and Search

MLP Simulation Operator. It can be observed that the unsupported operators on the NPU are primarily mathematical computation operators. A Multi-layer Perceptron (MLP) is a classical feedforward neural network consisting of an input layer, multiple hidden layers, and an output layer. Due to its universal approximation capability, nonlinear modeling ability, and parameter optimization characteristics, the MLP can approximate any continuous function, thereby enabling it to simulate the functionality of mathematical computation operators.

The Universal Approximation Theorem states that a single-hidden-layer MLP with a sufficient number of hidden neurons and an appropriate activation function can approximate any continuous function. Formally, let $f : R^n \to R$ be a continuous function defined on a compact subset $S \subseteq R^n$. For any $\varepsilon > 0$, there exists a single-hidden-layer feedforward neural network g such that:

$$|f(x) - g(x)| < \varepsilon, \quad \forall x \in S \tag{18}$$

where $g(x)$ is expressed as:

$$g(x) = \sum_{i=1}^{N} c_i \, \sigma\left(a_i^\top x + b_i\right) \tag{19}$$

Here, $\sigma(\cdot)$ denotes the activation function, $a_i \in R^n$ is the input weight vector, $b_i \in R$ is the bias, $c_i \in R$ is the output weight, and N is the number of neurons in the hidden layer.

Regardless of the complexity of the function $f(x)$, there is always a single hidden layer MLP capable of approximating $f(x)$ to an arbitrary degree of precision in the domain S.

Nair et al. [8] introduced the ReLU activation function, showing that it accelerates training and mitigates vanishing gradients, thereby enhancing the approximation power of neural networks. Accordingly, this work adopts ReLU as the activation function for the MLP.

Leveraging the universal approximation property of MLPs, we propose to simulate and replace unsupported mathematical operators in deep neural networks (DNNs). This facilitates model deployment on mobile NPU platforms, enabling efficient inference by fully utilizing NPU resources.

MLP Neuron Search. Due to its universal approximation capability, the MLP can simulate the functionality of unsupported operators. However, the simulation performance of the MLP varies depending on the number of hidden layers and neurons, particularly across different data ranges and operator types.

In this work, we investigate the ability of single-hidden-layer MLPs with varying numbers of neurons to simulate mathematical computation operators. As shown in Table 3, the evaluated data ranges are 10^{-4}–10^4 and -10^4–10^4, and the targeted operators include Abs, $Log10$, among others. The training data set consists of 10 million randomly generated samples and the test dataset consists of 100 thousand samples. The MLPs are trained for 100 epochs, and the accuracy of

Algorithm 1. Neuron Number Search for MLP

Require: Training dataset $D = \{(x_1, y_1), (x_2, y_2), \ldots, (x_n, y_n)\}$, Testing dataset $D' = \{(x'_1, y'_1), (x'_2, y'_2), \ldots, (x'_m, y'_m)\}$, Learning rate η, Initial MLP parameters θ, Array of candidate neuron numbers N, Number of training epochs E

Ensure: Optimal number of neurons n and updated MLP parameters θ'

1: Initialize dataset $D' \leftarrow D$, $p(x_i) = 1$
2: $best\_loss \leftarrow 0$
3: **for** each $neural\_num$ in N **do**
4:    **for** $i = 1$ to E **do**
5:       Forward pass: $output(D) \leftarrow \text{MLP}(D, \theta_{i-1})$
6:       Compute loss: $L(\theta_i) \leftarrow \text{Loss}(output(D))$
7:       Update parameters: $\theta_i \leftarrow \theta_{i-1} - \eta \nabla_\theta L(\theta_i)$
8:       **if** $L(\theta_i) < best\_loss$ **then**
9:          $best\_loss \leftarrow L(\theta_i)$
10:         Save MLP parameters θ'
11:       **end if**
12:    **end for**
13:    $correct \leftarrow 0$ (Number of correct predictions)
14:    $total \leftarrow m$ (Total number of test samples)
15:    **for** $i = 1$ to m **do**
16:       $\hat{y}_i \leftarrow \text{MLP}(x'_i, \theta')$
17:       **if** $|\hat{y}_i - y'_i| < 10^{-2}$ **then**
18:          $correct \leftarrow correct + 1$
19:       **end if**
20:    **end for**
21:    $accuracy \leftarrow correct/total$
22:    **if** $accuracy > 0.99$ **then**
23:       Save the MLP model parameters θ'
24:    **end if**
25: **end for**

Table 3. Simulation of Mathematical Operators by MLPs with Varying Neuron Counts

Op	2	4	8	16	32	64	96	128	224
Abs	47.44%	99.86%	99.99%	99.98%	99.93%	99.93%	99.88%	99.92%	99.82%
Celu	99.98%	99.98%	99.55%	99.10%	99.81%	99.56%	96.15%	98.99%	99.03%
Erf	25.85%	98.97%	94.59%	98.14%	99.79%	97.79%	95.60%	95.22%	95.98%
Selu	99.97%	98.91%	99.06%	98.24%	99.96%	99.89%	99.84%	99.36%	99.51%
Log10	8.69%	8.67%	16.46%	72.03%	71.49%	99.39%	98.57%	98.28%	99.37%

the simulation is considered acceptable if the MLP output matches the standard mathematical computation output within an absolute error tolerance of 10^{-2}. Experimental results show that MLPs with a single hidden layer and neuron counts ranging from 2 to 224 can accurately simulate the targeted operator functions.

We automate MLP neuron selection (2–224 range) by training on 1000k data, stopping when >99% accuracy is reached within 100 epochs. The optimal configuration replaces the original component (Algorithm 1) to balance performance/efficiency.

The computational complexity of the algorithm includes both the training and testing phases. The training phase has a time complexity of $O(N \cdot n)$, while the testing phase has a time complexity of $O(m)$. Therefore, the overall time complexity of the algorithm is $O(N \cdot n)$.

4 Experimental Evaluation

4.1 Experimental Setting

This study examines DNNs that include NPU-restricted operations. Unlike models such as Transformer or BERT, which rely on supported components such as matrix multiplications and activations, we target architectures like ConvMixer [13] and gMLP [6] (with large-stride convolutions), and ShortRes [14] and Musicnn [11], which involve unsupported functions (e.g., $\log_{10}$) or large kernels (Table 4).

The image models are fine-tuned in a 10 class subset of ImageNet [2]; audio models use a PUBG sound dataset that comprises 80K audio samples captured by Infinix GT20 devices, covering six categories of audio signals. We follow the MediaTek Dimensity 8200 NeuroPilot SDK for NPU constraints. For ConvMixer and gMLP, we replace the final 1000 class layer with a 10 output layer and fine-tune it using AdamW [7] earning rate 0.01, weight decay 2×10^{-5}) with a OneCyclePolicy. The MLP-based operator approximators are trained for 100 epochs with AdamW (learning rate 1×10^{-4}).

All preprocessing and training were conducted on an Intel Core i9-13900KF CPU with NVIDIA RTX 4080 GPU (16GB VRAM). Inference measurements used an Infinix GT20 Pro smartphone (MediaTek Dimensity 8200 Ultimate) under idle conditions. Precision, latency, and power were evaluated across four modes: (1) CPU-only (MNN), (2) GPU-only (MNN), (3) CPU+NPU fallback (NNAPI), and (4) full NPU (MDLA). Latency reports the average of 1000 runs; power consumption is the device's mean power increase.

Table 4. Unsupported or Constrained Operators in Selected Models

Model	Unsupported/Constrained Operator
ConvMixer	Convolution with large stride
gMLP	Convolution with large stride
ShortRes	$Log10$ (unsupported)
Musicnn	$Log10$ (unsupported), large convolution kernel

4.2 MLP Simulation Capabilities

To assess MLP-based operator approximation capability and training costs, we design a series of experiments that target the second category of unsupported operators. The MLP is configured with a single hidden layer containing 128 neurons. As shown in Fig. 4, we test the model under different input data distributions: $[-10^1, 10^1]$, $[-10^3, 10^3]$, and $[-10^5, 10^5]$, with the corresponding data ranges for the operator $\log_{10}$ being $[10^{-1}, 10^1]$, $[10^{-3}, 10^3]$, and $[10^{-5}, 10^5]$.

Furthermore, as shown in Fig. 5, we evaluate the effect of varying the size of the data set (2,000K, 5,000K and 10,000K) in a fixed data distribution of $[-10^3, 10^3]$ (for the operator $\log_{10}$: $[10^{-3}, 10^3]$). Results show that MLPs, when trained with sufficient epochs and provided with a large, well-distributed dataset, can achieve high-accuracy approximations of mathematical operators.

This study analyzed the training overhead of MLP-based mathematical operator approximation. Using an NVIDIA RTX 4080 GPU and a 10-million-sample dataset, single-GPU training required only 12 GB memory, achieving 8.6 min training time over 100 epochs. These results demonstrate that the MLP-based approximation imposes relatively low computational requirements and incurs minimal training overhead, making it practical for efficient operator simulation.

4.3 Validation of Operator Replacement Accuracy

To validate our operator replacement scheme, we compare both model accuracy and the mean squared error (MSE) between the original and replaced operators under identical inputs.

As shown in Table 5, ConvMixer and gMLP yield MSEs below 10^{-7} and accuracy drops under 0.5%, indicating that constrained operator replacement is effectively lossless. For ShortRes and Musicnn, MSEs stay below 3 with accuracy degradation under 0.5%, consistent with MLP-based approximations of unsupported functions. Minor accuracy differences are mainly due to hardware precision variations between GPU and NPU.

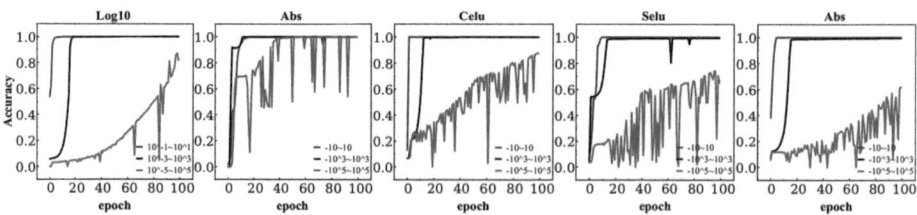

Fig. 4. MLP simulations lack operator support across data distributions.

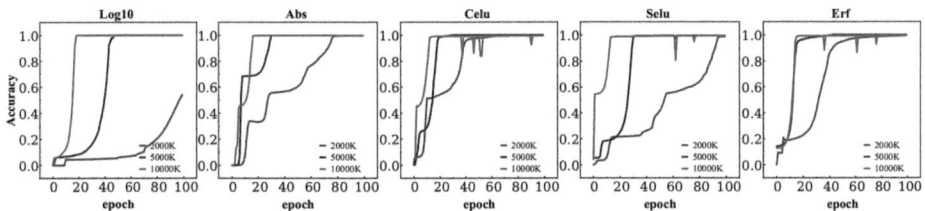

Fig. 5. MLP tests with different data scales lack operator support.

These results confirm that our replacement scheme introduces negligible precision loss and that MLP approximations maintain accuracy within acceptable bounds, validating the method's correctness and effectiveness.

Table 5. Model Accuracy and Operator Output MSE: Before/After Replacement

Model	Accuracy (Before)	Accuracy (After)	Operator MSE
ConvMixer	85.21%	85.18%	7.2×10^{-8}
gMLP	84.97%	84.92%	6.6×10^{-8}
ShortRes	89.43%	89.05%	2.73
Musicnn	90.26%	89.92%	2.81

4.4 Latency and Power Evaluation of Operator Replacement

To further validate the effectiveness of the proposed operator replacement strategy, we compare inference latency and power consumption across different execution modes. As shown in Fig. 6, NPU-CPU collaborative execution incurs significant latency due to frequent context switching, often resulting in higher latency than CPU-only execution.

For ConvMixer and gMLP, which have relatively large parameter sizes, NPU-only execution reduces latency by 3–4× over CPU, 2–30× over GPU, and 20–50× over NPU-CPU execution. In contrast, for the smaller ShortRes model, NPU-only execution offers minimal latency gain over CPU but still achieves 4× and 11× speedups over GPU and NPU-CPU modes, respectively.

For Musicnn, which includes both a constrained operator (large-kernel convolution) and an unsupported one ($\log_{10}$), the compact model causes NPU execution to be on par with CPU in latency, while still offering 2.5× and 4× improvements over GPU and NPU-CPU execution.

Although ShortRes and Musicnn show limited latency gains on the NPU, the next subsection demonstrates that their power efficiency improves significantly under NPU execution.

Figure 7 compares inference power consumption across four execution modes: CPU-only, GPU-only, NPU-CPU collaborative, and NPU-only. CPU-only shows the highest power draw, followed by GPU, while both NPU modes consume significantly less. Specifically, NPU-only reduces power consumption by 3–7× compared to CPU and 2–3× compared to GPU. Although power usage is similar between NPU-only and NPU-CPU modes, the former achieves much lower latency, as previously shown. For ShortRes and Musicnn, NPU-only offers latency on par with CPU but with much lower power usage, highlighting the NPU's energy efficiency.

Aggregating all experimental findings, the proposed operator replacement scheme maintains model accuracy within a 0.5% loss, while delivering substantial energy savings: 7–12× reduction versus CPU-only execution, 6–30× versus GPU-only execution, and approximately 20–40× versus NPU-CPU collaborative execution.

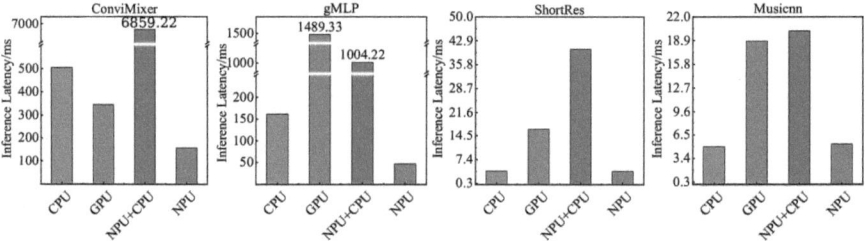

Fig. 6. Inference latency comparison.

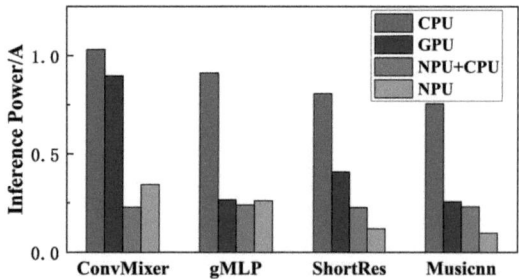

Fig. 7. Comparison of power consumption.

5 Conclusion

In this paper, we propose an operator optimization framework for NPU deployment constraints, addressing unsupported operators in DNN inference while maintaining model accuracy. Our approach replaces NPU-constrained operators with equivalent sub-operator combinations and substitutes NPU-incompatible mathematical operators with MLP-based emulation. Through evaluation of computational cost and accuracy-sensitive schemes, optimized models achieve deployment in mobile NPUs with less than 0.5% accuracy loss, reducing energy consumption by 7–12× versus CPU, 6–30× versus GPU and 20–40× versus hybrid execution of NPU-CPU.

Acknowledgments. This research was supported in part by the National Natural Science Foundation of China under Grant No. 62402270, the China Postdoctoral Science Foundation under Grant No. 2024M761684, and the Postdoctoral Fellowship Program of CPSF under Grant No. GZC20240832.

References

1. Google Deepmind (2023). https://deepmind.google/technologies/gemini/nano/
2. Deng, J., et al.: ImageNet: a large-scale hierarchical image database. In: 2009 IEEE Conference on Computer Vision and Pattern Recognition, pp. 248–255. IEEE (2009)
3. Ding, X., et al.: Diverse branch block: building a convolution as an inception-like unit. In: Proceedings of the IEEE/CVF Conference on Computer Vision and Pattern Recognition, pp. 10886–10895 (2021)
4. Jeong, J.S., et al.: Band: coordinated multi-DNN inference on heterogeneous mobile processors. In: Proceedings of the 20th Annual International Conference on Mobile Systems, Applications and Services, pp. 235–247 (2022)
5. Kim, S., Oh, S., Yi, Y.: Minimizing GPU kernel launch overhead in deep learning inference on mobile GPUs. In: Proceedings of the 22nd International Workshop on Mobile Computing Systems and Applications, pp. 57–63 (2021)
6. Liu, H., et al.: Pay attention to MLPs. Adv. Neural. Inf. Process. Syst. **34**, 9204–9215 (2021)
7. Loshchilov, I., Hutter, F.: Decoupled weight decay regularization. arXiv preprint arXiv:1711.05101 (2017)
8. Nair, V., Hinton, G.E.: Rectified linear units improve restricted Boltzmann machines. In: Proceedings of the 27th International Conference on Machine Learning (ICML-10), pp. 807–814 (2010)
9. Niu, W., et al.: DNNFusion: accelerating deep neural networks execution with advanced operator fusion. In: Proceedings of the 42nd ACM SIGPLAN International Conference on Programming Language Design and Implementation, pp. 883–898 (2021)
10. OpenAI (2024). https://github.com/Taewan-P/gptmobile
11. Pons, J., Serra, X.: musicnn: pre-trained convolutional neural networks for music audio tagging. arXiv preprint arXiv:1909.06654 (2019)
12. Shi, J., et al.: Pretrained quantum-inspired deep neural network for natural language processing. IEEE Trans. Cybern. (2024)
13. Trockman, A., Zico Kolter, J.: Patches are all you need?. arXiv preprint arXiv:2201.09792 (2022)
14. Won, M., et al.: Evaluation of CNN-based automatic music tagging models. arXiv preprint arXiv:2006.00751 (2020)
15. Xu, D., et al.: Mandheling: mixed-precision on-device DNN training with DSP offloading. In: Proceedings of the 28th Annual International Conference on Mobile Computing And Networking, pp. 214–227 (2022)

16. Zhang, J., et al.: MobiDepth: real-time depth estimation using on-device dual cameras. In: Proceedings of the 28th Annual International Conference on Mobile Computing And Networking, pp. 528–541 (2022)
17. Zhang, J., et al.: MobiPose: real-time multi-person pose estimation on mobile devices. In: Proceedings of the 18th Conference on Embedded Networked Sensor Systems, pp. 136–149 (2020)
18. Zhang, X., Zeng, H., Zhang, L.: Edge-oriented convolution block for real-time super resolution on mobile devices. In: Proceedings of the 29th ACM International Conference on Multimedia, pp. 4034–4043 (2021)

AIGC in Mobile Edge Networks: A Survey

Tingting Long and Deyu Zhang(✉)

Central South University, Changsha 410000, Hunan, China
{tingtinglong,zdy876}@csu.edu.cn

Abstract. Artificial Intelligence Generated Content (AIGC) refers to the automatic generation of various types of content—such as text, images, audio, and video—using artificial intelligence techniques. In recent years, AIGC has been widely applied across domains including software development, education, advertising, online customer service and beyond, significantly transforming the content creation paradigm and reshaping human–computer interaction. With the increasing computational capabilities of mobile devices, the proliferation of intelligent computing at the network edge, and growing user demands for low-latency, bandwidth-efficient, and privacy-preserving AIGC services, Mobile Edge Networks (MENs) have emerged as a promising infrastructure for delivering AIGC services efficiently and responsively. This survey provides an overview of the opportunities and challenges associated with deploying AIGC on mobile edge networks. We highlight the significance of enabling AIGC under the MENs paradigm and examine the key supporting technologies and system infrastructures—such as resource allocation, task offloading, and edge caching—that empower users to access AIGC services seamlessly at the edge. Finally, we outline future research directions and open issues in mobile AIGC networks to guide further investigation in this emerging field.

Keywords: AIGC · Generative AI · Mobile edge networks · AI training and inference · Internet technology

1 Introduction and Motivation

With the rapid advancement of generative AI models—such as large language models (LLMs) and diffusion models—Artificial Intelligence Generated Content (AIGC) has emerged as a transformative force in digital content creation [6,17,70]. AIGC enables the automatic generation of diverse content forms, including text, images, audio, and video. For example, Stable Diffusion [45] can generate high-quality images from textual prompts, while ChatGPT [1] facilitates intelligent dialog and content generation. Moreover, AIGC is extensively used in cross-modal tasks such as text-to-image, image-to-text, text-to-video, and video-to-text generation, demonstrating remarkable potential in a range of applications, including image enhancement and inpainting, video synthesis and extension, and automatic speech recognition and beyond. For example, KlingAI

[23,49] and Sora [41] offer functionalities such as AI-based image outpainting and video synthesis from a single image input, thus stimulating a new wave of user creativity.

In recent years, the performance of AIGC models has improved substantially. On the one hand, the development of more capable foundation models (e.g., Stable Diffusion XL [42]) and efficient inference techniques (e.g., one-step diffusion [35]) has significantly enhanced both quality and speed. On the other hand, the increasing computational capabilities of mobile devices have opened new opportunities for edge-side deployment. For instance, lightweight multimodal models like MobileVLM [7,8] and frameworks like MNN-Transformer [55] enable on-device execution of LLMs and content generation, allowing users to perform AIGC tasks directly at the network edge.

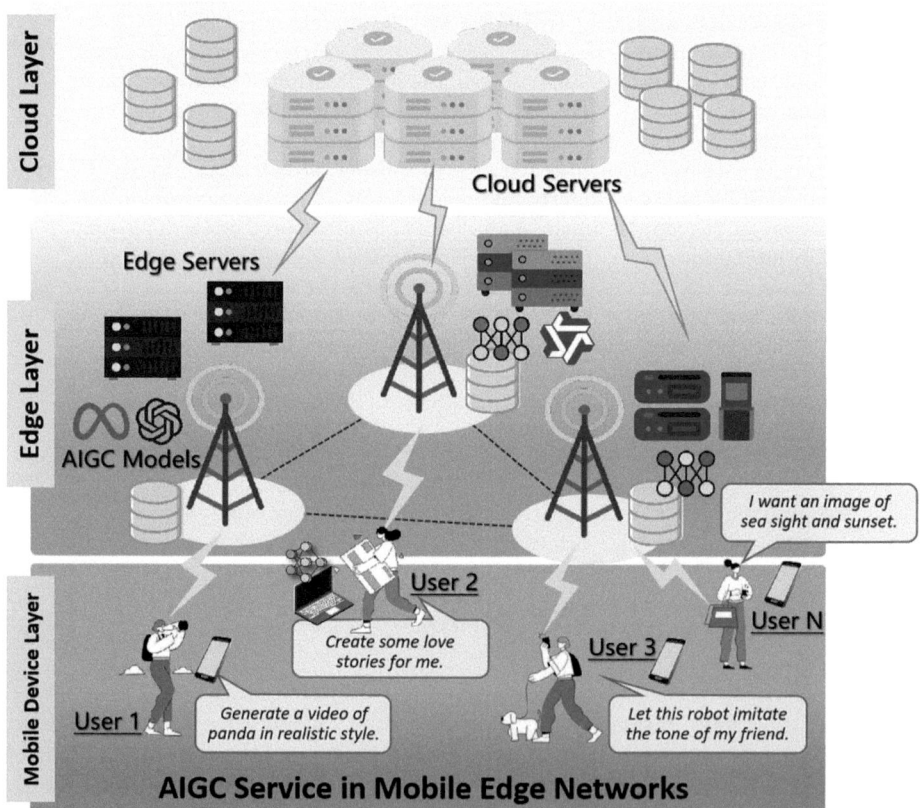

Fig. 1. Overview of AIGC service in mobile edge networks, illustrating the cloud, edge, and mobile device layers.

Mobile Edge Networks (MENs) are a distributed network architecture that pushes computing and storage resources from centralized cloud data centers to the edge of the network—closer to end users. By doing so, MENs can offer lower latency, higher bandwidth efficiency, improved privacy, and enhanced system reliability. It is being widely adopted in diverse scenarios such as content delivery networks (CDNs) [32], smart cities, the Internet of Things (IoT) [40], the Internet of Vehicles (IoV) [65], and immersive technologies like augmented reality (AR) and virtual reality (VR) [4].

In light of these trends, both academia and industry are actively exploring the deployment of AIGC within mobile edge networks (as shown in Fig. 1), driven by the following compelling benefits:

- **Latency reduction.** Placing AIGC models near the edge significantly reduces the time required for content generation and delivery, enabling near-instantaneous responses in latency-sensitive applications.
- **Bandwidth efficiency.** Generating and processing content locally reduces the volume of data that must traverse the core network, which is particularly beneficial for high-resolution images and video data.
- **Privacy and security.** By minimizing the need to transmit sensitive data to remote servers, edge-side AIGC enhances data security and reduces privacy risks.
- **System reliability.** The distributed nature of edge computing improves fault tolerance. Even in the event of local node failures, other edge nodes can continue to provide uninterrupted service.

With the continuous evolution of AIGC capabilities in mobile edge environments, both academia and industry are steadily advancing toward the realization of AI agents [38,61,72]. AI agents refer to autonomous intelligent systems deployed on personal or edge devices, capable of understanding users' multimodal inputs (such as text, images, audio, and video), performing context-aware reasoning, and generating intelligent responses or actions. This paradigm not only redefines the mode of human–computer interaction, but also marks a critical step toward achieving cognitive and decision-making intelligence on ubiquitous computing platforms, underscoring the practical value and strategic significance of deploying AIGC technologies in mobile edge networks.

Motivated by these trends, this survey aims to explore the emerging landscape of AIGC in mobile edge networks. We systematically review recent advancements, underlying system architectures, optimization techniques, and key challenges. Particular attention is paid to how AIGC models can be re-architected for edge execution, the trade-offs involved in model deployment, and how edge–device collaboration can unlock scalable and efficient generative AI applications.

2 Enabling Technologies for AIGC in MENs

The deployment of AIGC on mobile edge networks relies on a range of enabling technologies. This section begins with an introduction to the fundamental concepts, including commonly used generative models and the mobile edge computing paradigm. It then discusses recent advancements in AIGC optimization techniques tailored for mobile edge environments.

2.1 AIGC Models

AIGC models primarily refer to generative models that are capable of synthesizing data across various modalities. In recent years, several representative architectures have gained prominence, including Variational Autoencoders (VAEs), Generative Adversarial Networks (GANs), Diffusion models, Large Language Models (LLMs), and Multi-modal Large Language Models (MM-LLMS).

Variational Autoencoders were introduced by Kingma et al. in 2014 [11, 26], which are a class of deep generative models composed of an encoder and a decoder. The core idea of VAEs is to represent the latent variables output by the encoder as a probability distribution (typically Gaussian), rather than as deterministic values. By introducing variational inference, VAEs optimize the Evidence Lower Bound (ELBO) to learn an approximate posterior distribution of the latent variables. New samples can then be generated by sampling from this learned latent space and decoding them through the decoder. As a representative generative model, VAEs offer both interpretability and architectural stability. In recent years, they have been widely adopted in tasks such as image super-resolution and reconstruction, and have demonstrated promising performance [42,45,68].

Generative Adversarial Networks is a deep learning model proposed by Ian Goodfellow in 2014 [19]. GANs generate realistic data through an adversarial process, i.e. competition between generator and discriminator. The generator receives a random noise vector as input and generates samples that resemble real data. The discriminator receives a sample as input and determines whether the sample is real data or data generated by the generator. The training process of GANs is a zero-sum game, where the generator and discriminator compete with each other to improve their respective performance. Numerous GAN variants have since been developed and applied to various generation tasks, such as SRGAN [27], Real-ESRGAN [54] and GigaGAN [25].

Diffusion Models are a type of generative models that synthesize data by reversing a gradual noising process. A canonical method is the Denoising Diffusion Probabilistic Model (DDPM) [22], whose generation process typically involves two stages: (a) a forward diffusion process, where Gaussian noise is progressively added to the input data until it becomes pure noise; and (b) a reverse denoising process, in which a neural network learns to iteratively remove the noise and reconstruct the original data. DDPM relies on a stochastic sampling procedure that requires injecting Gaussian noise at every denoising step, resulting in slow generation. To address this inefficiency, the Denoising Diffusion Implicit Model (DDIM) [48] introduces a deterministic reverse process that eliminates the need for per-step noise sampling and enables "skip sampling," thereby significantly accelerating inference. In 2021, OpenAI advanced diffusion models by improving the architecture and introducing classifier guidance [10], achieving state-of-the-art performance in both unconditional and conditional image synthesis for the first time—surpassing GAN-based methods. Since then, diffusion models have rapidly proliferated in generative applications, powering models such as Palette [46], DALL-E 2 [44], Imagen [47], and the widely adopted Stable Diffusion series [42,45].

Large Language Models are large-scale deep neural networks, typically built on Transformer [52] or BERT [9] architectures, and are characterized by massive parameter counts, complex structures, and the need for extensive training data. When scaled beyond a certain threshold, LLMs exhibit emergent abilities [56], enabling them to autonomously learn abstract patterns, represent high-dimensional semantics, and generalize across diverse tasks. In 2022, OpenAI's release of ChatGPT [1] marked a turning point in the public adoption of LLMs and sparked a wave of open-source and commercial models, such as Qwen [66], LLaMA [50,51], and DeepSeek [21,33], which further advanced the development of generative AI.

Multi-modal Large Language Models leverage the powerful capabilities of LLMs as the core reasoning engine to perform multimodal tasks. Popular MM-LLMs include Flamingo [2], CLIP [43], the BLIP series [28,29], GPT-4o [24], Qwen-VL [3], and LLaVA [34], among others. These models are typically pretrained on large-scale paired cross-modal datasets—such as image-text, speech-text, or video-text pairs—to align representations across different modalities. A common pretraining strategy of MM-LLMs involves freezing the backbone components (e.g., vision encoders and LLMs) while learning a lightweight module that bridges the modalities. For instance, LLaVA-1 consists of a CLIP-based vision encoder, a linear projection layer, and an LLM model LLaMA [50]. The vision encoder extracts visual features from the input image, which are then projected into the same embedding space as the text tokens via a learnable linear mapping. The resulting image token embeddings are concatenated with the input text tokens and fed into the LLM to generate multimodal responses. During training, LLaVA employs a two-stage process: (1) only the lightweight projection module is trained on image-text instruction-following data to align visual and textual representations, and (2) the model is further fine-tuned using manually curated multimodal instruction datasets to enhance its ability to produce grounded and context-aware responses to visual inputs—all while keeping the backbone components fixed.

2.2 Collaborative Infrastructure for Mobile AIGC Networks

Mobile Edge Networks (MENs) refer to a distributed network architecture that extends cloud capabilities by integrating edge servers and mobile devices at the network periphery—closer to end users. Unlike traditional centralized cloud infrastructures, MENs leverage hierarchical collaboration among cloud, edge, and mobile layers to balance computational demands, bandwidth usage, latency constraints, and privacy concerns. This paradigm is particularly well-suited for AIGC, whose models are typically large, resource-intensive, and increasingly required to operate in real-time, interactive, and privacy-sensitive settings.

This section introduces the core components of the collaborative MENs infrastructure for AIGC: cloud computing provides centralized model training and resource provisioning; edge computing offers low-latency inference and context-aware services; and mobile computing enables on-device execution and last-mile personalization. Their interplay forms the backbone of scalable, adaptive AIGC deployments in mobile edge environments.

Cloud Computing is a traditional service model that delivers computing resources—such as servers, storage, databases, networking, software, and analytics—over the Internet, eliminating the need for users to manage or maintain physical infrastructure. Given the intensive data and computational demands of AIGC model training and inference, cloud platforms offer robust technical support and infrastructure, making them the most common medium for delivering AIGC services today. Cloud computing is typically categorized into three service patterns: Infrastructure as a Service (IaaS), Platform as a Service (PaaS), and Software as a Service (SaaS). In AIGC applications, IaaS offers access to virtualized computational resources such as GPU servers, storage, and databases; PaaS provides platforms for building and deploying AIGC applications; and SaaS allows end users to interact with generative models via web browsers or mobile applications over the Internet.

Edge Computing refers to the deployment of computing and storage resources closer to end users—at the edge of the network—enabling AIGC services to be delivered through wireless access infrastructure. This proximity drastically reduces latency, enhances data privacy, improves reliability, and alleviates core network bandwidth usage. Although edge servers may lack the capacity to support large-scale model training, they are well-suited for running real-time inference and lightweight fine-tuning tasks, which significantly enhance user experience. Moreover, edge computing facilitates context-aware, personalized content generation by leveraging local data, making it a key enabler for deploying adaptive AIGC services in mobile edge environments.

Mobile Computing provides a flexible and convenient way to provide AIGC services by leveraging the computing power and wireless network connections of mobile devices. Although mobile devices, such as smartphones and tablets, have limited computing power, with the continuous improvement of hardware performance, they already have the ability to run some lightweight AIGC models. Mobile devices can locally perform AIGC model inference tasks such as text generation, image processing, and speech synthesis. This reduces the need for data transfer, improving responsiveness and user experience. In cases where complex tasks or more computing resources are required, mobile devices can use nearby edge servers for collaborative computing to complete more complex AIGC tasks.

2.3 Enabling Technologies for AIGC in MENs

Delivering low-latency, personalized, and context-aware AIGC services in Mobile Edge Networks requires a suite of enabling technologies. This section provides an overview of recent advances across key areas, including unified framework design, resource allocation, task and computation offloading, edge caching, and mobility management.

Unified Framework Design. Running generative AI models—particularly diffusion-based AIGC models—demands substantial computational and storage resources. This issue is especially pronounced in diffusion models, where each denoising step incurs significant energy consumption [22]. In MENs, a common approach is to offload compute- and storage-intensive AIGC services to edge servers that act as AIGC Service Providers (ASPs) [15,53,57]. When a user initiates a generation request, the task is transmitted to a nearby ASP for processing and the output is returned to the user. This approach reduces the computational burden on user devices and enables scalable, flexible AIGC deployment.

However, delivering high-quality AIGC services in MENs poses new challenges. Since AIGC is inherently user-centric, users often exhibit diverse preferences, personalized needs, and task-specific expectations [36]. Meanwhile, ASPs differ in their deployed models, supported content modalities, and available computing resources. For example, some AIGC models are better suited for generating realistic human portraits, while others excel in producing natural scenes. As a result, selecting the most appropriate ASP for each user becomes a key optimization problem, requiring consideration of user interests, task types, model capabilities, and edge resource constraints.

To address this, several recent works have explored intelligent ASP selection mechanisms. Du [15] proposes the AI-Generated Optimal Decision (AGOD) algorithm, which leverages a diffusion model to recommend the best ASP for a given user request. Wen [57] introduces a contract theory-based model that incentivizes ASPs to provide high-quality AIGC services under asymmetric information. Lyu [39] formulates the ASP selection problem as a Markov Decision Process and proposes a Deep Q-Network-based solution, aiming to simultaneously maximize user utility and minimize energy consumption.

Another line of research explores collaborative distributed frameworks, which aim to improve the efficiency and accessibility of AIGC services through cooperation among heterogeneous devices. These frameworks are often tailored to the characteristics of specific AIGC models or application scenarios. For instance, Du [14] proposes a device collaboration strategy for diffusion-based AIGC, where shared denoising steps are jointly executed and intermediate results are transmitted for task-specific refinement—reducing latency and distributing computation. Zhang [69] designs a cloud–edge–end collaborative architecture for autonomous driving, developing resource allocation, communication, storage, and computation mechanisms to support AIGC service.

Resource Allocation. In mobile edge networks, efficient resource allocation is critical for balancing generation latency and output quality in AIGC services. Unlike conventional edge tasks, AIGC workloads—particularly those based on diffusion models—require careful coordination of model-specific parameters such as model size and the number of inference steps. To tackle these challenges, Gao [18] proposes a multi-stream queuing model and a cross-edge scheduling framework for text-to-image generation, along with a Monte Carlo Tree Search-based diffusion scheduler that dynamically adjusts computation based on realtime conditions. Feng [16] develops an online scheduling scheme that assigns AIGC requests to suitable servers while adapting the inference step count per request. They formulate the problem as a Markov Decision Process and solve it using deep reinforcement learning to optimize the trade-off between latency and generation quality under dynamic network conditions.

Beyond computational efficiency, AIGC services in mobile edge networks also face unique deployment challenges. While prior works advocate caching user inputs or generation results on edge servers to reduce response latency [12,67], such strategies may be insufficient for AIGC applications, where user requests are often highly personalized and content cannot be easily reused. Moreover, unlike cloud servers that can host all AIGC models, resource-constrained edge devices cannot support the simultaneous deployment of diverse models. When the required model is not preloaded at the edge, the system must fetch it from the cloud, introducing extra delay and bandwidth overhead. To address this, Liang [31] proposes a collaborative edge–cloud framework that formulates model heterogeneity as an optimization problem. They design a model-level selection algorithm that enables resource sharing while balancing deployment latency and resource consumption for AIGC model allocation.

Task and Computation Offloading. In mobile edge networks, AIGC users can offload computationally intensive generative tasks to nearby edge servers, significantly alleviating the burden on resource-constrained mobile devices. Some early studies [13,53,60] focus on single-server scenarios, where all mobile devices offload their AIGC tasks to a centralized edge server, which then jointly handles task offloading and resource allocation. However, under heavy workloads or large-scale user access, the limited computational capacity of a single edge server often leads to high latency, long task queues, and degraded service quality.

To overcome these limitations, recent works [30,57,62–64] have toured multi-server scenarios, where multiple edge servers are geographically distributed, and mobile devices can selectively offload tasks to different servers. These studies are generally categorized into homogeneous [30,63,64] and heterogeneous settings [57,62]. In homogeneous scenarios, all devices employ similar AIGC models and require comparable computational resources. In contrast, heterogeneous scenarios assume devices use different models for diverse services, such as image synthesis and music creation. Wu et al. [59] address task offloading and resource allocation for heterogeneous AIGC services under multi-server edge settings. To accommodate varying resource demands and the compute-intensive nature of AIGC tasks, they propose a queue-based multi-task flow architecture that enhances scheduling flexibility. Moreover, they introduce a multi-level asynchronous decision framework to decouple decision-making across stages, enabling more adaptive and responsive system behavior under dynamic conditions.

Edge Caching. Edge caching leverages the storage resources of edge servers and mobile devices to provide low-latency content and computation services. In the context of mobile AIGC, it enables users to access generative services by caching either the AIGC models or their outputs at the network edge, thereby reducing dependence on remote cloud data centers. Caching the outputs of AIGC models involves offloading inference tasks to the cloud and storing the generated results at edge servers for future reuse. While this approach reduces latency for repeated queries, it introduces new challenges, such as added transmission latency, bandwidth overhead, and potential privacy risks. Moreover, due to the highly personalized and interactive nature of AIGC services, pre-cached outputs may fail to satisfy user-specific requests.

Alternatively, caching the AIGC models themselves poses significant challenges due to their massive size and computational requirements. For example, GPT-3 [5] has up to 175 billion parameters, and LLaMA-3 [20] variants range from 8 to 70 billion. These models demand substantial storage and powerful GPU resources for real-time inference. As a result, edge servers must carefully balance storage capacity and computational capability. Intelligent caching strategies are needed—ones that dynamically account for model size, usage frequency, latency sensitivity, and hardware constraints—to support responsive and scalable AIGC services at the edge. To address these issues, Xu et al. [63] propose a joint optimization framework for model caching and inference in mobile edge networks, aiming to minimize service cost and accuracy degradation under limited GPU and compute resources. Liu et al. [37] further introduce a reinforcement learning-based approach using Deep Deterministic Policy Gradient (DDPG). DDPG effectively captures the dynamic state space—including user mobility and evolving AIGC request patterns—to derive optimal decisions for model caching and resource allocation, ultimately improving the quality and responsiveness of edge-deployed AIGC services.

Mobility Management. Due to their flexible deployment and low cost, unmanned aerial vehicles (UAVs) equipped with computing resources are increasingly used in mobile edge networks to provide reliable and wide coverage for edge computing services. In particular, UAV-mounted AIGC servers can offer ultra-low latency and highly reliable access to generative services, especially in scenarios where fixed edge servers are frequently overloaded in hotspot areas, or where deployment in remote regions is impractical due to high costs. UAVs are also well-suited for AIGC applications with mobility requirements, enabling dynamic service delivery as users move.

Some studies have explored using a single UAV to provide AIGC services. For example, Wu et al. [58] address the challenge of complex resource management in UAV-enabled mobile edge computing for Metaverse rendering. They propose a diffusion model-based algorithm that jointly optimizes rendering decisions and resource allocation, effectively reducing frame rendering time and improving the user experience. Building on this, Zhang et al. [71] investigate scenarios involving multiple UAVs. They model the complex problem of multi-UAV-assisted resource management and allocation as a Markov Decision Process, and leverage the generative capability of diffusion models along with their integration with reinforcement learning to derive efficient decision-making policies.

3 Future Research Directions and Open Challenges

The future research directions and open issues of AIGC in mobile edge networks cover many aspects, the following are some key research directions and open challenges:

3.1 Key Research Directions

Efficient Model Training and Inference. Develop more efficient AI model training and inference algorithms to accommodate the computing and storage constraints of edge devices. For example, lightweight models (e.g., model pruning, quantization), application and optimization of edge AI chips (e.g., TPU, NPU).

Model–Software–Hardware Co-design. Enable holistic optimization across AIGC models, system software, and hardware platforms to fully exploit the potential of inference acceleration on edge devices. By jointly considering model architecture, runtime scheduling, and hardware constraints (e.g., memory, energy, and compute units), co-design approaches can deliver substantial improvements in latency, energy efficiency, and model performance. This cross-layer optimization is essential to support complex generative tasks within the limited resources of mobile edge environments.

Distributed Computing and Collaborative Learning. Investigate distributed computing and cooperative learning methods between edge devices and between the edge and the cloud. For example, federated learning, distributed machine learning framework, heterogeneous computing architecture.

Intelligent Resource Management and Optimization. Develop intelligent resource management and optimization algorithms to dynamically adapt to changing network environments and user needs. For example, reinforcement learning, machine learning-driven resource scheduling, dynamic load balancing.

3.2 Open Challenges

Model and Data Heterogeneity. How to deal with the heterogeneity of edge devices and data sources of different types and sizes to achieve unified and efficient AIGC services is a major challenge. It is necessary to design a universal model architecture and adaptation mechanism to ensure compatibility and efficiency across devices and data sources.

Scalability and Portability. How to ensure that AIGC services have good scalability and portability in network environments of different sizes and types is one of the issues that need to be considered. It is necessary to develop a scalable system architecture and a portable software framework to support multiple application scenarios and deployment environments.

Energy Efficiency Optimization. Since the reasoning of the AIGC model requires a lot of computing resources, how to optimize energy efficiency on resource-limited edge devices and extend the running time of the devices is one of the urgent problems to be solved. It is necessary to study low-power models and algorithms to optimize the energy consumption of computing and communication.

Data Privacy and Compliance. Today, most AIGC services rely on cloud-edge or end-edge architecture, and data still needs to be uploaded. How to use and process user data legally and compliantly while ensuring data privacy will still be an issue that needs to be considered for a long time to come.

4 Conclusion

This survey centers on the deployment of AIGC services in mobile edge networks. We begin by outlining the background of AIGC and highlighting the importance of deploying generative AI within the mobile edge network paradigm. We then review the core infrastructure and enabling technologies that support AIGC services at the edge, including mainstream generative models, the MENs architecture, and key techniques such as framework design, resource allocation, task offloading, and caching strategies. Finally, we discuss future research directions and open challenges in mobile AIGC networks, aiming to guide further exploration and development in this rapidly evolving field.

References

1. Achiam, J., et al.: GPT-4 technical report. arXiv preprint arXiv:2303.08774 (2023)
2. Alayrac, J.-B., et al.: Flamingo: a visual language model for few-shot learning. Adv. Neural. Inf. Process. Syst. **35**, 23716–23736 (2022)
3. Bai, S., et al.: Qwen2. 5-VL technical report. arXiv preprint arXiv:2502.13923 (2025)
4. Bretos, M.A., Ibáñez-Sánchez, S., Orús, C.: Applying virtual reality and augmented reality to the tourism experience: a comparative literature review. Span. J. Mark.-ESIC **28**(3), 287–309 (2024)
5. Brown, T., et al.: Language models are few-shot learners. Adv. Neural. Inf. Process. Syst. **33**, 1877–1901 (2020)
6. Cao, Y., et al.: A survey of AI-generated content (AIGC). ACM Comput. Surv. **57**(5), 1–38 (2025)
7. Chu, X., et al.: MobileVLM2: faster and stronger baseline for vision language model. arXiv preprint arXiv:2402.03766 (2024)
8. Chu, X., et al.: MobileVLM: a fast, strong and open vision language assistant for mobile devices. arXiv preprint arXiv:2312.16886 (2023)
9. Devlin, J., et al.: BERT: pre-training of deep bidirectional transformers for language understanding. In: Proceedings of the 2019 Conference of the North American Chapter of the Association for Computational Linguistics: Human Language Technologies, Volume 1 (long and Short Papers), pp. 4171–4186 (2019)
10. Dhariwal, P., Nichol, A.: Diffusion models beat GANs on image synthesis. Adv. Neural. Inf. Process. Syst. **34**, 8780–8794 (2021)

11. Doersch, C.: Tutorial on variational autoencoders. In: arXiv preprint arXiv:1606.05908 (2016)
12. Drolia, U., et al.: Cachier: edge-caching for recognition applications. In: 2017 IEEE 37th International Conference on Distributed Computing Systems (ICDCS), pp. 276–286. IEEE (2017)
13. Du, H., et al.: Diffusion-based reinforcement learning for edge-enabled AI-generated content services. IEEE Trans. Mob. Comput. **23**(9), 8902–8918 (2024)
14. Du, H., et al.: Exploring collaborative distributed diffusion-based AI-generated content (AIGC) in wireless networks. IEEE Netw. **38**(3), 178–186 (2023)
15. Du, H., et al.: User-centric interactive AI for distributed diffusion model-based AI-generated content. arXiv preprint arXiv:2311.11094 (2023)
16. Feng, C., Zheng, Y., Yuedong, X.: Online AI-generated content-request scheduling with deep reinforcement learning. In: IEEE INFOCOM 2024-IEEE Conference on Computer Communications Workshops (INFOCOM WKSHPS), pp. 1–6. IEEE (2024)
17. Foo, L.G., Rahmani, H., Liu, J.: AI-generated content (AIGC) for various data modalities: a survey. ACM Comput. Surv. **57**(9), 1–66 (2025)
18. Gao, S., et al.: Characterizing and scheduling of diffusion process for text-to-image generation in edge networks. IEEE Trans. Mob. Comput. (2025)
19. Goodfellow, I.J., et al.: Generative adversarial nets. Adv. Neural Inf. Process. Syst. **27** (2014)
20. Grattafori, A., et al.: The LLaMA 3 herd of models. arXiv preprint arXiv:2407.21783 (2024)
21. Guo, D., et al.: Deepseek-RL: incentivizing reasoning capability in LLMs via reinforcement learning. arXiv preprint arXiv:2501.12948 (2025)
22. Ho, J., Jain, A., Abbeel, P.: Denoising diffusion probabilistic models. Adv. Neural. Inf. Process. Syst. **33**, 6840–6851 (2020)
23. Huang, Y., et al.: ConceptMaster: multi-concept video customization on diffusion transformer models without test-time tuning. arXiv preprint arXiv:2501.04698 (2025)
24. Hurst, A., et al.: GPT-4o system card. arXiv preprint arXiv:2410.21276 (2024)
25. Kang, M., et al.: Scaling up GANs for text-to-image synthesis. In: Proceedings of the IEEE/CVF Conference on Computer Vision and Pattern Recognition, pp. 10124–10134 (2023)
26. Kingma, D.P., Welling, M., et al.: Auto-encoding variational bayes. In: International Conference on Machine Learning, pp. 1109–1117. PMLR (2014)
27. Ledig, C., et al.: Photo-realistic single image super-resolution using a generative adversarial network. In: Proceedings of the IEEE Conference on Computer Vision and Pattern Recognition, pp. 4681–4690 (2017)
28. Li, J., et al.: BLIP-2: bootstrapping language-image pre-training with frozen image encoders and large language models. In: International Conference on Machine Learning, pp. 19730–19742. PMLR (2023)
29. Li, J., et al.: BLIP: bootstrapping language-image pre-training for unified vision-language understanding and generation. In: International Conference on Machine Learning, pp. 12888–12900. PMLR (2022)
30. Li, S., et al.: Multi-agent RL-based industrial AIGC service offloading over wireless edge networks. In: IEEE INFOCOM 2024-IEEE Conference on Computer Communications Workshops (INFOCOM WKSHPS), pp. 1–6. IEEE (2024)
31. Liang, Y., et al.: Resource-efficient generative AI model deployment in mobile edge networks. arXiv preprint arXiv:2409.05303 (2024)

32. Lin, S., et al.: ProAcher: secure and practical password pro-authentication by content delivery networks. In: 22nd USENIX Symposium on Networked Systems Design and Implementation (NSDI 2025), pp. 1399–1419 (2025)
33. Liu, A., et al.: Deepseek-v3 technical report. arXiv preprint arXiv:2412.19437 (2024)
34. Liu, H., et al.: Visual instruction tuning. Adv. Neural. Inf. Process. Syst. **36**, 34892–34916 (2023)
35. Liu, X., et al.: InstaFlow: one step is enough for high-quality diffusion-based text-to-image generation. In: The Twelfth International Conference on Learning Representations (2023)
36. Liu, Y., et al.: Towards multi-task generative-AI edge services with an attention-based diffusion DRL approach. In: 2024 9th IEEE International Conference on Smart Cloud (SmartCloud), pp. 60–65. IEEE (2024)
37. Liu, Z., et al.: Joint model caching and resource allocation in generative AI-enabled wireless edge networks. arXiv preprint arXiv:2411.08672 (2024)
38. Luo, J., et al.: Large language model agent: a survey on methodology, applications and challenges. arXiv preprint arXiv:2503.21460 (2024)
39. Lyu, X., Rani, S., Feng, Y.: Optimizing AIGC service provider selection based on deep Q-network for edge-enabled health-care consumer electronics systems. IEEE Trans. Consum. Electron. (2024)
40. Oliveira, F., et al.: Internet of intelligent things: a convergence of embedded systems, edge computing and machine learning. Internet Things 101153 (2024)
41. OpenAI. Sora/OpenAI (2024). https://openai.com/sora/
42. Podell, D., et al.: SdXL: improving latent diffusion models for high-resolution image synthesis. arXiv preprint arXiv:2307.01952 (2023)
43. Radford, A., et al.: Learning transferable visual models from natural language supervision. In: International Conference on Machine Learning, pp. 8748–8763. PMLR (2021)
44. Ramesh, A., et al.: Hierarchical text-conditional image generation with CLIP latents. arXiv preprint arXiv:2204.06125, vol. 1, no. 2, p. 3 (2022)
45. Rombach, R., et al.: High-resolution image synthesis with latent diffusion models. In: Proceedings of the IEEE/CVF Conference on Computer Vision and Pattern Recognition, pp. 10684–10695 (2022)
46. Saharia, C., et al.: Palette: image-to-image diffusion models. In: ACM SIGGRAPH 2022 Conference Proceedings, pp. 1–10 (2022)
47. Saharia, C., et al.: Photorealistic text-to-image diffusion models with deep language understanding. Adv. Neural. Inf. Process. Syst. **35**, 36479–36494 (2022)
48. Song, J., Meng, C., Ermon, S.: Denoising diffusion implicit models. arXiv preprint arXiv:2010.02502 (2020)
49. Kuaishou Technology. Kling AI: Next-Gen AI Video and AI Image Generator. https://app.klingai.com/cn/
50. Touvron, H., et al.: LLaMA 2: open foundation and fine-tuned chat models. arXiv preprint arXiv:2307.09288 (2023)
51. Touvron, H., et al.: LLaMA: open and efficient foundation language models. arXiv preprint arXiv:2302.13971 (2023)
52. Vaswani, A., et al.: Attention is all you need. Adv. Neural Inf. Process. Syst. **30** (2017)
53. Wang, J., et al.: A unified framework for guiding generative AI with wireless perception in resource constrained mobile edge networks. IEEE Trans. Mob. Comput. (2024)

54. Wang, X., et al.: Real-ESRGAN: training real-world blind super-resolution with pure synthetic data. In: Proceedings of the IEEE/CVF Conference on Computer Vision, pp. 1905–1914 (2021)
55. Wang, Z., et al.: MNN-LLM: a generic inference engine for fast large language model deployment on mobile devices. In: Proceedings of the 6th ACM International Conference on Multimedia in Asia Workshops, pp. 1–7 (2024)
56. Wei, J., et al.: Emergent abilities of large language models. arXiv preprint arXiv:2206.07682 (2022)
57. Wen, J., et al.: Diffusion model-based incentive mechanism with prospect theory for edge AIGC services in 6G IoT. In: IEEE Internet of Things J. (2024)
58. Wu, G., et al.: Diffusion model-based metaverse rendering in UAV-enabled edge networks with dual connectivity. In: 2024 IEEE Wireless Communications and Networking Conference (WCNC), pp. 1–6. IEEE (2024)
59. Wu, J., et al.: QoE-aware offloading and resource allocation for MEC-empowered AIGC services. IEEE Trans. Mob. Comput. (2025)
60. Wu, Y., et al.: Latency-aware resource allocation for mobile edge generation and computing via deep reinforcement learning. IEEE Netw. Lett. (2024)
61. Xi, Z., et al.: The rise and potential of large language model based agents: a survey. Sci. China Inf. Sci. **68**(2), 121101 (2025)
62. Xie, G., et al.: GAI-IoV: Bridging generative AI and vehicular networks for ubiquitous edge intelligence. IEEE Trans. Wireless Commun. (2024)
63. Xu, M., et al.: Cached model-as-a-resource: provisioning large language model agents for edge intelligence in space-air-ground integrated networks. arXiv preprint arXiv:2403.05826 (2024)
64. Xu, M., et al.: Joint foundation model caching and inference of generative AI services for edge intelligence. In: GLOBECOM 2023-2023 IEEE Global Communications Conference, pp. 3548–3553. IEEE (2023)
65. Yan, G., et al.: Edge intelligence for internet of vehicles: a survey. IEEE Trans. Consum. Electron. (2024)
66. Yang, A., et al.: Qwen3 technical report. arXiv preprint arXiv:2505.09388 (2025)
67. Yang, P., et al.: Dynamic mobile edge caching with location differentiation. In: GLOBECOM 2017-2017 IEEE Global Communications Conference, pp. 1–6. IEEE (2017)
68. Yang, T., et al.: Pixel-aware stable diffusion for realistic image super-resolution and personalized stylization. In: European Conference on Computer Vision, pp. 74–91. Springer (2024)
69. Zhang, J., et al.: Cloud-edge-terminal collaborative AIGC for autonomous driving. IEEE Wirel. Commun. **31**(4), 40–47 (2024)
70. Zhang, Y., et al.: Mobile generative AI: opportunities and challenges. IEEE Wirel. Commun. **31**(4), 58–64 (2024)
71. Zhang, Z., et al.: Diffusion-based reinforcement learning for cooperative offloading and resource allocation in multi-UAV assisted edge-enabled metaverse. IEEE Trans. Veh. Technol. (2025)
72. Zhao, A., et al.: Expel: LLM agents are experiential learners. In: Proceedings of the AAAI Conference on Artificial Intelligence, vol. 38, no. 17, pp. 19632–19642 (2024)

Large Language Models-Driven Personalized Adaptation Framework for Intelligent Agents

Zhelin Xu, Congle Fu, Nan Sun, Honglan Huang[✉], Bing He, and Xianyang Zhang

Rocket Force University of Engineering, Xi'an 710049, Shanxi, China
huanghonglan17@alumni.nudt.edu.cn

Abstract. In response to the challenge that the current intelligent wargaming system exhibits insufficient human-machine adaptability and struggles to meet diverse tactical requirements, this paper presents a personalized adaptation framework for agents underpinned by the Large Language Model (LLM). The framework adheres to a "cloud-based development and design combined with end-side autonomous training" paradigm. Specifically, the development team formulates the fundamental model and the structured prompt engineering module in the cloud environment. Subsequently, users leverage natural language interaction to convert their personalized tactical specifications into executable reward functions, which enables the autonomous training of agents. This study devises two representative preference-based requirements and conducts end-side experiments. The experimental findings reveal that, in comparison with the sparse feedback rewards in the traditional setting, the reward functions generated by LLM can substantially enhance the training efficiency and tactical performance of agents. Simultaneously, through the structured prompting design and the interactive optimization mechanism, the framework effectively mitigates issues such as the command misinterpretation of LLM during wargaming. Consequently, it offers a novel solution for addressing the personalized adaptation problem in intelligent wargaming systems.

Keywords: Wargaming · Reinforcement-Learning · Large Language Model · Preference-Based Requirements · Sparse Rewards

1 Introduction

As wargaming continues to advance in intelligence, human-computer collaboration has emerged as a pivotal developmental trajectory [1]. Effective cooperation between human commanders and AI agents serves a dual purpose: it not only mitigates the computational and multitasking constraints inherent to human operators but also preserves the distinctive strengths of human decision-makers, such as strategic acumen and the ability to make ad-hoc decisions in dynamic scenarios.

During the deployment of agents in wargaming environments, a central conflict has become increasingly evident: the adaptability of these agents. In an ideal scenario, human-machine collaboration requires AI agents to adaptively modify their behavior based on diverse user command patterns, tactical preferences, and task requirements.

However, in practice, most wargaming users lack the technical proficiency to independently develop or fine-tune agents. At the same time, professional AI development teams face significant challenges in providing customized support for a large number of users. From the users' perspective, standardized agents often fail to align with their individual tactical routines, thereby reducing the fluidity and efficacy of human-machine collaboration. For developers, the prohibitive cost of training specialized models for each user makes large-scale deployment unfeasible. This predicament ultimately restricts the widespread adoption of intelligent wargaming systems, hindering the full realization of their potential in the field.

The evolution of large language model (LLM) has opened a new technical pathway for the widespread deployment of customized wargaming agents [2]. Leveraging their strengths in natural language processing and code generation, LLM have effectively lowered the technical barriers for users in customizing and fine-tuning agents. However, numerous challenges persist in real-world implementation. First, the inherently complex program architecture of wargaming systems exceeds the generative capacity of current models. Second, a significant semantic gap exists when converting natural language instructions into executable symbolic representations [3]. Third, the reliability of LLM-generated outputs is compromised by the well-known issue of model hallucinations, and currently, no efficient automated verification mechanisms are available. Collectively, these technical hurdles severely hinder the practical effectiveness of applying LLMs in the wargaming domain.

To address these challenges, this paper proposes an LLM-based personalized adaptation framework for agents. The framework enables users to define their tactical preferences in a straightforward manner via a natural-language interactive interface. The LLM then automatically generates corresponding reward functions using user-engineered prompts. These functions are integrated with the Proximal Policy Optimization (PPO) algorithm to train agents that meet user-specified requirements. Furthermore, this study emphasizes the importance of validating and optimizing LLM outputs, effectively addressing the aforementioned issues and ensuring the accuracy and reliability of information processing within the framework. This approach not only enhances agent adaptability but also improves the overall performance and practicality of intelligent wargaming systems.

2 Related Work

2.1 Advancements in Intelligent Wargame Deduction

Research on the intelligence of wargames has undergone a transformative evolution from rule-driven to data-driven paradigms. In the early stages, investigations predominantly relied on expert systems [4], wherein intelligent decision-making was achieved through predefined rule sets. As reinforcement learning techniques advanced, contemporary wargame systems have increasingly embraced deep reinforcement learning approaches [5], demonstrating superhuman performance in intricate tactical decision-making scenarios. The AlphaDogfight test conducted by the Defense Advanced Research Projects Agency (DARPA) stands as a seinal milestone, exemplifying the practical utility of artificial intelligence in air combat for the first time [6].

In the domain of agent decision optimization, Zhang Qi introduced an innovative integration of rule-setting and autonomous learning. By embedding reinforcement learning into behavior trees and employing MAXQ hierarchical reinforcement learning, Zhang enabled online node optimization [7]. To address the limited adaptability of rule-based agents, Li Chen et al. applied the Reward Shaping technique to optimize the reward function, establishing a connection between the reward mechanism and target control points, thereby effectively alleviating the issue of sparse rewards stemming from expansive state spaces [8]. Zhang Zhen et al., focusing on the convergence challenges in reinforcement learning training, incorporated expert knowledge via supervised learning [9]. By combining the Proximal Policy Optimization (PPO) algorithm with a multi-reward mechanism, they significantly enhanced both the training efficiency and win rate of agents. Notwithstanding these numerous breakthroughs, existing intelligent wargame systems still commonly encounter adaptability bottlenecks, rendering them ill-equipped to satisfy the personalized requirements of diverse users. Consequently, there is an urgent need for further exploration in the areas of dynamic scene adaptation and user-customized interaction.

2.2 LLM in Decision-Support Applications

In recent years, LLMs have exhibited remarkable potential within the realm of military decision support. The GPT-4 Technical Report by OpenAI elucidates that LLMs facilitate data-driven decision-making by augmenting situational awareness, enabling effective comprehension of tactical commands, and generating rational operational plans [10].

The rapid progression of artificial intelligence has redefined the contours of military collaborative intelligence. Generative AI-based models, in particular, have the capacity to simulate human cognitive processes and thereby assist in generating multi-dimensional analysis outcomes and actionable recommendations within military decision-making contexts, fostering profound collaboration between commanders and intelligent systems [11].

LLMs are adept at processing vast volumes of multi-source data, including textual information, images, and sensor readings. Through the identification of latent patterns and extraction of critical insights, they enhance the comprehensiveness of battlefield situation awareness [12]. Additionally, by generating human-like text, LLMs promote transparent and efficient cross-hierarchical and cross-linguistic communication. They enable rapid data analysis, facilitate the generation of action plans, and thus expedite the decision-making process. Moreover, these models can construct highly realistic conflict scenarios, which prove invaluable for combat planning and training purposes.

Notwithstanding these advantages, extant research has also revealed that LLMs encounter significant challenges when applied to wargaming, such as biases in instruction comprehension and issues related to decision interpretability [13].

2.3 Reinforcement-Learning-Enabled Personalized Reward Generation

The generation of personalized reward functions constitutes a pivotal technology for realizing agent adaptation. The Proximal Policy Optimization (PPO) algorithm, introduced by Schulman et al., established an efficient framework for policy optimization

[14]. In subsequent academic discourse, a multitude of approaches to reward generation have been investigated, encompassing inverse reinforcement learning grounded in demonstration learning [15] and preference-based reward learning [16].

Typically, evaluation is conducted by computing the true reward of the optimized policy based on the learned reward [17]. Nevertheless, this approach is vulnerable to interference stemming from policy learning failures. Moreover, simplistic measures within the parameter space prove inadequate for accurately reflecting the functional disparities among reward functions [18]. This is attributable to the fact that transformations of reward functions frequently yield identical optimal policies and trajectory orderings. Recent research from Tsinghua University has highlighted that integrating the semantic understanding capabilities of large language models can substantially enhance the quality of reward function generation [19]. This integration effectively addresses the challenge of personalized adaptation in intelligent wargaming systems.

2.4 Research Gaps

Existing intelligent wargaming systems still suffer from several shortcomings. First, these systems lack the capability to adapt to diverse tactical preferences, often failing to align with commanders' personalized decision-making styles. Second, although LLMs demonstrate potential in human-computer interaction, the prevailing reward-generation mechanisms in such models exhibit demand comprehension biases—specifically, their insufficient grasp of professional military terminology and limited ability to resolve semantic ambiguities in complex battlefield scenarios. These deficiencies hinder the advancement of intelligent wargaming systems.

3 Methodology

In this section, the motivation and implementation of the proposed framework for agent-level personalized adaptation grounded in LLM are elucidated [20]. This framework synergistically integrates LLM with reinforcement-learning-based rewards to surmount the challenges associated with customized learning, as illustrated in Fig. 1. The discussion commences with an overview of the framework's overarching architecture, followed by an in-depth examination of its constituent core components.

3.1 Motivation

In the context of wargaming, assessing task completion efficacy typically requires a multi-dimensional evaluation framework [21]. However, contemporary AI agents trained solely on outcome-based reward mechanisms often overlook critical execution factors, such as resource consumption and intelligence collection, rendering them inadequate to meet users' diverse requirements.

To address this limitation, we propose the concept of User Self-Trained Preference Agents, supported by a "cloud-based development and terminal-side autonomous training" paradigm. Under this framework: The agent development team designs the core architecture and algorithms. End-users define their specific operational needs and leverage LLMs to autonomously train the agents, thereby tailoring their performance to individual decision-making preferences.

3.2 Framework

As depicted in Fig. 1, the framework is structured around three fundamental hierarchical components:

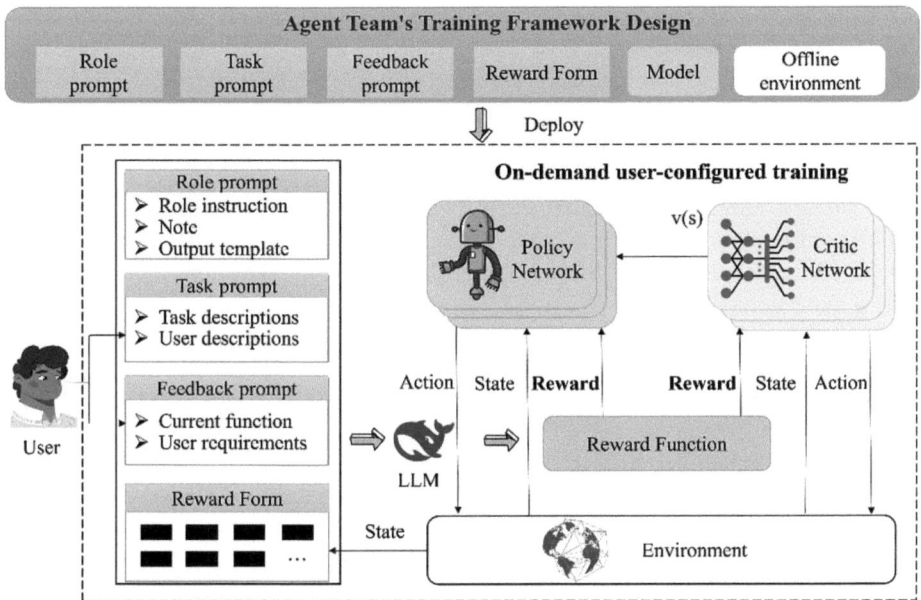

Fig. 1 LLM based agent personalized adaptation framework

Cloud-Based Foundation Layer. The agent foundation, meticulously constructed by the development team, comprises a policy model, a repository of pre-defined latent reward functions, and an LLM-based structured cue board system

Interactive Conversion Layer. This layer serves as the pivotal stratum for implementing user-specific personalized design, facilitating the precise mapping of user requirements from natural language to reward functions. The execution of this process predominantly hinges upon the LLM-based structured cue plate system [20], which comprises a total of four cue plates

We initiate the establishment of the Role Prompting segment, which is meticulously structured into three distinct components: Role Instruction, Note, and Output Template. The content within these modules is developed and finalized by the development team, thereby obviating the need for user modification. The Role Instruction component serves to unambiguously delineate the roles and responsibilities of LLM, thereby preventing any divergence from the core task at hand. The Note section provides a set of guidelines and constraints that govern the LLM's code-generation process. By adhering to these stipulations, the likelihood of generating hallucinatory or invalid outputs is significantly mitigated. The Output Template, on the other hand, prescribes a well-defined structured format for the agent's output, ensuring that the generated results align with the anticipated specifications. Through the strategic design of the Role Prompting mechanism, the decision-aiding function of the LLM is clearly demarcated, its response style is standardized, and potential risks associated with code generation are minimized. These elements collectively lay a robust foundation for the stable and reliable operation of the agent.

The Task Prompt component serves the purpose of infusing scene-environment knowledge into the system. It is primarily composed of two integral parts: Task Descriptions and User Descriptions. The Task Descriptions are meticulously crafted to elucidate the specific wargaming tasks at hand. This encompasses detailed specifications such as the task objectives, operational areas, participating action units, and scoring criteria. By providing such comprehensive information, the Task Descriptions act as a guiding framework for the output of LLM, enabling it to generate responses that are highly congruent with the requirements of the actual scenario. Conversely, the User Descriptions are designed to capture and convey the personalized requirements of the users. They serve as a directive for the LLM to perform demand-driven reasoning, thereby facilitating a more tailored alignment with the users' expectations. The content of the User Descriptions is contributed directly by the users, ensuring a high degree of customization and relevance.

The Feedback Prompt serves as a crucial conduit for users to provide feedback and effect adjustments to the training efficacy. It predominantly comprises two essential elements: Current Function and User Requirements. The Current Function meticulously delineates the reward function employed in the preceding training iteration, facilitating iterative refinement of the reward mechanism. Conversely, the User Requirements are contributed by the users themselves, encapsulating their evaluations and insights regarding the training outcomes. By integrating the Current Function and User Requirements, the Feedback Prompt emerges as the linchpin connecting the model's output with user expectations. This integration offers pivotal support for the iterative optimization of the agent, ensuring that its performance progressively aligns with the desired objectives.

The latent reward repository, painstakingly assembled by the agent development team, is capable of encapsulating the latent characteristics inherent in the paired-inference process. This repository enables the multi-dimensional assessment of an agent's behavior. The reward forms subsection within this repository comprehensively details the name, operational rules, and application scenarios of each function. The function names and usage guidelines primarily serve to elucidate the proper invocation protocols, while the rules explicitly define the scoring logics associated with each function. These rules play a foundational role in supporting the reasoning processes of LLM, providing it with a structured basis for generating informed outputs.

The Edge-Side Training Layer. Personalized reward functions, which are generated via natural-language-based interaction by the user, are employed for the training of agents

$$R = \sum_{i=1}^{n} \omega_i f_i(s, a) \qquad (1)$$

In the formula, ω_i is dynamically generated by LLM according to user requirements, and f_i is developed by the development team.

In the initial generation stage of the Reward Function, LLM needs to schedule the existing n Latent reward functions f_i in the agent to generate the reward function:

$$R_{init} \leftarrow M(role, task, f_1, f_2, \cdots, f_n) \qquad (2)$$

In the stage of reward function optimization generation, LLM adjusts the generation based on the generated reward function according to the user evaluation:

$$R_{final} \leftarrow M(role, task, feedback, f_1, f_2, \cdots, f_n) \tag{3}$$

The proposed framework integrates the semantic comprehension capabilities of LLM with pre-defined reward elements to generate hybrid rewards. This integration not only guarantees the stability and reliability of the algorithmic model but also endows ordinary users with a substantial degree of customization autonomy.

4 Experiments

We conducted experiments on the "MiaoSuan" wargaming platform to validate the feasibility and effectiveness of the LLM-based personalized agent adaptation framework.

4.1 Experimental Setups

We constructed a cooperative reconnaissance scenario on the "MiaoSuan" wargaming platform, where multiple heterogeneous units work together to rapidly locate hidden enemy units within a designated area. In this scenario, enemy units remain stationary in concealed positions but may ambush searching units when opportunities arise. Different types of reconnaissance units possess varied mobility, combat capabilities, and detection ranges. The reward mechanism is solely based on the number of detected enemy units and the time taken to locate them.

Using PPO as our baseline algorithm, we conducted comparative experiments with three reward schemes: environmental rewards, initial LLM-generated rewards, and optimized LLM-generated rewards after interactive feedback, thereby validating the framework's feasibility. Additionally, we designed two user preference requirements: (1) maximizing search coverage during reconnaissance operations, and (2) prioritizing unit survival rates. By comparing training results against standard reward conditions, we demonstrated the framework's effectiveness in accommodating personalized tactical needs.

4.2 Model Construction

The reinforcement learning agent is constructed using the PPO algorithm, with the following definitions for key RL components:

State Design. The state includes information about both friendly and enemy entities, as well as environmental data. Entity information (for both sides) comprises unit type, health, position and mobility status. Due to the fog of war, enemy entity states may be partially observable, with missing values are zero-padded. All data is normalized for dimensionality removal. Environmental information primarily consists of terrain features around friendly entities

Action Design. Decision-making is implemented by selecting movement target points for entities. Specifically, the environment is discretized using a hexagonal grid system, where each entity's movement target is set as the center point of an adjacent hexagon

Reward Design. The environmental reward consists of two components: a sparse detection reward that grants a fixed value upon identifying each enemy entity, and a mission completion bonus awarded when all entities are detected. The initial reward generated by LLM is automatically produced by the LLM based on the user's preliminary requirements. Subsequently, the optimized reward generated by LLM is generated through an iterative refinement process that incorporates both the reward function from the previous training round and the user's evaluative feedback on the corresponding results

Network Architecture. The Actor and Critic networks in our framework share a nearly identical architecture. The state information first passes through multiple normalization layers and convolutional layers for feature compression and extraction, followed by fully-connected layers for final output. Specifically, the Actor network generates probability distributions over possible actions, while the Critic network produces a 1-dimensional state value estimate

Hyperparameters. Complete experimental configurations are detailed in Tab. 1

Tab. 1 Experimental hyperparameter setting

Hyperparameters	Hyperparameter settings	Hyperparameters	Hyperparameter settings
Optimizer	Adam	GAE parameter	0.95
Discount factor	0.99	Policy epochs	8
Clip epsilon	0.2	Mini-batch size	2048
Learning rate	1e-4	Replay buffer size	3000

Tab. 2 Reconnaissance coverage rate statistics

Coverage rate	Environmental Reward	LLM-Reward	Preference reward
%	69	73	78

4.3 The Effectiveness of Algorithm

To validate the effectiveness of the framework, we conducted comparative experiments using three distinct reward schemes. As shown in Fig. 2, the three reward mechanisms demonstrated significant differences in agent training performance.

The environmental reward group exhibited notable instability during early training stages, with substantial fluctuations in reward values, and ultimately achieved significantly lower average rewards than the other two groups. These results confirm the inherent sparsity issue in native environmental reward signals.

While the initial LLM-generated reward group showed a stable upward trend, its convergence speed remained relatively slow during the first 1500 episodes. In contrast, the interactively optimized LLM reward group delivered the best performance: it surpassed the baseline significantly by episode 1000 and maintained highly stable training curves.

These findings demonstrate that the interactive optimization mechanism not only preserves the reliability of environmental rewards but also effectively improves convergence speed while addressing the generation bias in initial LLM rewards.

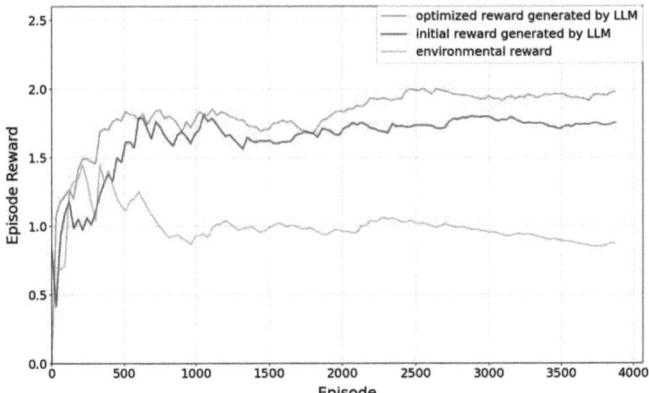

Fig. 2 Agent training performance under three reward schemes

4.4 The Feasibility of Algorithm

The comparative experiments with two typical tactical preferences validated that the LLM-based reward generation mechanism can effectively guide agents to develop differentiated strategies. Under the "maximize observation range" preference, the reward function generated by the LLM incorporated additional scores for exploring new areas, prompting the agents to adopt more proactive mobility strategies and successfully increasing the reconnaissance area coverage by 5 %. Training results showed good stability in the reward curve, with final performance reaching over 90 % of the interactively optimized LLM reward group. This demonstrates the LLM's capability to accurately comprehend and achieve the tactical intent of expanding reconnaissance scope Fig. 3.

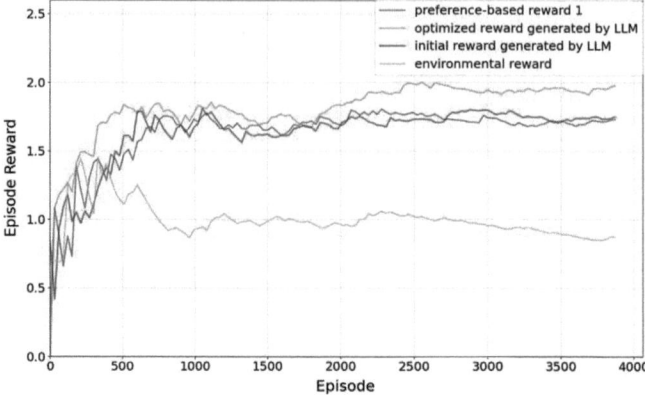

Fig. 3 Agent training performance under "maximize observation range" preference

For the "survival rate priority" preference, the hybrid reward function constructed by the LLM introduced penalty terms for receiving attacks and entering high-risk zones. During training, the agents exhibited strategic differentiation: high-survivability units primarily undertook enemy localization tasks, while low-survivability units focused on exploring new areas. This division of labor increased the average unit survival rate to 92.5 %, representing a 19 % improvement. Although the training convergence speed slightly decreased, the agents' demonstrated cautious path-planning capability fully verified the framework's effectiveness in balancing survival rates with mission performance. The experimental results indicate that through structured preference descriptions, the LLM can generate reward functions that align with user preference requirements Figs. 4 and 5.

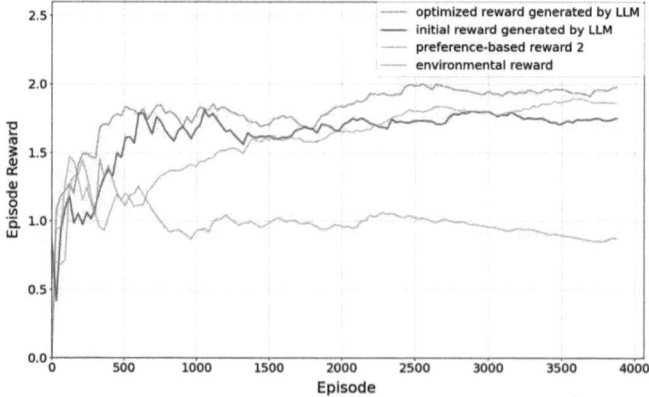

Fig. 4 Agent training performance under "survival rate priority" preference

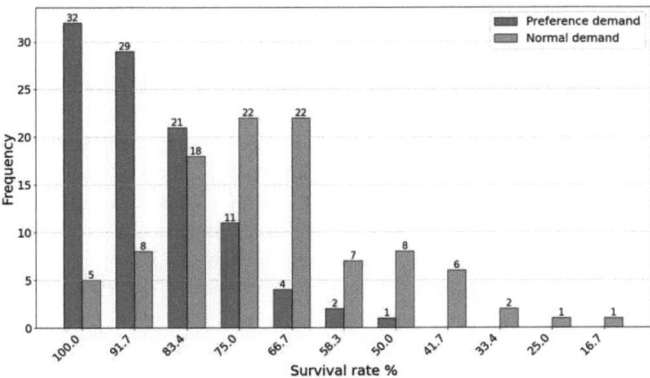

Fig. 5 Frequency distribution of unit survival rates: preference denmand vs normal demand

5 Conclusion

This paper addresses the challenge of personalized adaptation in intelligent wargaming systems by proposing a novel LLM-based personalized agent adaptation framework. The framework employs an innovative "cloud-based development and end-side autonomous training" paradigm that effectively resolves the precise mapping from wargaming-specific terminology to reward functions. It accurately transforms semantic expressions of user preferences into executable mathematical reward functions. Compared to sparse environmental feedback rewards, the LLM-generated rewards significantly enhance agent training efficiency and lower the skill threshold for human-machine collaborative wargaming. This breakthrough enables mass users to autonomously train agents that align with their individual tactical styles.

6 Discussions

The current experimental scenarios in this study are relatively simple, and the framework's adaptability to the extremely complex tactical environments in wargaming requires further exploration and improvement. Additionally, the on-device training mode demands significant computational resources, which somewhat limits the framework's broader application. Future research will focus on the following directions: (1) developing lightweight training algorithms to substantially reduce on-device computational resource consumption. (2) deeply integrating the LLM capability to automatically generate reward function code into the framework to enhance system automation. and (3) exploring a hybrid optimization model that combines automatic iterative optimization of reward functions with manual intervention to achieve an organic integration of algorithmic efficiency and human expertise.

Acknowledgments. A third level heading in 9-point font size at the end of the paper is used for general acknowledgments, for example: This study was funded by X (grant number Y).

Disclosure of Interests. We are deeply grateful to Professor Weijie Kang for his theoretical guidance and to Professor Peifeng Su for his expertise in experimental design, both of which were instrumental to this research.

References

1. Hu, X., Qi, D.: Intelligent wargaming system: change needed by next generation need to be changed. J. Syst. Simulation **33**(9), 1997–2009 (2021)
2. Yin, Q., Zhao, M., Ni, W., Zhang, J.-G., Huang, K.-Q.: Intelligent decision making technology and challenge of wargame. Acta Automatica Sinica **49**(5), 913–928 (2023)
3. Cao, Y., Zhao, H., Cheng, Y., Shu, T., Chen, Y., Liu, G., Liang, G., Zhao, J., Yan, J., Li, Y.: Survey on large language model-enhanced reinforcement learning: concept, taxonomy, and methods. IEEE Trans. Neural Netw. Learning Syst. **36**(6), 9737–9757 (2025)
4. Perla, P.P.: The Art of Wargaming: A Guide for Professionals and Hobbyists. Naval Institute Press, Annapolis (1990)
5. Silver, D., Hubert, T., Schrittwieser, J., Antonoglou, I., Lai, M., Guez, A., Lanctot, M., Sifre, L., Kumaran, D., Graepel, T., Lillicrap, T., Simonyan, K., Hassabis, D.: A general reinforcement learning algorithm that masters chess, shogi, and Go through self-play. Science **362**(6419), 1140–1144 (2018)
6. Johns Hopkins APL: AI Bests Human Fighter Pilot in AlphaDogfight Trial at Johns Hopkins APL (2025). https://www.jhuapl.edu/news/news-releases/200828-AI-bests-human-fighter-pilot-in-AlphaDogfight-trial-at-APL. Accessed 2025/6/13
7. Zhang, Q., Sun, L., Jiao, P., Yin, Q.: Combining Behavior Trees with MAXQ Learning to Facilitate CGFs Behavior Modeling. In: Proceedings of the 2017 4th International Conference on Systems and Informatics (ICSAI), pp. 525–531. IEEE, Hangzhou (2017).
8. Li, C., Huang, Y., Zhang, Y., Chen, T.: Multi-agent decision-making method based on Actor-Critic framework and its application in wargame. Syst. Eng. Electron. **43**(3), 755–762 (2021)
9. Zhang, Z., Huang, Y., Zhang, Y., Chen, T.: Combat entity game algorithm based on proximal policy optimization. J. Nanjing Univ. Sci. Technol. **45**(1), 77–83 (2021)
10. OpenAI, et al.: GPT-4 Technical Report. https://arxiv.org/abs/2303.08774. Accessed 2025/06/20
11. Nah, F.F.-H., Zheng, R., Cai, J., Siau, K., Chen, L.: Generative AI and ChatGPT: Applications, challenges, and AI-human collaboration. J. Inf. Technol. Case Appl. Res. **25**(3), 277–304 (2023)
12. Marasco, E., Bourlai, T.: Enhancing trust in Large Language Models for streamlined decision-making in military operations. Image Vision Comput. **158**, 105489 (2025)
13. Chen, Y., Chu, S.: Large Language Models in Wargaming: Methodology, Application, and Robustness. In: 2024 IEEE/CVF Conference on Computer Vision and Pattern Recognition Workshops (CVPRW), pp. 2894–2903 (2024)
14. Schulman, J., Wolski, F., Dhariwal, P., Radford, A., Klimov, O.: Proximal Policy Optimization Algorithms. https://arxiv.org/abs/1707.06347. Accessed 2025/06/20
15. Ho, J., Ermon, S.: Generative Adversarial Imitation Learning. In: Proceedings of the 30th International Conference on Neural Information Processing Systems (NIPS 2016), pp. 4572–4580. Curran Associates, Red Hook (2016)
16. Christiano, P.F., Leike, J., Brown, T., Martic, M., Legg, S., Amodei, D.: Deep Reinforcement Learning from Human Preferences. In: Advances in Neural Information Processing Systems 30 (NIPS 2017), pp. 4299–4307 (2017)
17. Wang, Y., Sun, Z., Zhang, J., Xian, Z., Biyik, E., Held, D., Erickson, Z.: RL-VLM-F: Reinforcement Learning from Vision Language Foundation Model Feedback. In: Proceedings of the 41st International Conference on Machine Learning (ICML 2024), vol. 2112, pp. 1–18. JMLR.org, Vienna (2024)
18. Jenner, E., Skalse, J.M.V., Gleave, A.: A General Framework for Reward Function Distances. In: NeurIPS ML Safety Workshop (2022)

19. Qu, Y., Jiang, Y., Wang, B., Mao, Y., Wang, C., Liu, C., Ji, X.: Latent reward: LLM-empowered credit assignment in episodic reinforcement learning. Proc. AAAI Conf. Artif. Intell. **39**(19), 20095–20103 (2025)
20. Shen, L., Yang, Q., Zheng, Y., Li, M.: AutoIOT: LLM-Driven Automated Natural Language Programming for AIoT Applications. In: Proceedings of the 31st Annual International Conference on Mobile Computing and Networking (MobiCom 2025), pp. 1–15. ACM, New York (2025)
21. Wei, Y., Shan, X., Li, J.: LERO: LLM-driven Evolutionary framework with Hybrid Rewards and Enhanced Observation for Multi-Agent Reinforcement Learning. https://arxiv.org/abs/2503.21807. Accessed 2025/06/20

Learning-Based Taxi Selection for Opportunistic Street Parking Sensing

Yongbin Huang[1,2,3], Yuezhong Wu[1,2,3](✉), Zhiyong Yu[1,2,3], and Fangwan Huang[1,2,3]

[1] College of Computer and Data Science, Fuzhou University, Fuzhou, China
{yuezhong.wu,yuzhiyong,hfw}@fzu.edu.cn
[2] Fujian Key Laboratory of Network Computing and Intelligent Information Processing, Fuzhou, China
[3] Engineering Research Center of Big Data Intelligence, Ministry of Education, Fuzhou, China

Abstract. With the rapid increase in urban parking demand, smart parking technologies have emerged as a key solution for improving parking experience. Traditional approaches that rely on fixed sensors to monitor parking availability often incur high deployment and maintenance costs. Mobile Crowdsensing (MCS) offers a cost-effective alternative by utilizing taxis to dynamically collect parking space information as they traverse city streets. However, intelligently selecting which taxis to equip with sensing devices is critical for ensuring high-quality data. The inherent uncertainty in taxi trajectories poses significant challenges to traditional participant selection strategies, leading to performance degradation and instability in real-world deployments. To address this issue, we propose E2RL, a taxi selection method based on embedding representation learning and reinforcement learning. E2RL captures the behavioral patterns of taxis from historical trajectory data and learns an optimal taxi selection strategy through reinforcement learning. Extensive experiments conducted on the San Francisco taxi trajectory dataset and a real-world parking availability dataset demonstrate that our method significantly outperforms several baselines in terms of sensing quality and robustness.

Keywords: smart parking · taxi selection · mobile crowdsensing (MCS) · parking space sensing · embedding representation learning · reinforcement learning(RL)

1 Introduction

With rapid urbanization, parking scarcity has become a common issue in major cities. Studies show that up to 30% of urban congestion is caused by drivers searching for parking spaces, leading to traffic inefficiency, increased emissions, and pollution [1]. Intelligent parking systems offer real-time guidance to reduce search time and ease traffic congestion. They also support environmental sustainability and improve urban livability. As a result, the smart parking market is expected to grow at a CAGR of 17.94% in the coming years [2].

Acquiring real-time parking availability information is essential for the effective implementation of intelligent parking systems. Traditional approaches typically rely on fixed sensors installed at individual parking spots to monitor occupancy. While these methods offer high accuracy, they also involve substantial deployment and maintenance costs [3, 4]. MCS offers robust sensing and communication capabilities [5], and has been widely adopted across various domains, including environmental monitoring [6], public safety [7], and intelligent transportation systems [8]. It serves as a flexible and cost-effective alternative to traditional fixed-sensor deployment schemes.

Taxis equipped with sensing devices—such as side-scan ultrasonic sensors [9, 10] or windshield-mounted cameras [11]—can serve as mobile agents to detect real-time parking occupancy while driving through city streets. This approach effectively leverages existing mobile resources, thereby significantly reducing infrastructure costs. Bock et al. [12] conducted a simulation study on random taxi selection to detect on-street parking availability but leaving optimal taxi selection strategies unexplored.

In the domain of Mobile Crowdsensing (MCS), participant selection is critical for the sensing quality and cost, which can be categorized into two types: online selection and offline selection [14]. Offline selection selects participants in advance before the sensing task starts, while online selection makes decisions during the task execution based on real-time participant arrivals. When selecting participants to maximize sensing quality under budget constraints, two typical scenarios arise. In the first, participants (e.g., private cars or buses) have stable daily routes, allowing effective selection based on expected coverage. In the second, participants like taxis show high trajectory variability, making historical-data-based selection unreliable due to future movement uncertainty, often resulting in suboptimal and unstable sensing performance. Consequently, selecting taxis in the offline scenario remains a major challenge.

In urban transportation systems, large-scale taxi trajectory data reveals rich mobility patterns. Although individual taxis differ in driving habits and service areas, they often exhibit shared group-level behaviors. This insight allows taxi selection systems to move beyond traditional approaches that rely solely on individual historical trajectories. Therefore, We propose a method named E2RL incorporating trajectory data from city-wide auxiliary taxis to effectively capture the underlying movement patterns of the candidate taxi fleet as a whole, thereby mitigating the challenges posed by trajectory uncertainty.

The main contributions of this paper are summarized as follows:

1. To the best of our knowledge, this is the first work to formulate the participant selection problem in the context of selecting taxis for sensing street parking availability.
2. We propose E2RL, a novel taxi selection framework combining embedding representation learning and reinforcement learning. It first learns embeddings from auxiliary taxi trajectories, then trains a classifier, and finally uses reinforcement learning to guide candidate taxi selection with the assistance of auxiliary taxi information. To our knowledge, E2RL is the first method to leverage auxiliary taxis for taxi selection. By modeling stable mobility patterns from uncertain trajectories, E2RL improves the stability of sensing quality.
3. We evaluate our approach using real-world taxi trajectory data and roadside parking data from San Francisco. Experimental results demonstrate that our proposed E2RL method significantly outperforms baseline methods in terms of coverage quality, both in grid-based evaluation and real-world parking street scenarios.

2 Related Work

In this section, related work on participant selection and reinforcement learning is introduced.

2.1 Participant Selection

In Mobile Crowdsensing (MCS), data collection faces the dual challenge of minimizing cost while ensuring high sensing quality [13, 15–17]. Many studies [16–19] assume the cost is proportional to the number of selected participants and evaluate sensing quality via spatiotemporal coverage. The most direct way to improve quality is to recruit more participants, but this leads to higher energy and communication costs [20]. To address this, Wang et al. [21] proposed a greedy bus route selection method for real-time waterlogging monitoring. He et al. [22] combined current and future positions with greedy and genetic algorithms for vehicle-based participant selection. Some studies [23–25] used trajectory prediction to estimate users' future utility. Liu et al. [26] introduced an improved greedy algorithm(CQO), and Hu et al. [27] proposed RL-Recruiter to maximize relative coverage through reinforcement learning from scratch.

Overall, existing participant selection methods fall into two types. One predicts future mobility and selects participants accordingly. However, prediction errors may propagate and worsen the final selection, especially for taxis with highly uncertain and variable routes over longer periods. The other type selects participants directly based on historical trajectories, typical in offline schemes, but these methods struggle to handle the inherent uncertainty in taxi movements.

2.2 Reinforcement Learning

Reinforcement Learning (RL), an important branch of machine learning, involves a process in which an agent learns the optimal behavior strategy through trial-and-error interactions with a dynamic environment [28]. In recent years, RL has achieved remarkable success in multiple fields, including robotics [29], autonomous control systems [30], game strategies [31], and so on.

RL has a wide range of applications in decision-making problems such as dynamic programming and combinatorial optimization. In recent years, it has also started to be applied to the participant selection problem [27, 32]. However, existing methods usually make combinatorial decisions only based on the historical trajectories of candidate participants, and it is difficult to effectively deal with the uncertainties of participants' daily trajectories. Such uncertainties may have a significant impact on the future coverage quality, but existing methods still fall short in addressing this issue.

3 Problem Definition and System Framework

Assume that a smart parking platform aims to equip a subset of taxis with sensors in order to sense citywide parking spaces as broadly as possible during their daily operations. The problem description is as follows:

Let the set of all sensing areas be A = $\{a_1, a_2, \cdots, a_m\}$. Taxis owned by the platform are defined as candidate taxis, and the set of candidate taxis is denoted as Z = $\{z_1, z_2, \cdots, z_{|Z|}\}$. The time period related to the historical trajectories of candidate taxis is evenly divided into g segments, and the set of these segments is T = $\{t_1, t_2, \cdots, t_g\}$. Taxis not included in the candidate set are called auxiliary taxis, and the set of auxiliary taxis is U = $\{u_1, u_2, \cdots, u_{|U|}\}$. The time period related to the historical trajectories of auxiliary taxis is evenly divided into l segments, and the set of these segments is $T\prime$ = $\{t_1, t_2, \cdots, t_l\}$. Unlike selecting a group of participants in each time slot, we will directly select a set of taxis P = $\{p_1, p_2, \cdots, p_{|P|}\}$ ($P \subseteq Z$) from the set of candidate taxis to continuously complete sensing tasks in a future period of time. This future period of time is evenly divided into w segments, and the set of these segments is T^* = $\{t_1, t_2, \cdots, t_w\}$. The sensing-coverage matrix of the selected set of taxis P in the future period of time is:

$$V^*(P) = \begin{bmatrix} v_{1,1} & \cdots & v_{1,w} \\ \vdots & \ddots & \vdots \\ v_{m,1} & \cdots & v_{m,w} \end{bmatrix} \quad (1)$$

If there is a taxi in the selected set of taxis P passing through the sensing area i in the j-th time slot in the future, then $v_{i,j} = 1$; otherwise $v_{i,j} = 0$, where $i \in \{1,2,\cdots,m\}$, $j \in \{1,2,\cdots,w\}$.

The objective of problem is to select n taxis from the candidate taxis to maximize the relative coverage of the selected set of taxis in the future time period. The relative coverage is defined as:

$$C^*(P) = \frac{|V^*(P)|}{|V^*(Z)|} \quad (2)$$

where,

$$|V^*(P)| = \sum_{i=1}^{m} \sum_{j=1}^{w} v_{i,j} \quad (3)$$

Relative coverage measures how well the selected taxis cover sensing areas relative to all candidate taxis. This objective is equivalent to maximizing the average number of sensing areas covered by the selected set of taxis over a future period of time, that is:

$$\max \frac{|V^*(P)|}{w} \quad s.t. |P| = n \quad (4)$$

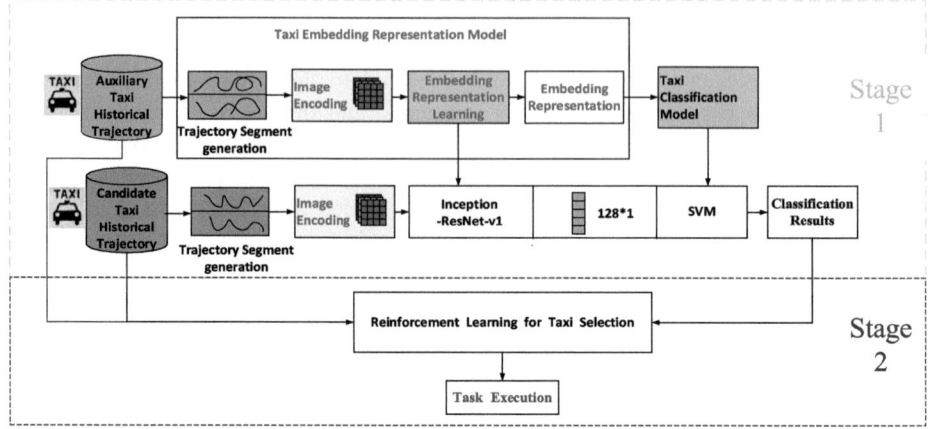

Fig. 1 The system framework of E2RL

4 Method

In this section, we detail the proposed E2RL method for taxi selection, combining embedding representation learning with reinforcement learning.

4.1 Overall Framework of the Proposed E2RL Method

The overall architecture of the E2RL system, illustrated in Fig. 1, is structured into two stages. In the first stage, the system trains a taxi embedding representation model using auxiliary taxis' historical trajectories, followed by a taxi classification model to categorize mobility patterns. Each candidate taxi is then processed through these two trained models to identify the auxiliary taxi with the most similar behavioral pattern. In the second stage, the system utilizes the outputs from the first stage to guide taxi selection through reinforcement learning. The reward function is designed to incorporate both the coverage of the selected candidate taxis and that of their corresponding auxiliary taxis. This dual-coverage objective captures both individual and group-level mobility behaviors, enhancing the stability and effectiveness of the selection process.

4.2 Taxi Embedding Representation Model

Inspired by advances in face recognition [33], we propose a taxi embedding representation learning approach using auxiliary taxis' historical trajectories. Each taxi's trajectory is segmented, similar to multiple facial images of a person. These segments are encoded into image-like representations that capture spatial-temporal features. A triplet loss is then used to learn embeddings. The complete pipeline is detailed as follows. **Trajectory segment generation.** To construct effective training samples from auxiliary taxis' historical data, we segment and concatenate their multi-day trajectories into two-day trajectory segments. For example, given a 10-day trajectory, two strategies can be used. The combinatorial splicing strategy combines any two days, generating $C_{10}^2 = 45$ segments with rich pattern diversity but potential spatiotemporal noise. The continuous splicing strategy uses only adjacent days, yielding 9 segments with better temporal consistency and lower complexity, though with limited diversity. To balance these trade-offs, we propose a hybrid strategy that integrates both adjacent and partially random combinations. This strategy maintains spatio-temporal continuity while enriching data diversity, thereby enhancing model robustness and generalization.

Trajectory-to-image encoding scheme. After generating the trajectory segments of taxis, each trajectory segment of every taxi can be encoded into a corresponding image. The time scope studied in this paper is from 6:00 to 24:00 every day. During the conversion from trajectory segments to images, the three primary color channels of red, green, and blue (rgb) in the image respectively map the coverage of the study area by the trajectory segments during the three time intervals of 6:00 - 12:00, 12:00 - 18:00, and 18:00 - 24:00. To achieve this goal, we implement a grid-based division of the study area to ensure that each pixel in the image can accurately correspond to a geographic grid, thereby establishing a direct mapping relationship between pixels and geographical space. When a taxi's trajectory segment traverses a particular geographic grid cell during a given time slot, we increment by 1 the corresponding pixel value in the rgb channel image associated with that time interval. This scheme intuitively reflects the spatial distribution characteristics of the trajectory segments during different time intervals.

To further clarify our encoding scheme, we provide an illustrative example. Suppose the study area consists of 16 sensing zones, and we have a taxi's trajectory segment data for one day (6:00 to 24:00), with each time slot set to 3 h. Figure 3 details the process of encoding this trajectory segment into image data. First, we initialize all pixel values in the RGB channels to 0. Then, based on the grids traversed by the taxi during each time slot, we increment the corresponding pixel values in the color channel image for the respective time interval. For instance, the time slot from 6:00 to 9:00 falls within the 6:00–12:00 interval, so the grids traversed during this slot will increment the corresponding pixel values in the red channel image. Similarly, data from other time slots are processed in the same manner.In the example, a time slot of 3 h is used for explaining the coding scheme in this paper. However, to more accurately capture a taxi's dwell time at specific locations and its repeated visits within a given interval, a finer time slot is required. Therefore, in our experiments, the time slot length in the encoding scheme is set to 20 min to better reflect mobility patterns.

Embedding representation learning for trajectory segments. Our method employs a deep convolutional neural network based on the inception architecture [34], specifically inception-v1, to learn embedding representations of trajectory segments. The embedding representation $f(u) \in R^d$ maps a trajectory segment u into a d-dimensional euclidean space. Since each taxi has its own unique movement pattern, we aim to ensure that the embedding representation of a specific trajectory segment u_i^a from a given taxi is as close as possible to the embeddings of all other trajectory segments u_i^p from the same taxi, while being as far as possible from the embedding of any trajectory segment u_i^n belonging to any other taxi. Here, u_i^a serves as the anchor taxi trajectory, u_i^p as the positive taxi trajectory, and u_i^n as the negative taxi trajectory. The pair u_i^a and u_i^p forms a positive taxi pair, and the pair u_i^a and u_i^n forms a negative taxi pair. Together, u_i^a, u_i^p and

u_i^n constitute a taxi triplet. As illustrated in Fig. 3, the taxi triplet loss function aims to learn a feature space where the distance between positive taxi pairs is minimized, and the distance between negative taxi pairs is maximized. The learned embeddings exhibit strong intra-class consistency and interclass discriminability, making this approach well-suited for learning taxi trajectory representations. Based on the above description, our objective can be summarized as follows:

$$||f(u_i^a) - f(u_i^p)||_2^2 + \delta < ||f(u_i^a) - f(u_i^n)||_2^2,$$

$$\forall (u_i^a, u_i^p, u_i^n) \in \Gamma \tag{5}$$

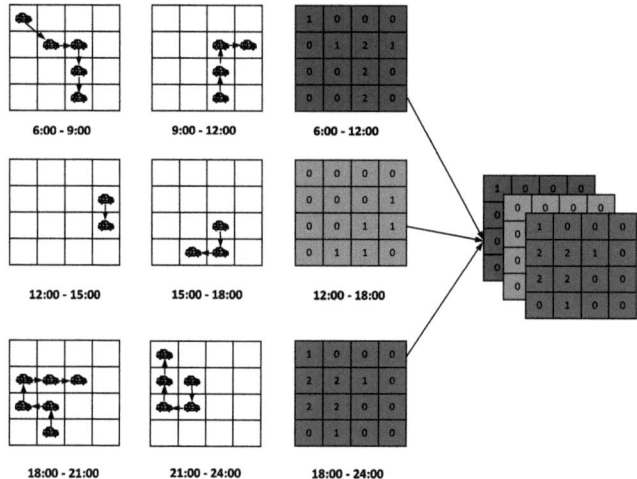

Fig. 2 Example of Encoding Scheme

Here, δ is a margin that is enforced between positive and negative taxi pairs. Γ represents the set of all possible taxi triplets in the training dataset, with a total count of N. The loss function to be minimized is:

$$L = \sum_i^N [||f(u_i^a) - f(u_i^p)||_2^2 + \delta - ||f(u_i^a) - f(u_i^n)||_2^2]_+ \tag{6}$$

where the operator $[x]_+ = \max(0, x)$.

Taxi triplets can be classified into three distinct categories based on their training characteristics. The "easy taxi triplets" are readily distinguishable but provide negligble contribution to model training, ultimately degrading training efficiency and compromising generalization performance; "hard taxi triplets" may introduce training instabil-

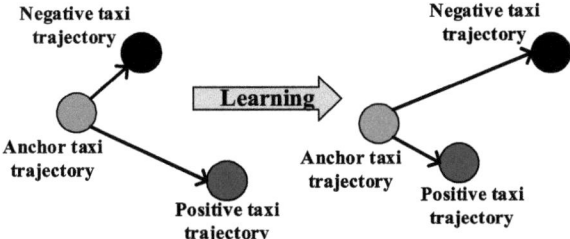

Fig. 3 Taxi Triplet Loss Learning

ity and increase the risk of overfitting; The "semi-hard taxi triplets" emerge as the optimal choice, as they effectively enhance model generalization while ensuring stable and efficient training convergence. Based on this analysis, our approach selectively employs "semi-hard taxi triplets" that satisfy specific criteria for model training. The "semi-hard taxi triplets" satisfy the following conditions:

$$\|f(u_i^a) - f(u_i^p)\|_2^2 < \|f(u_i^a) - f(u_i^n)\|_2^2 < \|f(u_i^a) - f(u_i^p)\|_2^2 + \delta \quad (7)$$

4.3 Taxi Classification Model

We utilized the Support Vector Machine (SVM) algorithm [35] to classify embedding representations of auxiliary taxi trajectory segments. Each trajectory segment was assigned a classification label corresponding to its respective taxi's unique identifier. By optimizing for maximal inter-class margin, this classification model effectively captures distinct driving pattern characteristics across different taxis.

After constructing the classification model, we used it to perform similarity matching on candidate taxis to identify the most similar auxiliary taxis, as follows: First, the trajectory segments of the candidate taxis are input into the pre-trained embedding representation model of auxiliary vehicles to generate their corresponding embedding representations. Then, these embedding representations are fed into the pre-trained classification model of auxiliary taxis to produce classification results, which are the IDs of the auxiliary vehicles.

4.4 Reinforcement Learning for Taxi Selection

We propose a reinforcement learning method for the final selection of taxis. We first define the state, action, and reward for our target problem as follows.

State: If each taxi's selection status were directly mapped as a binary bit in the state vector, the total number of possible states would be $2^{|Z|}$. This state representation would lead to an excessively large state space, consequently posing significant convergence challenges for Q-table training. To address this issue, we adopt the state aggregation strategy from RL-Recruiter [27], which redefines the state space by focusing on the count of currently selected taxis. This approach achieves effective state compression and has demonstrated superior performance in participant selection problems.

Action: At the beginning of each round, first initialize the set of available action spaces $A = \{a_1, a_2, \cdots, a_{|Z|}\}$. The action a_k represents selecting a taxi p_k in the current round. Since the same taxi cannot be selected repeatedly in the same round, after performing an action, the action will be removed from the available action space in the current round. We adopt the ε − greedy strategy to select actions, that is, with a probability of ε, a random available action is selected, or with a probability of $1 - \varepsilon$, the best available action is chosen.

Reward: The reward for the traditional selection of taxi p_k is defined as the coverage increment after taking this action. That is:

$$r = \frac{|V(P \cup p_k)| - |V(P)|}{g} \tag{8}$$

Each taxi in the selected set P corresponds to its most similar auxiliary taxi, collectively forming an auxiliary taxis subset H ($H \subseteq U$). When an action is taken to select taxi p_k, it simultaneously activates its corresponding auxiliary taxi h_j. Our reward not only considers the coverage gain brought by the selection of candidate taxi p_k, but also considers the coverage gain brought by the selection of corresponding auxiliary taxi h_j. Consequently, Eq. (8) for the reward should be modified to reflect this combined coverage benefit.

$$r = \frac{(1-\beta)(|V(P \cup p_k)| - |V(P)|)}{g} + \frac{\beta(|V\prime(H \cup h_j)| - |V\prime(H)|)}{l} \tag{9}$$

Among them, β is the weight factor of the auxiliary vehicle, where $0 \leq \beta \leq 1$. The larger this value is, the more important the information of the auxiliary taxi becomes.

Reinforcement learning based on the Q-table, where each row represents a state and each column represents an action. The Q-table is used to store the long-term return that can be obtained by performing an action in a certain state, that is, the Q-value. The update formula for the Q-value is:

$$Q(s, a_k) = Q(s, a_k) + \alpha[r + \gamma \max(Q(s\prime, a\prime)) - Q(s, a_k)] \tag{10}$$

Here, α is the learning rate, γ is the discount factor, and $\max(Q(s\prime, a\prime))$ represents the maximum Q-value among all possible actions in the next state $s\prime$.

The reinforcement learning training process is outlined in Algorithm 1. The model undergoes training for a total of *num* episodes. In each episode, taxis are sequentially selected according to the ε − greedy strategy. Each selection of a taxi triggers an immediate update of the Q-table. The episode terminates when the number of selected taxis reaches *n*. Upon completing *num* training episodes, we proceed to actual taxi selection using the trained Q-table. Since the ε − greedy strategy during training has achieved sufficient exploration and exploitation balance, the operational phase employs a purely greedy approach at each step selecting the action with the maximum Q-value for the current state, until *n* taxis are selected.

Algorithm 1. Reinforcement Learning Training Process

Input: Candidate taxi set Z, auxiliary taxi set U, correspondence relationship between candidate and auxiliary taxis
Output: Trained Q-table

1. Initialize Q-table;
2. **for** $epoch = 1,2,\cdots num$ **do**
3. $s = 0, P = \varnothing, H = \varnothing$, available actions set A;
4. **While** $|P| < n$ **do**
5.   Generate random number x;
6.   **if** $x > \varepsilon$ **then**
7.   Select the best action from A;
8.   **Else**
9.   Randomly select an action from A;
10. **End if**
11. Record: selected action a_k, chosen taxi p_k, corresponding auxiliary taxi h_j;
12. $s = s + 1, P = P \cup p_k, A = A - a_k, H = H \cup h_j$;
13. Compute reward using Eq. (9), update Q-table using Eq. (10);
14. **End while**
15. **end for**

5 Experiment

In this section, we conduct comprehensive experiments based on real-world datasets to validate the superiority of the proposed E2RL method and compare it with baseline approaches.

5.1 Dataset and Experimental Setup

In this paper, a taxi GPS trajectory dataset from [36] is used, which contains the GPS trajectory data of 536 taxis in San Francisco, USA, from May 17, 2008, to June 10, 2008. We set the study area as ranging from 122.370°W to 122.514°W in longitude and from 37.708°N to 37.813°N in latitude as shown in Fig. 4. In order to evaluate the coverage ability of the selected taxis, we divide the region into many small squares with length and width of 0.001° according to longitude and latitude. We use the advanced map matching algorithm [37] to interpolate taxi trajectory points and match them to the roads in the road network, thus improving the accuracy of the trajectory data. Taxi operation activities are relatively scarce in the early morning hours, so we set the study time period as 6:00–24:00 every day. We removed taxis only operated for a few days or had limited daily operation hours. Eventually, we selected 310 taxis as the experimental subjects.

We used 210 of these taxis as auxiliary vehicles, with a historical time set spanning 16 days and time slots of 1 h. Trajectory segments with a time slot of 20 min and a duration of two days were used as encoding units. Each taxi generated 30 images.

Fig. 4. Study area in San Francisco

for training the embedding representation model and classification model. The remaining 100 taxis served as candidate vehicles, with both the historical time set T and future time set T^* spanning two days and time slots of 1 h. We carried out 8 groups of independent experiments. The trajectories of candidate taxis in each group of experiments lasted for 4 days. Among them, the trajectories of the first two days were used for training, and the trajectories of the latter two days were used for testing. We assumed that each grid cell traversed by a taxi represents a sensing task and evaluated the relative coverage of various methods when selecting 10, 20, 30, and 40 taxis in a grid-based scenario.

We also use dataset of SFpark project [38] to evaluate the performance of the E2RL in practical scenarios. We matched the latitude and longitude information of 230 parking streets to the corresponding road network roads. A taxi is identified as traversing a particular parking street when its trajectory's road segment identifier matches that of the street. We further studied the coverage performance of the method we proposed on the parking streets. Similar to the grid scenario, we evaluated the relative coverage when 10, 20, 30, and 40 taxis were selected, which extensively verified the superiority of the E2RL model we proposed.

The parameter settings of the experiment are shown in Table 1. The exploration rate ε starts at 0.8 and is exponentially decayed toward 0.1 over the course of training using the following formula:

$$\varepsilon_t = \varepsilon_{end} + (\varepsilon_{start} - \varepsilon_{end})e^{-t/\tau} \tag{11}$$

where $\varepsilon_{start} = 0.8$ ensures extensive initial exploration, $\varepsilon_{end} = 0.1$ maintains a 10% random exploration probability late in training to avoid local optima, and t is the current episode. The decay rate is controlled by the time constant τ, which we set to 200.

Table 1. Parameter settings for E2RL

Paramater	Value
δ	0.1
num	1000
β	0.3
α	0.1
γ	0.9
ε	0.8–0.1
n	10/20/30/40

5.2 Comparison Methods

Random: Randomly select n taxis from the candidate taxis, and evaluate the relative coverage rate of the selected taxis in a certain period in the future.

CQO [26]: This method is an improved greedy algorithm applied to our problem, where at each step, select the taxi that can maximize the historical coverage quality gain until the number of selected taxis reaches n.

RL-Recruiter [27]: A participant selection method based on reinforcement learning, whose core idea is to construct a value function through Q-table learning. The immediate reward is defined as the historical coverage gain of a candidate. While the original method targets online selection with dynamic updates per time slot, we adapt it to an offline setting by leveraging the full historical trajectories of candidate taxis for training.

ERCQO: Replaces the second-stage reinforcement learning algorithm in E2RL with the CQO algorithm, while the first stage remains the same as our method, leveraging auxiliary information to learn embedding representations and classification models.

5.3 Experiment Results and Analysis

Coverage performance comparison. Figures 5 and 6 respectively show the comparison of the relative coverage rates of different methods for grids and parking streets when the number of taxis is 10, 20, 30, and 40. All the relative coverage here are the average test performance of 8 groups of experiments. As can be seen from Fig. 5, our method outperforms a series of baseline methods under different scales of taxi selection. This indicates that our method can well capture the movement patterns hidden in the complex trajectories of taxis, and can effectively deal with the problem of the uncertainty of taxi trajectories. As a result, the selected taxis can have higher coverage performance in the future. Figure 6 demonstrates the superiority of our taxi selection method in terms of the coverage performance on parking streets. This shows that our method can be applied to the selection of taxis for street parking space sensing, providing a low-cost and high-benefit solution for smart parking.

Figure 7 shows the comparison of the relative street coverage between our method and other methods in different experimental groups when 10, 20, 30, and 40 taxis are selected in the parking street scenario. Taking the selection of 20 taxis as an example,

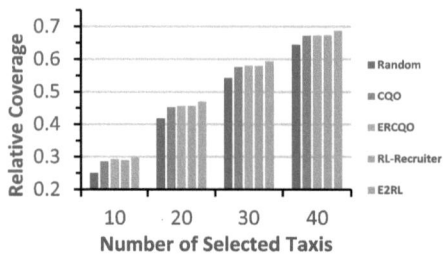
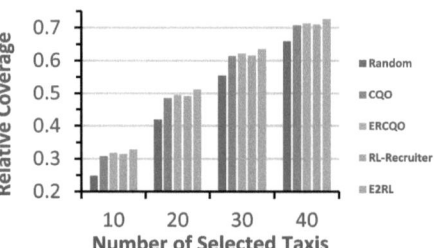

Fig. 5 Comparison of relative grid coerage varying with the number of selected taxis

Fig. 6 Comparison of relative parking streets coverage varying with the number of selected taxis

except for Group 5, our method has achieved the optimal relative coverage in the remaining 7 groups of experiments. Even in Group 5, the gap in relative coverage between our method and the best method is very small. The experimental results show that our method can not only achieve a higher average coverage quality, but also exhibit excellent stability. The following is an analysis and summary of the reasons why our method is superior to the baseline:

The random selection strategy (Random) performs the worst because it does not learn patterns from historical trajectory data. In contrast, both CQO and RL-Recruiter utilize the historical trajectory information of candidate taxis for taxi selection. However, CQO only focuses on maximizing the coverage gain at the current step in each selection step and lacks consideration for future returns. RL-Recruiter, through the learning mechanism of the Q-table, can more farsightedly evaluate the impact of the current selection on the coverage quality brought by subsequent taxi selections, thereby making better decisions. Nevertheless, RL-Recruiter only relies on the historical trajectories of candidate taxis and fails to effectively address the problem of unstable coverage sensing quality caused by the uncertainty of taxi trajectories. Furthermore, E2RL fully excavates taxi patterns by learning embedding representations from auxiliary taxis and conducting classification, and integrates the most similar auxiliary taxis for Q-table learning, thus improving the performance and stability of coverage sensing. Compared with CQO, ERCQO utilizes the information of auxiliary taxis and its performance is better than that of CQO, which demonstrates the effectiveness of the first stage of our method.

Although ERCQO conducts the learning in the first stage, using CQO in the second stage is not far-sighted enough and fails to fully utilize the potential of the embedding representations and taxi classification models learned in the first stage, resulting in its performance being inferior to that of E2RL. All in all, the excellent performance of E2RL is mainly attributed to the following key advantages:

1. Taxi embedding representation reveals latent trajectory features.
2. Taxi classification model selects well-matched auxiliary taxi.
3. Multi-source fusion reduces the uncertainty of single-source reliance.
4. Reinforcement learning foresight via Q-table optimizes long-term coverage decisions.

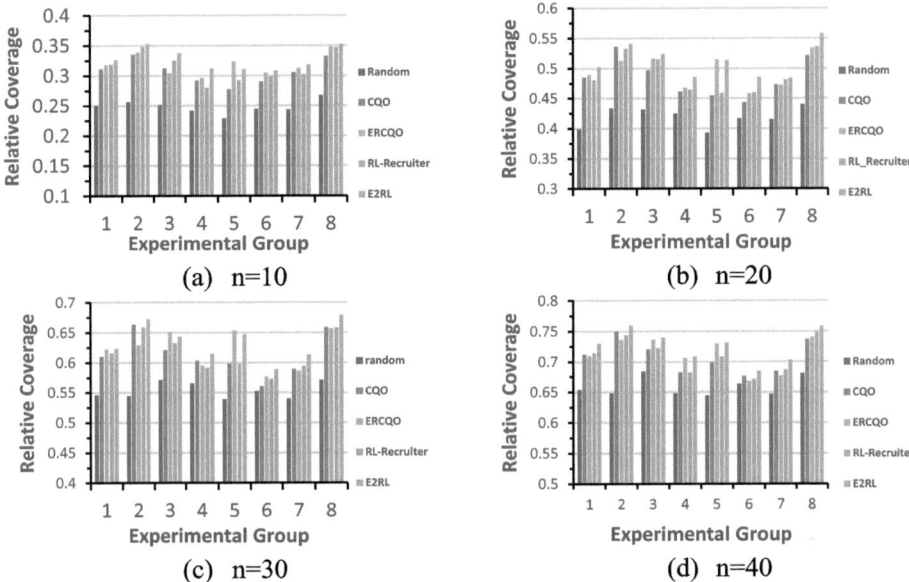

Fig. 7. Comparison of relative parking streets coverage with different experimental group. (a) n = 10. (b) n = 20. (c) n = 30. (d) n = 40

convergence analysis of the Q-table training process. Figure 8 presents the total reward using a 50-episode moving average with 95% confidence intervals across 1,000 training episodes for a Q-learning agent. In the early phase (0 - 200 episodes), the total reward climbs quickly, reflecting the agent's rapid acquisition of effective policies through exploration. During the transitional phase (200 - 600 episodes), the reward curve plateaus, indicating a shift from exploration to exploitation and a slowdown in learning progress. In the late phase (600 - 1000 episodes), the curve stabilizes with minimal fluctuations, demonstrating consistent, reliable performance—a strong signal that the algorithm has converged.

The impact of key parameters. We conducted a comprehensive parameter analysis by testing β (the weight factor of auxiliary taxis) at discrete values from 0 to 1.0 in 0.1 increments. As shown in Fig. 9, the experimental results show that neither a larger nor a smaller value of β leads to the best effect. When the value of β is 0.3, the model can achieve the optimal relative coverage.

As shown in Fig. 10, we also explored the impact of the setting of the number of channels in the image encoding scheme in the first stage on the results of parking street sensing coverage. The experimental results show that the 3-channel encoding scheme can achieve the best results, and this result can be explained by the time characteristics of taxi operations: during the operating period from 6:00 to 24:00, the behavior patterns of taxis can be naturally divided into three intervals with significant differences in the morning, afternoon, and evening. By aligning channels with these distinct operational.

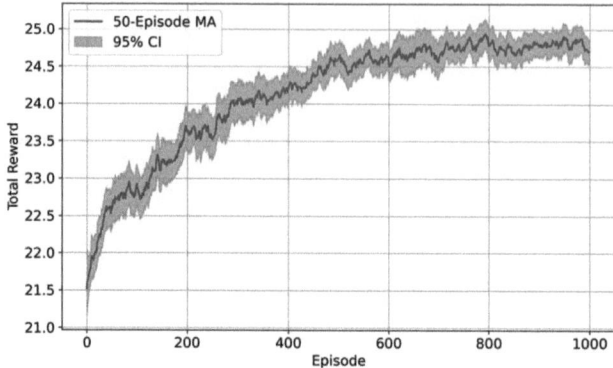

Fig. 8 Q-table training convergence curve of the E2RL method for selecting 20 taxis in the street parking scenario

intervals, the encoding captures temporal variations in taxi activity more precisely. Consequently, the taxi embedding representation model learns richer, period-specific behavior patterns, enhancing the guidance for taxi selection and ultimately delivering more robust coverage optimization.

Time complexity of various schemes. E2rl and ercqo are two-stage schemes, whereas cqo, rl recruiter, and random perform taxi selection in a single stage. Therefore, we analyze the runtimes of these single-stage methods together with the second stage of the two-stage schemes. Stage 1 requires about 15 h for training the taxi embedding representation model and 3 min for the taxi classification model; once trained, processing a batch of 100 candidate taxis takes just 26 s. In a realistic scenario with 230 parking streets, the average stage 2 runtimes are approximately 0 s for random, 0.63 s for cqo, 1.52 s for ercqo, 8.51 s for rl recruiter, and 58.72 s for e2rl. Since all methods are executed offline, even the slower two-stage approaches remain practical.

6 Conclusion

To address the challenges posed by taxi trajectory uncertainty in vehicle selection, this paper proposes a method named E2RL based on embedding representation learning and reinforcement learning. The method employs a deep convolutional network with triplet loss to learn discriminative trajectory embeddings, which subsequently train a classifier model to inform reinforcement learning decisions. Extensive experiments demonstrate that our solution achieves both high coverage quality and remarkable stability, providing an effective decision-making framework for street parking space monitoring through optimized taxi selection.

For future work, we will consider collecting high-quality parking space data through intelligent optimization of taxi selection strategies under the framework of Sparse Mobile Crowdsensing, and achieving high-precision reconstruction of the information about parking spaces on urban streets based on the data completion algorithm.

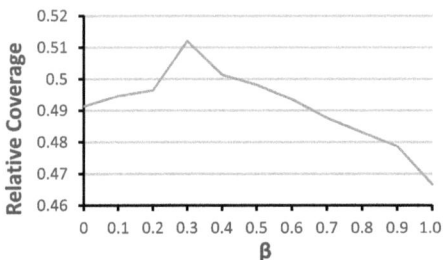
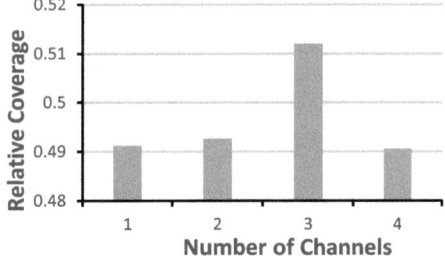

Fig. 9 Comparison of relative parking streets coverage with different β

Fig. 10 Comparison of relative parking streets coverage with different number of channels

References

1. Shoup, D.C.: Cruising for parking. Transp. Policy **13**(6), 479–486 (2006)
2. Misra, Prasant, et al. "The future of smart parking systems with parking 4.0." GetMobile: Mobile Computing and Communications 23.1 (2019): 10–15
3. Xu, Bo, et al. "Real-time street parking availability estimation." 2013 IEEE 14th International Conference on Mobile Data Management. Vol. 1. IEEE, 2013
4. Kotb, A.O., Shen, Y.-C., Huang, Yi.: Smart parking guidance, monitoring and reservations: a review. IEEE Intell. Transp. Syst. Mag. **9**(2), 6–16 (2017)
5. Guo, Wenzhong, et al. "A survey of task allocation: contrastive perspectives from wireless sensor networks and mobile crowdsensing." IEEE Access 7 (2019): 78406–78420
6. Dutta, Prabal, et al. "Common sense: participatory urban sensing using a network of handheld air quality monitors." Proceedings of the 7th ACM conference on embedded networked sensor systems. 2009
7. Lee, R., Wakamiya, S., Sumiya, K.: Discovery of unusual regional social activities using geo-tagged microblogs. World Wide Web **14**, 321–349 (2011)
8. Zhou, Pengfei, Yuanqing Zheng, and Mo Li. "How long to wait? Predicting bus arrival time with mobile phone based participatory sensing." Proceedings of the 10th international conference on Mobile systems, applications, and services. 2012
9. Mathur, Suhas, et al. "Parknet: drive-by sensing of road-side parking statistics." Proceedings of the 8th international conference on Mobile systems, applications, and services. 2010
10. Satonaka, Hisashi, et al. "Development of parking space detection using an ultrasonic sensor." PROCEEDINGS OF THE 13th ITS WORLD CONGRESS, LONDON, 8–12 OCTOBER 2006. 2006
11. Houben, Sebastian, et al. "On-vehicle video-based parking lot recognition with fisheye optics." 16th International IEEE Conference on Intelligent Transportation Systems (ITSC 2013). IEEE, 2013
12. Bock, Fabian, Sergio Di Martino, and Antonio Origlia. "Smart parking: Using a crowd of taxis to sense on-street parking space availability." IEEE Transactions on Intelligent Transportation Systems 21.2 (2019): 496–508
13. Restuccia, Francesco, et al. "Quality of information in mobile crowdsensing: Survey and research challenges." ACM Transactions on Sensor Networks (TOSN) 13.4 (2017): 1–43
14. Li, H.: Participant Selection and Task Assignment in Mobile Crowd Sensing. The University of North Carolina at Charlotte, Diss (2018)
15. Wang, Jiangtao, et al. "Task allocation in mobile crowd sensing: State-of-the-art and future opportunities." IEEE Internet of Things journal 5.5 (2018): 3747–3757

16. Zhang, Daqing, et al. "CrowdRecruiter: Selecting participants for piggyback crowdsensing under probabilistic coverage constraint." Proceedings of the 2014 ACM International Joint Conference on Pervasive and Ubiquitous Computing. 2014
17. Reddy, Sasank, Deborah Estrin, and Mani Srivastava. "Recruitment framework for participatory sensing data collections." Pervasive Computing: 8th International Conference, Pervasive 2010, Helsinki, Finland, May 17-20, 2010. Proceedings 8. Springer Berlin Heidelberg, 2010.
18. Guo, B., et al. "A Framework for Optimized Multitask Allocation in Mobile Crowdsensing Systems., 2017, 47." https://doi.org/10.1109/THMS (2016): 392–403
19. Wang, Jiangtao, et al. "PSAllocator: Multi-task allocation for participatory sensing with sensing capability constraints." Proceedings of the 2017 ACM Conference on Computer Supported Cooperative Work and Social Computing. 2017
20. Lane, Nicholas D., et al. "Piggyback crowdsensing (pcs) energy efficient crowdsourcing of mobile sensor data by exploiting smartphone app opportunities." Proceedings of the 11th ACM Conference on Embedded Networked Sensor Systems. 2013
21. Wang, Jingbin, et al. "Route selection for opportunity-sensing and prediction of waterlogging." Frontiers of Computer Science 18.4 (2024): 184503
22. He Z, Cao J, Liu X .High quality participant recruitment in vehicle-based crowdsourcing using predictable mobility[J].IEEE, 2015.https://doi.org/10.1109/INFOCOM.2015.7218644
23. Yang, Yongjian, et al. "A prediction-based user selection framework for heterogeneous mobile crowdsensing." IEEE Transactions on Mobile Computing 18.11 (2018): 2460–2473
24. Liu, Wenbin, et al. "Dynamic online user recruitment with (non-) submodular utility in mobile crowdsensing." IEEE/ACM Transactions on Networking 29.5 (2021): 2156–2169
25. Wang, En, et al. "Truthful incentive mechanism for budget-constrained online user selection in mobile crowdsensing." IEEE Transactions on Mobile Computing 21.12 (2021): 4642–4655
26. Liu, Ting, et al. "Mobile crowdsensing coverage degree-probability enhancement based on urban vehicles." GLOBECOM 2020–2020 IEEE Global Communications Conference. IEEE, 2020
27. Hu, Yunfan, et al. "Participants selection for from-scratch mobile crowdsensing via reinforcement learning." 2020 IEEE International Conference on Pervasive Computing and Communications (PerCom). IEEE, 2020
28. Kaelbling, Leslie Pack, Michael L. Littman, and Andrew W. Moore. "Reinforcement learning: A survey." J. Artif. Intell. Res. **4** 237–285 (1996)
29. Singh, Bharat, Rajesh Kumar, and Vinay Pratap Singh. "Reinforcement learning in robotic applications: a comprehensive survey." Artif. Intell. Rev. **55.2** 945–990 (2022)
30. Recht, B.: A tour of reinforcement learning: The view from continuous control. Ann. Rev. Control Robotics Auton. Syst. **2**(1), 253–279 (2019)
31. Wang, Fei-Yue, et al. "Where does AlphaGo go: From church-turing thesis to AlphaGo thesis and beyond." IEEE/CAA J. Automat. Sin. 3.2 (2016): 113–120
32. Xu, Ying, et al. "PSARE: A RL-based online participant selection scheme incorporating area coverage ratio and degree in mobile crowdsensing." IEEE Transactions on Vehicular Technology 71.10 (2022): 10923–10933
33. Schroff, Florian, Dmitry Kalenichenko, and James Philbin. "Facenet: A unified embedding for face recognition and clustering." Proceedings of the IEEE conference on computer vision and pattern recognition. 2015
34. Szegedy, Christian, et al. "Going deeper with convolutions." Proceedings of the IEEE conference on computer vision and pattern recognition. 2015
35. Chang, Chih-Chung. "A library for support vector machines." http://wwwcsie.ntu.edu.tw/~cjlin/libsvm. (2001)
36. M. Piorkowski, N. Sarafijanovic-Djukic, M. Grossglauser. (Feb. 2009). CRAWDAD Dataset Epfl/Mobility (v. 2009-02-24). http://crawdad.org/epfl/mobility/2009022

37. Gotrackit.[Online].Available:https://gotrackit.readthedocs.io/en/latest/index.html
38. SFMTA. (2014). SFpark: Putting Theory Into Practice. Pilot Project Summary and Lessons Learned. Accessed: Jun. 24, 2016. [Online]. http://sfpark.org/resources/docspilotsummary/

Eye-PPG: Remote PPG Signal Generation for Heart-Rate Estimation and User Identification in Virtual Reality

Rao Fu, Guangrong Zhao(✉), and Yiran Shen

School of Software, Shandong University, Jinan, China
guangrong.zhao@sdu.edu.cn

Abstract. As the awareness of maintaining health continues to rise, there is an increasing demand for advanced, convenient, and non-intrusive health monitoring solutions. Given the rapid advancements in and wide-spread adoption of virtual reality (VR) technology and devices, this paper introduces a contactless heart rate estimation and user identity recognition system specifically designed for real-world VR environments. We propose a video-based heart rate estimation model that utilizes remote photoplethysmography (rPPG) to reconstruct blood volume pulse (BVP). The rPPG signals are generated from the video recordings captured by the eye infrared cameras used for eye tracking in VR headset. Additionally, we investigate the potential application of the reconstructed rPPG signals in user identification tasks. The experiments are conducted using a self-constructed dataset, which comprises binocular video and fingertip photoplethysmography (PPG) signals captured from 11 participants while they naturally wore VR headsets. The results substantiate the feasibility of the proposed system in enabling unobtrusive heart-rate estimation and identity recognition within immersive virtual environments.

Keywords: Virtual Reality (VR) · Heart Rate Estimation · Remote Photoplethysmography (rPPG) · User Identity Recognition

1 Introduction

With immersive technologies increasingly integrated into daily life and professional practice, Virtual Reality has evolved into a comprehensive platform that combines perception, interaction, and feedback. Leveraging VR for personalized health monitoring and identity recognition has become a key direction for enhancing system intelligence and improving user experience. In this context, non-intrusive and efficient physiological sensing techniques are receiving increasing attention. A pressing challenge is how to utilize existing VR headset sensors extract physiological features for user health status assessment and identity recognition, without requiring additional hardware.

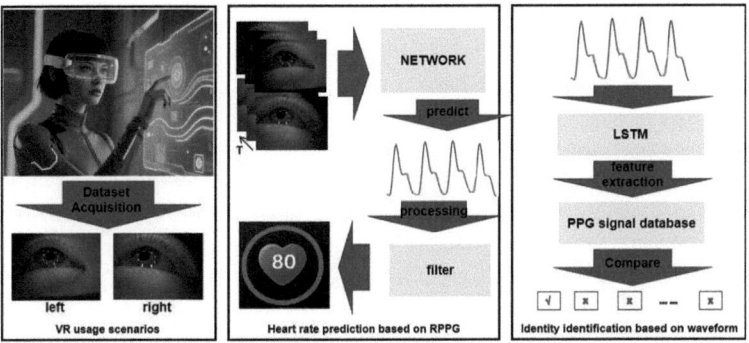

Fig. 1. Workflow of the Eye-PPG system for heart rate monitoring and user identification in VR. The system includes three stages: data acquisition, heart rate estimation, and identity identification.

Currently, mainstream physiological signal monitoring techniques for heart rate can be broadly categorized into contact-based and non-contact approaches, each with inherent limitations in immersive or mobile scenarios. For example, electrocardiography (ECG) detects the heart's electrical activity via electrode sensors, offering high accuracy but relying on specialized equipment, thus limiting its application to clinical settings. In contrast, wearable devices like smartwatches typically use photoplethysmography (PPG) to estimate pulse signals by detecting variations in light absorption through the skin, making them suitable for daily health management. Previous studies [12] have explored the feasibility of using PPG for heart rate monitoring in VR environments. However, wearing such PPG devices during prolonged immersive experiences still cause discomfort and interfere with user interaction. Additionally, integrating extra PPG sensors into resource-constrained VR devices can introduce additional system overhead. Despite recent efforts to estimate heart rate using non-contact methods such as WiFi sensing, these approaches often require dedicated infrastructure (e.g., specialized antennas or base stations) and are highly sensitive to environmental interference, motion artifacts, and occlusions [4]. As a result, their robustness and scalability in dynamic or immersive settings like VR remain limited.

These limitations have motivated the exploration of rPPG, a contactless technique that estimates heart rate by analyzing subtle color variations of the skin captured by CMOS based cameras. Many modern VR headsets are now equipped with embedded eye-tracking cameras, providing a promising opportunity to extract rPPG signals without requiring additional hardware. However, two major challenges remain: first, data privacy policies enforced by VR manufacturers often restrict access to raw eye-tracking images, resulting in a lack of publicly available datasets; second, conventional rPPG methods primarily rely on visible light, while camera in VR headsets typically operate in low-light or near-infrared (NIR) environments, which significantly limits the applicability and accuracy of such methods. Beyond heart rate estimation, rPPG signals have also shown promise in the field of biometric identification [17]. However, most existing studies rely on full-face RGB video data, which are not directly applicable to realistic VR settings where only infrared images of the eye region are available.

To address these challenges, this study proposes the Eye-PPG system, which leverages near-infrared eye-tracking cameras widely embedded in VR headsets to generate and utilize rPPG signals for contactless heart rate estimation and user identification. We design a deep neural network architecture that integrates multi-view feature fusion with long- and short-term temporal modeling to enhance the robustness and discriminability of signal extraction. Experimental validation was conducted by collecting eye-region videos and synchronized pulse reference signals from 11 participants, confirming the feasibility and effectiveness of the Eye-PPG system in real-world VR environments. The results demonstrate that the proposed system can accurately predict heart rate under near-infrared conditions and achieve effective user identity recognition, highlighting its potential for health monitoring and personalized identification in VR environments. The major contribution of this work are:

- We construct the first rPPG research dataset based on near-infrared eye-region videos in VR scenarios. The dataset includes synchronized binocular NIR eye-tracking videos and fingertip PPG signals collected from 11 participants wearing real VR headsets, providing valuable data support for contactless physiological signal research under infrared conditions.
- We propose the PhysVR method which is a deep neural network architecture that integrates multi-view feature fusion and long-short temporal modeling. This model enables robust extraction of rPPG signals from infrared eye videos for heart rate estimation and user identification, while satisfying the requirements for real-time deployment in VR scenarios.
- We conduct comprehensive experiments in realistic VR environments, this is the first attempt to apply rPPG-based heart rate estimation in VR scenarios. The results demonstrate that our system achieves accurate heart rate estimation and effective identity recognition under NIR conditions, validating the feasibility of Eye-PPG for immersive health monitoring and biometric identification in VR.

2 Related Work

2.1 Applications of Physiological Signals in Virtual Reality Environments

Photoplethysmography (PPG) is a non-invasive optical technique widely used to monitor blood volume changes. When a light source illuminates the skin surface, cardiac-induced pulsatile changes in blood vessel volume modulate the amount of transmitted or reflected light. These variations are captured by a photodetector and converted into electrical signals, from which physiological parameters such as heart rate and blood oxygen saturation can be inferred [6,26]. PPG signals are obtained by analyzing light absorption changes caused by variations in skin blood flow, offering advantages such as low cost and ease of integration. Consequently, PPG technology has been widely adopted in wearable devices, including smartwatches and fitness bands. These advantages have sparked interest in applying PPG to VR environments, where researchers have begun exploring its use for real-time interaction and emotional adaptation.

For example, Cui et al. [10] proposed a gamified biofeedback approach using a Cardboard VR headset. In their system, real-time PPG signals were captured via a smartphone camera to extract heart rate data, which was then used to influence gameplay in real time. This design aimed to increase user engagement in biofeedback therapy and make the experience more interactive and enjoyable. Some studies have explored the use of ear-clip PPG sensors to perform continuous HR monitoring, allowing the system to infer users' stress levels and emotional states during gameplay, training, or therapeutic scenarios [7,8]. These efforts not only enhance the intelligence and interactivity of human–computer interfaces in VR systems, but also lay the groundwork for emotionally adaptive and therapeutic VR content.

However, heart rate monitoring in VR using traditional PPG faces several challenges. Traditional PPG requires direct skin contact through devices like fingertip sensors or wristbands, which can reduce immersion and comfort, especially during extended sessions. Additionally, motion-induced artifacts and perspiration during physical VR activities can degrade signal quality, affecting measurement reliability. These issues highlight the need for contactless alternatives like rPPG, which can seamlessly integrate with VR systems without requiring extra hardware.

2.2 Advancements in Remote Photoplethysmography

To mitigate the discomfort and reduction in immersion caused by wearable sensors, rPPG has emerged as a promising contactless alternative for physiological signal monitoring. rPPG leverages commodity cameras—such as standard RGB webcams—to remotely detect subtle color variations in facial skin, which are then analyzed to estimate physiological parameters including heart rate and respiratory rate [26,31]. This technique offers several advantages, including high compatibility with existing hardware, ease of integration, and minimal user intrusiveness, making it particularly suitable for use in immersive VR environments.

Conventional rPPG approaches typically rely on color space transformations to extract chrominance signals from RGB video streams. A representative example is the CHROM method [11], which estimates pulse signals based on specific chrominance projections. Poh et al. [25] applied Independent Component Analysis (ICA) to normalized and spatially averaged color signals to extract the blood volume pulse (BVP). Verkruysse et al. [31] proposed a method based on skin-color tracking and spatial averaging for pulse estimation. Wang et al. [33] developed a technique that projects color signals onto a plane orthogonal to the skin tone vector, leveraging physiological and optical priors. Early rPPG methods predominantly focused on signal enhancement and blind source separation (BSS) techniques, such as Principal Component Analysis (PCA), which have demonstrated effectiveness in suppressing environmental noise and motion artifacts [26].

With technological advancements, an increasing number of rPPG approaches have been proposed to address challenges arising from varying illumination conditions, ethnic and skin tone differences, and subtle motion artifacts. Chen et al. [9] introduced DeepPhys, a dual-branch convolutional architecture that processes appearance and motion difference frames in parallel. The network learns to predict the PPG waveform by adaptively modeling skin color variations, and incorporates an attention mechanism to enhance measurement accuracy. Liu et al. [20] proposed an efficient on-device video-based system for physiological signal monitoring. Their method, MTTS-CAN (Multi-task Temporal Shift Convolutional Attention Network), enables real-time estimation of both cardiovascular and respiratory signals on mobile platforms. More recently, researchers have explored end-to-end learning-based models that directly regress the pulse signal from raw video frames. Traditional rPPG pipelines often rely on handcrafted preprocessing based on strong assumptions, which may fail under real-world conditions. To mitigate this, Niu et al. [22] developed RhythmNet, a deep learning-based framework that estimates heart rate in a fully data-driven manner, reducing dependence on fixed signal processing heuristics.

Overall, rPPG technology, a contactless and unobtrusive physiological monitoring method, is highly compatible with VR systems. It allows continuous, real-time physiological signal acquisition without disrupting the user's immersive experience, making it ideal for next-generation personalized VR platforms.

2.3 Identification in VR Environments

With the rapid advancement of VR technologies, user identification has become increasingly critical in immersive environments. Unlike conventional desktop systems, identification mechanisms in VR must not only ensure security but also preserve user experience by minimizing interruptions to immersion. Existing research on VR identification can be broadly categorized into approaches based on traditional passwords and behavioral biometrics, such as gestures and motion patterns [30,35].

Early studies on user identification in VR environments largely relied on traditional text-based or graphical passwords. However, such methods are often cumbersome to input and interrupt the immersive experience, making them less suitable for VR contexts. As a result, researchers have increasingly turned to more natural biometric identification approaches [18,23]. Behavioral biometric identification identifies users based on their interaction patterns in VR, such as hand gestures, head movements, or gait trajectories. These methods offer strong stealthiness and minimal user disruption. For instance, Mathis et al. [21] proposed an identification method called RubikAuth, where users interact with a virtual 3D cube using handheld controllers to select numerical digits for identification, enabling a fast and immersive login experience. In another study, Shen

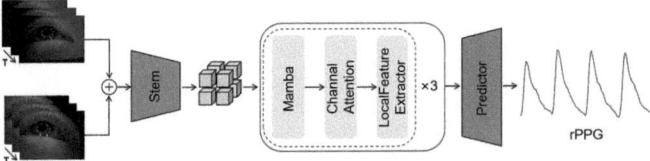

Fig. 2. Overview of PhysVR architecture. The Stem module extracts features from the fused binocular eye video input. These features are then processed through the Mamba, Channel Attention, and Local Feature Extractor modules in a three-stage cascade, followed by the predictor that outputs the rPPG signals.

et al. [29] introduced GaitLock, a gait-based identification system that leverages inertial measurement units (IMUs) embedded in VR/AR head-mounted displays to recognize users based on their walking patterns. This method enhances privacy and security by providing continuous and device-integrated identity verification. Beyond behavioral patterns, physiological signals offer another promising avenue for user identification in VR.

Due to the inherent inter-individual variability in cardiovascular signals, such features are difficult to replicate and thus can serve as robust auxiliary biometric cues for user verification, reducing the risk of biometric spoofing [5]. Building on this insight, heart rate and heart rate variability (HRV) have increasingly drawn attention in the context of VR-based user identification. Recent studies have explored the use of heart rate signals derived from PPG as an implicit and continuous identification modality [19].

In particular, Li et al. [16] demonstrated that rPPG can effectively approximate the signal characteristics of contact-based PPG. Through spoofing attack experiments, their findings revealed that rPPG signals encode individual-specific physiological traits and exhibit both temporal stability and user discriminability. These results provide strong empirical evidence supporting the theoretical feasibility and practical potential of rPPG as a biometric modality for continuous and unobtrusive identification in VR environments.

3 Method

The overall system workflow is illustrated in Fig. 1. Details regarding data acquisition will be introduced in the IV. This part focuses on the heart rate estimation and identity identification stages.

In the heart rate estimation stage, we propose a network architecture named physVR, as shown in Fig. 2. First, video data of the eye region from two different perspectives are combined (indicated by the plus sign). The combined data then enter the "Stem" module for initial processing, which outputs a feature representation. The resulting feature representation is subsequently processed through a composite block consisting of the Mamba module for capturing long-term temporal dependencies, the Channel Attention module for multi-view feature fusion by emphasizing informative channels, and the Local Feature Extractor module for fine-grained spatial feature extraction. This composite block is repeated three times to progressively refine the feature representation. Finally, the enhanced features are fed into the Predictor module, similar to that in [37], which projects them into a rPPG waveform through temporal upsampling and spatial averaging. We adopt the same loss function as in [36].

In the identity identification stage, features are extracted from the processed rPPG signals and modeled using a Bidirectional Long Short-Term Memory (BiLSTM) network, as shown in Fig. 3. The resulting feature obtained from the BiLSTM are then passed through a fully connected layer for final classification, completing the identity identification task.

3.1 Implicit Fusion Based on Channel Attention

The multi-view problem aims to observe, analyze, and interpret a subject from different viewpoints to achieve a more comprehensive and in-depth understanding. In the field of computer vision, a common scenario involves using multiple cameras to capture and monitor a target scene or object from various angles. Each camera captures information about the subject from a specific perspective, and these multi-view images provide complementary cues that together form a more complete and enriched representation of the target [24].

Due to the VR eye-tracking equipment being equipped with two cameras for the left and right eyes, the video signals processed by the model originate from dual-view inputs of the same subject. Specifically, the synchronized video streams captured by the two cameras for the left and right eye regions. This setup inherently forms a multi-view physiological signal modeling problem, providing richer spatiotemporal redundancy in the information.

Recent studies have demonstrated the outstanding performance of channel attention mechanisms in addressing multi-view problems, which has provided inspiration for the design of our system. [28,32]. As shown in Fig. 1, the input videos from the left and right eye regions, each represented as a single-channel video stream, are concatenated along the channel dimension to form a two-channel video sequence: $X_{\text{left}}, X_{\text{right}} \in \mathbb{R}^{B \times 1 \times T \times H \times W}$. Then it is fed into a shared 3D convolutional encoder for feature extraction. On top of this unified channel representation, a Channel Attention Module is incorporated to dynamically compute the response strength of each channel, thereby enabling implicit fusion of multi-view features.

3.2 Long-Term Temporal Modeling Module

Mamba is a sequence modeling framework based on the State Space Model (SSM), originally developed for natural language processing tasks, aiming to achieve efficient modeling of textual sequences [13]. Benefiting from its recurrent structure with linear complexity and superior capability in modeling low-frequency information, Mamba excels at capturing long-range dependencies while significantly reducing computational resource consumption [14]. However, the original Mamba model is optimized for one-dimensional language sequences. Directly applying it to high-dimensional video sequences—especially high-resolution and long-term video-based physiological signals—incurs substantial memory and computational costs. Recent studies [15] have successfully extended Mamba to the video domain, demonstrating the effectiveness and potential of state space models in video sequence modeling.

Inspired by previous works [34,37], we incorporate a three-stage 3D convolutional encoder (Stem) into our model to perform coarse-grained feature extraction on the raw video input prior to Mamba. Specifically, the input tensor $X \in \mathbb{R}^{B \times C_{in} \times T \times H \times W}$, where B denotes the batch size, $C_{in} = 2$ is the number of input channels, $T = 256$ is the number of frames, and $H = W = 128$ is the spatial resolution of the input, is sequentially passed through three 3D convolutional layers with kernel sizes of $(1 \times 5 \times 5)$, $(3 \times 3 \times 3)$, and $(3 \times 3 \times 3)$, respectively. After each convolutional block, a spatial max-pooling operation (i.e., only along the H and W dimensions) is applied to gradually downsample the spatial resolution, while the channel dimension is increased at each stage, thereby significantly reducing the dimensional burden of subsequent sequential modeling. Each stage includes Batch Normalization and ReLU activation. The output of the Stem is a feature tensor $X_{stem} \in \mathbb{R}^{B \times C_{out} \times T \times H' \times W'}$, where $H' = \frac{H}{4}$ and $W' = \frac{W}{4}$.

Through this feature transformation strategy, we significantly reduce the input sequence length to the Mamba module, effectively compressing computational overhead and improving model efficiency. This allows the state space model to handle long-range dependencies with substantially lower resource consumption, while preserving its ability to capture essential physiological information.

3.3 Local Feature Extractor Module

Physiological signals exhibit certain periodicity and regularity, and in addition to long-term temporal information, local temporal information along the time dimension should also be considered in signal computation. In related studies [20], the Temporal Shift Module (TSM) effectively captures local temporal features in video data by exchanging adjacent frame information, and its effectiveness has been experimentally validated.

Inspired by this, we employ 3D separable convolutions for local information extraction in our method. The module first uses depthwise $3 \times 3 \times 3$ convolutions (Depthwise Conv3D) to independently extract local spatiotemporal features along each channel. Then, a $1 \times 1 \times 1$ pointwise convolution (Pointwise Conv3D) is applied to fuse features along the channel dimension, enabling information reorganization and feature weighting.

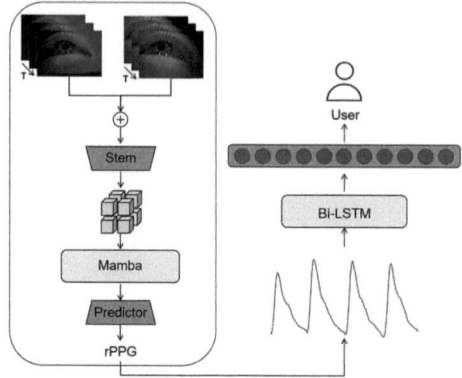

Fig. 3. Overview of identity identification architecture. At this stage, the predicted rPPG waveform is fed into the network to perform identity identification.

3.4 Identity Identification Based on rPPG

As shown in Fig. 3, during identity identification, we largely follow the PhysVR architecture but exclude the Channel Attention and Local Feature Extractor modules. Based on our observations, this may help avoid an overemphasis on heart rate-related features, which could potentially interfere with identity-specific representation. To effectively model and classify the predicted one-dimensional physiological waveform 1D signal, we employ a Bidirectional Long Short-Term Memory (BiLSTM) network to extract contextual dependencies from the temporal signal. This is followed by a dropout layer and a fully connected layer to perform the final classification.

The input to this module is a tensor $\mathbf{Y} \in \mathbb{R}^{B \times 1 \times 256}$, where B denotes the batch size, and each sample corresponds to a 1D rPPG signal of length 256. This input is first processed by a BiLSTM with two layers and hidden size 256. Due to its bidirectional nature, the output tensor becomes $\mathbf{H} \in \mathbb{R}^{B \times 1 \times 512}$.

A squeeze operation is then applied to remove the singleton temporal dimension, resulting in $\mathbf{H}' \in \mathbb{R}^{B \times 512}$. To alleviate overfitting, a dropout layer with a probability of 0.5 is used. Finally, the processed features are passed through a fully connected (FC) layer to obtain the final prediction $\mathbf{Y} \in \mathbb{R}^{B \times C}$, where $C = 11$ is the number of identity classes.

4 Evaluation

4.1 Dataset

To validate the effectiveness of our system in monitoring heart rate within VR environments, we conducted a user study approved by the Institutional Review Board (IRB). A total of 11 participants, aged between 18 and 25, were recruited. This age range was chosen for two main reasons: (1) young adults are the primary users of VR technology; and (2) existing rPPG datasets, such as UBFC-Phys [27], predominantly include participants from this age group.

Specifically, each subject contributed approximately 32 min of data across 8 sessions, resulting in a total data duration of 6.4 h. Data collection was conducted using an HTC VIVE PRO VR headset [3], equipped with a Droolon F1 eye-tracking module [2] from 7invensun. During data collection, a stable infrared light source provided by the Droolon F1 was used. The VR interface was standard, and participants were allowed to blink and perform natural, non-intensive head movements. A video data acquisition system was developed using the aSeeVRUserSDK, while a CMS50E pulse oximeter [1], placed on the subject's finger, was used to measure photoplethysmographic (PPG) signals, which served as the ground truth for the system. The video recordings were captured at a resolution of 320×240 pixels and a frame rate of 120 frames per second. The PPG signal was recorded at a frequency of 60 Hz.

To ensure reliable data quality and consistency between the CMS50E pulse oximeter signals and video frames, we applied strict synchronization filtering. Using microsecond-level timestamps, we resampled both signals to 30 Hz, achieving millisecond-level alignment.

4.2 Experimental Setup

In the heart rate estimation phase, we employed a two-fold cross-validation strategy to evaluate the system's performance. In the first fold, data from six subjects were used for training, while data from the remaining five subjects were used for testing. In the second fold, the roles of the two groups were swapped. This strategy ensures a strict separation between training and testing data in terms of both subjects and acquisition sessions, thereby effectively eliminating the risk of data leakage. Furthermore, it closely simulates real-world deployment scenarios where the individuals used to train the model differ from the end users, enabling a more realistic and robust assessment of the system's generalizability across unseen subjects.

In the identity identification phase, we adopted a user-specific session-based partitioning strategy. Specifically, for each subject, the first six data collections were used as the training set, while the subsequent 7th and 8th collections served as the test set. This setup ensures that the training and testing samples come from different acquisition sessions, thus evaluating the model's ability to generalize across sessions and over time.

4.3 Comparison Method

GREEN [31]: A traditional method for estimating PPG signals using linear algebra and conventional signal processing techniques. Since the data collected in the VR scenario is single-channel, RGB three-channel-based methods are not applicable. Therefore, this study adopts the GREEN method. This approach utilizes the green channel information as a proxy for the spatially averaged PPG signal from RGB video. Due to its reliance on a single channel, this approach can be readily adapted to the single-channel data setting of our system.

PhysNet [36]: PhysNet is an end-to-end spatiotemporal network designed for rapid and efficient recovery of remote photoplethysmographic (rPPG) signals from raw facial videos. This work proposed two network architectures for rPPG signal estimation: a 3D-CNN and a 2D-CNN combined with an RNN. The former leverages 3D convolutional kernels to extract spatiotemporal features, while the latter processes spatial and temporal information separately. Empirical comparisons have demonstrated that the 3D-CNN architecture yields superior performance. Therefore, in our comparative experiments, we adopt the 3D-CNN configuration. The model is trained using the negative Pearson correlation loss, which optimizes the trend alignment and pulse peak localization between the predicted and ground-truth signals. Unlike the original PhysNet, we replace the input with our system's single-channel data and treat the left and right eye regions as two separate channels. This modification preserves the core architecture and loss function, while enabling the investigation of how different input configurations affect rPPG signal estimation.

PhysFormer [37]: PhysFormer is an end-to-end video Transformer architecture designed for remote physiological measurement. Based on a video Transformer backbone, it adaptively aggregates both local and global spatiotemporal features to enhance the representation of rPPG-related signals. Prior studies have shown that it outperforms traditional and learning-based methods such as POS [33], DeepPhys [9], and PhysNet [36]. In terms of experimental configuration, similar to the modifications applied in the GREEN and PhysNet, we adapt the input format to match the characteristics of our system, which captures single-channel video data. While the core architecture of PhysFormer remains unchanged, the input configuration is adjusted to ensure compatibility with our single-channel recordings and to explore its impact on rPPG signal estimation performance.

PhysVR (w/o CA & LFE): We implement a simplified version of the Mamba framework. Unlike the full model, this variant excludes additional modules such as channel attention mechanisms and 3D separable convolutions. Instead, it retains only the Selective State Space Model (SSM) component of Mamba for temporal feature extraction. The model operates on grayscale video input to ensure consistency with the input modality used in the complete system. Due to its superior performance in identity identification, this simplified model is also adopted for the identity identification task.

4.4 Evaluation Metrics

Two commonly used evaluation metrics are employed in this work: Mean Absolute Error (MAE) and Root Mean Square Error (RMSE).

$$\text{MAE} = \frac{1}{N}\sum_{i=1}^{N}|\hat{y}_i - y_i|, \quad \text{RMSE} = \sqrt{\frac{1}{N}\sum_{i=1}^{N}(\hat{y}_i - y_i)^2} \tag{1}$$

Table 1. Comparison of Heart Rate Estimation across Different Methods.

Method	MAE↓	RMSE↓
GREEN [31]	21.34	28.07
PhysNet [36]	13.16	17.43
PhysFormer [37]	6.49	8.34
PhysVR (w/o CA & LFE)	5.89	7.37
PhysVR	5.11	6.51

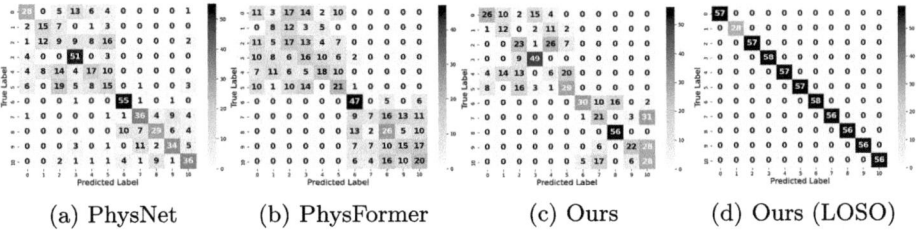

(a) PhysNet (b) PhysFormer (c) Ours (d) Ours (LOSO)

Fig. 4. Confusion matrices for identity identification. Classification results based on rPPG signals predicted by different methods. Darker diagonal elements indicate higher classification accuracy.

where N denotes the number of samples, y_i represents the ground-truth heart rate estimation calculated from the reference PPG signal, and $\hat{y}_i$ denotes the predicted heart rate estimation derived from the estimated rPPG signal.

4.5 Result

Heart Rate Estimation. The results of heart rate estimation task are summarized in Table 1. GREEN [31] and PhysNet [36] both rely on color information from the RGB space for pulse signal estimation. GREEN directly utilizes the green channel, while PhysNet projects the RGB input into a color subspace with enhanced representational capacity to improve feature extraction. However, in our study, single-channel infrared grayscale images are used, which differ significantly in imaging principles from RGB data. As a result, these color-dependent methods face limitations in feature representation, leading to degraded performance. As shown in Table I, GREEN achieves an MAE of 21.34 bpm and an RMSE of 28.07 bpm, while PhysNet yields 13.16 bpm and 17.43 bpm, respectively. These results clearly indicate that such models are not well-suited for infrared data and suffer from substantial accuracy loss in this setting.

PhysFormer [37], which adopts a Transformer-based architecture for spatiotemporal feature modeling, demonstrates stronger temporal modeling capability and overall better performance than traditional methods. It achieves an MAE of 6.49 bpm and an RMSE of 8.34 bpm, indicating a certain modeling advantage.

Our proposed PhysVR (w/o CA & LFE) model leverages a simplified Mamba-based Selective State Space Model (SSM) to capture long-range temporal dependencies in remote heart rate signals. Results show that, even under the condition of single-channel infrared video data, the model effectively captures temporal physiological information, achieving 5.89 bpm MAE and 7.37 bpm RMSE. This demonstrates the importance of long-term temporal information in rPPG signal prediction.

Building upon the PhysVR (w/o CA & LFE), we further introduce a channel attention mechanism and 3D depthwise separable convolutions to construct the final model i.e., PhysVR. These enhancements improve the fusion of multi-view features and strengthen the modeling of local dynamic variations. Experimental results indicate that this proposed architecture achieves an MAE of 5.11 bpm and an RMSE of 6.51 bpm, showcasing superior adaptability and feature modeling capabilities for this task. In addition, the average computation time for deriving a single heart rate estimate is approximately 0.1 s, confirming the model's suitability for real-time applications.

Identity Identification. To evaluate the potential of rPPG signals for identity identification tasks, we utilized the predicted rPPG waveforms as physiological representations for different users and introduced a classification strategy to assess their discriminability.

In Fig. 4(a)–(d) respectively show the confusion matrices for identity identification classification using rPPG waveforms obtained from PhysNet, Phys-Former, and Ours (refers to the model proposed in this paper based on Mamba i.e., PhysVR (w/o CA & LFE)), combined with BiLSTM. For each matrix, vertical labels are the actual classes and the horizontal labels are the predicted classes. The values in each cell indicate the number of samples where a given actual class was classified as a specific predicted class. The rPPG prediction in Fig. 4(a)–(c) was conducted using a two-fold cross-validation scheme, where users 0–5 and users 6–10 were processed by two independently trained models. In contrast, Fig. 4(d) presents the results obtained using a leave-one-subject-out (LOSO) strategy, where for each test subject, a distinct model was trained on data from the remaining ten users. Under the two-fold cross-validation setup, user identification achieved an accuracy of 34.56% on the waveforms predicted by Pherformer, 54.53% on the rPPG waveforms predicted by PhysNet, and 50.67% on the results from our proposed method. We achieved 100% accuracy when using the LOSO training strategy. Results indicate that training a dedicated model for each user can enhance the inter-user differences in rPPG signals.

Nevertheless, under the two-fold cross-validation training scheme, certain users (e.g., 55 out of 58 samples from user6 were correctly predicted, and 36 out of 56 samples from both user 7 and user 10 were correctly predicted, as shown in Fig. 4(a)) still exhibit strong identifiability, implying that rPPG signals do contain subject-specific features. This highlights the potential of rPPG signals as a biometric trait for identity identification.

Conclusion

This study makes an initial attempt to address the challenge of contactless physiological signal measurement and identity identification in real-world virtual reality scenarios. To this end, we propose a novel rPPG-based system tailored for virtual reality environments. This system leverages near-infrared video captured by head-mounted devices and processes the data using our proposed PhysVR model to estimate heart rate from the predicted rPPG waveforms. Furthermore, the predicted rPPG signals are fed into a BiLSTM network to perform identity identification. We conducted a comparative analysis between our method and existing approaches for heart rate estimation using rPPG waveforms. Additionally, we explored identity identification through a classification task. The physVR model achieved an MAE of 5.11 bpm and an RMSE of 6.51 bpm, while identity identification achieved 100% accuracy under the LOSO training. Experimental results validate the effectiveness of our proposed method and highlight the potential of VR-based rPPG signals in the field of biometric user identification.

Acknowledgments. This work is partially supported by the Natural Science Foundation of Shandong Province (Grant No. 2022HWYQ040 and Grant No. ZR2024ZD12, Major Basic Research), and supported by the Open Project Program of State Key Laboratory of Virtual Reality Technology and Systems, Beihang University (No. VRLAB2025C04).

Disclosure of Interests. The authors have no competing interests to declare that are relevant to the content of this article.

References

1. Cms50e introduction. https://www.contecmed.com.cn/productinfo/817903.html
2. Drollon F1 introduction. https://www.7invensun.com/droolonf1xqy
3. Vive-pro introduction. https://www.vive.com/cn/product/vive-pro/
4. Adib, F., Mao, H., Kabelac, Z., Katabi, D., Miller, R.C.: Smart homes that monitor breathing and heart rate. In: Proceedings of the 33rd Annual ACM Conference on Human Factors in Computing Systems, pp. 837–846 (2015)
5. Agrawal, V., Hazratifard, M., Elmiligi, H., Gebali, F.: Electrocardiogram (ECG)-based user authentication using deep learning algorithms. Diagnostics **13**(3), 439 (2023)
6. Allen, J.: Photoplethysmography and its application in clinical physiological measurement. Physiol. Meas. **28**(3), R1 (2007)
7. Buffington, S., Yu, S., Gersh, N., Elor, A.: A stress reactive immersive virtual reality survival game with biofeedback (2021)
8. Chauhan, U., Reithinger, N., Mackey, J.R.: Real-time stress assessment through PPG sensor for VR biofeedback. In: Proceedings of the 20th International Conference on Multimodal Interaction: Adjunct, pp. 1–5 (2018)
9. Chen, W., McDuff, D.: DeepPhys: video-based physiological measurement using convolutional attention networks. In: Proceedings of the European Conference on Computer Vision (ECCV), pp. 349–365 (2018)

10. Cui, F., Wang, Y., Lei, S., Shi, Y.: CardboardHRV: bridging virtual reality and biofeedback with a cost-effective heart rate variability system. In: Extended Abstracts of the CHI Conference on Human Factors in Computing Systems, pp. 1–6 (2024)
11. De Haan, G., Jeanne, V.: Robust pulse rate from chrominance-based rPPG. IEEE Trans. Biomed. Eng. **60**(10), 2878–2886 (2013)
12. Gnacek, M., et al.: Heart rate detection from the supratrochlear vessels using a virtual reality headset integrated PPG sensor, pp. 210–214 (2021). https://doi.org/10.1145/3395035.3425323
13. Gu, A., Dao, T.: Mamba: linear-time sequence modeling with selective state spaces. arXiv preprint arXiv:2312.00752 (2023)
14. Gu, A., Goel, K., Ré, C.: Efficiently modeling long sequences with structured state spaces. arXiv preprint arXiv:2111.00396 (2021)
15. Li, K., et al.: VideoMamba: state space model for efficient video understanding. In: Leonardis, A., Ricci, E., Roth, S., Russakovsky, O., Sattler, T., Varol, G. (eds.) ECCV 2024. LNCS, vol. 15084, pp. 237–255. Springer, Cham (2024). https://doi.org/10.1007/978-3-031-73347-5_14
16. Li, L., Chen, C., Pan, L., Zhang, J., Xiang, Y.: Video is all you need: attacking PPG-based biometric authentication. In: Proceedings of the 15th ACM Workshop on Artificial Intelligence and Security, pp. 57–66 (2022)
17. Li, L., Chen, C., Pan, L., Zhang, L.Y., Zhang, J., Xiang, Y.: SigA: rPPG-based authentication for virtual reality head-mounted display, pp. 686–699 (2023)
18. Li, M., Banerjee, N.K., Banerjee, S.: Using motion forecasting for behavior-based virtual reality (VR) authentication. In: 2024 IEEE International Conference on Artificial Intelligence and eXtended and Virtual Reality (AIxVR), pp. 31–40. IEEE (2024)
19. Liu, X., Wu, H., Ou, W.: Identity authentication via ECG and PPG signals: an innovative method incorporating singular spectrum analysis and feature integration. In: 2024 IEEE 7th Advanced Information Technology, Electronic and Automation Control Conference (IAEAC), vol. 7, pp. 505–511. IEEE (2024)
20. Liu, X., Fromm, J., Patel, S., McDuff, D.: Multi-task temporal shift attention networks for on-device contactless vitals measurement. Adv. Neural. Inf. Process. Syst. **33**, 19400–19411 (2020)
21. Mathis, F., Williamson, J., Vaniea, K., Khamis, M.: RubikAuth: fast and secure authentication in virtual reality. In: Extended Abstracts of the 2020 CHI Conference on Human Factors in Computing Systems, pp. 1–9 (2020)
22. Niu, X., Shan, S., Han, H., Chen, X.: RhythmNet: end-to-end heart rate estimation from face via spatial-temporal representation. IEEE Trans. Image Process. **29**, 2409–2423 (2019)
23. Olade, I., Fleming, C., Liang, H.N.: BioMove: biometric user identification from human kinesiological movements for virtual reality systems. Sensors **20**(10), 2944 (2020)
24. Peng, X., Cui, J., Kneip, L.: Articulated multi-perspective cameras and their application to truck motion estimation. In: 2019 IEEE/RSJ International Conference on Intelligent Robots and Systems (IROS), pp. 2052–2059. IEEE (2019)
25. Poh, M.Z., McDuff, D.J., Picard, R.W.: Advancements in noncontact, multiparameter physiological measurements using a webcam. IEEE Trans. Biomed. Eng. **58**(1), 7–11 (2010)
26. Poh, M.Z., McDuff, D.J., Picard, R.W.: Non-contact, automated cardiac pulse measurements using video imaging and blind source separation. Opt. Express **18**(10), 10762–10774 (2010)

27. Sabour, R.M., Benezeth, Y., De Oliveira, P., Chappe, J., Yang, F.: UBFC-phys: a multimodal database for psychophysiological studies of social stress. IEEE Trans. Affect. Comput. **14**(1), 622–636 (2021)
28. Salem, A., Ibrahem, H., Kang, H.S.: RCA-LF: dense light field reconstruction using residual channel attention networks. Sensors **22**(14), 5254 (2022)
29. Shen, Y., et al.: GaitLock: protect virtual and augmented reality headsets using gait. IEEE Trans. Dependable Secure Comput. **16**(3), 484–497 (2018)
30. Stephenson, S., Pal, B., Fan, S., Fernandes, E., Zhao, Y., Chatterjee, R.: SoK: Authentication in augmented and virtual reality. In: 2022 IEEE Symposium on Security and Privacy (SP), pp. 267–284. IEEE (2022)
31. Verkruysse, W., Svaasand, L.O., Nelson, J.S.: Remote plethysmographic imaging using ambient light. Opt. Express **16**(26), 21434–21445 (2008)
32. Wang, J., Peng, K.: A multi-view gait recognition method using deep convolutional neural network and channel attention mechanism. Comput. Model. Eng. Sci. **125**(1), 345–363 (2020)
33. Wang, W., Den Brinker, A.C., Stuijk, S., De Haan, G.: Algorithmic principles of remote PPG. IEEE Trans. Biomed. Eng. **64**(7), 1479–1491 (2016)
34. Xiao, T., Singh, M., Mintun, E., Darrell, T., Dollár, P., Girshick, R.: Early convolutions help transformers see better. Adv. Neural. Inf. Process. Syst. **34**, 30392–30400 (2021)
35. Yang, H., Fan, Y., Jin, Y., Shi, H., Li, T.: Pathword: a 3D identity authentication interface based on connection trajectory. In: 2023 IEEE Conference on Virtual Reality and 3D User Interfaces Abstracts and Workshops (VRW), pp. 951–952. IEEE (2023)
36. Yu, Z., Li, X., Zhao, G.: Remote photoplethysmograph signal measurement from facial videos using spatio-temporal networks. arXiv preprint arXiv:1905.02419 (2019)
37. Yu, Z., Shen, Y., Shi, J., Zhao, H., Torr, P.H., Zhao, G.: PhysFormer: facial video-based physiological measurement with temporal difference transformer. In: Proceedings of the IEEE/CVF Conference on Computer Vision and Pattern Recognition, pp. 4186–4196 (2022)

Temporal Decision-Making Optimization for Intelligent Agents with Gradually Clarified Objectives

Chen Gu[1], Guo Chen[2], Fangwan Huang[2(✉)], Xuanyun Liu[2], Zhiyong Yu[2], and Yuezhong Wu[2]

[1] Maynooth International Engineering College, Fuzhou University, Fujian, China
[2] College of Computer and Data Sciences, Fuzhou University, Fujian, China
hfw@fzu.edu.cn

Abstract. The intelligent decision-making system has significant application value in fields such as autonomous driving and industrial control. However, the performance of existing decision systems largely depends on the quality of information acquisition and analysis, which leads to the following limitations. Firstly, high-precision information processing requires the configuration of high-performance sensors and computing units, which increases the system deployment cost and makes it unsuitable for resource-constrained scenarios. Secondly, in dynamic scenarios where the granularity of information gradually becomes clearer over time, the early coarse-grained information is not effectively utilized, causing the overall decision-making benefits to fall short of the theoretical upper bound. Therefore, this paper first designs a temporal scenario with progressively clarifying objectives, which simulates the dynamic characteristic of information granularity gradually improving as the agent approaches the target. Additionally, a risk assessment mechanism is incorporated into the decision-making process. Based on this scenario, a Two-Stage Multi-Objective Dynamic Policy Network (TSMODPN) is proposed. In this model, the first stage involves offline training of a strategy pool with diverse preferences, while the second stage employs an online dynamic policy network. This network first employs a dynamic distillation module to efficiently extract features from initial fuzzy information. Thereafter, the temporal preference selection module dynamically selects optimal strategies from the offline strategy pool based on these extracted features. The experimental results explicitly show that TSMODPN surpasses other baseline models in its ability to approach the theoretical upper bound for both safety benefits and task duration across varying risk levels.

Keywords: Temporal Decision-Making · Risk Assessment · Reinforcement Learning · Multi-objective Optimization

1 Introduction

With the development of Internet of Things (IoT) technology, various sensing devices (such as smartphones, wearable devices, and contactless sensors) are able to collect large amounts of time-series data, including multiple indicators such as heart rate, blood pressure, and light intensity. By analyzing these time-series data, support can be provided for various intelligent systems. Sometimes, in the initial stage, only coarse-grained complete data (i.e., fuzzy data) can be obtained. As time goes by, the details of the data gradually become clearer, and the granularity of the information also gradually refines. For example, during the process of an autonomous vehicle moving towards a target, the resolution of the target image follows a pattern of progressive clarity [1].

Many existing applications adopt single-step decision-making based on the clearest available information [2]. The advantage of a single-step decision lies in its reliance on more complete and accurate information, which can reduce the decision-making error rate by utilizing a higher recognition rate. However, since the decision is made only when the information is fully available, it has poor timeliness and misses the opportunity for early decision-making. In contrast, multiple decision-making processes iteratively adjusts decisions as the data gradually becomes clearer. This strategy allows early decision-making when information is incomplete, making it suitable for applications requiring quick responses. To compare the characteristics of these two decision-making approaches, this paper first constructs a multi-objective optimization scenario, in which an agent performs sequential decision-making on progressively clarified images. In the end, the two approaches are appraised by determining the final decision-making benefits.

The scenario studied in this paper simulates the process in which an agent moves along a road from the starting point to the target. During this movement, the agent needs to make action decisions based on its judgment of the target. As illustrated in Fig. 1, when executing the m-th task, the agent moves toward the target image (marked as n) with an initial speed of V_0. During this movement, the agent receives fuzzy information about the target image, with smaller numbered blocks corresponding to more blurry images. As the agent gets closer to the target, the resolution of the process image gradually improves, simulating the process of progressively clarifying the target. For simplicity, this paper assumes that there are n process images evenly distributed along the distance D between the starting point and the target. Therefore, the distance between adjacent images is $d = D/n$. Based on this scenario, a single-step decision refers to the agent making a judgment and a single action decision (e.g., accelerate or decelerate) after reaching the target image (block n), while multiple decisions refer to the agent making judgments and action decisions for each block's process image.

To more reasonably evaluate the temporal decision-making rewards of an intelligent agent, this paper also considers the risk level of the target image in the given scenario. In real-world scenarios, the quality of an agent's decisions is closely related to its risk assessment of the environment [3]. This is because the consequences of misidentifying images with different risk levels vary. The consequences of misidentifying a safe image are usually minimal, whereas failing to recognize a dangerous one may cause severe or even irreversible losses. For example, when a person moves through a scene and needs to assess objects on the road to decide walking speed, mistakenly identifying a snake (dangerous category) as a rope (safe category) and choosing to speed up can pose a serious threat to his safety. By contrast, misidentifying a water pipe (safe category) as a rope or a snake has little impact on safety. Although these two decisions are made due to insufficient recognition abilities leading to decisions inconsistent with the actual situation, their consequences are greatly different. Therefore, the intelligent agent should fully consider the safety or risk level in their environment when making decisions, which is a key factor that cannot be overlooked in the decision-making process.

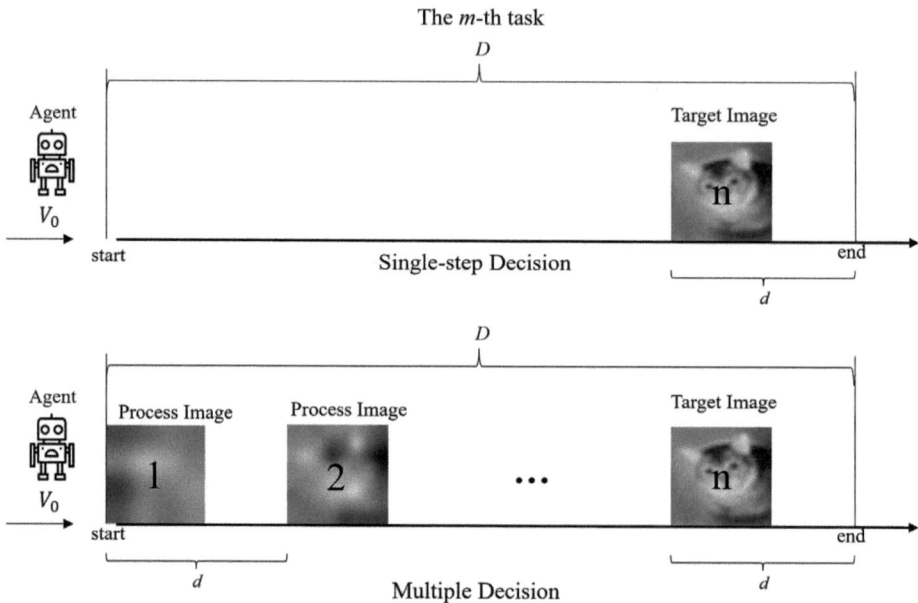

Fig. 1. A temporal decision-making scenario for agents with progressively clarified objectives, where smaller numbered blocks correspond to more blurred images, D represents the total distance, and d represents the distance between adjacent images

Based on the above scenario, the agent needs to make judgments on a series of target images with diverse risk levels and then make action decisions. The evaluation of the decisions must consider the risks brought by misidentifying the target images and the average duration of the agent's traversal. Therefore, the agent needs to strike a balance between the safety objective and the efficiency objective, which essentially constitutes a multi-objective optimization problem.

In view of the limitations of current multi-objective optimization methods, this paper proposes a two-stage multi-objective model based on dynamic policy network (TSMODPN), providing a new approach to solving the multi-objective optimization problem in sequential scenarios where the objectives gradually become clearer. The contributions of this paper are summarized as follows:

- This paper designs a temporal scenario with gradually clarified objectives, and integrates a risk assessment mechanism into the decision-making process, thus addressing a practical yet underexplored problem setting.
- TSMODPN is proposed not only to fully extract the features of early blurry images but also to dynamically select appropriate preferred strategies from the strategy pool according to the environment.

- Experimental data demonstrates that the multi-objective optimization of security rewards and task duration by TSMODPN in different risk scenarios gets the nearest to the theoretical upper bound.

2 Related Work

2.1 Early Classification

In the field of data mining research, early classification problems have received increasing attention, especially in applications in industries such as healthcare, manufacturing, energy, and transportation [4–7]. In these fields, some data are highly sensitive to time and environmental changes, making the real-time collection and accurate processing of data particularly critical in these application scenarios. During dynamic data collection, the main task of early classification is how to complete efficient classification as early as possible while the data is gradually refined. This optimization process typically involves a reasonable balance between the credibility of the data and the timeliness of the task. Based on different classification strategies, early classification methods can be divided into several types: prefix-based methods, shapelet-based methods, model-based methods, and miscellaneous methods [8].

As a typical classification technique, the minimum prefix method is based on the core idea of constructing a series of classifiers with increasing prefix lengths [9]. However, the computational consumption of this method grows linearly with the sequence length, limiting its application in long sequence. The shapelet-based methods focus on obtaining a set of key subsequences from the training dataset and using them as the class-distinguishing features of the time series [10]. However, the drawback of this method is that finding shapelets that meet the early classification requirements is a very complex process. Recently, an increasing number of studies have begun to apply deep learning methods to early classification tasks. For example, Sharma et al. [11] proposed a hybrid deep learning classification model using Convolutional Neural Networks (CNNs) and Recurrent Neural Networks (RNNs), which utilized mobile data for early transportation mode detection. Similarly, to solve the problem of early cardiovascular classification based on variable-length time series, Huang et al. [12] proposed a reinforcement learning framework based on long-term dependencies in time series. Martinez et al. [13] introduced a reinforcement learning method for early time series classification, enabling the model to dynamically adjust its decision strategy based on the characteristics of the time series and the difficulty of the task, introducing the concept of adaptive learning.

In summary, the existing research on early classification has the following two drawbacks. Firstly, most studies mainly focus on time-scale divisions, while the exploration of other scales (e.g., spatial scales) is insufficient. Secondly, the relevant studies frequently make single-step decisions based only on information from specific periods and do not continuously optimize decisions using subsequent data.

2.2 Multi-Objective Optimization

Many complex decision problems in the real world require the consideration of multiple objectives. Multi-objective optimization methods provide an important technical pathway for solving trade-off decisions in complex environments. Current multi-objective optimization methods can be mainly classified into three categories: evolutionary algorithms, weighted sum methods, and multi-objective reinforcement learning [14].

Firstly, evolutionary algorithms, represented by genetic algorithms, perform excellently in multi-objective optimization but have inherent defects in their adaptability in dynamic environments [15]. The target progressive clarity scenario discussed in this paper requires the establishment of a mapping relationship between environmental factors such as target images, danger coefficients, and the agent's actions. This can only be achieved if the solution of the genetic algorithm is presented in the form of a neural network, as otherwise, the solution set obtained under changing environments would not be applicable. If genetic algorithms are to be used to complete the multi-objective optimization task in the target progressive clarity scenario, the neural network weights of this strategy must be encoded as individuals and subjected to selection, crossover, mutation, and other operations [16]. However, encoding the neural network structure and parameters is complex, and as the network scale increases, the time complexity of genetic algorithms also rises sharply. Therefore, conventional evolutionary algorithms are not suitable for the target progressive clarity scenario proposed in this paper.

Secondly, the weighted sum method assigns different weights to each objective function and sums them up, thus converting a multi-objective problem into a single-objective optimization problem [17]. However, in the target progressive clarity scenario discussed in this paper, simply integrating multiple objectives into a single objective is not suitable. This is because the relative importance of safety and efficiency objectives changes with the variation in the danger coefficient. For instance, in high-risk situations, the importance of the security goal increases dynamically. But at the same time, the efficiency goal cannot be ignored either. Otherwise, there may arise the problem that multiple decisions tend to adopt conservative strategies. Therefore, for target progressive clarity scenarios with varying danger coefficients, traditional single-objective integration methods have obvious limitations.

Lastly, Multi-Objective Reinforcement Learning (MORL) methods can optimize multiple objectives simultaneously by vectorizing reward functions, but the Pareto front they generate faces a selection dilemma in sequential decision-making. According to the Pareto dominance relationship definition [18], in each decision step of the target progressive clarity scenario, the agent needs to select a specific solution from the Pareto front. However, existing MORL methods lack an effective dynamic selection mechanism. More importantly, as the granularity of information changes (from fuzzy to clear), the relative importance of each objective dynamically changes, making static Pareto front selection strategies unable to guarantee global optimality.

In summary, existing multi-objective optimization methods have insufficient dynamic adaptability in addressing the target progressive clarity scenario. This motivates the proposal of a two-stage model based on dynamic policy networks to solve the multi-objective optimization problem in multi-step decisions under the target progressive clarity scenario.

3 Optimization Objective

According to the scenario described in Fig. 1, this paper first defines the average duration G_t, which reflects the average traversal efficiency of the agent in multiple target image tasks under the progressive clarity scenario. Its formula is shown as (1).

$$G_t = \frac{1}{M} \sum_{m=1}^{M} T_m \qquad (1)$$

M is the number of tasks, and T_m represents the task duration for the agent to reach the destination in the m-th task, as shown in (2).

$$T_m = \sum_{k=1}^{n} \frac{d}{V_k^m} \qquad (2)$$

The velocity after going through the k-th process image V_k^m is given by (3), where Δv is the speed variation of the agent. For simplification, Δv can be set as a fixed value. To prevent the speed from becoming negative, it can be specified that $\Delta v < V_0/n$, thereby ensuring $V_k^m > 0$ at any moment. Clearly, a lower value of G_t indicates shorter task duration and higher efficiency for the agent.

$$V_k^m = V_{k-1}^m \pm \Delta v, k = 1, 2 \ldots, n \text{ and } V_0^m = V_0 \qquad (3)$$

Furthermore, this paper defines the average damage G_s, which reflects the cumulative harm inflicted on the agent by multiple target images in the environment. Its calculation formula is given in (4).

$$G_s = \frac{1}{M} \sum_{m=1}^{M} G_{sm} \qquad (4)$$

where G_{sm} denotes the damage when the agent reaches the destination in the m-th task, as defined in (5).

$$G_{sm} = \begin{cases} \alpha_i V_n^m, & \text{label} = l_{di} \\ 0, & \text{label} = l_{sj} \end{cases} \qquad (5)$$

When the agent passes a target image point belonging to the dangerous category l_{di} (e.g., obstacles or threats), the incurred damage is proportional to the product of its current speed V_n^m and the danger coefficient α_i. A higher product value indicates greater damage, and vice versa. Conversely, when passing a safe category l_{sj} (e.g., neutral or beneficial objects), the damage is 0, meaning no harm is sustained.

In summary, a smaller value of the average damage G_s indicates greater safety benefits. Therefore, minimizing G_s can be regarded as optimizing for safety objectives, effectively mitigating potential risks to the agent within the environment. Similarly, minimizing the average duration G_t optimizes for efficiency objectives by reducing the time required for the agent to complete a set of tasks. Thus, the dual optimization objectives of this paper are to maximize safety benefits (by minimizing G_s), and minimize task duration (by minimizing G_t), under the scenario where target clarity progressively improves.

4 Method

TSMODPN proposed in this paper is shown in Fig. 2, which includes an offline phase and an online phase. In the offline phase, a preference sampling mechanism is employed to assign different preferences to each target and train a separate reinforcement learning policy for each preference. This process constructs a diversified strategy pool with tailored preference orientations. In the online phase, a dynamic preference temporal network is introduced to real-time select the optimal policy from the pre-constructed strategy pool. This selection is based on comprehensive environmental information—including the agent's state, target image features, and danger coefficient. The offline phase evaluates the performance of each strategy under different target preferences using simulated environment data to ensure that the agent can respond quickly in the online phase. Meanwhile, the online phase combines the pre-trained strategy pool to quickly judge and switch to the optimal policy under high real-time requirements. Through this model, the agent can flexibly select the optimal policy at each decision-making step in a dynamic environment.

4.1 The Offline Stage

In the offline phase, by sampling different preferences, i.e., setting different preference weights for safety and efficiency goals, a separate reinforcement learning strategy is trained for each preference weight, thus constructing a pool of strategies with different preferences. This supports the agent in the online phase to quickly select the appropriate strategy in a dynamic environment.

For each strategy in the pool, a multi-decision method based on Deep Q Networks (DQN) is used. DQN can learn and optimize at multiple decision points, i.e., at different levels of information, through reinforcement learning, thus gradually improving decision quality. The multi-decision method based on Deep Q Networks is described as follows.

In a scenario where the target gradually becomes clearer, the agent moving toward the target on the road is defined as a reinforcement learning agent. The agent faces two action choices in different states: accelerate (acc) and decelerate (dec). In the environment set in this paper, the process of the agent moving from the most blurred position to the clearest position of a target image (i.e., Fig. 1) is an episode in reinforcement learning. In one episode, rewards are set for the decisions made by the agent based on different granularity of information obtained at different positions. In the offline phase, the approach used is to transform the multi-objective problem into a single-objective optimization problem. However, this paper does not simply sum up the multi-objective rewards but sets a different preference for each objective reward, obtaining multiple

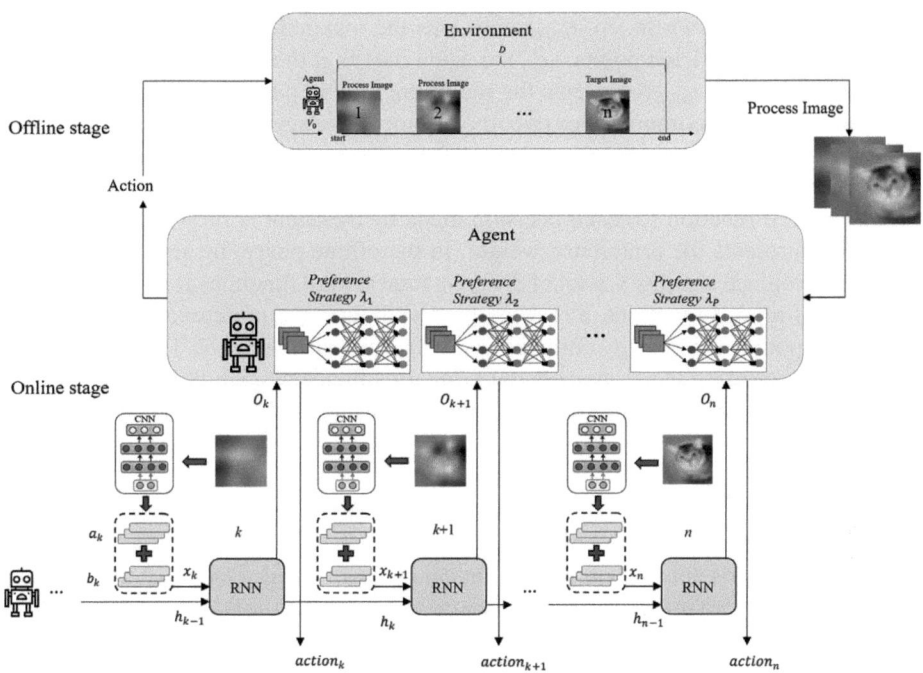

Fig. 2. A two-stage multi-objective overall model based on dynamic policy network, where the first stage involves offline training of a strategy pool with diverse preferences, while the second stage employs an online dynamic policy network

strategies with different preferences, thus avoiding overemphasis or underemphasis on any particular goal. The specific setup is described in (6–10).

$$R_k^s = \begin{cases} -1, & if label = l_{di} and action_k = acc \\ 0, & otherwise \end{cases} \quad (6)$$

$$R_k^t = -\frac{t_k - t_{min}}{t_{max} - t_{min}} \quad (7)$$

$$t_k = \frac{d}{V_k^m} \quad (8)$$

$$V_0 - n \times \Delta V \leq V_k^m \leq V_0 + n \times \Delta V \quad (9)$$

$$R_k = \lambda R_k^s + (1 - \lambda) R_k^t \quad (10)$$

R_k^s represents the safety penalty for the agent's decision at each decision point. If the current category is a dangerous category and the agent chooses to accelerate, it should be penalized to prevent the agent from accelerating in dangerous situations. R_k^t represents the time penalty for the agent's decision, where t_k is the time spent by the agent between two decision points, i.e., the time required from the (k-1)-th process image to the k-th process image, as shown in (8). t_{max} represents the maximum time the agent takes to pass between two decision points, i.e., the agent moves at the minimum speed over the distance. Conversely, t_{min} represents the minimum time to pass between them, i.e., the agent moves at the maximum speed over the distance. The agent's speed range is related to the number of decision points n the agent makes in the target progression clarity environment, as shown in (9). Combining the reward functions of the two objectives, the total reward function for each decision made by the agent is shown in (10), where $\lambda \in [0,1]$ represents the preference weight. In the offline phase, by setting a series of different preference weights λ, a set of different total reward functions is obtained. Based on this set of reward functions, a strategy pool with different preferences is trained. In the strategy pool, each strategy corresponds to a preference weight λ. The closer λ is to 0, the more aggressive the strategy is, favoring the efficiency goal; conversely, when λ is closer to 1, the strategy is more conservative, favoring the safety goal.

4.2 The Online Stage

In the target progression clarity environment designed in this paper, the strategy adopted by the agent depends not only on the current danger level but also on the extent to which the agent has obtained information about the target image and the agent's current state. If the agent is relatively confident that the target is safe and is far from the target point, the agent can initially adopt a strategy that favors efficiency. As time passes and the agent approaches the target, it obtains more complete information. If the probability of the target being classified as dangerous increases, the agent can then adopt a safety-focused strategy. Therefore, at each decision step, the agent should consider which preference strategy best aligns with the scenario proposed in this paper. Based on this, a dynamic preference temporal network as shown in Fig. 3 is proposed in the online phase of TSMODPN.

Specifically, at the k-th time step, the features of the k-th process image are first extracted using a CNN, and then a distilled category vector a_k is obtained with the help of a dynamic distillation sub-module. Based on the agent's current environmental state features, a vector b_k is obtained. Then, the distilled category vector a_k is concatenated with the state feature vector b_k forming the input x_k for the temporal preference selection module implemented using an RNN. At each decision step, the temporal preference selection module selects a strategy from the strategy pool obtained in the offline stage based on the output. The process image for the current time step is then passed to the selected strategy, allowing the agent to make a decision that aligns with the chosen preference. As the target becomes clearer over time, the dynamic preference network will dynamically select the strategy best suited to the current state. The various components of this network are explained in detail below.

Feature Representation Module. In order for the agent to make the most suitable temporal decisions dynamically based on the current environmental state, it is necessary to perform feature representation on the information that can currently be received. However, the feature extraction in this paper is not intended to directly provide the basis for the agent's decision-making, but rather to ensure that the agent fully receives the environmental information, thereby providing a basis for selecting preferences in the dynamic

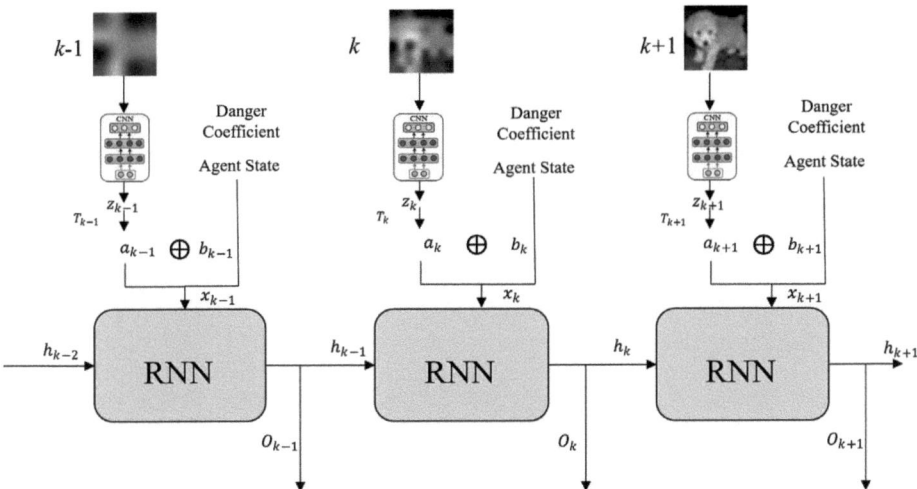

Fig. 3. The structure of the dynamic preference temporal network, which includes the feature representation module and the temporal preference selection module

preference network. Therefore, during the feature extraction phase, a CNN is used to process the image information, and a probability distribution vector for image classification is output through *Softmax*. This probability distribution vector can effectively reflect the agent's understanding of the current process image information. However, since the agent starts far from the target and the information obtained is blurry, this means the agent cannot overly trust the image classification probability distribution vector provided by the CNN in the early stages.

Based on this, this paper draws on the concept of soft targets from knowledge distillation and introduces a dynamic distillation sub-module for image feature extraction. Specifically, in the feature representation during the online phase, a soft target (i.e., a_k) for image classification is obtained using the distillation method. This soft target can serve as a means for data augmentation in the early stages when the information granularity is relatively coarse, and can also reduce the confidence of early process image predictions to some extent. By distilling the CNN output z_k, the category vector a_k is obtained, as calculated in (11).

$$a_{ki} = \frac{exp(z_{ki}/T_k)}{\sum_j exp(z_{kj}/T_k)} \tag{11}$$

where a_{ki} represents the i-th component of a_k, and T_k is the distillation temperature at the k-th time step, which controls the smoothness of the distribution (when $T_k > 1$, the distribution becomes smoother) [19].

In the target progression clarity environment, as the agent moves closer to the target image, the amount of information obtained from the image gradually increases, and the confidence in the prediction of the image recognition module (CNN) also increases. This indicates that the agent requires more "dark knowledge" from the soft target in the early stages, and this need decreases later. Based on this, the dynamic distillation sub-module provides different distillation temperatures for different time steps, with the distillation temperature changing from high to low over time. The specific distillation temperature settings will be detailed in the experimental evaluation section.

After completing the feature extraction of the image information, other factors that influence the agent's decision preference must be considered. These are represented by the vector b_k for the extraction of other features. Firstly, the danger coefficient α has a significant impact on the agent's temporal decisions. For example, as the danger coefficient increases, the agent is more likely to adopt a more conservative strategy. Therefore, the danger coefficient α is included in b_k. To ensure that each component of the input vector has comparable magnitudes, the representation component for the danger coefficient b_{k1} is given in (12).

$$b_{k1} = \frac{\lg\alpha - \lg\alpha_{min}}{\lg\alpha_{max} - \lg\alpha_{min}} \quad (12)$$

where α_{min} and α_{max} respectively represent the minimum and maximum danger coefficients observed in the environment.

Next, in the target progression clarity scenario set in this paper, the distance traveled by the agent $d_k^m = k \times d$ is also a reference factor for selecting the appropriate preference strategy. For example, if the agent's travel distance is short, the agent has more opportunities for adjustment and can adopt a slightly more aggressive strategy. Conversely, if the agent's travel distance is long, the agent has fewer opportunities for adjustment and needs to adopt a relatively conservative strategy. The representation of the agent's travel state b_{k2} is shown in (13).

$$b_{k2} = \frac{d_k^m}{D} \quad (13)$$

If there are other agent-related states or environmental factors that influence the agent's decision preference, these variables can be added as components to the feature vector b_k, depending on the specific application scenario. This also reflects the scalability of the model.

Finally, the distilled category vector a_k and the state feature vector $b_k = [b_{k1}, b_{k2}]^T$ are concatenated. At this point, the feature representation stage is complete, and the concatenated vector x_k becomes the input to the recurrent neural network for the current time step.

Temporal Preference Selection Module. For the target progression clarity scenario proposed in this paper, a decision needs to be made at each time step. Therefore, the temporal preference selection module needs to select the strategy that fits the current state from the strategy pool at each time step, and the preferences between adjacent time steps should be correlated. The choice of strategy preference will gradually evolve toward a certain trend as the target becomes clearer.

Based on these requirements, TSMODPN uses Gated Recurrent Units (GRU) as the core architecture for the temporal preference selection module. At each time step, the GRU needs to calculate the hidden state h_k and output O_k. The hidden state h_k represents the probability distribution of each preference, with its dimension corresponding to the number of different preferred strategies in the strategy pool. The output O_k can determine the preference weight λ for the strategy selected by the dynamic preference network from the strategy pool. For example, if the strategy pool in the offline phase has three strategies, with preference weights $\lambda \in \{0.3, 0.5, 0.7\}$, and if $Softmax(h_k) = [0.7, 0.2, 0.1]$, the temporal preference selection module will select the preference weight λ corresponding to the maximum component in $Softmax(h_k)$ as the output. In this case, $\lambda = 0.3$ will be chosen as the preference output for this time step. This is a classification task, so the loss function is calculated as shown in (14).

$$L = -\sum_{k=1}^{n} y_k \log(Softmax(h_k)) \tag{14}$$

where y_k is the ground-truth label, and $Softmax(h_k)$ is the probability of the preference strategy selected by the current state of the GRU.

5 Experimental Evaluation

This paper validates the effectiveness of TSMODPN in multi-objective tasks within the constructed target progression clarity scenario.

5.1 Dataset

This paper uses a modified CIFAR-10 dataset to evaluate the benefits of single-step decisions and multiple decisions. Below is a detailed introduction to the modified dataset. CIFAR-10 consists of 10 different categories of images, with each category providing 5000 training samples and 1000 test samples. All images are in RGB color format with a size of 32 × 32 pixels. In this paper, a subset of CIFAR-10 containing images of animals that move on land—namely cats, deer, dogs, frogs, and horses—is selected as the experimental dataset. When one of the categories is designated as the "dangerous" category, the other four categories are classified as "safe" categories.

To meet the requirements of the target progression clarity scenario for the agent's temporal decision-making, the down-sampling technique is applied to reduce the original images from 32 × 32 pixels to blurry images of sizes 4 × 4, 8 × 8, 16 × 16, and 24 × 24. This means that the agent needs to progressively identify and make decisions based on increasingly clearer information from the same image.

For the training dataset of the temporal preference selection module in the online phase of the proposed model, it needs to be constructed manually, which reflects the process by which the user provides preferences to the agent. Specifically, based on the vector x_k obtained from the feature representation stage in the online phase, the user selects the preference weight λ corresponding to the strategy in the strategy pool. A one-hot encoded vector y_k is generated, where the component corresponding to λ is 1, and all other components are 0. For example, suppose that it can be known from x_k that the image category is difficult to distinguish and the agent is about to reach the end, then the ground-truth label y_k should be set to allow the agent to adopt a conservative strategy that leans towards security. In this paper, a dataset of 5000 labels for the dynamic preference temporal network is created based on the number of tasks $M = 1000$ and the number of decision steps $n = 5$.

5.2 Experimental Scenario Setup

The problem studied in this paper is the multi-objective optimization of making sequential decisions for a series of targets with varying risk coefficients during their gradual clarification process. Given that this is a practical yet underexplored problem setting, the comparison schemes are self-proposed.

This paper uses n fixed resolutions to simulate the progressively clearer environmental images received by the agent. The agent makes a decision at each process image until it reaches the destination. This process constitutes one round of the experiment, repeated for M rounds using M different images for testing. Specifically, when $n = 5$, the resolutions of the process images are 4×4, 8×8, 16×16, 24×24, and 32×32. The distance D from the start to the endpoint in the environment is set to 100, so the decision interval for the agent is $d = 20$. The danger coefficient α is set to different values: 1, 10, 100, 500, and 1000, in order to simulate a scenario where the risk level of the dangerous category increases progressively. The agent moves towards the endpoint in the scenario shown in Fig. 1 with an initial speed $V_0 = 10$. After the agent makes a decision based on the current information, it alters its current speed by a change amount $\Delta v = 1$. In this experiment, the proportion of dangerous category images is set to β, so the number of images for each safe category is given by $(M - \beta \times M)/4$.

5.3 Benchmark Models and Configuration Explanation

To evaluate the effectiveness of TSMODPN for the multi-objective optimization problem in the target progression clarity scenario, the following models are designed for comparative analysis.

Upper Bound. This model makes multiple decisions based on the real labels, which means that the agent can make correct decisions at each step. Therefore, the decision benefit of this model is an ideal theoretical upper bound. Other models are considered more effective when their results approximate those of this model more closely.

Single-step Decision Based on Predictive Labels (SD-PL). In this model, the agent waits until the target image is completely clear (i.e., the resolution reaches 32×32) and makes a decision based on the recognition result of a CNN with three convolutional layers.

Multi-objective Weighted Sum Based on DQN (MWS-DQN). This model linearly combines the reward functions of the safety and efficiency objectives using fixed weights, converting the multi-objective problem into a single-objective problem. In this experiment, the reward function is given by (10), with a fixed preference weight $\lambda=0.5$. DQN first extracts features using the same CNN as SD-PL, and then makes decisions in an end-to-end manner.

Multi-objective Threshold Dictionary Based on DQN (MTD-DQN). This model optimizes the strategy for each objective in order of priority and uses a threshold to ensure that the expected return for each objective meets the minimum requirement, ultimately finding a balanced optimal strategy. Since the order of the objectives affects the performance of MTD-DQN, the safety and efficiency objectives are set to the highest priority separately, resulting in two models: MTD-DQN-S and MTD-DQN-E.

TSMODPN (Ours). In the offline phase, different preference weights $\lambda \in \{0.1, 0.3, 0.5, 0.7, 0.9\}$ are sampled to construct the strategy pool. The multiple decisions corresponding to each preference weight are obtained by the same DQN as MWS-DQN and MTD-DQN. In the online phase, the dynamic distillation module processes image features. For different levels of clarity in the process images ($1 \leq k \leq 5$), corresponding distillation temperature parameters $[T_1, T_2, T_3, T_4, T_5]$ are set to $[4, 2, 2, 1, 1]$. This draws on the setting of knowledge distillation in the task of image classification [20].

5.4 Evaluation of Safety and Efficiency Objectives for Different Models

In this section, the multi-objective optimization effects of different models under the target progression clarity scenario are evaluated. These experiments are conducted under the settings of $M = 1000$ and $\beta = 0.1$. Tables 1 and 2 show the average damage G_s and the average duration G_t of various models when the dangerous category is set to "dogs". G_s and G_t are both dimensionless quantities, and it is preferable for their values to be as small as possible.

From the experimental results, it can be seen that TSMODPN proposed in this paper is closest to the theoretical upper bound (indicated in italics) in terms of both safety and efficiency objectives. Compared to other models, this model can effectively balance the two objectives.

Firstly, TSMODPN can dynamically select strategies from the strategy pool generated in the offline phase based on different danger coefficients α. In contrast, the other models can only adopt one strategy for different danger coefficients. Therefore, the average damage G_s for these models is proportional to the danger coefficient α, while the average duration G_t remains unchanged with the danger coefficient.

Secondly, TSMODPN can effectively balance safety and efficiency objectives in the environment with different danger coefficients α. For example, although in the high danger coefficient scenario (α = 1000), the duration of MTD-DQN-E is slightly shorter than that of TSMODPN, the average damage of TSMODPN is 11.3% less than that of MTD-DQN-E. This is because MTD-DQN-E pursues efficiency priority, while the dynamic preference temporal network of TSMODPN can select different strategies at each time step, thereby balancing safety and efficiency objectives.

Table 1. Average damage G_s for different models (dangerous category-dogs)

Model	α=1	α=10	α=100	α=500	α=1000
Upper Bound	*0.5*	*5*	*50*	*250*	*500*
SD-PL	0.97	9.7	97	485	970
MWS-DQN	0.79	7.9	79	395	790
MTD-DQN-S	0.65	6.48	64.8	324	648
MTD-DQN-E	0.69	6.92	69.2	346	692
TSMODPN	**0.62**	**6.34**	**62.2**	**308**	**614**

Table 2. Average duration G_t for different models (dangerous category-dogs)

Model	$\alpha=1$	$\alpha=10$	$\alpha=100$	$\alpha=500$	$\alpha=1000$
Upper Bound	8.50	8.50	8.50	8.50	8.50
SD-PL	9.87	9.87	9.87	9.87	9.87
MWS-DQN	9.83	9.83	9.83	9.83	9.83
MTD-DQN-S	10.08	10.08	10.08	10.08	10.08
MTD-DQN-E	9.78	9.78	9.78	9.78	**9.78**
TSMODPN	**9.47**	**9.54**	**9.62**	**9.69**	9.81

Table 3. Average damage G_s for different models (dangerous category-horses)

Model	$\alpha=1$	$\alpha=10$	$\alpha=100$	$\alpha=500$	$\alpha=1000$
Upper Bound	0.5	5	50	250	500
SD-PL	0.95	9.46	94.6	473	946
MWS-DQN	0.76	7.58	75.8	379	758
MTD-DQN-S	0.63	6.34	63.4	317	634
MTD-DQN-E	0.70	7.04	70.4	352	704
TSMODPN	**0.61**	**6.27**	**61.5**	**302**	**598**

Table 4. Average duration G_t for different models (dangerous category- horses)

Model	$\alpha=1$	$\alpha=10$	$\alpha=100$	$\alpha=500$	$\alpha=1000$
Upper Bound	8.50	8.50	8.50	8.50	8.50
SD-PL	9.86	9.86	9.86	9.86	9.86
MWS-DQN	9.82	9.82	9.82	9.82	9.82
MTD-DQN-S	9.85	9.85	9.85	9.85	9.85
MTD-DQN-E	9.76	9.76	9.76	9.76	**9.76**
TSMODPN	**9.42**	**9.46**	**9.58**	**9.69**	9.78

Tables 3 and 4 show the average damage G_s and the average duration G_t of various models when the danger category is set to "horses". Even if the dangerous category changes, the same conclusion can still be drawn that TSMODPN performs closest to the theoretical upper bound in terms of safety and efficiency objectives.

5.5 Ablation Experiment

To verify the effectiveness of each module in TSMODPN, ablation experiments are conducted in this section under the setting of the danger category being "dogs". The performance of TSMODPN is compared with that of the model without dynamic preference temporal network (referred to as w/o-DPN) and the model without dynamic distillation module (referred to as w/o-DDM). In the online stage of w/o-DPN, randomly select strategies with different preferences from the strategy pool to replace the adaptive selection of the dynamic preference temporal network. In the online stage of w/o-DDM, the distillation temperature is set to $T = 1$ at each time step, so the probability distribution of the image classification is undistilled.

According to the experimental results in Table 5 and 6, compared with w/o-DPN, the safety performance of TSMODPN has increased by 8.65% to 10.76%, and the efficiency performance has increased by 1.32% to 3.37% at different danger coefficients. This reflects the effectiveness of the dynamic preference network (DPN), which can select a preference strategy suitable for the current state from the strategy pool. Similarly, compared with w/o-DDM, the safety performance of TSMODPN has increased by 3.13% to 3.75%, and the efficiency performance has increased by 0.51% to 1.13% at different danger coefficients. This reflects that the dynamic distillation module, compared to directly using the *Softmax* output of the image classification probability distribution, is able to extract more information from early-stage blurred process images, fully utilizing early information and providing a solid basis for selecting strategy preferences in the online phase.

Table 5. Comparison of average damage G_s (dangerous category-dogs)

Model	$\alpha=1$	$\alpha=10$	$\alpha=100$	$\alpha=500$	$\alpha=1000$
w/o-DPN	0.68	6.94	68.8	342	688
w/o-DDM	0.64	6.46	64.4	320	636
TSMODPN	**0.62**	**6.34**	**62.2**	**308**	**614**

Table 6. Comparison of average duration G_t (dangerous category-dogs)

Model	$\alpha=1$	$\alpha=10$	$\alpha=100$	$\alpha=500$	$\alpha=1000$
w/o-DPN	9.80	9.84	9.87	9.82	9.96
w/o-DDM	9.56	9.65	9.72	9.76	9.86
TSMODPN	**9.47**	**9.54**	**9.62**	**9.69**	**9.81**

Figures 4 and 5 further demonstrate the performance optimization ranges of TSMODPN in terms of security and efficiency compared to w/o-DPN and w/o-DDM. It can be found from Fig. 4 that for the average damage G_s, the optimization range brought by DPN shows an upward trend with the increase of the danger coefficient α, while that brought by DDM is relatively stable. This is mainly because DDM pays more attention to the smoothing processing of the classification probability distribution of early blurred images, while DPN places greater emphasis on the influence of the danger coefficient on decision-making. Similarly, it can also be verified from Fig. 5 that the efficiency optimization brought by DPN is higher than that brought by DDM. Furthermore, for the average duration G_t, the optimization ranges induced by DPN and DDM exhibit a downward trend as the danger coefficient α increases. This implies that TSMODPN predominantly takes safety goals into account in high-risk scenarios, whereas in low-risk situations, it mainly concerns itself with efficiency goals.

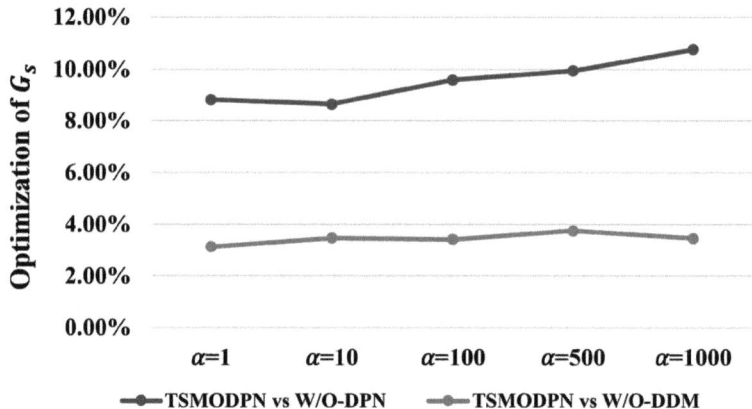

Fig. 4. The average damage optimization range of TSMODPN compared to w/o-DPN and w/o-DDM

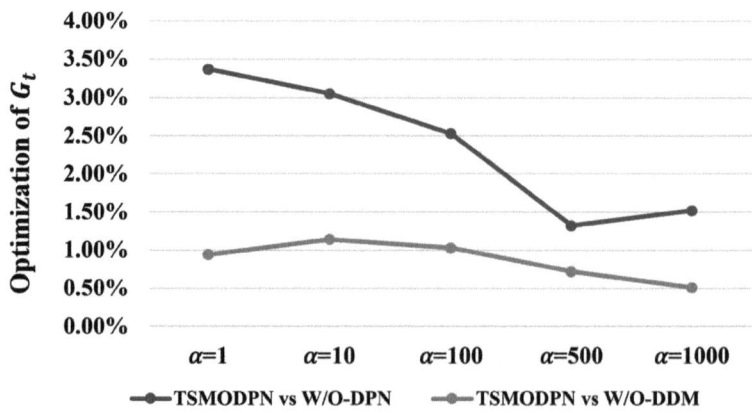

Fig. 5. The average duration optimization range of TSMODPN compared to w/o-DPN and w/o-DDM

6 Conclusion

This paper proposes a temporal scenario with gradually clarified objectives and introduces the concept of danger coefficient to model this scenario according to real-world conditions. This scenario simulates the dynamic characteristic of gradually increasing information granularity as the agent approaches the target, and embeds a risk assessment mechanism into the decision-making process. The goal of designing this scenario is to explore a perception-decision integration strategy that allows the agent to fully utilize early-stage, blurred, coarse-grained information to make decisions in advance.

For the multi-objective optimization task in the above scenario, this paper proposes a two-stage multi-objective model based on dynamic policy networks. The experimental results show that the multi-objective optimization of security rewards and task duration by this model under different risk levels is closest to the theoretical upper bound. This provides a new approach for handling multi-objective optimization problems in dynamic environments.

Undoubtedly, the model presented in this paper can still be refined and enhanced. For example, the current design of the target progression clarity scenario uses a fixed resolution change mechanism, which does not effectively reflect the feedback effect of decision actions on the information. A more realistic design would involve the granularity of the information the agent acquires changing as decisions are made. In future work, the changes in the process image resolution can be modeled by considering both the target image resolution and the agent's distance from the target. This increases the complexity and realism of the scenario, helping to explore more powerful temporal decision models.

Acknowledgments. This study was funded by the National Natural Science Foundation of China (grant No. 62332014), and the Open Fund of National Engineering Research Center of Geographic Information System (grant No. 2023KFJJ11).

Disclosure of Interests. The authors have no competing interests to declare that are relevant to the content of this article.

References

1. Teng, S., Hu, X., Deng, P., Li, B., Li, Y., Ai, Y., et al.: Motion planning for autonomous driving: The state of the art and future perspectives. IEEE Trans. Intell. Veh. **8**(6), 3692–3711 (2023). https://doi.org/10.1109/TIV.2023.3274536
2. Yoo, S.B.M., Hayden, B.Y., Pearson, J.M.: Continuous decisions. Philos. Trans. R. Soc. B **376**(1819), 20190664 (2021). https://doi.org/10.1098/rstb.2019.0664
3. Chia, W.M.D., Keoh, S.L., Goh, C., Johnson, C.: Risk assessment methodologies for autonomous driving: A survey. IEEE Trans. Intell. Transp. Syst. **23**(10), 16923–16939 (2022). https://doi.org/10.1109/TITS.2022.3163747
4. El-Assy, A.M., Amer, H.M., Ibrahim, H.M., Mohamed, M.A.: A novel CNN architecture for accurate early detection and classification of Alzheimer's disease using MRI data. Sci. Rep. **14**(1), 3463 (2024). https://doi.org/10.1038/s41598-024-53733-6
5. Chen, Z., Chen, Y., Xiao, T., Wang, H., Hou, P.: A novel short-term load forecasting framework based on time-series clustering and early classification algorithm. Energy Build. **251**, 111375 (2021). https://doi.org/10.1016/j.enbuild.2021.111375

6. Alinezhad, H.S., Shang, J., Chen, T.: Early classification of industrial alarm floods based on semisupervised learning. IEEE Trans. Industr. Inf. **18**(3), 1845–1853 (2021). https://doi.org/10.1109/TII.2021.3081417
7. Chen, W., Lyu, F., Wu, F., Yang, P., Xue, G., Li, M.: Sequential message characterization for early classification of encrypted internet traffic. IEEE Trans. Veh. Technol. **70**(4), 3746–3760 (2021). https://doi.org/10.1109/TVT.2021.3063738
8. Gupta, A., Gupta, H.P., Biswas, B., Dutta, T.: Approaches and applications of early classification of time series: A review. IEEE Trans. Artif. Intell. **1**(1), 47–61 (2020). https://doi.org/10.1109/TAI.2020.3027279
9. Gupta, A., Gupta, H.P., Biswas, B., Dutta, T.: A divide-and-conquer–based early classification approach for multivariate time series with different sampling rate components in iot. ACM Trans. Internet Things **1**(2), 1–21 (2020). https://doi.org/10.1145/3375877
10. Wan, X., Cen, L., Chen, X., Xie, Y., Gui, W.: Memory shapelet learning for early classification of streaming time series. IEEE Trans. Cybern. **54**(5), 2757–2770 (2023). https://doi.org/10.1109/TCYB.2023.3337550
11. Sharma, A., Singh, S.K., Udmale, S.S., Singh, A.K., Singh, R.: Early transportation mode detection using smartphone sensing data. IEEE Sens. J. **21**(14), 15651–15659 (2020). https://doi.org/10.1109/JSEN.2020.300931
12. Huang, Y., Yen, G.G., Tseng, V.S.: Snippet policy network for multi-class varied-length ECG early classification. IEEE Trans. Knowl. Data Eng. **35**(6), 6349–6361 (2022). https://doi.org/10.1109/TKDE.2022.3160706
13. Martinez, C., Ramasso, E., Perrin, G., Rombaut, M.: Adaptive early classification of temporal sequences using deep reinforcement learning. Knowl.-Based Syst. **190**, 105290 (2020). https://doi.org/10.1016/j.knosys.2019.105290
14. Sharma, S., Kumar, V.: A comprehensive review on multi-objective optimization techniques: Past, present and future. Arch. Comput. Methods in Eng. **29**(7), 5605–5633 (2022). https://doi.org/10.1007/s11831-022-09778-9
15. Tian, Y., et al.: Evolutionary large-scale multi-objective optimization: A survey. ACM Comput. Surv. (CSUR) **54**(8), 1–34 (2021). https://doi.org/10.1145/3470971
16. Ma, H., Zhang, Y., Sun, S., Liu, T., Shan, Y.: A comprehensive survey on NSGA-II for multi-objective optimization and applications. Artif. Intell. Rev. **56**(12), 15217–15270 (2023). https://doi.org/10.1007/s10462-023-10526-z
17. Wang, R., Zhou, Z., Ishibuchi, H., Liao, T., Zhang, T.: Localized weighted sum method for many-objective optimization. IEEE Trans. Evol. Comput. **22**(1), 3–18 (2016). https://doi.org/10.1109/TEVC.2016.2611642
18. Dou, J., Wang, X., Liu, Z., Sun, Q., Wang, X., He, J.: Towards Pareto-optimal energy management in integrated energy systems: A multi-agent and multi-objective deep reinforcement learning approach. Int. J. Electr. Power Energy Syst. **159**, 110022 (2024). https://doi.org/10.1016/j.ijepes.2024.110022
19. Lin, Y.E., Yin, S., Ding, Y., Liang, X.: ATMKD: Adaptive temperature guided multi-teacher knowledge distillation. Multimedia Syst. **30**(5), 292 (2024). https://doi.org/10.1007/s00530-024-01483-w
20. Belinga, A. G., Tekouabou Koumetio, C. S., El Haziti, M., El Hassouni, M.: Knowledge distillation in image classification: the impact of datasets. Computers, **13**(8), 184. https://doi.org/10.3390/computers13080184

A LoRa Positioning Algorithm Based on the Integration of Kalman Filtering and Neural Networks and Its Implementation

Yiwei Li[1], Zhanjun Hao[1(✉)], Yuejiao Wang[2], Guowei Wang[2], Jiang Zhang[1], and Fenfang Li[1]

[1] Northwest Normal University, Lanzhou 730000, China
{haozhj,2023222169,lifenfang}@nwnu.edu.cn
[2] Lanzhou University, Lanzhou 730000, China
{wyuejiao2024,wanggw2024}@lzu.edu.cn

Abstract. This paper proposes a LoRa-based positioning and tracking method designed for large-scale outdoor environments. We develop a positioning algorithm that integrates Kalman filtering with a neural network. The algorithm first applies Kalman filtering to preprocess the RSSI and SNR values of LoRa signals, effectively reducing noise and smoothing the data to enhance signal reliability. For position prediction, we construct a multi-input LSTM neural network model that fuses static features (such as RSSI and SNR) with temporal features (such as historical location and time differences) to accurately estimate the device's location. We evaluate the proposed algorithm on a public dataset collected by three LoRaWAN gateways, conducting a comprehensive performance analysis in terms of positioning accuracy and error distribution. Experimental results demonstrate that the algorithm reduces the average positioning error to 50 m, significantly outperforming existing state-of-the-art neural network-based positioning methods.

Keywords: LoRa positioning · Kalman filter · LSTM neural network · outdoor tracking · oenergy-efficient Internet-of-Things (IoT) localization

1 Introduction

As an emerging wireless communication solution, LoRa has shown significant advantages in the IoT field, especially in outdoor positioning. It features wide coverage, long communication range, low power consumption, and low cost [1], making it an ideal choice for outdoor IoT device positioning. LoRa uses spread-spectrum modulation to achieve long-distance communication. Compared to traditional wireless technologies, it can reach further distances at the same power level, allowing broader outdoor coverage and meeting IoT devices' needs for long-distance communication and positioning.

Moreover, LoRa technology boasts extremely low power consumption. By optimizing communication protocols and hardware design, LoRa devices can drastically reduce power usage while maintaining long-distance communication, thus extending their service life. This is crucial for IoT devices requiring long-term operation, especially in outdoor settings where battery replacement or charging is inconvenient. LoRa technology also offers low cost and easy deployment advantages [2]. Its relatively inexpensive hardware and simple network topology make LoRa-based outdoor positioning systems highly cost-effective for large-scale deployment. Additionally, the ease of deployment reduces system maintenance complexity and costs, enhancing overall reliability and stability.

Thus, researching LoRa - based outdoor positioning systems is highly significant and valuable. It can overcome traditional outdoor positioning technology limitations, enhance positioning accuracy and stability, reduce system power consumption and cost, and offer a more efficient and reliable outdoor positioning solution for IoT devices.

As the Internet of Things (IoT) technology develops and spreads rapidly, the demand for outdoor positioning technology is surging and its applications are expanding into more fields. As an emerging positioning technology, LoRa-based outdoor positioning systems, with their unique advantages, show great application potential and market opportunities [3]. It has been initially applied in search and rescue [4,5], monitoring and warning [6], smart agriculture [7], industrial production [8], smart cities [9], and more. However, in actual deployment, factors like gateway clock precision, signal propagation errors, gateway deployment strategies, and specific application scenarios can affect the accuracy of LoRa positioning, making high-precision positioning challenging [10]. Therefore, in depth research on the key technologies of LoRa-based outdoor positioning systems and their diverse applications can greatly support and boost the development and implementation of IoT technologies.

To address the issue of positioning errors in large-scale outdoor environments using LoRa, this paper proposes a LoRa positioning algorithm that combines Kalman filtering with neural networks. The algorithm first applies Kalman filtering to preprocess the received signal strength indicator (RSSI) and signal-to-noise ratio (SNR) of LoRa signals, effectively eliminating noise interference and enhancing signal smoothness and accuracy. Subsequently, we design a multi-input LSTM neural network that integrates static features (such as RSSI and SNR) with temporal features (such as historical locations and time intervals) to achieve more accurate position prediction. Experimental results on a public dataset demonstrate that the proposed method outperforms existing state-of-the-art neural network-based positioning techniques in terms of both positioning accuracy and error control. The main contributions of this paper are summarized as follows:

- We introduce the Kalman filter to process the received signal strength indicator (RSSI) and signal-to-noise ratio (SNR) data of LoRa signals, and investigate its effectiveness in improving positioning accuracy. This preprocess-

ing provides more reliable input data for the subsequent position prediction model.
- We design a multi-input LSTM network that integrates static features (such as RSSI and SNR) with temporal features (such as historical locations and time intervals), enhancing the model's ability to capture dynamic changes in positional information.
- We evaluate the performance of the proposed positioning algorithm on a public dataset. Experimental results show that our algorithm reduces the average positioning error to 50 m, outperforming current mainstream neural network-based positioning methods.

2 Related Work

The LoRa positioning system primarily consists of three common methods: Time Difference of Arrival (TDOA) positioning, Time of Flight (TOF) positioning, and Received Signal Strength Indicator (RSSI) positioning. The following sections provide an overview and summary of these three positioning techniques and their related research.

2.1 LoRa Positioning Based on TDOA

TDOA positioning is a method that determines the target's location based on the time difference of signal arrival. The core principle involves measuring the time differences at which the same signal reaches multiple receiving nodes, and then calculating the relative distance differences between the signal source and each node. These differences form hyperbolas with the receiving nodes as foci, and the intersection point of multiple hyperbolas indicates the actual position of the signal source [11].

In LoRa networks, TDOA is widely adopted in positioning strategies due to its energy-efficient backend computation and decent accuracy. One study applied TDOA technology to locate targets in rural areas; although the positioning error could be reduced to around 100 m in certain scenarios, the overall accuracy remained unstable, with errors sometimes exceeding 1 km. Moreover, the requirement to process more than 10,000 data packets resulted in prolonged total runtime, rendering the method unsuitable for real-time positioning [12]. Another work combined TDOA-based localization with road and mobility trajectory data, which significantly improved the positioning accuracy of LoRa for outdoor target tracking. The results demonstrated a median error of 75 m, with 90% of the errors below 180 m [13]. To improve accuracy, some researchers have proposed compensating for time-related errors. One approach used a deep neural network (DNN) to learn and correct timing deviations, showing that as the number of reference nodes increased and the mobile node approached them, the average localization error was notably reduced to approximately 61 m [14].

Although the TDOA-based positioning method shows considerable potential, current LoRa localization systems leveraging TDOA still face several challenges.

On one hand, some approaches suffer from significant positioning errors, making them unsuitable for high-precision application scenarios. On the other hand, certain methods rely on additional hardware support, which increases the complexity and cost of system deployment. As a result, TDOA-based positioning systems built on standard LoRa devices are currently insufficient to meet the practical requirements of high-accuracy outdoor localization.

2.2 LoRa Positioning Based on TOF

In recent years, localization technologies based on the LoRa 2.4 GHz frequency band have attracted increasing attention from researchers. These technologies, which rely on the Time of Flight (TOF) ranging principle, demonstrate significantly improved positioning accuracy. Studies have shown that LoRa 2.4 GHz transceivers can accurately measure distances up to 1.7 km, with a maximum deviation of only 0.6% from actual ground distances [15]. A method combining TOF ranging with trilateration and Kalman filtering has been used to estimate the position of mobile targets, effectively reducing the positioning error to less than 5 m [16]. To address the issue of LoRa ranging accuracy being susceptible to non-line-of-sight (NLOS) propagation, a weighted multi-sample data fusion algorithm combining TOF and RSSI was proposed [17]. Additionally, a positioning algorithm was developed to reduce the posterior RSSI error in multi-anchor collaborative estimation scenarios. The research team constructed and tested a wide-area positioning system called LoRaWAPS (Long Range Wide Area Positioning System) in a real campus environment, demonstrating its feasibility and effectiveness.

Although LoRa 2.4 GHz exhibits excellent ranging performance, current error mitigation methods still primarily rely on traditional approaches. These methods are mostly based on fixed models or parameters and lack adaptability to dynamically changing channels in complex environments, making it difficult to effectively control ranging errors in certain scenarios. Moreover, these approaches often require manual parameter tuning when applied to different environments, lacking sufficient robustness and adaptability.

In the field of LoRa-based identification and error mitigation, current research remains limited. In particular, there has been no work found applying machine learning or deep learning techniques for identification and error suppression in the LoRa 2.4 GHz frequency band. This presents a promising direction for future research.

2.3 LoRa Positioning Based on RSSI

RSSI-Based Fingerprint Localization. Fingerprint localization is a positioning method based on signal feature matching. The core idea is to collect RSSI values from different locations during the offline phase to build a fingerprint database, with each location having a unique RSSI fingerprint. During online localization, the RSSI value at the target location is measured in real-time and matched with the fingerprints in the database to determine the target's location [18].

A multilayer perceptron - based neural network positioning algorithm can accurately determine measurement point locations [19]. Interpolation techniques reduce fingerprint data collection time and cost. Studies show the minimum Euclidean distance algorithm is more accurate and precise than classic pattern matching, while the random forest algorithm is superior in minimizing maximum estimation errors [20]. Stochastic neural networks can predict unknown location coordinates on Antwerp's public LoRaWAN dataset, achieving high - precision outdoor urban positioning [21]. SF and RSSI are used as fingerprints in a proposed deep reinforcement learning model [22].

However, RSSI signals are susceptible to multipath effects, shadowing, and interference, causing significant signal strength fluctuations and making it hard to steadily reflect the target location. Also, the high costs of building and maintaining fingerprint databases increase the practical application difficulty of this algorithm.

RSSI-Based Signal Strength Distance Localization. This method relies on the correlation between signal strength and distance. In ideal conditions, signal strength decreases with distance in a predictable manner. By measuring RSSI values, we can estimate the distance between a target and a receiver. Furthermore, using RSSI values from multiple receivers, we can create multiple distance constraints to determine the target's position.

Some studies have modeled RSSI signals in both indoor and outdoor environments using Gaussian processes, and proposed a corresponding localization approach based on maximum likelihood estimation [23]. Other works introduced Kalman filtering to smooth RSSI data and combined it with trilateration to enhance localization accuracy [24]. To improve the precision of path loss estimation, a framework called DeepLoRa was proposed, which uses remote sensing images to identify land cover types along the transmission path and applies deep learning techniques to model the impact of land cover on signal path loss [25]. In addition, an artificial neural network model was developed to accurately predict the propagation loss of LoRa communication links, showing improved performance in terms of RSSI prediction and root mean square error (RMSE) when compared with traditional path loss models [26].

Overall, although RSSI-based LoRa positioning methods offer clear advantages in terms of low cost and ease of deployment, their performance remains highly susceptible to interference caused by multipath effects and other environmental factors. The number and placement of anchor nodes play a critical role in determining the final positioning accuracy. In ideal radio environments, RSSI-based positioning can achieve high levels of distance estimation and localization precision. However, in practical applications, wireless channels are often affected by unpredictable external factors such as weather changes and physical obstructions, which cause significant fluctuations in RSSI values and, consequently, reduce the reliability and accuracy of the positioning results. Therefore, enhancing the robustness and precision of RSSI-based positioning algorithms under complex and dynamic conditions remains a major challenge in current research.

To address the aforementioned challenges, this study focuses on location recognition and error suppression of LoRa devices in large-scale, complex outdoor environments. It explores strategies that integrate efficient filtering algorithms with deep learning models, aiming to propose a positioning method that offers high accuracy, wide coverage, and low power consumption—thereby meeting the dual demands of performance and resource efficiency in practical applications.

3 Methodology

3.1 Dataset Description

We evaluate the performance of the proposed algorithm on a publicly available dataset hosted on the Zenodo repository [27]. This time-series dataset contains measurement data collected by a mobile device equipped with LoRa communication and GPS positioning capabilities. The data was captured by LoRa gateways deployed on the island of Sálvora, located in Galicia, Spain. The team installed three LoRa gateways at the following locations (latitude, longitude, altitude): Gateway 1 (42.46972, −9.01345, 73), Gateway 2 (42.49955, −9.00654, 5), and Gateway 3 (42.50893, −9.04902, 31).

The dataset is provided in the form of a single comma-separated values (CSV) file, where each row contains the following fields: device identifier (#device_id), received signal strength indicator from Gateway 1 (rssi_1), signal-to-noise ratio from Gateway 1 (snr_1), rssi_2 and snr_2 from Gateway 2, rssi_3 and snr_3 from Gateway 3, LoRa spreading factor (spreading_factor), timestamp (ts), GPS-derived latitude (lat), longitude (long), and altitude (alt) of the device. An example of the dataset structure is shown in Table 1.

Table 1. Dataset entry example.

#device_id	rssi_1	snr_1	rssi_2	snr_2	rssi_3	snr_3	spreading_factor	ts	lat	long	alt
1	−109	4.2	−107	5.2	−120	−13.5	12	1676461237.919161	42.47060775756836	−9.00178813934326	68.9000015258789
1	−109	4.2	−107	6	−119	−12.5	11	1676461312.525769	42.47065353393555	−9.001777649	57
...	...	...	...	...	...	...	...	...	...	...	...
7	−119	−5.8	−118	−3.2	−119	−15.2	12	1589595495.438987	42.47781753540039	−9.013547079	22.500000381459727

3.2 Dataset Preprocessing

In real-world environments, the measured data inevitably contain noise caused by various interference factors. These noises arise not only from the system itself but also from numerous unpredictable disturbances in the indoor environment, resulting in significant fluctuations in the measured RSSI and SNR values. Therefore, effective filtering techniques are necessary to reduce measurement errors induced by such interferences.

The Kalman filter is a recursive filtering algorithm that provides optimal estimation of a system's state based on linear state equations and observed input-output data. It performs effectively in both fixed-end and mobile device localization scenarios, demonstrating strong adaptability. To reduce the instantaneous fluctuations in RSSI and SNR sequences caused by multipath effects, obstructions, and device errors, we apply a one-dimensional Kalman filter to smooth each time series. This provides more stable inputs for the subsequent neural network modules used for localization and modeling.

Based on this, in the preprocessing stage, we apply a Kalman filter to the six signal measurements received by the three gateways: rssi_1, snr_1, rssi_2, snr_2, rssi_3, and snr_3. Each of these one-dimensional time series is processed using a separate one-dimensional Kalman filter with identical models and parameters. The system is defined by the following equations:

State Transition Model:

$$x_k = Ax_{k-1} + w_{k-1}, \quad w_{k-1} \sim \mathcal{N}(0, Q) \tag{1}$$

Observation Model:

$$z_k = Hx_k + v_k, \quad v_k \sim \mathcal{N}(0, R) \tag{2}$$

where:

- x_k is the hidden state (true signal) at time k,
- z_k is the observed RSSI or SNR at time k,
- $A = 1$ is the state transition coefficient,
- $H = 1$ is the observation matrix,
- Q is the process noise covariance,
- R is the observation noise covariance.

The Kalman filter performs two steps iteratively: prediction and update.

Prediction Step:

$$\hat{x}_{k|k-1} = A\hat{x}_{k-1} \tag{3}$$

$$P_{k|k-1} = AP_{k-1}A^\top + Q \tag{4}$$

Update Step:

$$K_k = \frac{P_{k|k-1}H^\top}{HP_{k|k-1}H^\top + R} \tag{5}$$

$$\hat{x}_k = \hat{x}_{k|k-1} + K_k(z_k - H\hat{x}_{k|k-1}) \tag{6}$$

$$P_k = (I - K_kH)P_{k|k-1} \tag{7}$$

where:

- $\hat{x}_{k|k-1}$ is the prior estimate at time k,
- $\hat{x}_k$ is the posterior estimate after measurement update,
- $P_{k|k-1}$, P_k are prior and posterior error covariances respectively,
- K_k is the Kalman gain determining the weight of the observation.

Algorithm 1. Kalman Filter for RSSI/SNR Denoising

Input: Observation sequence $Z = \{z_1, z_2, \ldots, z_T\}$
Output: Filtered sequence $\hat{X} = \{\hat{x}_1, \hat{x}_2, \ldots, \hat{x}_T\}$
Initialize state estimate $\hat{x}_0 \leftarrow z_1$
  Initialize error covariance $P_0 \leftarrow 1$
  Set model parameters: $A \leftarrow 1$, $H \leftarrow 1$, $Q \leftarrow 0.01$, $R \leftarrow 1$

for $k = 1$ **to** T **do**
  Predict Step:
  $\hat{x}_{k|k-1} = A \cdot \hat{x}_{k-1}$
  $P_{k|k-1} = A \cdot P_{k-1} \cdot A^\top + Q$

  Update Step:
  $K_k = \frac{P_{k|k-1} \cdot H^\top}{H \cdot P_{k|k-1} \cdot H^\top + R}$
  $\hat{x}_k = \hat{x}_{k|k-1} + K_k \cdot (z_k - H \cdot \hat{x}_{k|k-1})$
  $P_k = (I - K_k \cdot H) \cdot P_{k|k-1}$

return $\hat{X} = \{\hat{x}_1, \ldots, \hat{x}_T\}$

The initial state $\hat{x}_0$ is set to the first observed value, and all covariances Q, R, and P_0 are tuned empirically for optimal filtering results. The pseudocode block diagram for the Kalman filter data algorithm is shown in Algorithm 1.

We present the filtering results of the RSSI and SNR data received by Device 3 from three gateways. As shown in Fig. 1, the blue line represents the raw data, the yellow line shows the data after Butterworth filtering, and the green line indicates the data processed by Kalman filtering. We compare the effects of the two filtering methods and validate the effectiveness of the proposed Kalman filtering approach. Therefore, the Kalman-filtered data is selected for use in the subsequent neural network-based positioning model.

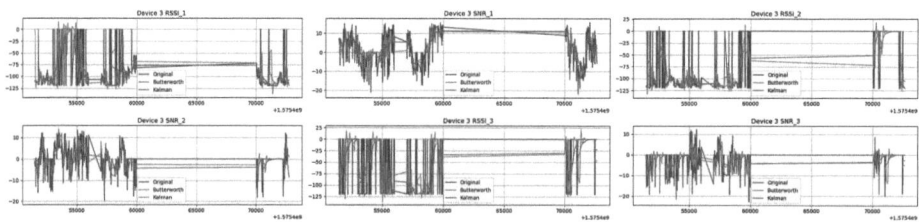

Fig. 1. Comparison between original and filtered data.

3.3 LSTM Localization Model

Although LoRa offers advantages such as low power consumption and long-range communication, most existing studies focus on ideal indoor or outdoor environments—such as equidistant antennas and clear line-of-sight—where positioning errors are relatively small. However, in wide open-field environments, the accuracy of LoRa-based localization significantly decreases, making it difficult to compete with GNSS. Traditional trilateration methods are limited in accuracy, and their performance deteriorates as the distance to gateways increases or environmental conditions change.

The data collected from the public dataset covers an area of approximately 2.5 km^2, located in an environment with limited communication infrastructure. The system is required to meet the demands of low cost, low power consumption, fine-grained tracking, and sufficient positioning accuracy. Although LoRa transmission is free, its accuracy decreases with increasing distance from gateways, requiring high-density deployment. On the other hand, GNSS offers high accuracy but suffers from high energy consumption and limited session cycles, which restrict its real-time capabilities and battery life.

In recent years, research has increasingly shifted from geometric algorithms to the use of artificial intelligence to enhance positioning performance. Accordingly, we propose a hybrid positioning method based on LoRa signal features (RSSI, SNR), historical trajectory information, and a small number of GNSS measurements. This approach achieves a balance among energy consumption, sampling rate, and positioning accuracy. By combining past location data with current LoRa features, the method improves the prediction performance of neural networks while maintaining low power consumption.

The device moves within the target area, and its position changes over time slots. In the dataset, GNSS data is not sampled in every time slot but only in selected ones. Therefore, the key task is to estimate the current latitude $\text{lat}_i(t)$ and longitude $\text{lon}_i(t)$ of the terminal device using the current time slot's LoRa features ($\text{RSSI}_{ij}(t)$, $\text{SNR}_{ij}(t)$) and the most recent available GNSS location. This enables a low-cost and efficient positioning system. To this end, we propose an LSTM-based positioning model that fuses both static features and temporal coordinate history.

The input variables of the LSTM model include static features, which are the RSSI and SNR values received from three LoRa gateways at the current time slot t, as well as the spreading factor (SF) used in LoRa signal modulation, resulting in a 7-dimensional feature vector. In addition, historical location sequences are used, extracted from the previous $n_{\text{prev\_ts}}$ time steps (default is 3) in each sample. These include latitude, longitude, altitude (optional), and time intervals. These values are converted into Cartesian coordinates (x, y) for temporal modeling. Each time step includes 4 features, thus the temporal input has a shape of $(n_{\text{prev\_ts}}, 4)$, as shown in Table 2(a). The output of the LSTM model for historical position sequences is concatenated with the static input features to form the final fused feature representation. The output variables of the LSTM model are Cartesian coordinates, representing the predicted user position at the current time step $(\hat{x}_i(t), \hat{y}_i(t))$, which is finally converted to the estimated latitude and longitude $(\hat{\text{lat}}_i(t), \hat{\text{lon}}_i(t))$, as shown in Table 2(b). The localization error is measured as the Euclidean distance between the estimated and the ground-truth position, calculated as follows:

$$Error_i(t) = \sqrt{(x_i(t) - \hat{x}_i(t))^2 + (y_i(t) - \hat{y}_i(t))^2} \qquad (8)$$

Table 2. Inputs and outputs of the LSTM model.

Static Features	$RSSI_{ij}(t), \forall j$	RSSI received by end-device i from each LoRa gateway j at time slot t.
	$SNR_{ij}(t), \forall j$	SNR received by end-device i from each LoRa gateway j at time slot t.
	Spreading_factor	The spreading factor of LoRa signal modulation.
Temporal Features	Latitude, longitude, altitude (optional), and time interval	Extract the latitude, longitude, altitude (optional), and time interval of the previous n_prev_ts (default: 3) time steps in each sample.

(a) Input data.

$\widehat{lat}_i(t), \widehat{lon}_i(t)$	Latitude and longitude of end-device i at the current time slot t.

(b) Output variables.

To improve localization accuracy using wireless signal data, we design an LSTM-based neural network model that fuses both static and temporal features. The input consists of two components: (1) static features extracted from RSSI, SNR, and spreading factor values (a total of 7 dimensions), and (2) a temporal sequence of location information over the previous n time steps, including historical coordinates and time differences. Latitude and longitude values are converted into planar (x, y) coordinates, and altitude is optionally included when predicting in 3D space. All input features are normalized using Min-Max scaling to eliminate differences in magnitude and ensure training stability.

The temporal sequence is processed by a two-layer LSTM network with dropout regularization, which captures high-level temporal dependencies. Meanwhile, the static input is fed directly into the network through a separate dense pathway. The two feature representations are concatenated and passed through a fully connected regression layer to produce the final coordinate output. The model is trained using the Adam optimizer and mean absolute error (MAE) loss function, with a learning rate scheduler (ReduceLROnPlateau) to dynamically adjust the learning rate during training. This hybrid model effectively leverages both current signal characteristics and historical motion patterns to predict accurate 2D or 3D positions. The model is shown in Fig. 2, and the algorithm pseudocode is presented in Algorithm 2.

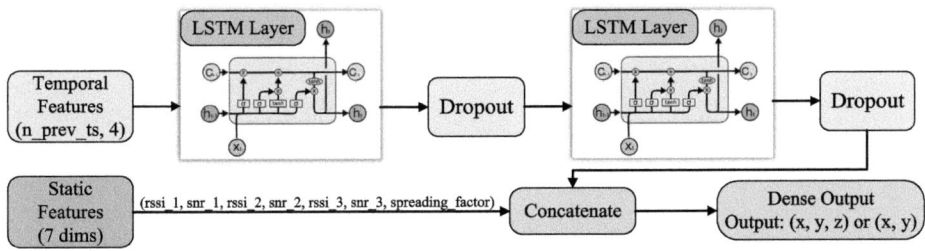

Fig. 2. LSTM localization model diagram.

Algorithm 2. LSTM-based Localization with RSSI, SNR, SF, and Historical Coordinates

Input: Training data $\mathcal{D}_{\text{train}}$, test data $\mathcal{D}_{\text{test}}$, epochs e, LSTM units u, history length n, flag $predict\_alt$
Output: Predicted coordinates $\hat{Y}$, true coordinates Y
Merge data $\mathcal{D} = \mathcal{D}_{\text{train}} \cup \mathcal{D}_{\text{test}}$;
foreach *sample l in $\mathcal{D}$* **do**
    Extract static features $s = [\text{rssi}_1, \text{snr}_1, \text{rssi}_2, \text{snr}_2, \text{rssi}_3, \text{snr}_3, \text{SF}]$;
    Initialize sequence buffer $seq = []$;
    for $i = 0$ *to* $n-1$ **do**
        if *no historical coordinate available* **then**
            Append $[0,0,0,0]$ to seq;
        else
            Retrieve lat, lon, alt, Δt from history;
            $(x,y) \leftarrow$ `get_xy_from_latlon`(lat, lon);
            Append $[x, y, alt, \Delta t]$ or $[x, y, 0, \Delta t]$ to seq depending on $predict\_alt$;
    Compute target (x, y, z) from current lat, lon, alt;
    Append $(s, seq, target)$ to input buffer;
Normalize static, sequence, and target data with MinMaxScaler;
Split into training and test sets;
Build LSTM model:
  – LSTM layers for sequence input
  – Dense layers for static input
  – Concatenate outputs
  – Final Dense layer to predict (x, y) or (x, y, z)

Train model using MAE loss and Adam optimizer with ReduceLROnPlateau;
Predict on test set and inverse transform result;
return predicted and true coordinates;

4 Evaluation

We analyzed the prediction results of the improved LSTM model in comparison with the true device positions. In the figure, the green dots represent the true positions, while the red dots represent the predicted positions. As observed from the distribution, the predicted points generally follow the true trajectory well, demonstrating strong consistency and stability. Most predicted points are close to their corresponding true positions, indicating that the model has good spatial perception capability. Statistical analysis shows that the average positioning error across the entire test set is approximately 50 m, which outperforms traditional neural network methods. This verifies the effectiveness of the proposed LSTM model that integrates static and temporal features for LoRa-based positioning tasks. The experimental results are illustrated in Fig. 3.

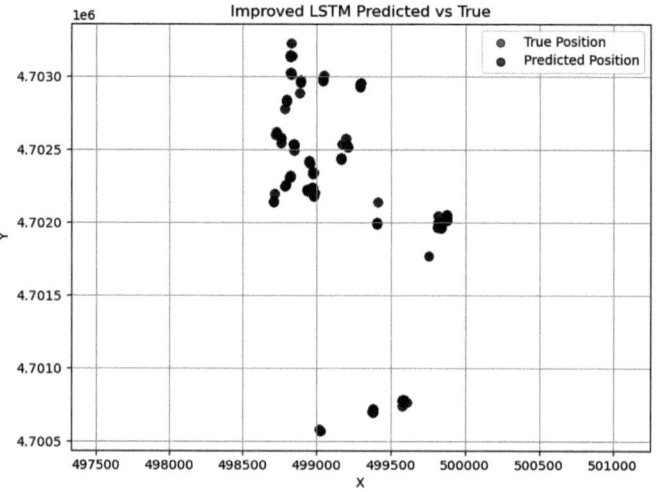

Fig. 3. Comparison between true and predicted positions.

We analyzed the error distribution of the improved LSTM positioning model on the test set. The x-axis shows positioning error (in meters), and the y-axis shows the number of samples within corresponding error ranges. Most errors cluster between 0 and 50 m, peaking at around 10 m. This indicates the model achieves high positioning accuracy in most cases. As errors increase, sample numbers drop quickly, showing a long - tail distribution and suggesting occasional larger deviations. Overall, the error distribution is concentrated, proving the model's robustness and reliability in LoRa positioning tasks. The trend curve in the figure further reflects error statistical characteristics, aiding in assessing model suitability under different precision requirements. The positioning error distribution is shown in Fig. 4.

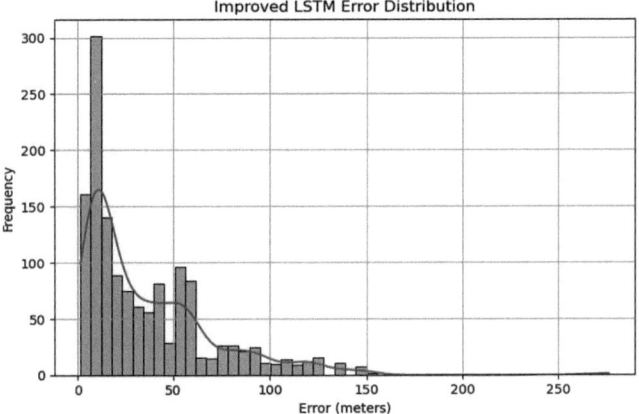

Fig. 4. Positioning error distribution chart.

This paper analyzes the loss variation of an enhanced multi-input LSTM positioning model during training. The blue curve denotes training loss, and the orange curve stands for validation loss. From the figure, it can be seen that the model's loss drops rapidly in the initial training stages and then stabilizes, indicating fast convergence. Both training and validation losses remain low without significant divergence, suggesting no overfitting and good generalization ability. This further confirms the effectiveness and robustness of the proposed LSTM model for LoRa positioning. The model's training loss variation is shown in Fig. 5.

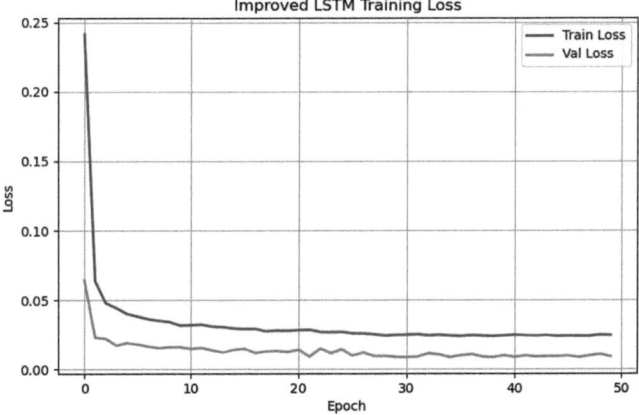

Fig. 5. Training loss variation of the positioning model.

We conducted a comparative evaluation of several positioning algorithms, including a baseline method, RSSI-based trilateration using the Least Squares

(LS) method, RSSI-based trilateration using a Neural Network (NN), and the proposed improved multi-input LSTM positioning model. The baseline method estimates the device's current position by simply using the most recent Global Navigation Satellite System (GNSS) location, assuming minimal movement within that time interval. The RSSI-LS method performs trilateration using RSSI values received from multiple LoRa gateways and applies the LS algorithm for position estimation. In contrast, the RSSI-NN method also relies on trilateration but employs a neural network to model the nonlinear relationship between RSSI and distance to improve accuracy.

Experimental results demonstrate that the proposed improved LSTM model outperforms all other methods, achieving the lowest average positioning error of approximately 50 m. In comparison, the baseline and RSSI-LS methods exhibit significantly larger errors. While the RSSI-NN method shows some improvement, its accuracy remains inferior to that of our proposed model. Overall, the results confirm the effectiveness and superiority of the improved LSTM model in enhancing LoRa-based outdoor positioning accuracy. The performance comparison of different algorithms is shown in Fig. 6.

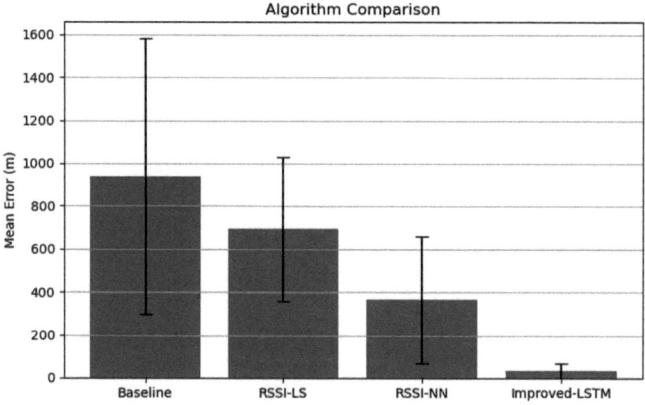

Fig. 6. Comparison of the performance of different positioning algorithms.

5 Conclusion

This study focuses on an intelligent positioning system based on LoRaWAN, demonstrating its feasibility for large-scale coverage under low-power conditions. Through simple LoRaWAN deployment, the system efficiently supports wide-area positioning tasks with high energy efficiency. In the positioning model, we incorporate a neural network trained on LoRa signal features such as RSSI and SNR, as well as historical location information, enabling accurate position estimation while minimizing reliance on GNSS. Considering the trade-off between power consumption and coverage in practical applications, LoRa communication offers an ideal solution. In future work, we plan to integrate time difference of arrival (TDoA) technology using available LoRaWAN gateways to enhance the spatial resolution and accuracy of the positioning system, thereby improving overall performance.

Acknowledgments. This work was supported by the National Natural Science Foundation of China (Grant 62262061), Major Science and Technology Special Program of Gansu Province (23ZDGA009), Central Government Guided Local Science and Technology Development Fund Project (25ZYJA007), Industrial Support Project of Gansu Provincial Department of Education (2022CYZC-12).

References

1. Janssen, T., BniLam, N., Aernouts, M., Berkvens, R., Weyn, M.: LoRa 2.4 GHz communication link and range. Sensors **20**(16), 4366 (2020)
2. Chen, N., Hu, A., Fu, H.: LoRa radio frequency fingerprint identification based on frequency offset characteristics and optimized LoRaWAN access technology. In: Proceedings of the IEEE 5th Advanced Information Technology, Electronic and Automation Control Conf. (IAEAC). IEEE (2021)
3. Müller, P., Stoll, H., Sarperi, L., Schüpbach, C.: Outdoor ranging and positioning based on LoRa modulation. In: Proceedings of the International Conference on Localization and GNSS. IEEE (2021)
4. Bianco, G.M., Giuliano, R., Marrocco, G., Mazzenga, F., Mejia-Aguilar, A.: LoRa system for search and rescue: path-loss models and procedures in mountain scenarios. IEEE Internet Things J. **8**(3), 1985–1999 (2021)
5. Chen, L., et al.: WideSee: towards wide-area contactless wireless sensing. In: Proceedings of the 17th Conference on Embedded Networked Sensor Systems, pp. 258–270 (2019)
6. Baghel, L.K., Gautam, S., Malav, V.K., Kumar, S.: TEMPSENSE: LoRa enabled integrated sensing and localization solution for water quality monitoring. IEEE Trans. Instrum. Meas. **71**, 1–11 (2022)
7. Yang, K., Chen, Y., Du, W.: OrchLoc: in-orchard localization via a single lora gateway and generative diffusion model-based fingerprinting. In: Proceedings of the 22nd Annual International Conference on Mobile Systems, Applications and Services, pp. 304–317 (2024)
8. Tang, L., Huang, Y.: Design of accurate positioning system based on LoRa for mine. J. Hunan Ind. Polytech. **20**(2), 4 (2020)
9. Zhuang, K.: Design of LoRa-based vehicle positioning system applied to underground parking lots. Guangxi University for Nationalities (2020)
10. Marquez, L.E., Calle, M.: Understanding LoRa-based localization: foundations and challenges. IEEE Internet Things J. **10**(13), 11185–11198 (2023)
11. Al-Odhari, A.H.A., Fokin, G., Kireev, A.: Positioning of the radio source based on time difference of arrival method using unmanned aerial vehicles. In: Proceedings of the Systems of Signals Generating and Processing in the Field of on Board Communications. IEEE (2018)
12. Fargas, B.C., Petersen, M.N.: GPS-free geolocation using LoRa in low-power WANs. IEEE (2017)
13. Podevijn, N., et al.: TDoA-based outdoor positioning with tracking algorithm in a public LoRa network. Wireless Commun. Mob. Comput. **2018**, 1–9 (2018)

14. Cho, J.S., Hwang, D.Y., Kim, K.H.: Improving TDoA based positioning accuracy using machine learning in a LoRaWAN environment (2019)
15. Andersen, F.R., Ballal, K.D., Petersen, M.N., Ruepp, S.: Ranging capabilities of LoRa 2.4 GHz. In: Proceedings of the IEEE 6th World Forum on Internet of Things (WF-IoT), pp. 1–5 (2020)
16. Müller, P., Stoll, H., Sarperi, L., Schüpbach, C.: Outdoor ranging and positioning based on LoRa modulation. In: Proceedings of the International Conference on Localization and GNSS. IEEE (2021)
17. Li, B., Xu, Y., Liu, Y., Shi, Z.: LoRaWAPS: a wide-area positioning system based on LoRa mesh. Appl. Sci. **13**(17), 9501 (2023)
18. Wang, P., Feng, Z., Tang, Y., Zhang, Y.: A fingerprint database reconstruction method based on ordinary Kriging algorithm for indoor localization. In: Proceedings of the International Conference on Intelligent Transportation, Big Data & Smart City (ICITBS). IEEE (2019)
19. Merhej, D., Ahriz, I., Garcia, S., Terré, M.: LoRa based indoor localization. In: Proceedings of the IEEE 95th Vehicular Technology Conference (VTC2022-Spring), pp. 1–5. IEEE (2022)
20. Qureshi, A.U.H.: Outdoor node localization using random neural networks for large scale urban IoT LoRa networks. Algorithms**14** (2021)
21. Suroso, D.J., Rudianto, A.S., Arifin, M., Hawibowo, S.: Random forest and interpolation techniques for fingerprint-based indoor positioning system in un-ideal environment. IJCDS J. **10**(1), 701–713 (2021)
22. Etiabi, Y., Jouhari, M., Burg, A., Amhoud, E.M.: Spreading factor assisted LoRa localization with deep reinforcement learning. In: Proceedings of the IEEE 97th Vehicular Technology Conference (VTC2023-Spring), pp. 1–5. IEEE (2023)
23. Savazzi, P., Goldoni, E., Vizziello, A., Favalli, L., Gamba, P.: A wiener-based RSSI localization algorithm exploiting modulation diversity in LoRa networks. IEEE Sensors J. (99), 1 (2019)
24. Anugrah, T.W., Rakhmatsyah, A., Wardana, A.A.: Non-line of sight LoRa–based localization using RSSI-Kalman-filter and trilateration. Int. J. Inf. Commun. Technol. (IJoICT) 1 (2020)
25. Liu, L., Yao, Y., Cao, Z., Zhang, M.: DeepLoRa: learning accurate path loss model for long distance links in LPWAN. In: Proceedings of the IEEE INFOCOM 2021. IEEE (2021)
26. Rofi, A.S.M., Habaebi, M.H., Islam, M.R., Basahel, A.: LoRa channel propagation modelling using artificial neural network. In: Proceedings of the 8th International Conference on Computer and Communication Engineering (ICCCE), pp. 58–62. IEEE (2021)
27. Escobar, J.J.L., Fondo-Ferreiro, P., González-Castaño, F.J., Gil-Castiñeira, F.: LoRa signal quality and GPS positioning time series dataset. https://doi.org/10.5281/zenodo.13835721. Accessed 01 Sept 2024

MSGR-DCM: Multi-Scale Graph Relational Learning and DC-Mamba for Multivariate Time Series Anomaly Detection in Industrial Control Systems

Xinjie Wang[1,2], Kaixiang Liu[1,2], Shijie Li[1,2], Zhiwen Pan[1,2], Shichao Lv[1,2], and Limin Sun[1,2(✉)]

[1] Institute of Information Engineering, CAS, Beijing, China
{8208210330,kaixiangliu}@csu.edu.cn, {lishijie,panzhiwen,
lvshichao,sunlimin}@iie.ac.cn
[2] School of Cyber Security, UCAS, Beijing, China

Abstract. Industrial Control Systems (ICSs) are critical to infrastructure security, but detecting anomalies in their multivariate time series remains challenging due to significant limitations in existing methods for modeling hierarchical multi-scale spatio-temporal dependencies and capturing long-range temporal dynamics. This paper introduces a novel framework integrating a Multi-Scale Graph Relational Learner and a DC-Mamba Block to address these gaps. The former employs convolutional feature extraction and self-attention to model hierarchical dependencies across different resolutions, while the latter fuses diffusion convolution with Mamba to efficiently capture long-range temporal dynamics in industrial processes. Experiments on two benchmark ICS dataset, Secure Water Treatment (SWaT) and Water Distribution (WADI), demonstrate the framework's superiority. On SWaT, it achieves a precision of 88.64% and an F1-score of 0.86; on the more challenging WADI dataset, it yields a precision of 94.41% and an F1-score of 0.56. Ablation studies validate the indispensable roles of both modules in enhancing detection accuracy. This work provides a robust solution for ICS anomaly detection, emphasizing the effectiveness of multi-scale graph learning and adaptive temporal modeling in industrial cyber-physical systems.

Keywords: Multi-scale · DC-Mamba · Industrial control systems · Multivariate time series · Anomaly detection

1 Introduction

Industrial Control Systems (ICSs) serve as the linchpin of crucial infrastructure sectors, encompassing water treatment facilities, power grids, and manufacturing industries [1].The digital transformation driven by Industry 4.0 has exponen-tially increased ICS connectivity and automation, while simultaneously exposing them to escalating cyber risks [2]. Security breaches in these systems can trigger cascading societal and economic impacts, as demonstrated by high-profile attacks like BlackEnergy 3 (2015), and Triton (2017), which revealed systemic vulnerabilities and caused significant operational disruptions[3–5].

Despite their critical role, current multivariate time series anomaly detection approaches for ICS environments face two significant gaps. First, few methods systematically address the multi-scale nature of industrial data. For example, a chemical plant's temperature sensors may exhibit high-frequency vibration patterns at the component level while also showing diurnal variations in reactor performance. Existing models often fail to integrate these hierarchical temporal and spatial scales, leading to missed anomalies that manifest across multiple resolutions. Additionally, most techniques struggle to capture long-range temporal dependencies inherent in ICS systems. Due to mechanical inertia and complex feedback loops, industrial processes like power grid voltage regulation or water distribution networks exhibit delayed responses spanning minutes to hours. Conventional recurrent models such as LSTM, however, suffer from gradient degradation when processing sequences with long time steps [6]. This limitation becomes particularly problematic in detecting incipient cyber-physical attacks, where adversarial perturbations may propagate slowly through the system before causing observable failures [7].

To address these limitations, we introduce two novel modules that enhance the modeling capabilities of graph and sequential neural networks. First, a multi-scale graph relational learner is proposed to explicitly capture hierarchical dependencies across temporal dimensions. This module integrates a convolutional feature extractor to capture local patterns and adaptively aggregates them into global representations via self-attention, enabling detection of anomalies that span different levels of granularity. Second, the DC-Mamba block is developed to tackle long-range temporal dependencies, combining diffusion convolution to model spatial graph structures with Mamba's adaptive gating mechanism for efficient long-sequence temporal modeling, thereby bridging spatial-temporal correlations in industrial data.

We evaluate our method on two widely used ICS benchmark datasets: the Secure Water Treatment (SWaT) and the Water Distribution (WADI) dataset. On SWaT, our approach achieves a precision of 88.64% and an F1-score of 0.86. For the more challenging WADI dataset, the model delivers a precision of 94.41% and an F1-score of 0.56. These results collectively validate the method's robustness across diverse ICS scenarios, from moderate to large-scale systems.

The main contributions of this work are summarized as follows:

1. Multi-scale Graph Relational Learner: A novel module integrating convolutional feature extraction and self-attention to model hierarchical spatio-temporal dependencies, capturing fine-grained local patterns and coarse-grained global trends in industrial data.
2. DC-Mamba Block: Integrates graph neural networks (GNNs) via diffusion convolution with Mamba to capture spatio-temporal information, addressing long-range temporal dependencies.
3. We conducted experiments on two ICS benchmark datasets (SWaT and WADI) and achieved superior anomaly detection performance compared to conventional approaches.

2 Related Works

2.1 Multivariate Time Series Anomaly Detection

Traditional time series anomaly detection approaches primarily consist of three categories: statistical modeling [8], distance-based methods [9], and clustering algorithms [10]. Statistical models, such as autoregressive integrated moving average, predict normal patterns by capturing temporal dependencies through parametric equations. Distance-based methods identify outliers by measuring dissimilarity between data points using metrics like Euclidean distance. Clustering-based approaches group similar data instances and flag samples deviating from dominant clusters as anomalies. However, these methods often fail to capture complex non-linear temporal features and temporal corre-lations, making them unsuitable for real-world datasets.

2.2 Multi-Scale Relationship Mining in Multivariate Time Series

The analysis of multivariate time series data in industrial control systems requires a hierarchical understanding of temporal and spatial dependencies across multiple scales. Traditional multiscale modeling approaches mainly rely on signal processing techniques. For example, Wang et al. [11] proposed a framework that combines discrete wavelet transform (DWT) with dynamic graph convolution to decompose the data into frequency components and construct adaptive graphs at each scale to capture variable dependencies in multivariate industrial data. Another approach is to start with data periodicity segmentation. For example, Guo et al. [12] introduced a 3D convolutional neural network to model the temporal dependencies of traffic data at daily, weekly, and monthly scales by dividing the data into time periods. These methods excel at capturing local patterns but struggle to integrate hierarchical depen-dencies across scales [13].

2.3 Long-Range Temporal Dependency Modeling

Another critical challenge in ICS anomaly detection is capturing long-range temporal dependencies, which arise from mechanical inertia, thermal delays, and complex feedback loops in industrial processes. Conventional recurrent neural networks (RNNs) such as LSTM and GRU are widely used for temporal modeling but suffer from gradient degradation when processing sequences with long time steps. This issue is particularly problematic in ICSs, where attack propagation may span hours (e.g., stealthy sensor spoofing in water treatment plants).

Transformer-based architectures have achieved break-throughs in long-range dependency modeling through self-attention mechanisms, but their quadratic time complexity makes them computationally infeasible for real-time ICS applications. Recent efforts to address this include sparse attention variants like Longformer [14], which reduce complexity to. However, these methods still require significant memory resources and are challeng-ing to deploy on edge devices.

3 Methodology

The overall architecture of our model, MSGR-DCM (shown in Fig. 1), follows the VAE structure and consists of three interconnected modules for spatio-temporal dynamics capture and anomaly detection optimization. The Multi- scale Graph Relational Learner (Fig. 2) uses a hierarchical convolutional architecture with different window sizes to extract local and global spatial features from industrial time series, and integrates them via self-attention for sensor interdependency modeling. The DC-Mamba Block (Fig. 3) fuses diffusion-convolved graph features (from bidirectional random walks) with the Mamba model, enabling efficient modeling of long-range spatio-temporal dependencies. The VAE-based loss function module combines reconstruction and KL divergence losses to optimize the model for normal pattern learning and anomaly detection. Collectively, these modules allow MSGR-DCM to systematically model hierarchical spatio-temporal features and achieve robust anomaly detection in industrial control systems.

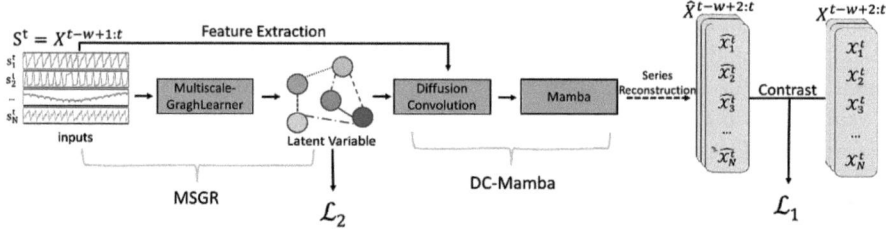

Fig. 1. Overall architecture of MSGR-DCM

3.1 Multi-Scale Graph Relational Learner

In order to capture multi-scale information in the construction of the initial graph, we introduce a Multi-layer convolution module that seamlessly integrates into conventional graph relational learners. As shown in Fig. 1, this module employs a three-stage convolutional architecture with kernel sizes [3] to fuse multi-scale information into the graph construction process. By applying successive convolutional layers with specific window sizes, the module extracts fine-grained local patterns and coarse-grained global trends in a hierarchical manner. These multi-scale features are then aggregated through a connection mechanism, ensuring that both detailed and contextual information are preserved in the final graph representation.

After the data is processed by the Multi-scale Graph Relational Learner, we apply the self-attention-based mechanism. The input to this part is the processed data, denoted as $\mathbf{H}^t = \{\mathbf{h}_i^t, i = 1, \ldots, N\}$, where N is the number of sensors or time series.

Self-attention-based mechanism is applied to calculate Θ^t, which represents the distribution of dependence relation-ships. More specifically, the high-level extracted feature H^t is projected into latent query, key subspaces Q^t and K^t with learnable parameters W_q and W_k respectively. Then the dependence relationship distribution parameters can

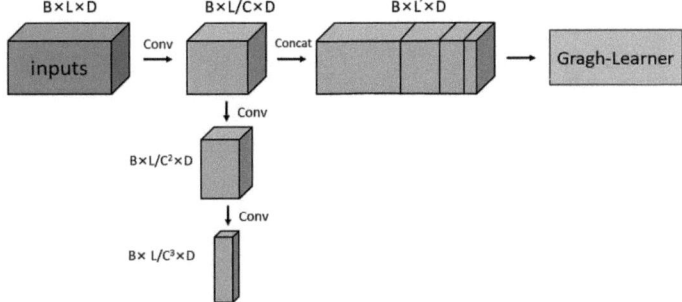

Fig. 2. Architecture of the Multi-Scale Graph Relational Learner

be obtained by dot-product attention. The formulas are as follows:

$$\mathcal{Q}^t = H^t W_q \tag{1}$$

$$\mathbf{K}^t = \mathbf{H}^t \mathbf{W_k} \tag{2}$$

$$\Theta^t = \mathbf{Q}^t \mathbf{K}^T = [\theta_{i,j}^t] \quad 1 \leq i, j \leq N \tag{3}$$

Here, $\theta_{i,j}^t$ represents the dependence relationship para-meter between the ith and j^{th} time series at time step t. Each pair-wise parameter $\theta_{i,j}^t$ is a key factor in determining the dependence structure among different time series. By using the self-attention mechanism, the encoder can effectively capture the complex relationships between different time series, which is essential for accurately learning the latent dependence relationships in the data. This information will be further used in subsequent parts of the model.

3.2 DC-Mamba Block

To integrate graph-structured information into the Mamba model, we adopt the diffusion convolution framework from DCRNN, which systematically captures multi-scale spatial dependencies through bidirectional random walk processes. The adjacency matrix $\Theta \in \mathbb{R}^{N \times N}$, representing node connections in the industrial system, is first normalized to construct symmetric random walk matrices that facilitate feature propagation across the graph. Specifically, we add self-loops to Θ (i.e., $\Theta + \mathcal{I}$) to ensure numerical stability and then perform symmetric normali-zation:

$$\widehat{\Theta} = \mathcal{D}^{-1/2}(\Theta + \mathrm{I})\mathcal{D}^{-1/2} \tag{4}$$

$$\widehat{\Theta}^T = \mathcal{D}^{-1/2}(\Theta^T + \mathcal{I})\mathcal{D}^{-1/2} \tag{5}$$

where $\mathcal{D}$ is the diagonal degree matrix with $\mathcal{D}_{ii} = \sum_j \Theta_{ij}$. These matrices model forward and backward information flow on the graph, respectively.

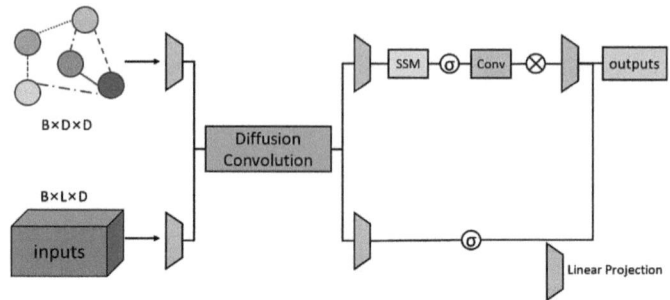

Fig. 3. Architecture of DC-Mamba Block

For each time-step feature matrix $X \in \mathbb{R}^{B \times L \times D}$, we compute K-order diffusion sequences along both directions. The forward diffusion sequence is generated as:

$$\mathbf{X}_t^{(0)} = \mathbf{X}_t, \quad \mathbf{X}_t^{(k)} = \widehat{\Theta} \mathbf{X}_t^{(k-1)} \text{ for } k = 1, 2, \ldots, K \tag{6}$$

capturing dependencies up to K-hop neighbors in the original graph. Similarly, the backward diffusion sequence follows:

$$\mathbf{X}_t'^{(0)} = \mathbf{X}_t, \quad \mathbf{X}_t'^{(k)} = \widehat{\Theta}^T \mathbf{X}_t'^{(k-1)} \text{ for } k = 1, 2, \ldots, K \tag{7}$$

modeling reverse-directional interactions. These sequences aggregate local and global spatial features by propagating node states across multiple scales.

The forward and backward diffusion features at all orders are concatenated as $\mathbf{X}_t^{\text{fwd}} = \left[\mathbf{X}_t^{(0)}, \mathbf{X}_t^{(1)}, \ldots, \mathbf{X}_t^{(K)}\right]$ and $\mathbf{X}_t^{\text{bw}} = \left[\mathbf{X}_t'^{(0)}, \mathbf{X}_t'^{(1)}, \ldots, \mathbf{X}_t'^{(K)}\right]$, The final graph-modulated feature combines both directions as

$$X_t^{graph} = ReLU(X_t^{fwd} W_0^T) + ReLU(X_t^{bw} W_1^T) \tag{8}$$

where W is a linear layer with ReLU activation.

This graph-modulated feature is fed into the Mamba model to enhance its temporal dynamics with structural context. The state transition mechanism of Mamba is adapted to incorporate the graph-aware inputs through gated operations, where the update gate $\tilde{z}_t$ and reset gate r_t modulate the information flow between the current graph features and the previous hidden state h_{t-1}:

$$z_t = \sigma(\mathbf{W}_z \mathbf{X}_t^{\text{sraph}} + \mathbf{U}_z \mathbf{h}_t - 1) \tag{9}$$

$$\mathbf{r}_t = \sigma(\mathbf{W}_r \mathbf{X}_t^{\text{sraph}} + \mathbf{U}_r \mathbf{h}_t - 1) \tag{10}$$

where σ denotes the sigmoid activation. The candidate state $\tilde{h}_t$ is computed using a non-linear transformation (e.g., tanh) that integrates the reset hidden state and the current graph features:

$$\tilde{\mathbf{h}}_t = \phi\left(\mathbf{W}_{\tilde{h}} \mathbf{X}_t^{\text{graph}} + \mathbf{U}_{\tilde{h}}(\mathbf{r}_t \odot \mathbf{h}_{t-1})\right) \tag{11}$$

The final hidden state is updated as a weighted combination of the previous state and the candidate state:

$$\mathbf{h}_t = (1 - \mathbf{z}_t) \odot \mathbf{h}_{t-1} + \mathbf{z}_t \odot \tilde{\mathbf{h}}_t \qquad (12)$$

This formulation seamlessly merges graph-structured spatial dependencies with Mamba's efficient long-range temporal modeling, enabling the model to capture intricate spatio-temporal patterns in industrial control systems.

3.3 Objective Function and Training

We employs a Variational Autoencoder (VAE) framework, with its objective function designed to maximize the Evidence Lower Bound (ELBO). The ELBO is formulated as:

$$\mathcal{L} = \mathbb{E}_{q_\phi(\mathcal{Z}^t|\mathcal{S}^t)}[\log p_\psi(\mathcal{S}^t|\mathcal{Z}^t)] - \mathrm{KL}[q_\phi(\mathcal{Z}^t|\mathcal{S}^t)||p_\psi(\mathcal{Z}^t)] \qquad (13)$$

This equation consists of two key components. The first term, $\mathbb{E}_{q_\phi(\mathcal{Z}^t|\mathcal{S}^t)}[\log p_\psi(\mathcal{S}^t|\mathcal{Z}^t)]$, represents the reconstruction loss. It quantifies the model's capability to reconstruct input data $\mathcal{S}^t$ using the learned latent dependencies. The second term, $\mathrm{KL}[q_\phi(\mathcal{Z}^t|\mathcal{S}^t)||p_\psi(\mathcal{Z}^t)]$ is the Kullback-Leibler divergence loss. This term measures the discrepancy between the learned dependency distribution from the encoder and the predefined prior distribution.

In the decoder, the reconstruction loss is decomposed recursively:

$$p_\psi(S^t|Z^t) = \prod_{t'=t-w+1}^{t-1} p_\psi(X^{t'+1}|X^{t'}, \ldots, X^{t-w+1}, Z^t) \qquad (14)$$

For each input sample S^t, the total reconstruction loss is calculated as:

$$\mathcal{L}_1 = -\sum_{i=1}^{N} \sum_{t'=t-w+2}^{t} \frac{\left\|x_i^{t'} - \mu_i^{t'}\right\|^2}{2\sigma^2} \qquad (15)$$

The KL divergence loss is computed as:

$$\mathcal{L}_2 = \mathrm{KL}[q_\phi(\mathcal{Z}^t|\mathcal{S}^t)||p_\psi(\mathcal{Z}^t)] = \sum_{k=1}^{h_0} \overline{p}_k^t \log\left(\frac{\overline{p}_k^t}{\overline{q}_k}\right) \qquad (16)$$

where $\overline{p}_k^t = \sum_{i=1}^{N} \sum_{j=1}^{N} z_{i,j}^t(k)$ aggregates the probabi-lities of dependencies assigned to type k.

4 Experiments and Analysis

4.1 Dataset

We select two commonly used datasets for industrial process anomaly detection: Secure Water Treatment (SWaT) and Water Distribution (WADI). Among them, the WADI dataset has a larger data volume for detection and is more challenging, while the SWaT dataset provides a relatively moderate-scale scenario for analysis. Table 1 details the two benchmark datasets, including their sizes and Anomaly Rates (%).

Table 1. Statistics of datasets.

Dataset	Train	Test	Dimensions	Anomalies (%)
SWaT	496800	449919	51	11.98
WADI	1048571	172801	123	5.99

4.2 Evaluation Metrics

We evaluate the performance of our method using three standard metrics: Precision (Pre), Recall (Rec), and F1-score(F1). The corresponding formulas are defined as follows:

$$\text{Precision} = \frac{TP}{TP + FP} \tag{17}$$

$$\text{Recall} = \frac{TP}{TP + FN} \tag{18}$$

$$F1 - \text{score} = \frac{2 \times TP}{2 \times TP + FP + FN} \tag{19}$$

where, TP (True Positive) and TN (True Negative) represent the number of correctly identified positive and negative samples, respectively. FP (False Positive) refers to negative samples incorrectly classified as positive, while FN (False Negative) refers to positive samples incorrectly classified as negative.

4.3 Baseline

- GDN[15]:learns sensor dependency graphs with em-beddings and graph attention for interpretable anomaly scoring.
- USAD[16]:uses dual autoencoders with adversarial training to combine reconstruction errors for anomaly detection.
- GreLeN[17]:uses VAE, GNN, and self-attention to model sensor dependencies via latent graph struc-tures and dynamic degree changes for multivariate time series anomaly detection.
- AnomalyBERT[18]:uses a Transformer with 1D relative position bias and a self-supervised data degradation schemeto detect anomalies by predicting degraded parts.
- TopoGDN[19]: uses multi-scale temporal convolution, dynamic graph construction, and topological-temporal feature fusion with anomaly scoring to enhance GAT for multivariate time-series anomaly detection.

4.4 Overall Performance

Table 2 presents the performance results of all methods on public datasets. To comprehensively assess the performance of different models in the anomaly detection task, we compare them using three key metrics: Precision (Pre), Recall (Rec), and F1-score (F1). Several critical observations can be drawn as follows:

Table 2. Comparison of our method with baseline methods on the SWaT and WADI dataset.

Method	SWaT			WADI		
	Prec	Rec	F1	Prec	Rec	F1
GDN	84.35%	52.63%	0.65	64.95%	38.71%	0.48
USAD	74.89%	59.46%	0.66	75.83%	15.00%	0.25
GreLeN	81.14%	**86.91%**	0.84	75.98%	**48.92%**	**0.6**
AnomalyBERT	72.26%	78.62%	0.75	92.42%	28.81%	0.44
TopoGDN	87.93%	71.91%	0.79	58.08%	44.85%	0.51
Ours	**88.64%**	84.22%	**0.86**	**94.41%**	39.67%	0.56

On the SWaT dataset, the proposed method demon-strates superior overall performance, achieving the highest F1-score of 0.86. This is attributed to its balanced precision (88.64%) and recall (84.22%), indicating strong capability in both correctly identifying anomalies (recall) and minimizing false positives (precision).

The WADI dataset presents a different challenge, where most models struggle to balance precision and recall. But our method still stands out with the highest F1-score of 0.56, despite a lower recall (39.67%). Its precision (94.41%) is the highest among all models, demonstrating strong reliability in identifying true anomalies with minimal false positives. This is particularly valuable in scenarios where false alarms are costly. The recall and F1-score are relatively lower compared to SWaT. This discrepancy can be attributed to two key factors: (1) Data imbalance: WADI features a lower anomaly rate, making rare anomalies harder to detect; (2) High-dimensional complexity: With 123 sensor dimensions, WADI introduces more intricate cross-feature dependencies that may exceed the model's current multi-scale fusion capacity. In the future, we hope to explore anomaly-sensitive data augmentation and hierarchical attention mechanisms to address the above issues.

4.5 Ablation Analysis

To investigate the contribution of key architectural components, we conducted ablation studies by removing the multi-scale convolution module and DCmamba module respectively: removing the multi-scale convolution module involved replacing it with a standard self-attention mechanism to assess the role of multi-scale feature extraction in capturing fine-grained anomaly patterns, while removing the DCmamba module entailed substituting it with a traditional DCRNN (Dynamic Convolutional Recurrent Neural Network) to evaluate the necessity of DCmamba's adaptive temporal modeling for sequence-based anomaly detection, with the impact of these modifications on model performance analyzed accordingly. The results are shown in Table 3.

Table 3. Results of ablation study

Method	SWaT			WADI		
	Prec	Rec	F1	Prec	Rec	F1
w/o DCMamba	74.06%	**94.92%**	0.83	65.92%	**55.66%**	**0.6**
w/o MultiConv	80.92%	88.10%	0.84	83.80%	35.71%	0.5
MSGR-DCM	**88.64%**	84.22%	**0.86**	**94.41%**	39.67%	0.56

Ablating the DC-Mamba module (replacing it with DCRNN) results in a 3.5% F1-score decline on SWaT, primarily driven by a significant precision drop to 74.06% despite high recall (94.92%). This trade-off indicates that without DC-Mamba's adaptive temporal modeling, the model produces more false positives (incorrectly flagging normal instances as anomalies), likely due to DCRNN's limited ability to distinguish complex temporal patterns between normal and abnormal sequences. On WADI, the imbalance between precision (65.92%) and recall (55.66%) further reveals that DC-Mamba is critical for maintaining a stable detection threshold in noisy, large-scale data—its absence leads to both missed anomalies (low precision) and false alarms (moderate recall), undermining the model's reliability in real-world industrial settings.

Ablating the multi-scale convolution module (replacing it with standard self-attention) causes a 2% F1-score drop on SWaT, with relatively balanced precision (80.92%) and recall (88.10%). This suggests self-attention alone can capture some scale-invariant dependencies, but the loss of multi-scale feature fusion reduces the model's sensitivity to fine-grained anomalies, as reflected by the marginal F1 decline. On WADI, however, the severe recall drop to 35.71% (despite high precision at 83.80%) indicates a critical failure to detect true anomalies, likely because the lack of multi-scale features prevents the model from identifying subtle, multi-resolution patterns characteristic of complex attacks. This highlights the module's role in ensuring comprehensive anomaly coverage (recall) across diverse data resolutions.

Summary: The DC-Mamba module is indispensable for maintaining high precision in noisy environments by modeling long-term temporal consistency, while the multi-scale convolution module enhances recall by enabling multi-resolution anomaly feature capture. Their combined effect balances precision (reducing false positives) and recall (minimizing false negatives), as demonstrated by the full model's superior F1-scores. This synergy underscores that both components are essential for robust anomaly detection in industrial systems, where neither over-penalizing normal operations nor missing stealthy attacks can be compromised.

5 Conclusion

In this study, we address the challenges of multi-scale spatio-temporal dependency modeling and long-range temporal dependency capture in multivariate time series anomaly detection for industrial control systems (ICSs). We propose two novel modules: a Multi-Scale Graph Relational Learner to explicitly model hierarchical dependencies across different granularities and a DC-Mamba Block to efficiently capture long-range temporal dynamics via diffusion convolution and Mamba's adaptive gating mechanism. Experimental results on two benchmark ICS datasets (SWaT and WADI) demonstrate that our approach achieves superior anomaly detection performance, outperforming conventional approaches in terms of precision, recall, and F1-score. Ablation studies validate the critical contributions of both modules, highlighting their synergistic role in enhancing feature representation and temporal modeling. This work provides a robust solution for ICS anomaly detection, particularly in scenarios requiring high precision and multi-scale dependency analysis, and paves the way for future research on adaptive spatio-temporal modeling in industrial cyber-physical systems.

Acknowledgments. This work was supported by National Natural Science Foundation of China (Grant No.92467201), the Beijing Natural Science Foundation (Grant No.L234033). This work was supported in part by the computational resources provided by the School of Computer Science and Engineering, Central South University.

References

1. Raman, M.G., Mathur, A.P.: A hybrid physics-based data-driven framework for anomaly detection in industrial control systems. IEEE Trans. Syst. Man and Cybern.: Systems **52**(9), 6003–6014 (2021)
2. Liu, J., et al.: ShadowPLCs: a novel scheme for remote detec-tion of industrial process control attacks. IEEE Trans. Dependable and Secure Comput. **19**(3), 2054–2069 (2020)
3. Liu, K., et al.: SecureSIS: Securing SIS Safety Functions With Safety Attributes and BPCS Information. IEEE Trans. Inf. Forensics Secur. **20**, 3060–3073 (2025)
4. Liang, G., et al.: The 2015 Ukraine blackout: Implications for false data injection attacks. IEEE transactions on power systems **32**(4), 3317–3318 (2016)
5. Di Pinto, A., Dragoni Y., Carcano A.: TRITON: The first ICS cyber attack on safety instrument systems. Proc. Black Hat USA **2018**, 1–26 (2018)
6. Al-Selwi, S.M., et al.: LSTM inefficiency in long-term dependencies regression problems. J. Adv. Res. Appl. Sci. Eng. Technol. **30**(3), 16–31 (2023)
7. Zizzo, G., et al.: Adversarial attacks on time-series intrusion detection for industrial control systems. 2020 IEEE 19th international conference on trust, security and privacy in computing and communications (TrustCom). IEEE (2020)
8. Zhai, S., et al.: Deep structured energy based models for anomaly detection. International Conference on Machine Learning. PMLR, (2016)
9. Sarmadi, H., Karamodin, A.: A novel anomaly detection method based on adaptive Mahalanobis-squared distance and one-class kNN rule for structural health monitoring under environmental effects. Mech. Syst. Signal Process. **140**, 106495 (2020)

10. Syarif, I., Prugel-Bennett, A., Wills, G.: Unsupervised clustering approach for network anomaly detection. Networked Digital Technologies: 4th International Conference, NDT 2012, Dubai, UAE, April 24–26, 2012. Proceedings, Part I 4. Springer Berlin Heidelberg, (2012)
11. Wang, J., et al.: Multiscale wavelet graph autoencoder for multivariate time-series anomaly detection. IEEE Trans. Instrum. Meas. **72,** 1–11 (2022)
12. Guo, Shengnan, et al. Deep spatial–temporal 3D convolutional neural networks for traffic data forecasting. IEEE Transactions on Intelligent Transportation Systems **20**(10), pp. 3913–3926 (2019)
13. Nawaz, M., et al.: Review of multiscale methods for process monitoring, with an emphasis on applications in chemical process systems. IEEE Access **10**, 49708–49724 (2022)
14. Beltagy, I., Peters, M.E., Cohan, A.: Longformer: the long-document transformer. arXiv preprint arXiv:2004.05150 (2020)
15. Deng, A., Bryan H.: Graph neural network-based anomaly detection in multivariate time series. Proceedings of the AAAI conference on artificial intelligence. Vol. 35. No. 5. (2021)
16. Audibert, J., et al.: Usad: Unsupervised anomaly detection on multivariate time series. Proceedings of the 26th ACM SIGKDD international conference on knowledge discovery & data mining. (2020)
17. Zhang, W., Chen Z., Fugee T.: GRELEN: Multivariate Time Series Anomaly Detection from the Perspective of Graph Relational Learning. IJCAI. (2022)
18. Jeong, Y., et al.: Anomalybert: Self-supervised transformer for time series anomaly detection using data degradation scheme. arXiv preprint arXiv:2305.04468 (2023)
19. Liu, Z., et al.: Multivariate time-series anomaly detection based on enhancing graph attention networks with topological analysis. Proceedings of the 33rd ACM International Conference on Information and Knowledge Management. (2024)

A Review on Temporal Knowledge Graph Completion in the Context of Internet of Things and Industrial Security

Runze Li[1], Sha Xiang[2], Shuo Zhu[1], and Banglie Yang[2(✉)]

[1] College of Software Engineering, Sichuan University, Chengdu 610000, China
{2022141461090,2022141461135}@stu.scu.edu.cn
[2] College of Computer Science, Sichuan University, Chengdu 610000, China
{xiangsha,yangbanglie}@stu.scu.edu.cn

Abstract. As Internet of Things (IoT) grow rapidly, security concerns have escalated due to the massive interconnection of industrial devices. Traditional static knowledge graphs fall short in representing evolving threats and real-time security events. To address this limitation, Temporal Knowledge Graphs (TKGs) have been proposed, enabling the modeling of time-dependent relations and entities that reflect dynamic security incidents. However, issues such as sparse threat intelligence and incomplete event data remain, necessitating the development of Temporal Knowledge Graph Completion (TKGC) techniques to predict potential threats and update security knowledge bases in real time. In this review, we provide a comprehensive overview of TKGC approaches, categorizing them into interpolation-based and extrapolation-based methods. We analyze recent advancements in the field, emphasizing their technological improvements over earlier techniques. Furthermore, we explore the distinct challenges presented by IoT and industrial security contexts and outline promising future directions to enhance the applicability and effectiveness of TKGC in safeguarding connected systems.

Keywords: Temporal Knowledge Graph · Temporal Knowledge Graph Completion · Internet of Things and Industrial Security

1 Introduction

Graphs, as structured multi-relational knowledge bases, store rich real-world events in the form of structured facts. In the context of IoT and Industrial Security, these graphs can represent complex interactions among industrial devices, control systems, and monitoring sensors. Each fact is represented as a triple (head entity, relation, tail entity, i.e., s, r, o), such as (Firewall, protects, Industrial Network), where "Firewall" and "Industrial Network" are the head and tail entities respectively, and "protects" is the relation. The structured and dynamic nature of knowledge graphs makes them highly useful in IoT security applications, enabling intrusion detection systems [1], vulnerability analysis [2], and threat intelligence integration across industrial infrastructures [3].

However, despite the large volume of IoT-generated security data, knowledge graphs derived from such environments often suffer from incompleteness and outdated threat information. To address this, Knowledge Graph Completion (KGC) techniques have been proposed to automatically infer missing security facts. With the rise of machine learning and deep neural networks [4], KGC methods have become effective at link prediction by learning low-dimensional embeddings of entities and relations, then applying scoring functions to validate potential connections. A major limitation of traditional KGC methods is their assumption of static facts [5], which conflicts with the dynamic and real-time nature of industrial security scenarios. In this domain, temporal context is essential—security events evolve quickly, and systems must react in real time to threats, anomalies, and attack patterns. This need has led to the emergence of Temporal Knowledge Graph Completion (TKGC).

TKGC incorporates temporal information into the learning process through quadruples (s,r,o,t), for example, (Lawrence Page, founder, google, 1998-9-4). This temporal modeling enhances link prediction performance and reflects changes in security states over time. TKGC techniques bridge the gap between static knowledge and dynamic threat landscapes by enabling timely updates and integration of diverse, often heterogeneous, data sources. We classify TKGC methods into two groups: interpolation-based methods that learn timestamp embeddings along with entity and relation embeddings (e.g., TNTComplEx [6], TimeIE), and extrapolation-based methods that predict future or unknown events beyond the observed range (e.g., Long Short-Term Memory (LSTM) [7], Temporal Graph Networks (TGN) [3]). TKGC still faces key challenges. On one hand, security knowledge graphs must be updated in real time to capture emerging threats and vulnerabilities. On the other hand, integrating timestamped information from multiple security data sources remains complex, with challenges in data format compatibility and temporal reasoning.

Recently, there has been increasing innovation in TKGC research. For example, frequency attention mechanisms [8] and multi-task learning frameworks [9] are being employed to improve temporal modeling. Frequency attention can detect periodic patterns in attack behaviors and better represent evolving threats, as seen in methods like TuckER-FA [10]. Multi-task learning, incorporating tasks like entity alignment [11], improves TKGC across heterogeneous IoT security datasets. Despite these developments, many surveys still provide only broad overviews of TKGC, with limited focus on recent advances and its growing relevance in industrial cybersecurity. The contributions of this paper are summarized as follows:

1) We provide a systematic summary of the existing literature on Temporal Knowledge Graph Completion (TKGC), with a particular focus on recent advancements.
2) We categorize TKGC methods based on their ability to predict future security events into interpolation and extrapolation types. We further introduce new methods, compare their strengths, and summarize key techniques.
3) We propose future research directions aimed at improving TKGC performance and applicability in the context of IoT and Industrial Security.

This paper is divided into five sections: first, we introduces the topic and motivation in section one, and presents the background of TKGC in section two. The third section categorizes and overviews current TKGC methods, and then the fourth section discusses future research directions in the context of IoT and Industrial Security. Finally, we provide a conclusion in section five.

2 Background

2.1 Preliminaries

Temporal Knowledge Graphs (TKGs) represent a sophisticated evolution of traditional Knowledge Graphs, specifically designed to capture and represent dynamic, time-sensitive information. Unlike traditional knowledge graphs that capture relationships between entities at a single moment in time, TKGs are capable of modeling the evolution of facts and connections and then providing a temporal perspective that reflects how the real world changes. In formal terms, a TKG is defined as $G = Q - E, R, T$, where E is the set of entities that includes nodes such as people, organizations, places, or even abstract concepts; R is the set of relationships that serve as the connective tissue between these entities, describing interactions such as employment, geographical location, or ownership; and T is the set of timestamps that imbue the graph with a temporal dimension, allowing for the representation of either discrete time points (for example, a specific date like "2024–10-09") or extended time intervals (such as "2020–2025") during which certain facts remain valid.

In a temporal knowledge graph, each fact is formally represented as a quadruple (s, r, o, t), , where temporal dimension is essential for accurate representation. For instance, the relation (Barack Obama, held_position, President of the United States, "2009–2017") precisely captures the temporal scope of this political fact, unlike static representations that lack chronological context. Outside that window, he might be labeled as an author or activist. These time-specific labels are critical in fields like political history or corporate strategy, where context changes constantly. Without time stamps, a knowledge graph could misleadingly suggest Obama is still president today—a clear disconnect from reality. By embedding time into the graph's structure, TKGs create a living snapshot of the world, letting researchers track relationships as they evolve.

2.2 Loss Functions

Loss functions are a fundamental component of TKGC models, playing a critical role in determining how effectively the model can predict missing quadruples. Given the dynamic nature of temporal relationships, recent loss function designs have increasingly emphasized temporal consistency and prediction accuracy. Here are some commonly used loss functions and regularization techniques in TKGC tasks:

1) Cross-Entropy Loss: Cross-entropy loss [13] is commonly used in classification tasks and has also been widely applied in TKGC models, especially those based on neural networks, such as Transformer-based TKGC models. The formula is:

$$-\sum_{i=1}^{n}\left[y_i \log(x_i) + (1 - y_i)\log(1 - x_i)\right] \quad (1)$$

where y_i is the true label (either 0 or 1) for the sample, and x_i is the predicted probability that the sample belongs to the positive class. For each sample, the loss is $-\log(x_i)$ if the true label is 1, and $-\log(1 - x_i)$ if the true label is 0.

2) Self-Adversarial Negative Sampling Loss: Self-adversarial negative sampling [14] introduces a strategic approach to negative sample selection. Unlike conventional methods that randomly select negative samples, this technique assigns higher weights to samples that are semantically close to positive instances but temporally inconsistent. For instance, the model would heavily penalize temporally impossible scenarios like "Company A acquired Company B in 2025" when records show Company B dissolved in 2020. The loss function is defined as:

$$-\log(\sigma(f(s, r, o, t))) - \sum_{i=1}^{n} p_i \log(\sigma(-f(s_i, r_i, o_i, t_i))) \qquad (2)$$

where p_i denotes the adversarial weight calculated based on the scoring function $f(s, r, o, t)$. Benchmark studies on ICEWS political event data show this method improves temporal discrimination accuracy compared to random negative sampling [14].

3) Cosine Similarity-based Temporal Regularization Loss: Recent work uses cosine similarity between consecutive timestamp embeddings to ensure gradual temporal transitions. This penalizes abrupt changes in temporal representations. The loss function is formulated as:

$$-\frac{1}{|T| - 1} \sum_{i=1}^{|T|-1} cos(t_i, t_{i+1}) \qquad (3)$$

where t_i and t_{i+1} are embeddings of sequential timestamps. This approach effectively reduces inconsistencies in long-term temporal reasoning tasks like tracking entity evolution in financial knowledge graphs [15].

4) Adaptive Margin-based Ranking Loss: The adaptive margin-based ranking loss [16] dynamically adjusts the margin parameter $\lambda(t)$ based on temporal characteristics like relationship duration. For example, in modeling executive tenure (e.g., "CEO of Company X, 2010–2020"), a larger margin is allocated to reflect the sustained significance of such relationships compared to transient events (e.g., "temporary partnership, 2021"). The loss function is expressed as:

$$\sum_{(s,r,o,t) \in Q} \left[\lambda(t) + g(s, r, o, t) - \sum_{o' \in E} g(s, r, o', t) \right]^+ \qquad (4)$$

where $g(s, r, o, t)$ is the scoring function. This adaptive mechanism is valuable in scenarios with imbalanced event distributions, such as political event prediction.

2.3 Benchmark Datasets

To effectively evaluate TKGC models, researchers employ several benchmark datasets, each presenting unique challenges in terms of temporal complexity, entity relationships, and data sparsity. The most widely-used datasets include:

1) ICEWS [17]: The Integrated Crisis Early Warning System (ICEWS) is one of the most popular temporal knowledge graph datasets, consisting of global political event data. The events are represented in the form of triples (subject, relation, object) and include timestamps detailing when events occurred.

2) GDELT [18]: The Global Database of Events, Language, and Tone (GDELT) is a large-scale global news event database that has tracked events since 1979. It is often used for temporal event prediction and modeling entity changes.
3) YAGO-T [19]: YAGO-T is a temporal extension of the YAGO knowledge graph, linking triples with time information. It is commonly used in long-term knowledge reasoning tasks and modeling historical records of entities.
4) Wikidata-T [20]: This is the temporal extension of Wiki data, enriched with temporal annotations, making it suit able for time-based reasoning in knowledge graph research.
5) DE-KG: The DE-KG dataset focuses on exploring delayed effects in temporal knowledge graphs, recording both event occurrence time and effect time. It is primarily used for evaluating reasoning tasks involving time dependent relationships.
6) Flickr30k-T [21]: This dataset combines temporal knowledge graphs with visual content, containing images from Flickr30k along with their time annotations. It mainly focuses on research that integrates temporal reasoning with multimodal data.

3 Methods

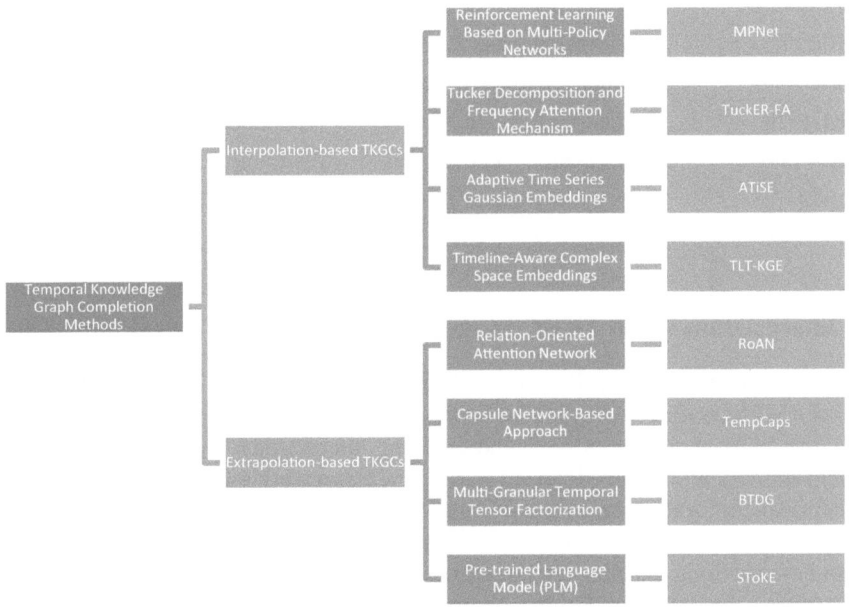

Fig. 1. Classification of Temporal Knowledge Graph Completion methods

We categorize TKGC methods into two types: Interpolation-based TKGC Methods and extrapolation-based TKGC Methods. We summarize the latest methods and techniques and classify them into these two categories, as shown in Fig. 1. And we also list TKGC methods in recent years, as shown in Table 1.

Table 1. Summary of the Temporal Knowledge Graph Completion methods.

Methods	Category	Technique
MPNet [23]	Interpolation-based TKGCs	Multi-Policy Networks
TITer [24]	Interpolation-based TKGCs	Reinforcement Learning
TuckER-FA [25]	Interpolation-based TKGCs	Frequency Attention Mechanism
TuckERTNT [27]	Interpolation-based TKGCs	Tucker Decomposition
ATiSE [28]	Interpolation-based TKGCs	Time Series Decomposition
TTransE [29]	Interpolation-based TKGCs	Time Series Decomposition
HyTE [30]	Interpolation-based TKGCs	Time Series Decomposition
TLT-KGE [31]	Interpolation-based TKGCs	Timeline-traced Embedding
TeAST [40]	Interpolation-based TKGCs	Archimedean SpiralTimeline
SiMFy [41]	Interpolation-based TKGCs	Fixed-frequency Strategy
RoAN [32]	Extrapolation-based TKGCs	Relation-Oriented Attention
DEGAT [33]	Extrapolation-based TKGCs	Attention-based
MRGAT [33]	Extrapolation-based TKGCs	Attention-based
TempCaps [34]	Extrapolation-based TKGCs	Capsule Network-Based
BTDG [35]	Extrapolation-based TKGCs	Block Term Decomposition
K-BERT [38]	Extrapolation-based TKGCs	Pre-trained Language Model
SToKE [37]	Extrapolation-based TKGCs	Pre-trained Language Model
ECOLA [39]	Extrapolation-based TKGCs	Pre-trained Language Model
SPA [42]	Extrapolation-based TKGCs	Neural Architecture Search
TLogic [43]	Extrapolation-based TKGCs	Rule-based

3.1 Interpolation-Based TKGC Methods

1) Reinforcement Learning Based on Multi-Policy Networks: The Multi-Policy Network (MPNet) [23] deploys three dedicated policy networks to capture static entity relationships, dynamic entities, and evolving interactions. By leveraging reinforcement learning with tailored reward signals, the framework refines decision-making via historical exploration enhanced by innovative edge types. Unlike traditional single-policy methods such as TITer [24], MPNet employs a multi-perspective evaluation of candidate actions, offering a more robust characterization of both static and dynamic properties. Its modular design effectively tackles challenges related to incomplete action spaces and historical data processing, while traceable exploration paths and an expanded action space further empower the model to conduct sophisticated temporal reasoning. This framework elevates predictive performance and improves the interpretability of the decision process, making it highly suitable for complex dynamic environments.
2) Tucker Decomposition and Frequency Attention Mechanism: The Tucker Decomposition with Frequency Attention (TuckER-FA) framework [25] enhances temporal

knowledge graph completion by smoothly combining Tucker decomposition [26] with frequency domain processing techniques. It maintains the core tensor factorization technique for learning compact representations while innovatively incorporating Discrete Cosine Transform (DCT) analysis of timestamp embeddings. This dual strategy effectively captures both periodic patterns and irregular temporal variations. This method employs a frequency attention mechanism that simultaneously captures global temporal dependencies, detects periodic patterns, and accommodates irregular fluctuations. As demonstrated in financial market predictions [27], TuckER-FA reduces computational complexity compared to TuckERTNT while improving temporal stability, particularly when processing streaming IoT data with variable sampling rates.

3) Adaptive TimeSeries Gaussian Embeddings: Adaptive Time Series Gaussian Embeddings (ATiSE) framework [28] advances temporal modeling by integrating Gaussian embeddings with additive time series decomposition. In this approach, entities and relations are represented as multivariate Gaussian distributions whose parameters explicitly capture trend, seasonal, and residual components - for instance, enabling the separation of long-term stock price movements from quarterly earnings patterns in financial forecasting. By employing KL divergence between predicted and observed distributions as a scoring function, ATiSE effectively quantifies the temporal uncertainty inherent in dynamic systems. Compared to earlier methods like TTransE [29] and HyTE [30], this dual modeling of time intervals and fluctuations reduces computational complexity in benchmark tests [28], making it particularly effective for applications requiring real-time updates, such as IoT sensor networks.

4) Timeline-Aware Complex Space Embeddings: The Timeline-traced Knowledge Graph Completion (TLT-KGE) model [31] introduces an innovative embedding architecture using hypercomplex number systems to decouple temporal dynamics from semantic features. By establishing orthogonal temporal axes in complex vector spaces, the model distinctly separates time-dependent attributes (e.g., duration of executive positions) from time-invariant properties (e.g., industry classifications). This separation proves particularly valuable in corporate relationship analysis, where it enables clear differentiation between persistent ownership structures and transient partnerships. Empirical validation on Wikidata-T [20] demonstrates significant improvements in temporal link prediction accuracy compared to conventional approaches, while the structured decomposition cybersecurity enhances interpretability of relationship evolution patterns [31].

3.2 Extrapolation-Based TKGC Methods

1) Relation-Oriented Attention Network: The Relation-Oriented Attention Network (RoAN) [32] represents a paradigm shift in temporal knowledge graph modeling by focusing on relational dynamics rather than entity attributes. Unlike traditional graph attention networks such as GAT [33] that primarily analyze node characteristics, RoAN specifically examines sequences of connected relations (e.g., "acquired → merged → divested") through specialized temporal attention mechanisms. In supply chain analytics, this approach improves prediction of multi-step partnership trajectories compared to RGAT [33]. Its modular architecture enables integration with

existing TKGC frameworks, offering scalability for large-scale industrial knowledge graphs.
2) Capsule Network-Based Approach: Temporal Capsule Networks (TempCaps) [34] leverage dynamic routing mechanisms to assign time-aware weights for information aggregation. For example, in epidemiological knowledge graphs, capsule vectors encode temporal proximity of disease spread events (e.g., COVID-19 variant emergence dates), allowing the model to prioritize temporally relevant connections. A multi-layer perceptron (MLP) decoding layer estimates event probabilities, achieving a higher F1-score than static capsule models on the ICEWS dataset [17]. The hierarchical architecture also reduces parameter redundancy by 33%, addressing scalability challenges in real-time applications [34].
3) Multi-Granular Temporal Tensor Factorization: The Block Term Decomposition with Distinct Time Granularities (BTDG) model [35] introduces a two-level tensor decomposition strategy. A core tensor captures macro-level trends (e.g., decade-long economic cycles), while factor matrices model micro-level variations (e.g., quarterly sales fluctuations). In financial fraud detection, this approach improves anomaly detection in transaction timelines compared to Tucker decomposition [36]. BTDG also supports hybrid temporal data (discrete events and adaptable to domains like healthcare, where both episodic patient visits and longitudinal vital signs coexist).
4) Context-Aware Pre-trained Language Integration: Learning Joint Structural and Temporal Contextualized Knowledge Embeddings (StoKE) [37] introduces a groundbreaking integration of pre-trained language models (PLMs) with structured temporal knowledge through its Event Evolution Tree (EET) architecture. These trees systematically encode temporal relationships between events while preserving the rich semantic embeddings from PLMs. A BERT-based [2] masked prediction mechanism ensures robust maintenance of temporal dependencies in the model's representations. Unlike previous PLM-based approaches such as K-BERT [38] and ECOLA [39] that focused primarily on static text-graph alignment, SToKE achieves seamless unification of three critical dimensions: temporal dynamics, structural relationships, and textual context. This leads to state-of-the-art performance in temporal knowledge reasoning tasks by enabling superior context utilization and efficient knowledge transfer from large-scale pretraining models.

4 Prospect

4.1 Future Direction

1) More Efficient Temporal Information Modeling: Many current TKGC methods rely on linear time embeddings [44] to capture temporal information. However, in industrial security scenarios, time dependencies among events—such as threat propagation, delayed system responses, or gradual infiltration—are often non-linear. For instance, a vulnerability may only be exploited after a certain latency, or the impact of a breach may intensify over time. Future research could investigate non-linear time modeling techniques, such as temporal convolutional networks or recurrent attention mechanisms, to better capture these intricate dynamics. Moreover, industrial security events may occur at varying temporal granularities—ranging from milliseconds in

intrusion detection to years in system upgrades. Multi-scale modeling strategies like hierarchical time embeddings and multi-resolution temporal convolutions can help address this challenge by accurately modeling event patterns across different time scales. Furthermore, since both periodic and non-periodic events coexist in industrial environments [45], specialized embedding strategies that differentiate between these event types can significantly improve the robustness and adaptability of TKGC models.
2) Enhanced Representation by Integrating Pre-trained LanguageModels (PLMs): Pre-trained language models [46] such as BERT and RoBERTa [48] have demonstrated significant potential for security-related NLP tasks, including threat intelligence extraction and log analysis. In industrial security contexts, incorporating PLMs into TKGC frameworks could substantially improve the semantic representation of security entities and their relationships. For instance, merging PLM-derived security context with traditional graph embeddings may enable more accurate modeling of complex attack patterns or system misconfigurations. These models could continuously process real-time text streams from security bulletins, vulnerability reports, and system logs [49], allowing dynamic updates to the knowledge graph that reflect emerging threats and newly discovered vulnerabilities.
3) Multi-modal Representation of Security Events: Industrial security environments generate diverse data modalities including textual logs, visual surveillance footage, firmware images, and configuration files that collectively describe security events. Future TKGC systems could leverage multi-modal learning approaches [50] to create comprehensive representations of security entities. For example, correlating visual data from physical security systems with textual alert logs could provide a more complete understanding of security incidents. Pre-trained models such as CLIP offer promising capabilities for aligning visual and textual modalities, particularly for analyzing physical-cyber interactions. A key challenge involves developing effective cross-modal reasoning mechanisms [51] to synthesize complementary information when different data sources describe the same event (e.g., matching video evidence of a breach with corresponding network intrusion alerts). Multi-modal attention and GNNs could be explored to effectively model temporal and semantic dependencies across modalities.
4) Model Explainability and Transparency: Enhancing explainability is critical in industrial security, where model outputs must often be audited by analysts. Future TKGC research could develop visualization tools that show how temporal embeddings evolve, helping stakeholders understand how the model interprets event timelines. Attention-based mechanisms [8] that highlight influential temporal patterns or critical entity relationships can provide intuitive and interpretable insights into model decisions. Transparent modeling not only aids in trust-building but is also necessary for compliance with industry regulations and security standards.
5) Self-supervised and Unsupervised Methods: Given the scarcity of labeled industrial security data, future research could benefit from self-supervised learning techniques that leverage inherent graph structure. Tasks like predicting masked entities or reconstructing time-aware subgraphs [52] could serve as effective pre-training strategies. Additionally, contrastive learning, which has demonstrated success in other domains [54], can be adapted to distinguish between similar and dissimilar threat behaviors

across time [53]. Such methods can enhance the model's ability to capture nuanced temporal security patterns without requiring extensive annotation.
6) Cross-domain and Cross-lingual Temporal Knowledge Graph Completion: Industrial environments often operate across sectors (e.g., energy, manufacturing, transportation) and geographies, leading to diverse but structurally similar knowledge graphs [56]. Future research could explore domain adaptation methods [55] to transfer insights from one industrial sector to another, especially where labeled data is limited. Similarly, cross-lingual knowledge graphs built from multilingual threat reports and advisories [57] could be fused using temporal alignment methods. This could facilitate a more holistic understanding of global threat dynamics and support multilingual industrial security monitoring.
7) Efficient Computation and Distributed Training for Large-scale Knowledge Graphs: Security knowledge graphs in industrial settings can be massive, containing numerous assets, vulnerabilities, and event logs. This scale introduces challenges for training and inference. Future work could focus on computational optimization via graph partitioning, mini-batch sampling, or memory-efficient architectures. In addition, distributed TKGC models designed for cloud or edge environments [58] can support real-time analytics and reduce latency. Leveraging scalable infrastructures will be essential for deploying TKGC solutions across large-scale industrial systems.
8) Real-time Updates for Dynamic Knowledge Graphs: Industrial security environments are inherently dynamic, with new threats, alerts, and policy changes occurring constantly. Future TKGC research could develop incremental learning mechanisms [59] that allow knowledge graphs to be updated in real time without retraining from scratch. Integrating event stream processing and online learning [59] could enable timely updates to the graph, preserving consistency and relevance. This would help TKGC models swiftly react to emerging incidents, provide up-to-date threat assessments, and support rapid response in industrial environments.

4.2 Application

As data volume and the optimization of algorithms grow continuously, the application prospects of Temporal Knowledge Graph Completion (TKGC) are becoming increasingly broad. At the same time, IoT and Industrial Security faces the challenge of processing massive, time-varying, and multi-source data generated by countless connected devices and industrial systems. Integrating TKGC with IoT and Industrial Security not only enables more effective detection and analysis of evolving threats but also enhances situational awareness and proactive defense in various security scenarios. Specifically, this integration can be observed in the following areas:

1) Natural Language Processing (NLP) and Industrial Log Analysis: In the field of natural language processing [60], TKGC enhances tasks such as entity recognition and linking, semantic role labeling [61], and threat report classification and sentiment analysis [62]. In industrial contexts, unstructured textual data (e.g., equipment error logs, incident reports) can be integrated with structured sensor data via NLP, enabling fine-grained interpretation of operational status and potential security risks. For example, correlating abnormal vibration readings with maintenance logs can help identify mechanical tampering or early signs of sabotage.

2) Threat Intelligence and Anomaly Detection: TKGC enhances threat intelligence systems by modeling the evolution of threat actors, tactics, and vulnerabilities over time [48]. In smart factory settings, analyzing temporal patterns of access logs and machine behavior through TKGC supports early detection of anomalous or unauthorized activities, significantly improving detection accuracy and response time compared to static models.
3) Medical IoT and Cyber-Physical Security: In medical cyber-physical systems, TKGC techniques help track and predict patterns of device communication failures or abnormal patient data transmissions, enabling timely identification of potential cybersecurity breaches. For example, constructing temporal KGs from heart monitor signals and access records can reveal patterns indicative of spoofing or man-in-the-middle attacks on wearable health devices.
4) Social Engineering and Insider Threat Detection: In cybersecurity, TKGC can uncover hidden communication links and behavioral shifts indicative of insider threats [65]. In industrial environments, combining time-aware modeling with IoT device access patterns enables detection of irregular user activities, contributing to defense-in-depth strategies for protecting critical infrastructure.
5) Knowledge Fusion and Security Asset Management: TKGC allows knowledge from different systems (e.g., SCADA logs, IT security alerts) to be merged into a unified knowledge base. This cross-domain integration supports comprehensive security asset tracking, threat correlation, and event prioritization across complex industrial environments.
6) Intelligent Response and Decision Support: Security response systems rely on complete and timely information to recommend mitigation actions [66]. By incorporating TKGC, security knowledge graphs can be dynamically updated based on real-time alerts, sensor data, and environmental changes, improving threat identification and enhancing decision-making precision. For instance, systems can advise optimal shutdown sequences or isolation strategies based on evolving network states. As technology advances, TKGC is expected to be applied in even more scenarios, including industrial threat hunting, secure automation, and resilience modeling.

5 Conclusion

We provide a comprehensive analysis of Temporal Knowledge Graph Completion (TKGC), covering its fundamental concepts, state-of-the-art advancements, and its integration with IoT & Industrial Security. By examining various TKGC methodologies, we demonstrate their effectiveness in modeling dynamic relationships within knowledge graphs and highlight their advantages over traditional static approaches. Furthermore, we discuss future research directions and potential application scenarios for TKGC in enhancing situational awareness, anomaly detection, and intelligent decision-making across complex industrial environments. In conclusion, although TKGC has made notable strides, key challenges remain—particularly in real-time adaptability, multimodal data fusion, and model explainability. Through continued innovation, TKGC holds the potential to not only improve the security and resilience of IoT and industrial systems, but also to unlock new opportunities for analyzing and forecasting temporal dynamics in critical infrastructure.

References

1. Gou, L., Zhou, R., Wan, J., Zhang, J., Yao, Y., Yang, C.: Temporal gate-attention network for meta-path based explainable recommendation. Proc. IEEE Int. Conf. Knowl. Graph, 219–226 (2023)
2. Liu, Y., et al.: Time-aware multiway adaptive fusion network for temporal knowledge graph question answering. Proc. IEEE Int. Conf. Acoust Speech Signal Process, 1–5 (2023)
3. Li, Y., Sun, S., Zhao, J.: Tirgn: Time-guided recurrent graph network with local-global historical patterns for temporal knowledge graph reasoning. Proc. Int. Joint Conf. Artif. Intell., 2152–2158 (2022)
4. Qu, L., Huang, A., Pan, J., Dai, C., Garg, S., Hassan, M.M.: Deep reinforcement learning-based multireconfigurable intelligent surface for mec offloading. Int. J. Intell. Sys. **2024**(1), 2960447 (2024)
5. Xu, Y., Ou, J., Xu, H., Fu, L.: Temporal knowledge graph reasoning with historical contrastive learning. Proc. AAAI Conf. Artif. Intell. **37**(4), 4765–4773 (2023)
6. Zhang, F., Chen, H., Shi, Y., Cheng, J., Lin, J.: Joint framework for tensor decomposition-based temporal knowledge graph completion. Inf. Sci. **654**, 119853 (2024)
7. Wu, C., et al.: Knowledge graph-based multi-context-aware recommendation algorithm. Inf. Sci. **595**, 179–194 (2022)
8. Liu, K., Zhao, F., Chen, H., Li, Y., Xu, G., Jin, H.: Da-net: Distributed attention network for temporal knowledge graph reasoning. Proc. Int. Conf. Inf. Knowledge Manage, 1289–1298 (2022)
9. Ma, T., Lin, X., Song, B., Philip, S.Y., Zeng, X.: Kg-mtl: Knowledge graph enhanced multi-task learning for molecular interaction. IEEE Trans. Knowl. Data Eng. **35**(7), 7068–7081 (2022)
10. Xiao, L., Zhang, R., Chen, Z., Chen, J.: Tucker decomposition with frequency attention for temporal knowledge graph completion. Proc. Annu. Meet. Assoc. Comput. Linguist., 7286–7300 (2023)
11. Liu, X., Wu, J., Li, T., Chen, L., Gao, Y.: Unsupervised entity alignment for temporal knowledge graphs. Proc. World Wide Web Conf., 2528–2538 (2023)
12. Guo, H., Mao, Y.: Interpolating graph pair to regularize graph classification. Proc. AAAI Conf. Artif. Intell. **37**(6), 7766–7774 (2023)
13. Mao, A., Mohri, M., Zhong, Y.: Cross-entropy loss functions: Theoretical analysis and applications. Proc. Mach. Learn. Res. 803–828 (2023)
14. Zhou W., et al.: Self-adversarial learning with comparative discrimination for text generation. Proc. Int. Conf. Learn. Represent. (2020)
15. Chuah, W., Tennakoon, R., Bab-Hadiashar, A.: Enhanced online test-time adaptation with feature-woss-entropy loss funeight cosine alignment. arXiv (2024)
16. Li, J., Xiao, D., Lu, T., Wei, Y., Li, J., Yang, L.: Hamface: Hard- ness adaptive margin loss for face recognition with various intra-class variations. Expert Syst. Appl. **240**, 122384 (2024)
17. Messner, J., Abboud, R., Ceylan, I.I.: Temporal knowledge graph completion using box embeddings. Proc. AAAI Conf. Artif. Intell. **36**(7), 7779–7787 (2022)
18. Bai, L., Ma, X., Zhang, M., Yu, W.: Tpmod: A tendency-guided prediction model for temporal knowledge graph completion. ACM Trans. Knowl. Discov. Data **15**(3), 1–17 (2021)
19. Suchanek, F.M., Alam, M., Bonald, T., Chen, L., Paris, P.H., Soria, J.: Yago 4.5: A large and clean knowledge base with a rich taxonomy. Proc. Int. ACM SIGIR Conf. Res. Dev. Inf. Retr. 131–140 (2024)
20. Diefenbach, D, Wilde, M.D., Alipio, S.: Wikibase as an infrastructure for knowledge graphs: The eu knowledge graph. Proc. Lect. Notes Comput. Sci. 631–647 (2021)

21. Xie, X., Li, Z., Tang, Z., Yao, D., Ma, H.: Unifying knowledge iterative dissemination and relational reconstruction network for image– text matching. Inf. Process. Manage. **60**(1), 103154 (2023)
22. Shen, T., Zhang, F., Cheng, J.: A comprehensive overview of knowledge graph completion. Knowl. Based Syst. **255**, 109597 (2022)
23. Wang, J., Wu, R., Wu, Y., Zhang, F., Zhang, S., Guo, K.: Mpnet: Temporal knowledge graph completion based on a multi-policy network. Appl. Intell. **54**(3), 2491–2507 (2024)
24. Mirtaheri M. et al.: Tackling long-tail entities for temporal knowledge graph completion. Proc. Word Wide Web Conf. 497–500 (2024)
25. Xiao, L., et al.: Tucker decomposition with frequency attention for temporal knowledge graph completion. Proc. Annu. Meet. Assoc. Comput Linguist. 7286–7300 (2023)
26. Dai, C., et al.: Compressing deep model with pruning and tucker decomposition for smart embedded systems. IEEE Internet Things J. **9**(16), 490–500 (2021)
27. Shao, P., Zhang, D., Yang, G., Tao, J., Che, F., Liu, T.: Tucker decomposition-based temporal knowledge graph completion. Knowledge-Based Syst. **238**, 107841 (2022)
28. Zhang, L., Zhou, D.: Temporal knowledge graph completion with approximated gaussian process embedding. Proc. Main Conf. Int. Conf. Comput. Linguist., 4697–4706 (2022)
29. Hou, X., Ma, R., Yan, L., Ma, Z.: T-gae: A timespan-aware graph attention-based embedding model for temporal knowledge graph completion. Inf. Sci. **642**, 119225 (2023)
30. Wang, Z., Li, L., Zeng, D.D. Time-aware representation learning of knowledge graphs. Proc. Int. Jt. Conf. Neural. Networks. IEEE 1–8 (2021)
31. Zhang, F., Zhang, Z., Ao, X., Zhuang, F., Xu, Y., He, Q.: Along the time: Timeline-traced embedding for temporal knowledge graph completion. Proc. Int. Conf. Inf. Knowledge Manage, 2529–2538 (2022)
32. Bai, L., Ma, X., Meng, X., Ren, X., Ke, Y.: Roan: A relation-oriented attention network for temporal knowledge graph completion. Eng. Appl. Artif. Intell. **123**, 308–308 (2023)
33. Dai, G., Wang, X., Zou, X., Liu, C., Cen, S.: Mrgat: Multi-relational graph attention network for knowledge graph completion. Neural Netw. **154**, 234–245 (2022)
34. Fu, G. et al. Tempcaps: A capsule network-based embedding model for temporal knowledge graph completion. Proc. Workshop Struct. Predict. NLP. 22–31 (2022)
35. Lai, Y., Chen, C., Zheng, Z., Zhang, Y.: Block term decomposition with distinct time granularities for temporal knowledge graph completion. Expert Sys. Appl. **201**, 117036 (2022)
36. Dai, C., Liu, X., Li, Z., Chen, M.Y.: A tucker decomposition based knowledge distillation for intelligent edge applications. Appl. Soft Comput. **101**, 107051 (2021)
37. Y. Gao, Y. He, Z. Kan, Y. Han, L. Qiao, and D. Li, "Learning joint structural and temporal contextualized knowledge embeddings for temporal knowledge graph completion," in Proc. Annu. Meet. Assoc. Comput. Linguist., 2023, pp. 417–430
38. Liu, W., et al.: K-bert: Enabling language representation with knowledge graph. Proc. AAAI Conf. Artif. Intell. **34**(03), 2901–2908 (2020)
39. Han, Z., et al.: Ecola: Enhanced temporal knowledge embeddings with contextualized language representations. arXiv (2023)
40. Li, J., Su, X., Gao, G.: Teast: Temporal knowledge graph embedding via archimedean spiral timeline. Proc. Annu. Meet. Assoc. Comput Linguist. 460–474 (2023)
41. Liu, Z., Tan, L., Li, M., Wan, Y., Jin, H., Shi, X.: Simfy: A simple yet effective approach for temporal knowledge graph reasoning. Proc. Find. Assoc. Comput. Linguist. 3825–3836 (2023)
42. Wang, Z., Du, H., Yao, Q., Li, X.: Search to pass messages for temporal knowledge graph completion. Proc. Find. Assoc. Comput. Linguist. 6160–6172 (2022)
43. Liu, Y., Ma, Y., Hildebrandt, M., Joblin, M., Tresp, V.: Tlogic: Temporal logical rules for explainable link forecasting on temporal knowledge graphs. Proc. AAAI Conf. Artif. Intell. **36**(4), 4120–4127 (2022)

44. Zhang, Z., Wang, J., Ye, J., Wu, F.: Rethinking graph convolutional networks in knowledge graph completion. Proc. Word Wide Web Conf., 798–807 (2022)
45. Liang, K., et al.: Learn from relational correlations and periodic events for temporal knowledge graph reasoning. Proc. Int. ACM SIGIR Conf. Res. Dev. Inf. Retr. 1559–1568 (2023)
46. Hu, L., Liu, Z., Zhao, Z., Hou, L., Nie, L., Li, J.: A survey of knowledge enhanced pre-trained language models. IEEE Trans. Knowl. Data Eng. (2023)
47. Shen, J., Wang, C., Gong, L., Song, D.: Joint language semantic and structure embedding for knowledge graph completion. arXiv (2022)
48. Zhong, Q., Ding, L., Liu, J., Du, B., Jin, H., Tao, D.: Knowledge graph augmented network towards multiview representation learning for aspect-based sentiment analysis. IEEE Trans. Knowl. Data Eng. **35**(10), 98–111 (2023)
49. Sun, H., Zhong, J., Ma, Y., Han, Z., He, K.: Timetraveler: Reinforce-ment learning for temporal knowledge graph forecasting. arXiv (2021)
50. Liang, K., et al.: "A survey of knowledge graph reasoning on graph types: Static, dynamic, and multi-modal. IEEE Trans. Pattern Anal. Mach. Intell. (2024)
51. Chen, X., et al.: Hybrid transformer with multi-level fusion for multi-modal knowledge graph completion. Proc. Int. ACM SIGIR Conf. Res. Dev. Inf. Retr. 904–915 (2022)
52. Liu, X., et al.: Self-supervised learning: Generative or contrastive. IEEE Trans. Knowl. Data Eng. **35**(1), 857–876 (2021)
53. Liu, K., Zhao, F., Chen, H., Li, Y., Xu, G., Jin, H.: Knowledge graph contrastive learning based on relation-symmetrical structure. IEEE Trans. Knowl. Data Eng. **36**(1), 226–238 (2023)
54. Yang, B., Zhu, L., Dai, C., Garg, S., Kaddoum, G.: An improved reconstruction-based multi-attribute contrastive learning for digital-twin-enabled industrial system. IEEE Internet Things J. **12**(4), 3670–3679 (2025)
55. Zhong, L., Wu, J., Li, Q., Peng, H., Wu, X.: A comprehensive survey on automatic knowledge graph construction. ACM Comput. Surv. **56**(4), 1–62 (2023)
56. Zhu, R., Zhao, Y., Qu, W., Liu, Z., Li, C.: Cross-domain product search with knowledge graph. Proc. Int. Conf. Inf. Knowledge Manage, 3746–3755 (2022)
57. Wang, R., et al.: Mutually-paced knowledge distillation for cross-lingual temporal knowledge graph reasoning. Proc. World Wide Web Conf. 2621–2632 (2023)
58. Mitropoulou, K., Kokkinos, P., Soumplis, P., Varvarigos, E.: Anomaly detection in cloud computing using knowledge graph embedding and machine learning mechanisms. J. Grid Comput. **22**(1), 6 (2024)
59. Wu, J., Xu, Y., Zhang, Y., Ma, C., Coates, M., Cheung, J.C.K.: Tie: A framework for embedding-based incremental temporal knowledge graph completion. Proc. Int. ACM SIGIR Conf. Res. Dev. Inf. Retr. 428–437 (2021)
60. Wu, T., Khan, A., Yong, M., Qi, G., Wang, M.: Efficiently embedding dynamic knowledge graphs. Knowledge-Based Syst. **250**, 109124 (2022)
61. Wang, L., Zhao, W., Wei, Z., Liu, J.: Simkgc: Simple contrastive knowledge graph completion with pre-trained language models. arXiv (2022)
62. Xiong, H., Wang, S., Tang, M., Wang, L., Lin, X.: Knowledge graph question answering with semantic oriented fusion model. Knowledge-Based Syst. **221**, 106954 (2021)
63. Zhao, Y., et al.: Time-aware path reasoning on knowledge graph for recommendation. ACM Trans. Inf. Syst. **41**(2), 1–26 (2022)
64. Fernańdez-Torras, A., Duran-Frigola, M., Bertoni, M., Locatelli, M., Aloy, P.: Integrating and formatting biomedical data as pre-calculated knowledge graph embeddings in the bioteque. Nat. Commun. **13**(1), 5304 (2022)
65. Vilela, J., et al.: Biomedical knowledge graph embeddings for personalized medicine: Predicting disease-gene associations. Expert. Syst. **40**(5), e13181 (2023)
66. Zhang, J.C., Zain, A.M., Zhou, K.Q., Chen, X., Zhang, R.M.: A review of recommender systems based on knowledge graph embedding. Expert Sys. Appl. 123876, (2024)

67. Huang, R., Wei, W., Qu, X., Xie, W., Mao, X., Chen, D.: Joint multi-facts reasoning network for complex temporal question answering over knowledge graph. Proc. IEEE Int. Conf. Acoust Speech Signal Process Proc. 331–335 (2024)
68. Wang, R., et al.: Rete: Retrieval-enhanced temporal event forecasting on unified query product evolutionary graph. Proc. Word Wide Web Conf. 462–472 (2022)
69. Yang, B., et al.: Cross-city transfer learning for traffic forecasting via incremental distribution rectification. Knowl.-Based Syst. **315**, 113336 (2025)

Federated Learning for AIoT Security: Current Advances and Future Challenges

Hao Yin[1(✉)] and Peng Wang[2]

[1] College of Software Engineering, Sichuan University, Chengdu 610000, China
2022141461088@stu.scu.edu.cn
[2] College of Computer Science, Sichuan University, Chengdu 610000, China
2021141461157@stu.scu.edu.cn

Abstract. Although Artificial Intelligence (AI) has advanced rapidly in the Internet of Things (IoT), security in distributed model training(whether in industrial applications or other domains) is still a challenge. Federated Learning (FL), a method that enables devices to collaboratively train a shared model while minimizing centralized data exposure, demonstrates superior security properties compared to other methods. However, simply adopting FL cannot guarantee the security of IoT during model training. Therefore, this paper provides a survey on existing approaches to addressing security threats in FL to facilitate its application in IoT. This paper comprehensively categorizes the FL-based approaches dealing with the unique security challenges in different scenarios. Finally, this paper concludes with an analysis of the difficulty when integrating FL with IoT and provide insights for future directions.

Keywords: Embedded systems security · Internet of things · Federated learning frameworks

1 Introduction

Recent breakthroughs in hardware resources such as the proliferation of edge AI chips have made it feasible to deploy AI models efficiently on resource-constrained IoT devices. Traditionally, IoT devices upload their own data to their data center or a server where AI models are trained and deployed back to the IoT devices after training. However, uploading their data to third-party servers raises privacy concerns and security problems [1]. Therefore, it is necessary for developing a new approach to realize security-enhanced intelligent IoT environments.

FL is a distributed collaborative AI method that enables IoT devices to train collaboratively without sharing their own data [2]. Since only model parameters are uploaded, even if a data breach occurs during their communications, attackers can't get sensitive information from the exchanged parameters easily. Specifically, instead of transferring their own data to the center server, all clients trains on data locally and uploads model parameters for a global aggregation. Then, the center server combines these updates to generate a global model and redistributed to the devices for further local training. This stops until convergence. However, security in FL extends beyond data leakage, threats such as Byzantine attacks also have risks. Therefore, this paper comprehensively categorizes the FL's approaches and implementation styles with the unique security challenges in different scenarios, aiming to equip IoT researchers and practitioners with insights into FL security, fostering safer FL-based IoT deployments.

While research already exists in IoT discussing about FL, sufficient survey has not been made about FL's security in IoT. This survey hopes to make a survey on existing approaches to addressing security threats in FL to facilitate its application in IoT. The work provides a blueprint for IoT experts to address FL-related security challenges in IoT. The contributions of this survey are summarized as follows:

- Identifying the security threats in FL environments.
- Providing an overview of defensive techniques in FL.
- Identifying the mitigation techniques to solve the security threats in FL.
- Providing insights into future directions for FL security in IoT (Fig. 1).

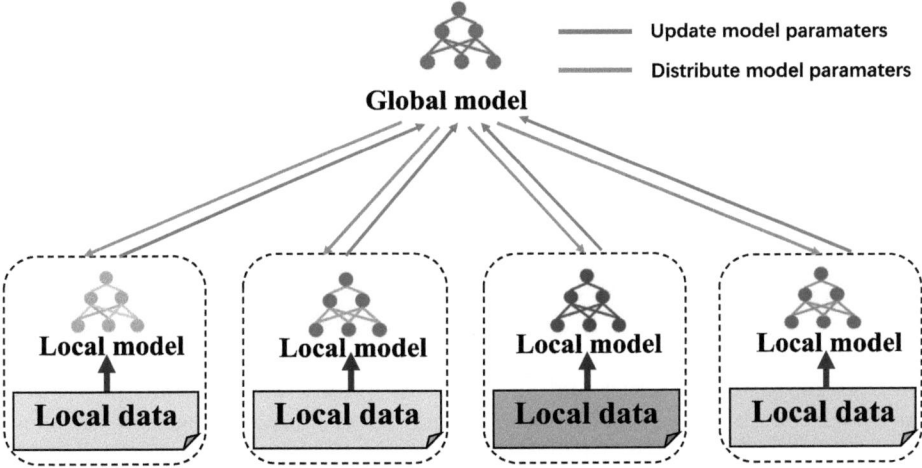

Fig. 1. Federated Learning Diagram

2 Related Work

2.1 Federated Learning

Since FL's introduction in 2016 [3], FL offers new solutions with its security-enhancing and distributed characteristics. In FL, data owned by the client never share with other clients, enabling a cooperative training without get each other's data. Its arrival may reshape the current IoT system. The concept of FL is described as follows (Fig. 2):

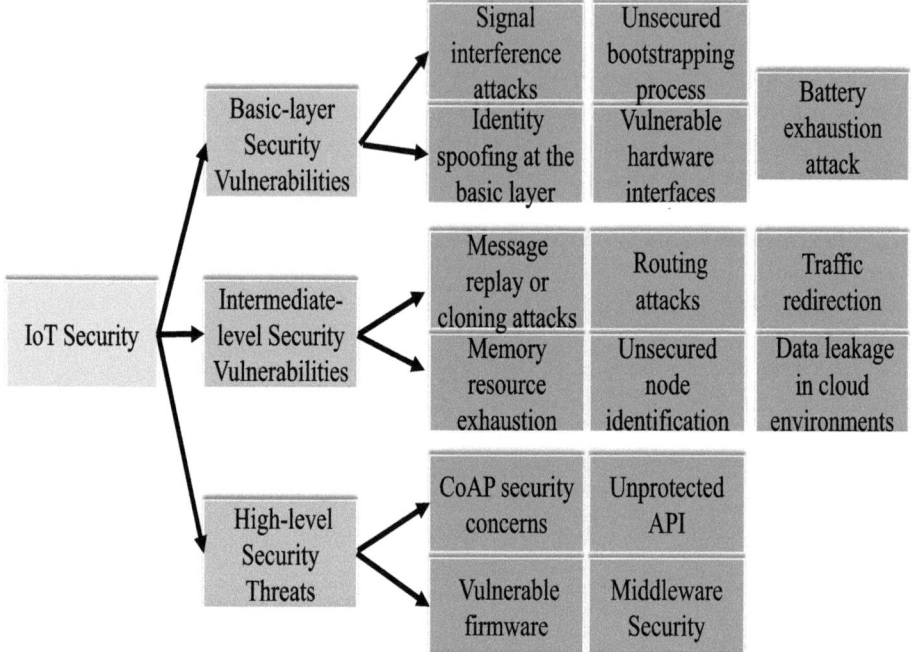

Fig. 2. IoT Security Diagram

FL Concept: Together, the clients(IoT devices) and server (Base station)make up FL. Let $C = 1, 2, ..., c$ denote the set of clients which collavoratively train a global model for a IoT task. FL enables devices to collaboratively train a shared model while minimizing centralized data exposure. In a FL process, each client c collaboratively train a shared model with their data D_c. After each local training their model w_k on their own data D_k, clients transfer their local model parameters w_k to the server and aggregate to build a shared model w_G. Then, the server can obtain a globally optimized model without knowledge of the client data. The specific process of Federated Learning (FL) is described as follows:

Client Selection: The server employs specific strategies to select users for the next round of federated learning and initializes a global model w_G^0.

Clients' Local Training: The server transmit w_G^0 to clients after this round of selection. Each client selected by the server trains locally with local data. Then, each client transfers its weight to the global server for a global aggregation.

Model aggregation: After receiving all the model parameters sent by the selected clients, the server The server aggregates them into a global model that distributed to all clients and start the next epoch.

FL Classifications: FL is categorized by data partitioning and network structure. [4,5]:

Data Partitioning: This category can be devided into: horizontal federated learning (HFL), federated transfer learning [5] (FTL) and vertical federated learning (VFL).

(a) bVFL: Vertical Federated Learning (VFL) involves participants sharing identical sample IDs but distinct feature sets. During VFL training, aligned samples from different clients are aggregated to collaboratively optimize a shared model, typically secured through cryptographic methods.

(b) FTL: FTL operates in scenarios where participants possess non-overlapping samples and heterogeneous feature spaces. FTL employs feature transformation techniques to project disparate data representations into a unified latent space, enabling joint model training across selectively aligned clients.

(c) HFL: HFL assumes that clients share identical feature dimensions but contribute distinct local datasets. Under this framework, participants collaboratively train a unified model by iteratively aggregating locally computed updates.

Network Architecture: FL can becategorized into two classes based on network structure:

(a) Centralized FL(CFL): CFL architecture contains a central server. Each epoch, the clients selected by the central server update their model parameters to the central server and the central server using algorithm such as FedAvg [8] to aggregate the parameters. Then, the parameters are sent back to for the next round of training.

(b) Decentralized FL (DFL) [4]: DFL architecture doesn't contain a central server in the training porcoss. Instead, all selected clients are connected together in P2P manner. During the FL process, each client get the model parameters from neighbor clients through the P2P method.

2.2 IoT Security Requirements

For a secure IoT environment, many factors need to be considered (Table 1), Two primary directions are outlined below:

Data Privacy and Integrity: Due to the need for multiple transmissions of IoT data in network nodes, a encryption method is required to ensure the privacy of the data. Attackers may modify the data for malicious purposes to impact the data integrity (Fig. 3).

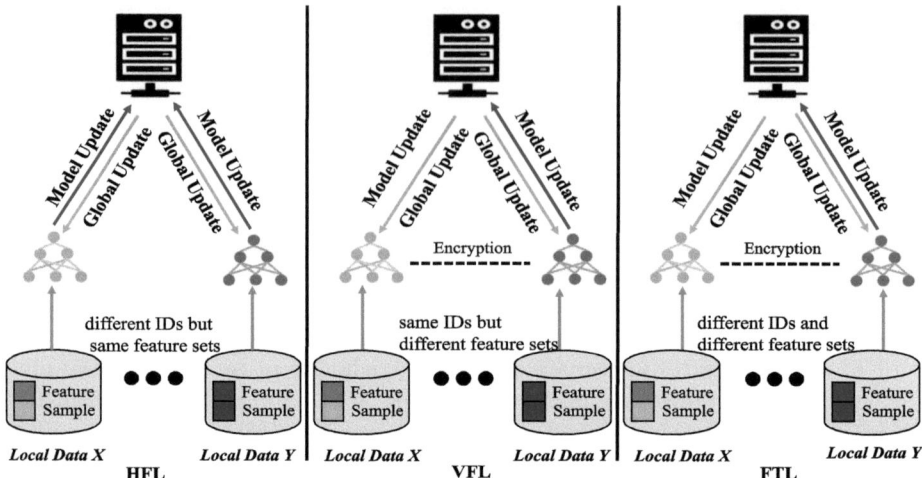

Fig. 3. FL Classification Chart

Authentication, authorization and Accounting: To maintain secure communication in IoT, the authentication is needed. Devices seeking privileged service access must undergo authentication. Authorization mechanisms control system access, granting permissions only to verified entities. When properly implemented, these measures establish a reliable and secure operational environment.

3 Security Vulnerabilities in FL

Vulnerabilities refer to system weaknesses that adversaries may exploit to execute harmful attacks. Detecting these vulnerabilities is crucial for establishing a protected environment and mitigating potential threats. For a clearer analysis, this section classifies the origins of vulnerabilities in federated learning (FL). The following four key sources are identified as potential vulnerabilities:

Communication Protocol: FL involves extensive communication between selected clients and the server. Existing FL frameworks [9] employ mixed networks based on public-key cryptography to anonymize messages. However, the frequent training rounds in FL still pose potential security risks.

Client Operations: FL has multiple clients, which means that they can become attackers. They can access the global model and compromise training outcomes by submitting malicious data.

Weak Aggregation Algorithm: A proper aggregation algorithm is supposed to intelligently identify abnormal client behavior and block malicious participants.

Table 1. Security Threats Classification in FL

Threat Type	Severity	Vulnerability Source
Poisoning Attacks	High	Client-side data manipulation, Compromised server
Backdoor Attacks	High	Malicious model parameter updates
GAN-based Attacks	High	Adversarial sample generation
Malicious Server	Critical	Central server compromise
Free-riding	Medium	Non-participating clients
Eavesdropping	Medium	Unencrypted communication channels
System Downtime	Low	Infrastructure vulnerabilities

The Breached Central Server: This core system orchestrates the distribution of initial model parameters, consolidates localized model updates, and synchronizes universal model iterations. The infrastructure must accurately differentiate between authenticated clients and hostile entities, implementing safeguards to thwart malicious infiltration attempts.

4 Security Threats in FL

Threats are actions where attackers exploit vulnerabilities to compromise FL systems. In FL, attackers uses vulnerabilities to disrupt model training outcomes or maliciously obtain private information. This section categorizes and describes these attacks as follows:

Poisoning: In a FL process, the most likely attack is poisoning. Due to all clients can get access to their own training data, they may tamper with their data to manipulate parameter weights a_k and model parameters w_k to influence the global model. Poisoning attack targets many aspects of the FL process. We further classify poisoning attacks into:

Data Poisoning: The attackers inject malicious samples into training data, aiming to degrade model performance.

Model Poisoning: The attackers alter local model parameters before uploading them to the server for aggregation, thereby corrupting the global model.

Data Modification: Data Modification include feature collision (merging two classes), adding interference patterns (e.g., noise), random label swapping, etc., all designed to induce model misclassification.

Backdoor Attacks: A backdoor attack embeds malicious functionality into the model without affecting its accuracy on primary tasks, making detection challenging. Studies in [10,11] investigate backdoor attack implementations. [32, 33] propose fine-tuning-based solutions to mitigate this threat.

GANs: Recent studies have examined generative adversarial network (GAN)-powered exploits in federated learning frameworks [12]. Such attacks present critical risks since they enable malicious data contamination, undermining the learning system's reliability. As detailed in [13], these adversarial approaches can extract sensitive training datasets and subsequently corrupt the model's input pipeline.

System Disruption IT Downtime: Production system downtime is unavoidable in Information Technology industry. While this issue exists across various domains, it requires special attention in FL as downtime periods may be strategically exploited to steal information from the federated learning process.

Malicious Server: In FL processes,the central server performs most of the work. However, a compromised server can create severe problems, as it can easily manipulate the global model and get private client data.

Free-Riding Attacks: A separate challenge in FL systems involves non-contributing participants who solely access aggregated models without participating in collaborative training efforts. To address this free-riding vulnerability, [14] developed a neural autoencoder-enhanced detection mechanism for improved identification and mitigation.

Eavedropping: In FL, attackers may ntercept and extract data during communication rounds between clients and the central server.

5 Defensive Strategies Against Security Vulnerabilities in FL

To address the security threats faced by FL systems, this section elaborates on state-of-the-art defensive techniques and summarizes their applications and effectiveness:

Data Sanitization: Data sanitization refers to the process of cleaning harmful or anomalous data. In FL, it acts as the first line of defense against attacks.

Though data sanitization techniques can filter out poisoned data to preserve usability and validity, they require access to users' local data and compromise privacy. Therefore, data sanitization is difficult to implement in FL widely and is only used in specific scenarios (Table 2).

Table 2. Defense Mechanism in FL

Defense Mechanism	Description	Associated Threats
Model Distillation	Compressing learned patterns from complex to compact models	Eavesdropping, Inference attacks, GAN-based reconstruction
Backdoor Defense	Detecting hidden trigger patterns in model parameters	Targeted misclassification, Stealthy model manipulation
Free-riding Prevention	Identifying clients with fake local updates	False participation claims, Resource draining
Adversarial Training	Robustifying models against crafted inputs	Gradient-based attacks, Evasion attacks
Differential Privacy	Adding noise to gradients during transmission	Membership inference, Data reconstruction
Secure Aggregation	Cryptographically protecting parameter exchanges	Man-in-the-middle attacks, Parameter tampering
Byzantine Resilience	Tolerating malicious participant behaviors	Model poisoning, Gradient flipping
Blockchain Verification	Decentralized validation of model updates	Consensus manipulation, Fake node injection

Data Poisoning Defense: Normally, to prevent malicious attackers from interfering with training using incomplete or contaminated data, data poisoning defense verifies the reality of parameters before model aggregation. [15] proposes the Sinper scheme, which identifies benign users by inspecting parameter update directions during the aggregation process to address this issue.

Model Poisoning Defense: For model poisoning defense, servers need to select clients that beneficiard against clients that degrade it. [16] introduces malicious update detection techniques and alternating minimization strategies to defend against model poisoning attacks executed by malicious users.

Model Distillation: Model distillation learns a new model from the original model to reducing attackers' ability to exploit the model [17,18]. [19] introduce a defense mechanism incorporating distillation techniques to mitigate the effectiveness of adversarial attacks during training.

Free-Riding Attacks Defense: [20,21] design a blockchain-based federated learning architecture for free-riding attack defense scenarios. This architecture leverages blockchain to evaluate users' contributions to model updates and incentivizes or restricts users based on their contributions.

GAN Defense: [22] proposes Anti-GAN to address adversarial perturbations in dataset samples. It automatically distinguishes between real and perturbed samples and demonstrates excellent performance across multiple datasets.

Model Pruning: Model pruning defends against poisoning attacks by removing parameters with minimal contributions or those that are contaminated, while also reducing model complexity and improving computational efficiency. Yu et al. [23] propose pruning techniques to eliminate poisoned parameters in backdoor attacks, significantly reducing attack success rates. Jiang et al. [24] introduce adaptive distributed pruning methods, reducing training time and communication overhead while maintaining accuracy. This technique effectively mitigates data/backdoor poisoning attacks and optimizes computational and communication performance.

Malicious Collusion Defense: Malicious collusion refers to many clients collaboratively uploading malicious parameters. It is covert and difficult to detect. Besides, model ownership abuse also has garnered significant attention.

Multi-party Collusion Defense: [34,35] propose defense methods based on the diversity of parameter updates, which are effective against label flipping and backdoor attacks. [36] solve this by monitoring participants' average losses and design predictive cost reports to defend against this attack.

Ownership Protection: [37] designs the WAFFLE framework to prevent collusion from overwriting watermarks through watermark retraining. [25,26] uses unique fingerprints to trace malicious attackers. [27,28] uses digital signatures and watermarking techniques to verify user identities and protect model security.

Secure Aggregation Protocols: The aggregation algorithm is important in the process of aggregating local parameters. A FL aggregation algorithm should adaptively detect anomalous parameters uploaded by malicious clients, or autonomously adjust the aggregation method and integrate encryption methods to ensure the security.

Federated Stochastic Controlled Averaging (SCAFFOLD): To solve promblems such as gradient tampering by malicious attackers and instability during aggregation in FL, [38] proposed the Federated Stochastic Controlled Averaging (SCAFFOLD) aggregation algorithm. It aims to maintain gradient parameter variables of clients and make sure that all clients update toward the right direction.

Secure Multi-Party Computation (SMPC) Aggregation Protocol: [29] designed a secure aggregation protocol based on secure multi-party computation (SMPC) which aggregates parameters from mutually distrustful clients without leaking private information. Besides, the aggregation method can achieve secure aggregation even when some malicious attackers refuse to participate.

Personalized Federated Aggregation: Due to the heterogeneity of non-centrally aggregated data, FL algorithms should address statistical heterogeneity across user devices. [30] proposed a secure aggregation method which combines base personalized layers with personalized layers. For heterogeneous data, personalized methods are used for alignment aggregation without leaking any information. [31] proposed an Adaptive Personalized Federated Learning (APFL) algorithm to get a generalized boundary for the mixture of local and global models.

6 Challenges in Integrating FL with IoT

Due to the characteristics of IoT and the nature of FL, the integration of FL and IoT presents some difficulties:

Resource Constraints: IoT devices are typically resource-constrained. However, the frequent model updates in FL processes need substantial bandwidth.

Heterogeneity of Data and Devices: IoT environments involve diverse devices generating non-IID data. Therefore, FL should address the issue of data heterogeneity.

Dynamic Network Conditions: IoT networks are dynamic, with devices frequently joining, leaving, or experiencing intermittent connectivity. This may disrupt the training process of FL.

7 Future Research Directions

To address the challenges outlined in Section VI and advance FL's applicability in IoT, future research should focus on the following directions:

Resource-Efficient FL Algorithms: Develop lightweight neural networks for IoT devices using techniques such as pruning and knowledge distillation.

Enhanced Security Methods: Integrate robust aggregation rules to detect malicious attackers or combine FL with encryption methods.

Adaptive Communication Protocols: Design FL algorithms that reduce update sizes via gradient sparsification without compromising the final performance. Alternatively, assign different priorities to clients to alleviate the issue of insufficient bandwidth.

Integration with Other Technologies: For example, introduce blockchain and other decentralized technologies to prevent tampering with model update records.

8 Conclusion

In conclusion, due to FL's advantages in protecting data privacy, it holds significant potential in IoT. This paper provides a systematic survey on FL's security, categorizing its approaches, implementation styles, and scenario-specific challenges. The analysis concludes with difficulties in FL-IoT integration and insights for future research directions.

References

1. Granjal, J., Monteiro, E., Silva, J.S.: Security for the internet of things: a survey of existing protocols and open research issues. IEEE Commun. Surveys Tuts. **17**(3), 1294–1312 (2015)
2. Konečný, J., McMahan, H.B., Yu, F.X., et al.: Federated learning: strategies for improving communication efficiency. arXiv:1610.05492 (2016)
3. McMahan, B., Moore, E., Ramage, D., et al.: Communication-efficient learning of deep networks from decentralized data. In: Proceedings of the AISTATS, pp. 1273–1282 (2017)
4. Li, Q., Wen, Z., Wu, Z., et al.: A survey on federated learning systems: vision, hypes and realities for data privacy and protection. arXiv:1907.09693 (2019)
5. Yang, Q., Liu, Y., Chen, T., Tong, Y.: Federated machine learning: concept and applications. ACM Trans. Intell. Syst. Technol. **10**(2), 1–19 (2019)
6. Park, J., Samarakoon, S., Bennis, M., et al.: Wireless network intelligence at the edge. arXiv:2008.02608 (2020)
7. Feng, S., Yu, H.: Multi-participant multi-class vertical federated learning. arXiv:2001.11154 (2020)
8. Reisizadeh, A., Mokhtari, A., Hassani, H., et al.: FedPAQ: a communication-efficient federated learning method with periodic averaging and quantization. In: Proceedings of the AISTATS, pp. 2021–2031 (2020)
9. Chaum, D., Gritzalis, D.A.: Anonymous electronic communications using digital pseudonyms. In: Secure Electronic Voting, pp. 211–219. Springer (2003)
10. Sun, Z., Kairouz, P., Suresh, A.T., et al.: Can you really backdoor federated learning?. arXiv:1911.07963 (2019)
11. Bagdasaryan, E., Veit, A., Hua, Y., et al.: How to backdoor federated learning. In: AISTATS, pp. 2938–2948 (2020)
12. Wang, Z., Song, M., Zhang, Z., et al.: Beyond inferring class representatives: user-level privacy leakage from federated learning. In: IEEE INFOCOM, pp. 2512–2520 (2019)
13. Zhang, J., Chen, J., Wu, D., et al.: Poisoning attack in federated learning using generative adversarial nets. In: IEEE TrustCom, pp. 374–380 (2019)
14. Lin, J., Du, M., Liu, J.: Free-riders in federated learning: attacks and defenses. arXiv:1911.12560 (2019)
15. Cao, D., Chang, S., Lin, Z., et al.: Understanding distributed poisoning attack in federated learning. In: IEEE ICPADS, pp. 233–239 (2019)

16. Bhagoji, A.N., Chakraborty, S., Mittal, P., et al.: Analyzing federated learning through an adversarial lens. In: Proceedings of the ICML, pp. 634–643 (2019)
17. Li, D., Wang, J.: FedMD: heterogenous federated learning via model distillation. In: NeurIPS FL Workshop (2019)
18. Wang, A., Zhang, Y., Yan, Y.: Heterogeneous federated transfer learning for defect prediction. IEEE Access **9**, 29530–29540 (2021)
19. Papernot, N., McDaniel, P., Wu, X., et al.: Distillation as a defense to adversarial perturbations against deep neural networks. In: IEEE SP, pp. 582–597 (2016)
20. Kim, H., Park, J., Bennis, M., et al.: On-device federated learning via blockchain and its latency analysis. arXiv:1808.03949 (2018)
21. Weng, J., Weng, J., Zhang, J., et al.: DeepChain: auditable and privacy-preserving deep learning with blockchain-based incentive. IEEE TDSC **18**(5), 2438–2455 (2019)
22. Zhang, X., Luo, X.: Defending against the label leakage attacks in federated learning. arXiv:2004.12571 (2020)
23. Yu, S., Nguyen, P., Anwar, A., et al.: Dynamic network pruning for non-IID federated learning. arXiv:2106.06921 (2021)
24. Jiang, Y., Wang, S., Valls, V., et al.: Model pruning enables efficient federated learning on edge devices. IEEE TNNLS **33**(4), 1–13 (2022)
25. Li, F., Wang, S., Liew, A.W.: Watermarking in federated learning: a novel framework for model ownership protection. arXiv:2105.03167 (2021)
26. Rouhani, B.D., Chen, H., Koushanfar, F.: DeepSigns: an end-to-end watermarking framework for ownership protection of deep neural networks. In: Proceedings of the ASPLOS (2019)
27. Chen, Y., Luo, F., Li, T., et al.: A privacy-preserving federated learning framework via TEE-based trusted execution environment. Inf. Sci. **522**, 69–79 (2020)
28. Uchida, Y., Nagai, Y., Sakazawa, S., et al.: Embedding watermarks into deep neural networks. In: Proceedings of the ICMR, pp. 269–277 (2017)
29. Mohammadi, M., Al-Fuqaha, A.: Enabling cognitive smart cities using big data and machine learning: trends and challenges. IEEE Commun. Mag. **56**(2), 94–101 (2018)
30. Wang, S., Cao, J., Yu, P.S.: Deep learning for spatio-temporal data mining: a survey. IEEE TKDE **34**(8), 3681–3700 (2022)
31. Roh, Y., Heo, G., Whang, S.E.: A survey on data collection for machine learning: a big data - AI integration perspective. IEEE TKDE **33**(4), 1328–1347 (2019)
32. Liu, K., Dolan-Gavitt, B., Garg, S.: Fine-pruning: defending against backdooring attacks on deep neural networks. In: RAID, pp. 273–294 (2018)
33. Jiang, Y., Wang, S., Ko, B.J., et al.: Model pruning enables efficient federated learning on edge devices. arXiv:1909.12326 (2019)
34. OpenMined. PySyft: a library for encrypted, privacy-preserving machine learning (2018). https://github.com/OpenMined/PySyft
35. Nasr, M., Shokri, R., Houmansadr, A.: Comprehensive privacy analysis of deep learning: stand-alone and federated learning under passive and active white-box inference attacks. In: IEEE SP, pp. 739–753 (2019)
36. Jiang, Y., Li, Y., Zhou, Y., et al.: Mitigating sybil attacks on differentially private federated learning. arXiv:2010.10572 (2020)
37. Jiang, Y., Li, Y., Zhou, Y., et al.: Differential privacy protection against sybil attacks in federated learning. arXiv:2010.10572 (2020)
38. Li, T., Sahu, A.K., Talwalkar, A., et al.: Federated learning: challenges, methods, and future directions. IEEE Signal Process. Mag. **37**(3), 50–60 (2020)

Lightweight Chaos-Based Image Encryption Algorithm for the Internet of Things

Wangcan Liu[1], Dawei Ding[1(✉)], Yuanyuan Wang[2], Zongli Yang[1], Chaoma Qian[2], and Jingwen Zhao[3]

[1] School of Electronic Information Engineering, Anhui University, Hefei, China
dwding@ahu.edu.cn
[2] China Railway Fourth Bureau Group Shanghai Engineering Co., Ltd., Shanghai, China
[3] Anhui Digital Intelligent Construction Research Institute Co., Ltd., Hefei, China

Abstract. With the rapid development of the Internet of Things (IoT), the secure transmission and storage of image data in resource-constrained devices has become an urgent challenge. Traditional cryptographic algorithms are often computationally intensive and unsuitable for IoT devices that require lightweight and efficient security solutions. To address this issue, this paper proposes a lightweight image encryption algorithm based on a two-dimensional cosine-coupled chaotic map (2D-CCM). By constructing a high-complexity chaotic system and generating pseudo-random sequences, the algorithm performs pixel-level scrambling and multi-round diffusion to enhance encryption strength and resistance to attacks. Experimental results demonstrate that the proposed method exhibits strong robustness against cropping and noise interference, while significantly reducing encryption and decryption time compared to existing methods. These features make it well-suited for secure image processing in IoT applications such as smart sensing and edge computing.

Keywords: Image Encryption · Chaotic System · IoT Security · Lightweight Encryption · 2D Cosine Map · Embedded Devices

1 Introduction

With the rapid expansion of the Internet of Things (IoT) [1] technologies such as intelligent sensing, edge computing, and ubiquitous connectivity have become integral to modern information systems. IoT devices are widely deployed in smart cities [2], industrial automation [3], healthcare [4], and home environments [5], where they continuously collect and transmit image data for real-time processing and decision-making. However, the open and resource-constrained nature of IoT environments introduces significant security and privacy concerns [1].

As shown in Fig. 1, a typical IoT ecosystem consists of numerous sensing nodes and communication modules, where image data is frequently transmitted over wireless channels. This makes it vulnerable to eavesdropping, tampering, and other forms of cyber attacks. Although traditional cryptographic techniques such as AES and RSA are effective, their high computational overhead makes them impractical for lightweight, real-time applications on embedded IoT devices [6].

To meet these challenges, chaos-based encryption has emerged as a promising alternative due to its inherent randomness, sensitivity to initial conditions, and low computational complexity [7]. In this paper, we propose a lightweight image encryption scheme based on a two-dimensional cosine-coupled chaotic map (2D-CCM). The scheme utilizes chaotic sequences for pixel permutation and multi-round diffusion, achieving a good balance between encryption strength and computational efficiency. The proposed method is specifically designed for secure image protection in low-power IoT devices and real-time edge computing environments.

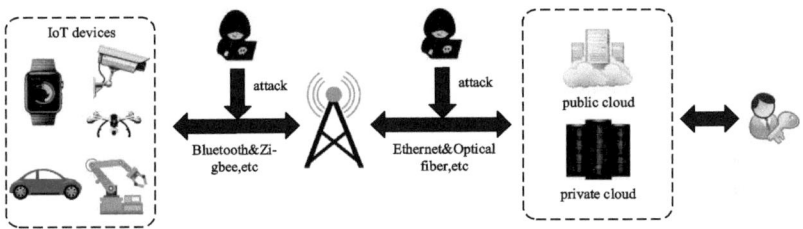

Fig. 1. Internet of Things(IoT)ecosystem.

2 Analysis of Chaotic System

2.1 Proposed Chaotic System

In this section, based on the one-dimensional Cosine Map (CM)[8], we propose a novel two-dimensional cosine-coupled chaotic map (2D-CCM) through coupling, as shown in Eq. (1).

$$\begin{cases} x_{n+1} = a\cos(\omega y_n)\cos(b/x_n), \\ y_{n+1} = a\cos(\omega x_{n+1})\cos(b/y_n). \end{cases} \quad (1)$$

The control parameters $\{a, b, \omega\} \in R$ and coupling structure diagram are shown in Fig. 2 The two-dimensional cosine map consists of two iterative channels: the x-channel and the y-channel. In the x-channel, the state x_n is input into a divider and a delay unit, where it is multiplied by pa-rameters b and w, respectively, before being fed into the input of the cos(·) module. Meanwhile, the input of the y-channel is multiplied by ω and the cos(·) module, then coupled into the x-channel via a multiplier. After being multiplied by parameter a, it is used as the output of the x-channel. The coupling pattern of the y-channel is similar to that of the x-channel and will not be elaborated further here.

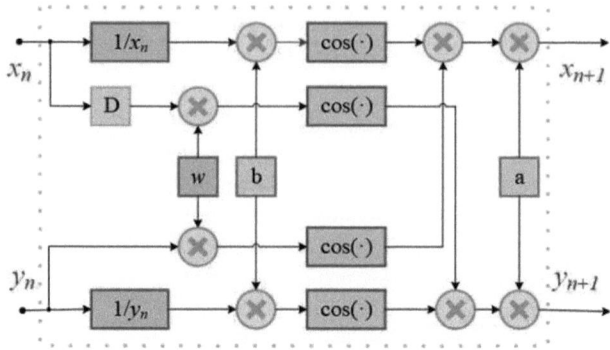

Fig. 2. The coupling structure of the proposed 2D-CCM.

2.2 Bifurcation and Lyapunov Diagram

A bifurcation diagram is an important tool for studying nonlinear dynamical systems [9]. It describes the stable states or periodic behaviors of a system as a control parameter varies. By plotting a bifurcation diagram, we can observe the evolution of the system from stability to period-doubling and eventually into a chaotic state.

When the initial values (x0,y0) = (0.3,0.4), Fig. 3 depicts the three-dimensional bifurcation diagram and the Lyapu-nov exponent plots with respect to parameters a,b,ω of the 2D cosine-coupled chaotic map. From the three-dimensional bifurcation diagram and the Lyapunov exponent plot, it can be observed that the bifurcation diagram exhibits a symmetric slanted structure. The Lyapunov exponent plot clearly shows that the system has two positive Lyapunov exponents over a large range, and the three parameters are mostly in chaotic and hyperch-aotic states. A periodic window exists near the parameter $\omega = 0.8126$.

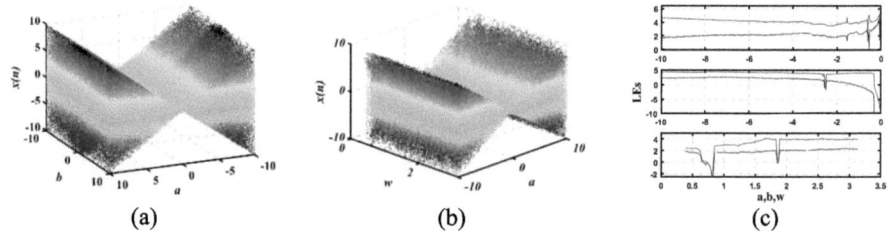

Fig. 3. The bifurcation diagram and Lyapunov exponent diagram of the system. (a) The bifurcation diagram of parameters a and b; (b) The bifurcation diagram of parameters a and w; (c) Regarding the Lyapunov exponent diagrams of a, b, and w.

2.3 Attractors

An attractor is a set to which the state of a dynamical system tends in the long-term evolution [10]. The complexity and ergod-icity of the discrete chaotic map can be intuitively observed through the attractor. In Fig. 4, various attractor traj-ectories of the 2D-CCM are plotted by fixing $\omega = \pi$ and b = 5, while varying the value of parameter a.

As seen in the figure, the chaotic map proposed in this paper exhibits an attractor with complex trajectories. Additionally, it is not difficult to observe that if the two halves of the attractor's sidelobes are considered as a complete sidelobe, the number of attractor sidelobes here equals twice the absolute value of a.

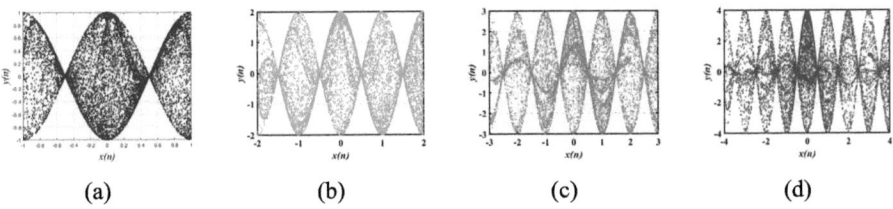

Fig. 4. Attractor diagram of parameter a. (**a**) a = -1; (**b**) a = -2; (**c**) a = -3; (**d**) a = -4.

Similarly, by fixing a = 1, b = 5, and varying the value of parameter w, the attractor of the system is plotted, as shown in Fig. 5. From the figure, it can be seen that the number of attractor sidelobes follows the same pattern with respect to ω.

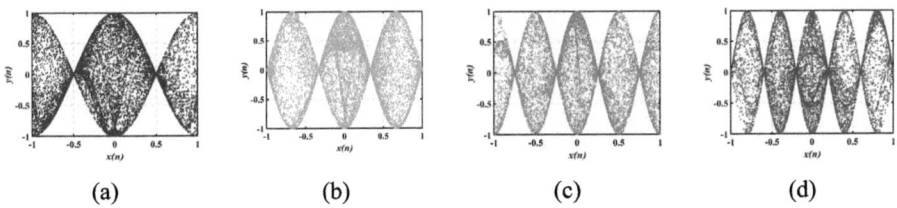

Fig. 5. Attractor diagram of parameter ω. (**a**) ω = π; (**b**) ω = 1.5π; (**c**) ω = 2π; (**d**) ω = 2.5π.

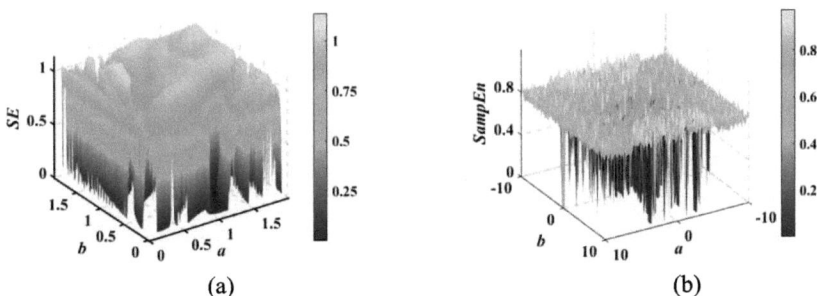

Fig. 6. Spectral entropy and sample entropy of coupled systems. (**a**) Spectral entropy; (**b**) Sample entropy.

2.4 SE and SampEn Complexity

The primary purpose of entropy is to determine the disorder of a time series, mainly due to the presence of random noise in the analyzed sequence. Spectral Entropy (SE) based on Fourier Transform and Discrete Wavelet Transform has advantages such as fast computation, simple implementa-tion, and robust calculation. Within the range of [0,1], a higher SE value indicates a higher complexity of the chaotic system. Similarly, signals with lower Sample Entropy (SampEn) exhibit more regularity, whereas signals with higher entropy values are considered disordered or more complex [11].

When the initial values(x0,y0,ω) = (0.3,0.4,π), Fig. 6 prese-nts the three-dimensional complexity maps of spectral entropy and sample entropy for the two-dimensional cosine-coupled chaotic map. From Fig. 6(a), it can be observed that when (a,b) ∈ (0, 2), the SE complexity remains close to 1 over a large parameter space. From Fig. 6(b), it can be seen that when (a,b) ∈ (-10, 10), the SampEn values are mostly above 0.8, and even close to 1. This indicates that the proposed 2D-CCM exhibits high uncertainty and complexity.

3 Proposed Image Encryption Methods

Based on the two-dimensional coupled chaotic map proposed in this paper, a new lightweight image encryption algorithm is introduced. The algorithm only uses simple mathematical operations, without the need for complex matrix transformations, making it suitable for embedded and AIoT devices. The basic flowchart of the algorithm is shown in the figure. The encryption process consists of three main parts. The first step is to generate the key stream using the chaotic system. The second step is to perform row and column interleaving permutation. The third part involves five rounds of diffusion. Figure 7 shows the encryption process flowchart.

3.1 Random Sequence Generation

In the encryption process, generating high-quality random sequences is crucial to ensuring the security of the encryption system. To achieve this, the system employs a random sequence generation method based on a chaotic system. First, by setting initial conditions and system parameters, a discretized chaotic map is used to generate a pseudo-random sequence. Specifically, the system's initial conditions are $x(1) = 0.3$ and $y(1) = 0.4$, and iterations are performed using the chaotic map formula, where a, b, and ω are preset parameters that control the chaotic behavior of the system.

After a certain number of warm-up iterations, the chaotic system reaches a stable state, generating a more random sequence. In this system, the number of warm-up iterations is set to 500, ensuring that the generated sequence does not contain initial non-randomness for subsequent processing. Then, based on the outputs $x(n)$ and $y(n)$ of the chaotic system, sequences for permutation and diffusion are extracted. These sequences are used to perform corresponding permutation and diffusion operations on the image pixel values, successfully generating pseudo-random sequences that play a key role in the encryption process.

It is worth noting that the permutation and diffusion operations are applied independently to each RGB channel of the image, but they both use the same chaotic sequence. This approach effectively increases the complexity and security of the encryption system, making the encrypted image difficult to break using traditional methods.

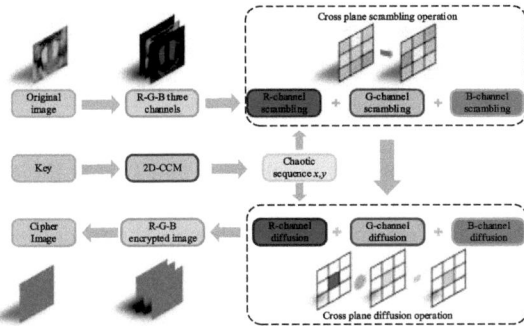

Fig. 7. Encryption processing flowchart based on 2-D -CCM.

3.2 Encryption Algorithm

In the 2D-CCM encryption scheme proposed in this study, the encryption process mainly consists of two core modules: pixel scrambling and diffusion operations. First, a high-quality pseudo-random sequence is generated through a chaotic system for subsequent image encryption processing. Then, the pixel positions and pixel values of the image are scrambled separately, thereby achieving encryption protection for the image content.

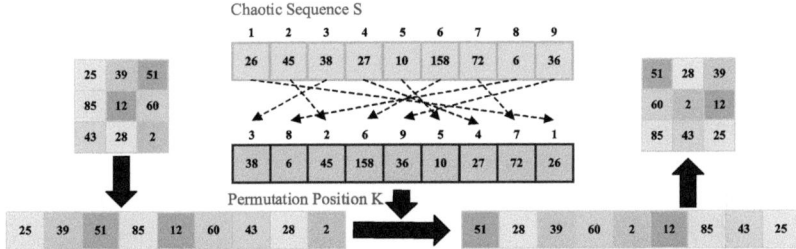

Fig. 8. Diagram of the scrambling process.

Figure 8 shows the process of image transformation during the scrambling phase. Specifically, the system first uses the chaotic sequence s(i) generated by the chaotic system to scramble the rows and columns of the image pixels. The sequence is processed using Eq. 1 and then mapped to integers within the pixel index range, resulting in a scrambled index sequence Ix(i), from which the final shuffled order G(i) is obtained through sorting. This process is performed separately for each color channel of the image (R, G, B), and the scrambling is done across planes, causing the spatial structure of the original image to be thoroughly disrupted, significantly enhancing the unrecogniz-ability of the encrypted image.

$$Ix(i) = \mod([|s(i)| - [|s(i)|]] \times \exp(1)^{15}, n) + 1 \qquad (2)$$

After completing the pixel position scrambling, the system further performs a diffusion operation using the chaotic sequence q(i) to generate the diffusion sequence Z(i), and applies a bitwise XOR operation between each pixel value and the diffusion sequence, as shown in Equation 3.

$$D(i) = B(i) \oplus Z(i) \qquad (3)$$

where B(i) represents the scrambled pixel value and C(i) is the final encrypted pixel value after diffusion. This operation effectively breaks the original grayscale distribution characteristics of the image and further enhances the randomness and anti-analysis capability of the encryption system.

Algorithm 1 represents encrypted pseudocode.

Algorithm 1 Encryption Algorithm

Input: Color image P (M × N × 3)
    Chaotic sequences s(i) and q(i) of length n = M × N
1.   For i ← 1 TO n DO
2.   Ix(i) ← mod(floor((abs(s(i))-floor(abs(s(i))))× e^15),n) + 1
3.   End For
4.   [F, G] ← sort(Ix)
5.   For ch ← 1 TO 3 DO
6.   A ← flatten(P(:,:,ch))
7.   For i ← 1 TO n DO
8.   B(i) ← A(G(i))
9.   End For
10.   C_permuted(:,:,ch) ← reshape(B, M, N)
11. End For
12. For i ← 1 TO n DO
13.   Z(i) ← mod(floor((abs(q(i))-floor(abs(q(i))))×e^15), 256)
14. End For
15. For ch ← 1 TO 3 DO
16.   B ← flatten(C_permuted(:,:,ch))
17.   For i ← 1 TO n DO
18.   D(i) ← B(i) XOR Z(i)//Bitwise XOR with diffusion sequence
19.   END FOR
20.   E(:,:,ch) ← reshape(D, M, N)
21. End For

Output: Display the encrypted image E

---

4 Experimental Simulation and Performance Analysis

In this chapter, the color image are tested by the encryption algorithm, and the effect and security of the algorithm are analyzed from many aspects.

4.1 Experimental Simulation

In this section, tests were conducted using 256 × 256 color chili images and two 512 × 512 industrial equipment images. Figure 9 shows the original image, enc-rypted image, and decrypted image.

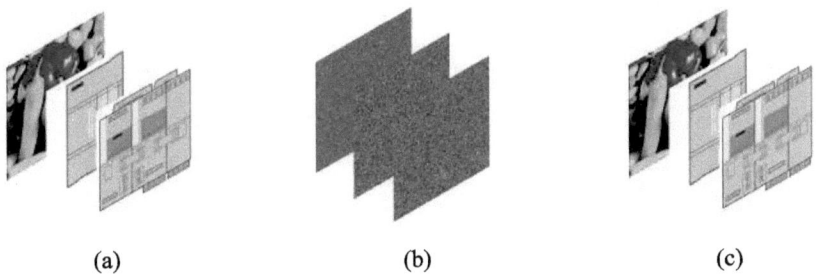

Fig. 9. Encryption and Decryption Results. (a) Plain Image; (b) Cipher Image; (c) Decrypted Image.

From Fig. 9, it can be seen that the encrypted image no longer contains any information from the original image. Calculate the difference between encrypted and decrypted images using Eq. 4.

$$diffference\_image = sum(difference\_image, 3) \qquad (4)$$

The final difference result is 0, indicating that this encryption algorithm has good feasibility.

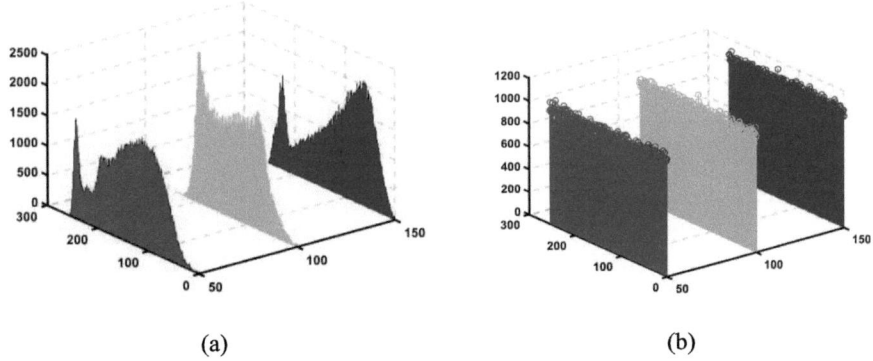

Fig. 10. Histogram analysis results of RGB channels. (a) original image (b) cipher image.

4.2 Histogram Analysis

A histogram is used to observe the distribution of pixel values between the original image and the encrypted image. As shown in Fig. 10, the results of histogram analysis of the original image and the encrypted image show that the pixel value distribution of the encrypted image is very uniform.Therefore, the encryption algorithm is better.

4.3 Correlation Between Adjacent Pixels

The correlation coefficient of adjacent pixels is quantified by calculating the correlation coefficient of the image [12]. As shown in Fig. 11, Table 1 shows the pixel correlation of the plaintext and ciphertext images. The comparison between the two correlation diagrams clearly demonstrates the advantage of the encryption algorithm.

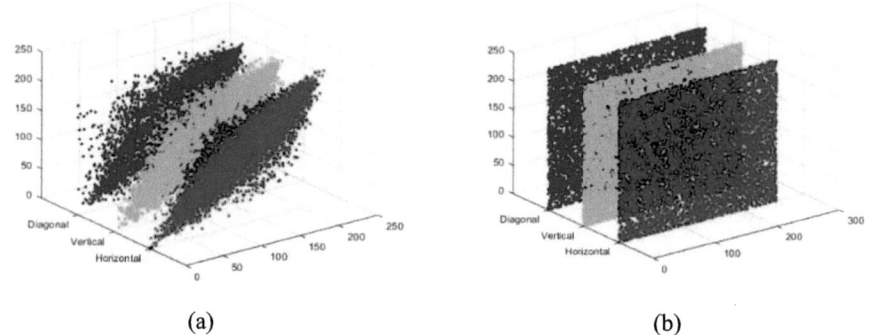

Fig. 11. Adjacent pixel correlation of the R channel. (**a**) plaintext image (**b**) encrypted image.

In the original image, adjacent pixels exhibit strong correlation, forming a concentrated distribution along the diagonal in horizontal, vertical, and diagonal directions. In contrast, after encryption, pixel pairs are scattered randomly, indicating that the correlation is significantly reduced. This decorrelation effectively conceals the inherent structure of the original image, thereby enhancing security against statistical and differential attacks. Table 1 shows the correlation between adjacent pixels in plaintext and ciphertext images.

Table 1. Correlation between adjacent pixels in plaintext and ciphertext images

Correlation coefficient	Original image			Encryption image		
	R	G	B	R	G	B
Horizontal	0.9218	0.8643	0.9071	-0.0016	-0.0009	-0.0008
Vertical	0.8624	0.7591	0.8782	-0.0005	0.0002	0.0020
Diagonal	0.8531	0.7299	0.8411	-0.0004	-0.0001	-0.0026

4.4 Robustness Analysis

A cropping attack is a method used to evaluate the robustness of image encryption systems [13]. In this attack, the encrypted image is partially cropped, and the resulting cropped ciphertext is decrypted to see if the main content of the original image can still be recovered. If the encryption method is not strong enough, an attacker may be able to retrieve significant image information even from a partial ciphertext.We introduced cropping attacks with magnitudes of 1/16 and 1/4 to evaluate the encryption scheme. Figure 12 shows the decryption results when cropping occurs simultaneously at the top-left and bottom-right corners. It can be observed that when the cropped area is 1/8, the decrypted image is nearly unaffected. Even when the cropped area reaches 1/2, the decrypted image still reveals recognizable information of the original image. This demonstrates that the proposed encryption algorithm is effective in resisting cropping attacks.

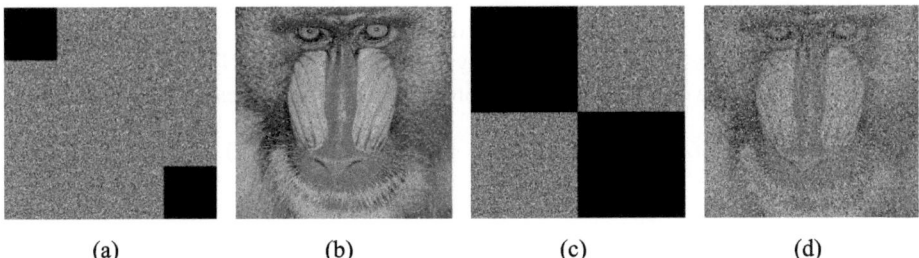

(a)　　　　　　　(b)　　　　　　　(c)　　　　　　　(d)

Fig. 12. Cropping attack. (**a**) Crop 1/16; (**b**) Crop 1/16 decrypted image; (**c**) Crop 1/4; (**d**) Crop 1/4 decrypted image.

To evaluate the robustness of the encryption algorithm, this paper also introduces salt-and-pepper noise and Gaussian noise attacks. The former simulates impulsive interference, while the latter represents continuous transmission noise, aiming to test the algorithm's stability under noisy conditions.During image transmission, they are often affected by channel noise. In the simulation, to mimic the actual channel degradation and noise interference, we artificially introduced 1% and 10% salt-and-pepper noise and Gaussian noise. Subsequently, the decryption results are shown in Fig. 13.

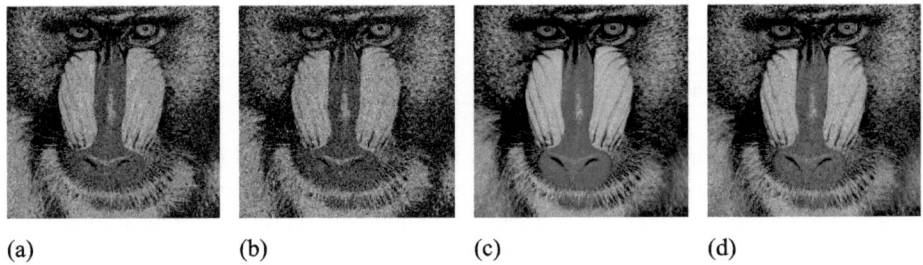

(a)　　　　　　　(b)　　　　　　　(c)　　　　　　　(d)

Fig. 13. Noise attack. (**a**) Salt-and-pepper noise with densities of 1%; (**b**) Salt-and-pepper noise with densities of 10%; (**c**) Gaussian noise with variances of 1%; (**d**) Gaussian noise with variances of 10%.

4.5 Complexity and Running Efficiency Analysis

Running efficiency is as essential as security when assessing lightweight encryption algorithms. This section evaluates the scheme from two perspectives: theoretical time complexity and actual execution time.

For a color image of size M × N, the proposed 2D-CCM lightweight encryption algorithm first generates two chaotic sequences s(i) and q(i) of length M × N, giving a complexity of O(MN). Next, a single sorting operation maps the permutation indices, after which the pixels of all three RGB channels are permuted, resulting in approximately O(3MN) operations.

The chain diffusion stage includes one inverse diffusion and one forward diffusion pass; both scans across the three channels, yielding about O(6MN).

Consequently, the overall theoretical complexity can be expressed as

$$O((1+3+6)MN) = O(10MN) \approx O(MN) \qquad (5)$$

which grows linearly with the number of pixels and thus meets real-time requirements in IoT scenarios.

Table 2 compares the encryption and decryption times of our scheme with the methods in Refs. [14] and [15] at two image resolutions. Regardless of image size, the proposed scheme outperforms the references in both stages, confirming its lightweight nature.

Table 2. Comparison of running time (unit: s)

Image size	Scheme	Encryption	Decryption	Total
256 × 256	Ref. [14]	0.4049	1.4724	1.8773
	Ref. [15]	0.4642	1.4950	1.8400
	Proposed	0.1369	0.1200	0.2569
512 × 512	Ref. [14]	0.9279	10.5830	11.5109
	Ref. [15]	1.7743	2.1113	3.8856
	Proposed	0.5525	0.3120	0.8645

In summary, the linear-time complexity and the experimental speed-up corroborate each other, demonstrating that the 2D-CCM algorithm achieves a sound balance between security and efficiency, making it suitable for deployment on resource-constrained edge and embedded devices.

4.6 Differential Attacks

NPCR (Number of Pixels Change Rate) and UACI (Unified Average Changing Intensity) are two crucial metrics used in differential cryptanalysis. These metrics are primarily used to measure the responsiveness of the encrypted image to minor changes in the original image.

Table 3. NPCR and UACI between different algorithms

Algorithm	Ref. [16]	Ref. [17]	Ref. [18]	Proposed
NPCR	99.6060%	99.6207%	99.6093%	99.6105%
UACI	33.4689%	33.4125%	33.4798%	33.4615%

Table 3 presents a comparative analysis of the NPCR and UACI values between the proposed method and several representative schemes reported in Refs. [16]–[18]. As shown, the NPCR values of all methods are close to the ideal value of 99.6094%, indicating strong sensitivity to pixel-level changes. The proposed method achieves an NPCR of 99.6105%, which is slightly higher than those in previous works, demonstrating excellent resistance against differential attacks. Similarly, the UACI value of 33.4615% is consistent with theoretical expectations and is comparable to or slightly better than the values in Refs. [16]–[18]. These results confirm that the proposed encryption scheme can effectively diffuse small changes in the plaintext image across the entire ciphertext, ensuring robust cryptographic security.

5 Conclusion

In this paper, a lightweight image encryption algorithm based on a two-dimensional cosine-coupled chaotic map (2D-CCM) was proposed to meet the security and efficiency requirements of IoT environments. The method integrates pixel-level scrambling and multi-round diffusion, leveraging the high complexity and sensitivity of the designed chaotic system. Experimental results confirm its robustness against noise and cropping attacks while maintaining significantly reduced computational cost.Compared with existing schemes, such as those based on compressive sensing or DNA encoding, the proposed algorithm achieves faster runtime and comparable or improved cryptographic strength, as evidenced by NPCR, UACI, and correlation tests. This makes it particularly suitable for real-time applications on resource-constrained edge devices.However, practical deployment in IoT systems also requires consideration of hardware-level constraints, such as memory usage, power consumption, and integration with embedded processors. Future work will focus on implementing the algorithm in hardware platforms such as FPGA or MCU, conducting large-scale deployment tests, and exploring its applicability to video encryption and secure federated learning scenarios.

References

1. Abuserrieh, L., Alalfi, M.H.: A survey on verification of security and safety in IoT systems. IEEE Access **12**, 138627–138645 (2024)
2. Zahan, H., Al Azad, M.W., Ali, I., Mastorakis, S.: IoT-AD: A Framework to Detect Anomalies Among Interconnected IoT Devices. IEEE Internet of Things Journal, **11**(1), 478–489 (2024)
3. Tan, L., et al.: Energy-efficient tactile-driven rule configuration and anomaly detection in industrial IoT Systems. IEEE Internet of Things Journal https://doi.org/10.1109/JIOT.2025

4. Bera, B., Das, A.K., Sikdar,B.: Quantum-resistant secure communication protocol for digital twin-enabled context-aware iot-based healthcare applications. IEEE Transactions on Network Science and Engineering https://doi.org/10.1109/TNSE.2025
5. Verma, P., Sood, S.K., Kaur, H., Kumar, M., Wu, H., Gill, S.S.: Data driven stochastic game network-based smart home monitoring system using IoT-enabled edge computing environments. IEEE Transactions on Consumer Electronics, https://doi.org/10.1109/TCE.2024
6. Jain, K., Titus, B., Krishnan, P., Sudevan, S., Prabu, P., Alluhaidan, A.S.: A Lightweight multi-chaos-based image encryption scheme for IoT networks. IEEE Access **12**, 62118–62148 (2024)
7. Liu, W., Sun, K., Wang, H., Li, B., Chen, Y.: Inverse proportional chaotification model for image encryption in IoT scenarios. IEEE Trans. Circuits Syst. I Regul. Pap. **72**(1), 254–264 (Jan.2025)
8. Tang, J., Zhang, Z., Huang, T.: Two-dimensional cosine-sine interleaved chaotic system for secure communication. IEEE Trans. Circuits Syst. II Express Briefs **71**(4), 2479–2483 (April 2024)
9. Min, F., Yin, S., Cheng, Y.: Hybrid-diode-based Shinriki circuit: Coexisting oscillations and bifurcation trees. IEEE Trans. Circuits Syst. I Regul. Pap. **72**(2), 896–906 (Feb.2025). https://doi.org/10.1109/TCSI.2024
10. Yang, Y., Huang, L., Kuznetsov, N.V., Chai, B., Guo, Q.: Generating multiwing hidden chaotic attractors with only stable node-foci: Analysis, implementation, and application. IEEE Trans. Industr. Electron. **71**(4), 3986–3995 (April 2024)
11. Mou, J., Han, Z., Cao, Y., Banerjee, S.: Discrete second-order memristor and its application to chaotic map. IEEE Trans. Circuits Syst. II Express Briefs **71**(5), 2824–2828 (May2024)
12. Huang, H., Wang, Y., Niu, L., Chen, J., Zhang, H., Zhang, M. Chaotic brillouin optical correlation-domain analysis based on differential correlation demodulation. IEEE Sensors Journal **25**(8), 13033–13038 (2025)
13. Zhang, G., Zheng, L., Su, Z., Zeng, Y., Wang, G.: M-sequences and sliding window based audio watermarking robust against large-scale cropping attacks. IEEE Trans. Inf. Forensics Secur. **18**, 1182–1195 (2023)
14. Wang, H., Xiao, D., Li, M., Xiang, Y.P., Li, X.Y.: A visually secure image encryption scheme based on parallel compressive sensing. Signal Process. **155**, 218–232 (2019)
15. Zhu, L.Y., et al.: A robust meaningful image encryption scheme based on block compressive sensingand SVD embedding. Signal Process. **175**, 107629 (2020)
16. Du, L., Teng, L., Liu, H., Lu, H.: Multiple face images encryption based on a new non-adjacent dynamic coupled mapping lattice. Expert Syst. Appl. **238**, 121728 (Mar.2024)
17. Wang, X., Li, Y.: Chaotic image encryption algorithm based on hybrid multi-objective particle swarm optimization and DNA sequence. Opt. Lasers Eng., **137**, 106393 (2021)
18. Lai, Q., Hu, G., Erkan, U., Toktas, A.: A novel pixel-split image encryp-tion scheme based on 2D Salomon map. Expert Syst. Appl. **213**, 11845 (2023)

Image Watermarking Encryption Algorithm for IoT Utilizing Chaotic Neural Network

Yuan Zhu[1], Dawei Ding[1(✉)], Yuanyuan Wang[2], Zongli Yang[1], Chaoma Qian[2], and Jingwen Zhao[3]

[1] School of Electronic Information Engineering, Anhui University, Hefei, China
dwding@ahu.edu.cn
[2] China Railway Fourth Bureau Group Shanghai Engineering Co., Ltd., Shanghai, China
[3] Anhui Digital Intelligent Construction Research Institute Co., Ltd., Hefei, China

Abstract. This paper addresses the need for image copyright protection in the Internet of Things (IoT) by proposing a watermarking algorithm based on chaotic neural networks. The algorithm employs a three-dimensional memristive Rulkov model to select strong chaotic parameters for generating encryption sequences, combines cross-offset scrambling and Fibonacci matrix diffusion to achieve dual encryption, and embeds the watermark into the high-frequency sub-bands of the image through secondary discrete wavelet transform. Experimental data show that the watermarked images achieve an average PSNR of 47.8 dB, with the SSIM index consistently ranging between 0.9974 and 0.9990. By leveraging the sensitivity of chaotic systems to globalize the impact of local attacks, the watermark can be fully extracted even under cropping and noise interference, demonstrating the algorithm's strong concealment and attack resistance. Real-time processing is enabled through edge computing, providing a reliable copyright management solution for IoT scenarios such as drone inspections.

Keywords: Rulkov neuron · discrete wavelet transform · IoT security · watermark embedding

1 Introduction

With the advancement of IoT technology, the application scenarios of images continue to expand, but copyright security issues have become increasingly prominent [1] Unauthorized copying, tampering, and misuse occur frequently, harming the rights of creators and the authenticity of images. In the IoT environment, the real-time transmission and sharing of massive images further complicate copyright protection. How to effectively safeguard image copyrights, ensuring their integrity and traceability, has become a critical issue that urgently needs to be addressed [2].

Digital watermarking is a key technology for protecting image copyrights in the IoT era[3]. It embeds invisible information into images, creating a unique "digital fingerprint" that remains traceable even after compression or modification[4]. When combined with blockchain technology, digital watermarking enables decentralized verification, enhancing the transparency and reliability of copyright protection, and providing an efficient solution for securing digital content.

The integration of digital watermarking and edge computing provides an efficient copyright protection solution for drone applications[5]. By embedding watermarks in real-time through onboard edge devices, it not only avoids the bandwidth consumption of transmitting high-definition images back but also ensures data timeliness and traceability. During inspection operations, the system can perform real-time watermark verification on video streams. Once unauthorized tampering or interception is detected, security mechanisms such as data encryption, alarms, or automatic return are immediately triggered, achieving end-to-end protection from data collection to transmission. This end-to-end security architecture significantly enhances the data reliability and copyright protection capabilities of drone operations[6]. Figure 1 describes the proposed drone watermarking framework for image data protection.

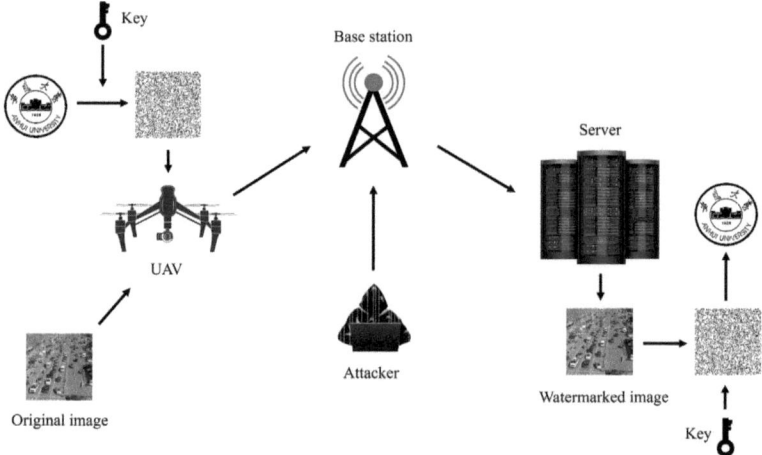

Fig. 1. UAV watermark image data protection framework

Meanwhile, to understand how the brain works, neurobiological analysis and experiments have investigated the dynamic mechanisms behind neuronal firing activity[7], this paper introduces a chaotic neural network to encrypt watermark information, thereby enhancing its security and reliability[8]. Research on digital watermarking based on chaos theory primarily focuses on improving the robustness of watermarks and enhancing system security. Behni et al. applied coupled map lattices to watermarking schemes, significantly enhancing the security of the scheme by leveraging the sensitivity to initial conditions of coupled map lattices[9]. Gu et al. proposed a robust watermarking algorithm that embeds watermarks in the low-frequency coefficients of the integer wavelet transform domain of images. This method uses chaotic pseudo-random sequences to select embedding positions, ensuring the watermark information is evenly distributed in the low-frequency part of the image. This enhances robustness against compression attacks and improves the security of the embedding locations[10].

2 Analysis of Chaotic System

2.1 Discrete Memristor

In this study, we employ the 2-D Rulkov model as our neuron model. Consequently, the memristor used for simulating magnetic induction should also be discretized. By modifying the flux-controlled memristor described in the literature[11], we have derived its mathematical model as follows:

$$\begin{cases} i_n = W(\varphi_n)v_n = \dfrac{\varphi_n}{\varphi_n + 1}v_n \\ \varphi_{n+1} = \varphi_n + v_n \end{cases} \quad (1)$$

where φ_{n+1} represents the value of φ_n after n + 1 iterations. To demonstrate the voltage-current trajectory of the discrete memristor in Eq. (1), we implemented the model on the MATLAB platform with the input voltage set as $v_n = A\sin(\omega_n)$, where A denotes the amplitude of the discrete voltage and w represents the angular frequency. The hysteresis loop of the memristor is shown in Fig. 2.

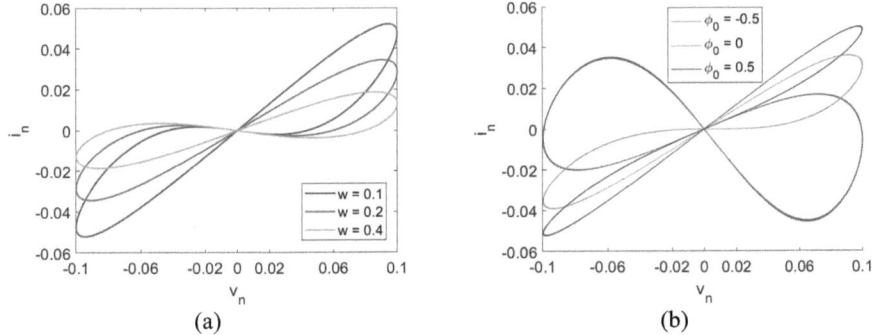

Fig. 2. Fingerprints of the discrete memristor with A = 0.1. (a) Frequency-relied voltage-current loci for A = 0.1, φ_0 = -0.1,and ω = 0.1,0.2,0.4.(b) Initial state-relied voltage-current loci for A = 0.1, ω = 0.2, φ_0 = -0.5,0,0.5.

2.2 The 3D Memristive Rulkov Neuron Model

Inspired by existing literature[12], this paper proposes a novel 3D discrete Rulkov neuron model as follows:

$$\begin{cases} x_{n+1} = \dfrac{a}{1+|x_n|} + y_n + b\dfrac{z_n}{z_n+1}x_n \\ y_{n+1} = y_n - cx_n \\ z_{n+1} = z_n + x_n \end{cases} \quad (2)$$

In the proposed model, x_n, y_n and z_n represent the outputs of the neuron at the nth iteration. The additive term $b(z_n/z_n + 1)x_n$ corresponds to the magnetic induction current, where b serves as a parameter controlling the intensity of induction. Additionally, a and c are system control parameters that govern the dynamical behavior of the model.

2.3 Dynamical Analysis

Bifurcation diagrams visually reveal transitions from periodic to chaotic states, while Lyapunov exponent plots quantify chaos: positive for chaos, zero for periodic or quasi-periodic, and negative for stability.

With the initial conditions set to $(0.1, 0, 0)$ and parameters $b = 0.6$ and $c = 0.01$, Fig. 3 illustrates the bifurcation diagram and Lyapunov exponent plot of the chaotic system as parameter a varies within the range $[0,5]$. The results demonstrate that the chaotic behavior of the system becomes increasingly pronounced as a increases. Additionally, with fixed parameters $a = 1$ and $b = 0.6$, Fig. 4 depicts the dynamical evolution of the system as parameter c increases within the interval $[0.8,1]$. In this case, intermittent periodic windows emerge within the chaotic regime, highlighting the significant influence of parameter c on the system's dynamics. In summary, while the chaotic characteristics of the system are enhanced with increasing a, variations in c introduce transient periodic behaviors, revealing a complex interplay between the parameters in shaping the system's dynamical properties.

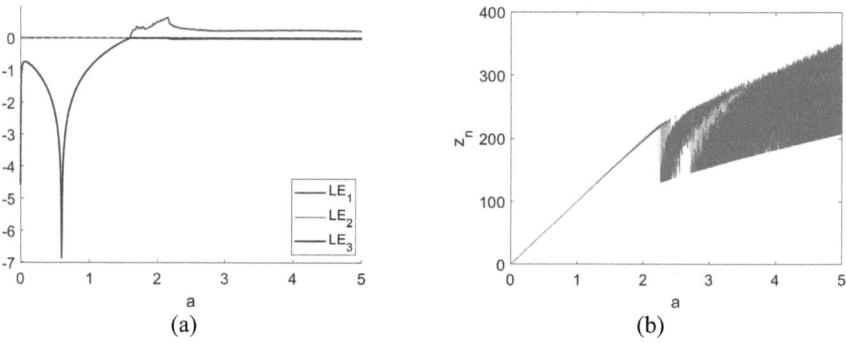

Fig. 3. With the initial state set to $(0.1, 0, 0)$ and parameters $b = 0.6$ and $c = 0.01$, the Lyapunov exponent (a) and the bifurcation diagram (b) are presented as parameter varies within the range $[0,5]$.

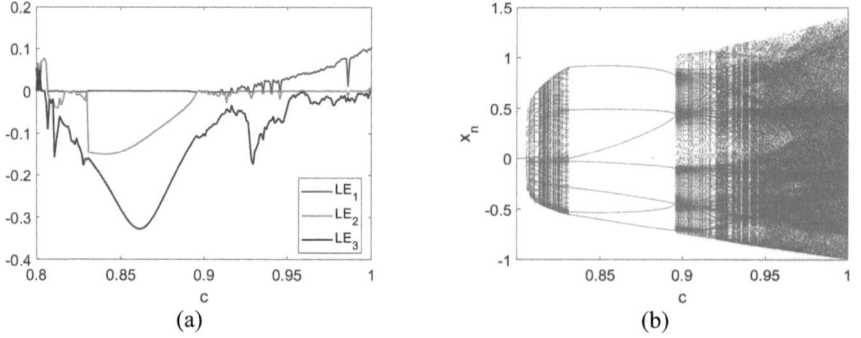

Fig. 4. Dynamical evolution of the map (2) with $b = 0.6$, $a = 1$ and c varies in $[0.8,1]$ under ($x_0, y_0, z_0) = (0.1,0,0)$, where (a) Lyapunov exponents, (b) bifurcation diagram.

3 Digital Image Watermarking Methods

The digital watermarking scheme proposed in this paper is illustrated in detail in Fig. 5.

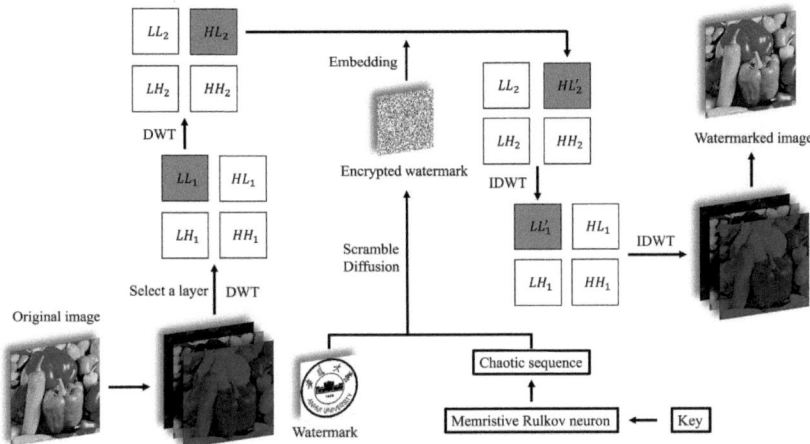

Fig. 5. Watermark embedding process

3.1 Chaotic Sequence Generation

By combining the neuron model from the second part with appropriate system parameters, the chaotic sequence required for subsequent encrypted watermarking can be generated. First, the size of the watermark needs to be read. Assuming the watermark image has a size of $M \times N$, the generated chaotic sequence K should correspond to the image size. Then, let the key K corresponding to the neuron model in the second part be $K = (a, b, c, x_0, y_0, z_0)$. The values of key K can be selected based on the corresponding Lyapunov exponent diagram. For example, let $K = (9, 0.6, 0.01, 0.1, 0, 0)$. After iterative selection, the chaotic sequence S is obtained.

3.2 Scrambling Algorithm

To encrypt the watermark, the method of cross-offset scrambling is employed, which confuses the entire pixel arrangement by altering the positions of rows and columns, thereby effectively disrupting the correlation between pixels. The specific operations are as follows:

Step 1. Read an image of size $M \times N$, and convert the chaotic sequence S into a matrix S_1 of size $M \times N$.

Step 2. Use S1 to construct the row index matrix I and the column index matrix J, and then interleave them according to (3).

$$row = \mod(I_{i,j} + 2 \times J_{i,j}, M) + 1$$
$$col = \mod(J_{i,j} - I_{i,j}, N) + 1 \quad (3)$$

Step 3. The watermark image is reordered according to the determined sorting sequence to obtain the scrambled watermark image.

3.3 Diffusion Algorithm

This paper employs a Fibonacci matrix diffusion technique, which offers a larger key space and more uniform pixel distribution compared to traditional Arnold or Logistic scrambling, improving security and resistance to attacks.

The Fibonacci sequence can be defined as:

$$C_n = C_{n-1} + C_{n-2}, n \geq 2 \tag{4}$$

The Fibonacci matrix can be derived from the following equation:

$$Q = \begin{bmatrix} C_2 & C_1 \\ C_1 & C_0 \end{bmatrix} \tag{5}$$

where Q is the Fibonacci matrix.

From this, the matrix Q^n is as follows:

$$Q^n = \begin{bmatrix} C_{n+1} & C_n \\ C_n & C_{n-1} \end{bmatrix} \tag{6}$$

Divide the scrambled watermark image into 2×2 sub-blocks, and multiply each sub-block by the Fibonacci matrix to obtain the final encrypted watermark image. The calculation process is as follows:

$$\begin{bmatrix} P_{i,j} & P_{i,j+1} \\ P_{i+1,j} & P_{i+1,j+1} \end{bmatrix} = \begin{bmatrix} S_{i,j} & S_{i,j+1} \\ S_{i+1,j} & S_{i+1,j+1} \end{bmatrix} \begin{bmatrix} C_{n+1} & C_n \\ C_n & C_{n-1} \end{bmatrix}^n \mod 256 \tag{7}$$

where P is the final encrypted image, and S is the scrambled image.

3.4 Watermark Embedding and Extraction

The algorithm adopted in this paper embeds a binary image watermark to be embedded into the high-frequency subband of the carrier image after discrete wavelet transform, achieving the purpose of hiding the watermark. The specific steps are as follows:

Step 1. Encrypting the watermark image using chaos-based index scrambling and Fibonacci matrix diffusion

Step 2. Perform a second-level discrete wavelet transform on one channel of the carrier image using Haar wavelets to obtain a series of second-level subbands. (LL_2, HL_2, LH_2, HH_2).

Step 3. Embed the encrypted watermark image into the HL_2 subband to obtain the subband containing watermark information.

Step 4. Combine the subbands embedded with the watermark with other subbands, apply a 2-level IDWT, and obtain the image containing the watermark.

The embedding strength α of the watermark can be inversely determined by the PSNR value of the image after embedding the watermark. The calculation formula between PSNR and embedding strength α is as follows:

$$PSNR = 10\log_{10}(255^2 / 16\alpha^2\gamma) \tag{8}$$

where γ is the attenuation factor of capability. Since high-frequency energy attenuates during the inverse transformation, its range is [0.1, 0.3]. The embedding strength α can be derived inversely from Eq. (8) to balance the robustness and invisibility of the watermark.

Watermark extraction is the inverse process of embedding. It involves applying IDWT twice to the watermarked image, then extracting the encrypted watermark image from the HL_2 subband, and finally decrypting the encrypted watermark image to obtain the final watermark image.

4 Analysis of Experimental Results

4.1 Experimental Simulation

Using a carrier image of size 512 × 512 and a watermark image of size 128 × 128, the carrier image is taken from the VisDrone dataset. Figure 6 shows the simulation results, which are satisfactory.

(a)　　　　　(b)　　　　　(c)　　　　　(d)　　　　　(e)

Fig. 6. (a) Watermark image (b) Encrypted watermark (c) Carrier image (d) Watermarked image (e) Extracted watermark image

4.2 Performance Analysis

Peak Signal-to-Noise Ratio (PSNR), Normalized Cross-Correlation (NC), Structural Similarity Index (SSIM), and Bit Error Rate (BER) are commonly used metrics to evaluate the similarity between watermarked images and original images, as well as the quality of the watermark. Generally, a PSNR value exceeding 40dB ensures the invisibility of the watermark, while NC and SSIM values closer to 1 and lower BER values indicate higher similarity between the two images. Table 1 compares the PSNR and SSIM of the proposed algorithm with those of algorithms from other literature.

Table 2 lists the PSNR, NC, and BER values of the watermarked images.

Table 1. Comparison of the PSNR and SSIM value between our method with related algorithms

	Image	Barbara	Baboon	Peppers
our	PSNR	47.89	47.88	47.87
	SSIM	0.9978	0.9990	0.9974
[13]	PSNR	39.83	37.03	44.07
	SSIM	0.9984	0.9619	0.9976
[14]	PSNR	39.92	47.77	40.70
	SSIM	0.9778	0.9730	0.9637
[15]	PSNR	34.17	32.98	36.11
	SSIM	0.9971	0.9951	0.9921

Table 2. PSNR, NC, and BER of watermarked image

Watermarked image	PSNR(dB)	NC	BER
Baboon	47.8836	1	0
Peppers	47.8747	1	0
Barbara	47.8863	1	0

4.3 Key Analysis

Key Space Analysis: The size of the key space is generally positively correlated with the ability to resist brute-force attacks. The encryption algorithm proposed in this paper uses six parameters of a neuron model as the key. Assuming the floating-point precision is 10^{16}, the key space size is approximately $10^{96} \approx 2^{319} > 2^{256}$.

Key sensitivity analysis: The key sensitivity of the algorithm is tested by modifying the key values. The initial *key* is used for encryption, and a new decryption key_1 is obtained by modifying $x_0 + 10^{-15}$ in the key. Figure 7 shows the test results of the algorithm's key sensitivity, demonstrating that when the key undergoes minor changes, the algorithm fails to restore the original image.

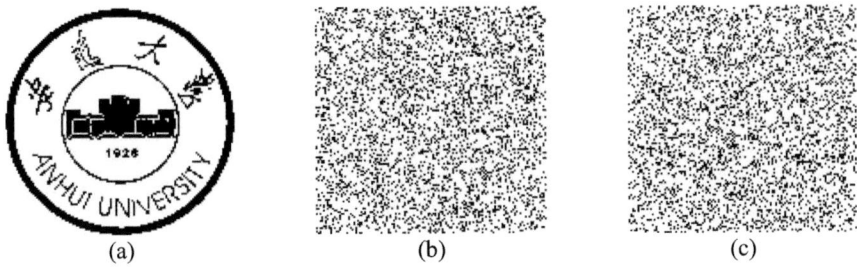

Fig. 7. (a) Original watermark image (b) Encrypted watermark (c) Decryption failed watermark

4.4 Computational Complexity Analysis

In addition, the running speed of the algorithm must also be considered. Using this algorithm to encrypt the watermark image requires generating $M \times N$ random numbers, so the complexity of generating $M \times N$ random numbers is $O(M \times N)$. The total time complexity of the entire encryption algorithm is $O(M \times N \log(M \times N))$. The computational complexity of performing DWT decomposition and embedding on the carrier image is $O(4M \times 4N)$, and the overall computational complexity of the entire algorithm is $O(4M \times 4N)$.

4.5 Histogram Analysis

Histogram analysis is used to evaluate the concealment and robustness of watermarks. By comparing the histogram distributions before and after watermark embedding, statistical anomalies can be detected to assess visual concealment; changes in the histogram after attacks reflect robustness. As shown in Fig. 8, the histograms of the carrier and watermarked images are highly similar, proving that the proposed algorithm ensures the invisibility of the watermark.

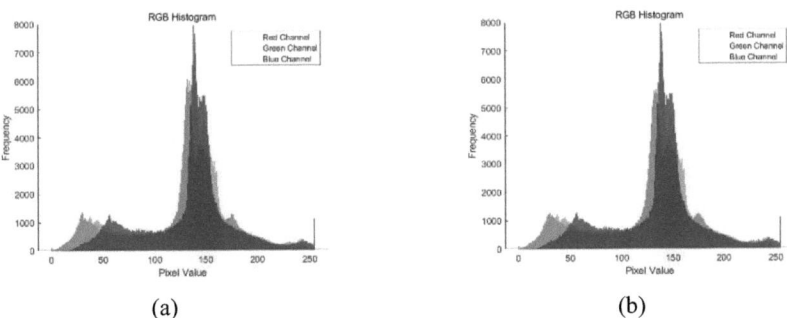

Fig. 8. Histogram comparison between the carrier image and the watermarked image: (a) original carrier image, (b) watermarked image.

4.6 Robustness Test

To test the robustness of the embedding algorithm, this paper employs cropping and noise as separate attack methods.

First, the watermarked image is processed using a cropping attack, and then the watermarked information is extracted from the cropped image, as shown in Fig. 9. It can be concluded that the cropping attack has a certain interference effect on the watermarked image. However, the proposed algorithm can diffuse the impact of the cropped area across the entire image, thereby reducing the influence on local regions. The information of the watermarked image can still be obtained, indicating that the algorithm in this paper can effectively resist cropping attacks.

Secondly, noise attacks were applied to the carrier images. This paper uses salt-and-pepper noise with noise values of 0.001 and 0.01 to test the image's noise resistance. The experimental results, as shown in Fig. 10, indicate that the proposed scheme exhibits a certain level of noise resistance and demonstrates high reliability.

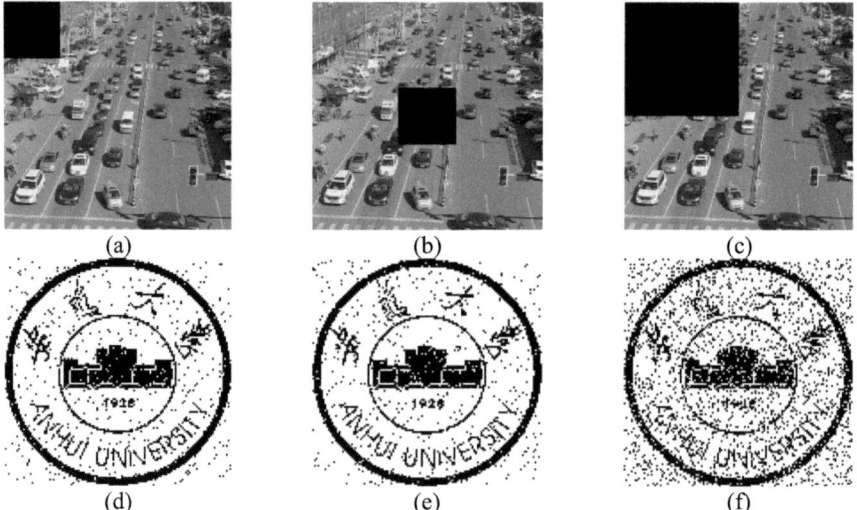

Fig. 9. Cropping attack results: (a) 6.25%, (b) 6.25%, (c) 25%, (d)-(f) the corresponding recovery watermark

Fig. 10. Salt-and-pepper noise attack results: (a) 0.001, (c) 0.01, (b) (d) the corresponding recovery watermark

5 Conclusion

In this paper, a three-dimensional memristive chaotic neural network watermarking encryption method for IoT image security is proposed. This method constructs a chaotic system with biological neuron characteristics, verifying its high complexity and parameter sensitivity through bifurcation diagrams and Lyapunov exponents. Dynamic keys are generated based on image features to form a mechanism resistant to chosen-plaintext attacks. The encryption adopts a hierarchical strategy: local watermark embedding employs cross-offset scrambling to achieve pixel-level dynamic perturbation; global encryption integrates Fibonacci matrix diffusion, constructing a dual-interleaved diffusion network in spatial-numerical domains. The algorithm excels in histogram uniformity, pixel correlation, and resistance to cropping attacks. Combined with lightweight edge computing, it provides a copyright protection solution for IoT images that integrates chaotic dynamic characteristics, multi-level encryption, and real-time processing capabilities.

References

1. Bounceur, A., Kara, M., Ferik, B., et al.: An adaptive ACM watermarking technique based on combined feature extraction and non-linear equation. Expert Syst. Appl. **275**, 126954 (2025)
2. Li, C., Zhang, Y., Xie, E.Y.: When an attacker meets a cipher-image in 2018: A year in review. Journal of Information Security and Applications **48**, 102361 (2019)
3. Gao, G., Han, T., Feng, Z., et al.: A data integrity authentication scheme in WSNs based on double watermark. IEEE Internet Things J. **10**(9), 8245–8256 (2022)
4. Yin, Z., She, X., Tang, J., et al.: Reversible data hiding in encrypted images based on pixel prediction and multi-MSB planes rearrangement. Signal Process. **187**, 108146 (2021)
5. Wei, Z., Zhu, M., Zhang, N., et al.: UAV-assisted data collection for Internet of Things: A survey. IEEE Internet Things J. **9**(17), 15460–15483 (2022)
6. Wei, Z., Feng, Z., Zhou, H., et al.: Capacity and delay of unmanned aerial vehicle networks with mobility. IEEE Internet Things J. **6**(2), 1640–1653 (2018)
7. Deng, Z., Wang, C., Lin, H., et al.: A memristive spiking neural network circuit with selective supervised attention algorithm. IEEE Trans. Comput. Aided Des. Integr. Circuits Syst. **42**(8), 2604–2617 (2022)
8. Makbol, N.M., Khoo, B.E.: Robust blind image watermarking scheme based on redundant discrete wavelet transform and singular value decomposition. AEU-International Journal of Electronics and Communications **67**(2), 102–112 (2013)
9. Behnia, S., Ahadpour, S., Ayubi, P.: Design and implementation of coupled chaotic maps in watermarking. Appl. Soft Comput. **21**, 481–490 (2014)
10. Gu, Q., Gao, T.: A novel reversible robust watermarking algorithm based on chaotic system. Digital Signal Processing **23**(1), 213–217 (2013)
11. Bao, H., Hu, A., Liu, W., et al.: Hidden bursting firings and bifurcation mechanisms in memristive neuron model with threshold electromagnetic induction. IEEE transactions on neural networks and learning systems **31**(2), 502–511 (2019)
12. Li, K., Bao, H., Li, H., et al.: Memristive Rulkov neuron model with magnetic induction effects. IEEE Trans. Industr. Inf. **18**(3), 1726–1736 (2021)
13. Devi, K.J., Singh, P., Bilal, M., et al.: Enabling secure image transmission in unmanned aerial vehicle using digital image watermarking with H-Grey optimization. Expert Syst. Appl. **236**, 121190 (2024)

14. Ye, G., Wu, H., Liu, M., et al.: Reversible image-hiding algorithm based on singular value sampling and compressive sensing. Chaos Solitons Fractals **171**, 113469 (2023)
15. Wang, X., Liu, C., Jiang, D.: An efficient double-image encryption and hiding algorithm using a newly designed chaotic system and parallel compressive sensing. Inf. Sci. **610**, 300–325 (2022)

BluePLP: Dynamic Vulnerability Patching for Heterogeneous BLE Devices

Xupu Hu[1], Zhongfeng Jin[2], Tongjie Wei[1], Peng Zhang[1], Chonghua Wang[3], and Ming Zhou[1(✉)]

[1] School of Cyber Science and Engineering, Nanjing University of Science and Technology, Nanjing 210094, China
{huxupu,weitongjie,zhang_peng,mingzhou}@njust.edu.cn
[2] National Computer Network Emergency Response Technical Team, Coordination Center of China, Beijing 100029, China
[3] China Industrial Control Systems Cyber Emergency Response Team, Beijing 100040, China
chonghuaw@live.com

Abstract. Bluetooth Low Energy (BLE) is a wireless communication protocol based on the Bluetooth standard, specifically developed to enable short-range data exchange between low-power, resource-constrained embedded devices. Attacks that exploit packet-based vulnerabilities in the BLE protocol stack are common. Rapidly patching these vulnerabilities faces two primary challenges. First, the standard firmware update process typically necessitates taking devices offline, resulting in service interruptions that may disrupt critical operations. Second, real-time operating system vendors must develop and maintain separate patches for each device architecture, significantly increasing both development time and engineering workload. This paper presents BluePLP, a lightweight live patching framework designed for BLE devices. BluePLP leverages the system's built-in exception handling mechanisms to redirect execution from vulnerable code to corresponding patch code, enabling vulnerability mitigation without requiring device downtime. In addition, it utilizes hardware breakpoints to support live patching across multiple processor architectures. We evaluated BluePLP on three BLE devices based on Cortex M3, Cortex M4, and Xtensa LX7 cores. The framework successfully remediated 25 vulnerabilities in BLE protocol stacks and was deployed in a commercial heart rate monitor. Experimental results show that BluePLP can apply patches in as little as two microseconds, providing rapid, architecture-independent runtime protection.

Keywords: Bluetooth Low Energy · Live Patching · Patch Generation

1 Introduction

Bluetooth Low Energy (BLE) is a fundamental wireless communication technology that underpins the Internet of Things (IoT). Its low power consumption and modest hardware requirements have made it the de facto standard for connectivity in resource-constrained embedded devices [18]. However, improper or incomplete implementations of the BLE protocol can introduce critical security vulnerabilities, which may be exploited through maliciously crafted packets. Several real-world incidents highlight the severity of such threats. For instance, Mantz et al. [9] showed a remote code execution attack on Broadcom BLE stacks by exploiting malformed link-layer frames. Similarly, the Bleeding-Bit vulnerabilities affecting Texas Instruments chips enabled attackers to bypass authentication and execute arbitrary code via a sequence of specially crafted advertising packets. These cases underscore the urgent need for effective mitigation strategies to address packet-based vulnerabilities in BLE-enabled IoT devices.

Although several defense mechanisms have been proposed for Linux-based BLE implementations, their applicability to deeply embedded platforms remains limited. For example, LBM [14] enforces stateless payload filtering within the BlueZ stack, while ProFactory [15] automatically generates secure low-level protocol code from formal specifications. However, both approaches require computational and storage resources that are not available on resource-constrained embedded devices [5]. In real-world deployments, vendors of real-time operating systems (RTOS) typically rely on over-the-air (OTA) firmware updates to address security flaws. Yet, OTA updates are often associated with considerable latency and require a full system reboot, leading to service interruptions. Such disruptions are especially problematic in mission-critical applications, such as medical monitoring and industrial control, where continuous device availability is essential. This underscores the need for lightweight, real-time mitigation strategies suitable for embedded BLE devices.

Dynamic live-patching techniques offer the potential for in-place vulnerability remediation without disrupting device operation. HERA [12] was among the first to implement this approach on Cortex-M3/M4 devices, utilizing the Flash Patch and Breakpoint (FPB) unit to redirect execution at runtime. However, this method is limited to architectures that provide FPB support and cannot be generalized to platforms lacking such hardware features. To overcome these architectural constraints, solutions such as BlueSWAT [5] and RapidPatch [8] embed an eBPF virtual machine (VM) within the device firmware, enabling dynamic patching across a broader range of hardware. However, the additional overhead introduced by the VM can significantly impact performance in resource-limited embedded systems. As a result, there remains a critical need for efficient, architecture-agnostic live-patching frameworks that impose minimal resource requirements and are suitable for deployment in deeply embedded BLE devices.

To overcome these limitations, we introduce BluePLP, a lightweight dynamic patching framework specifically designed for BLE devices. BluePLP intercepts program execution at the entry points of vulnerable routines by leveraging standard exception-handling mechanisms that are widely supported across modern processors. Patch code is stored in RAM and triggered using hardware breakpoints, which are available on platforms such as Cortex-M3, Cortex-M4, and Xtensa LX7 cores. This approach removes the need for embedding a full virtual machine or relying on architecture-specific debugging units. By adopting this design, BluePLP enables immediate and non-disruptive patch deployment, while imposing minimal overhead on CPU and memory resources. As a result, BluePLP provides an efficient and generalizable solution for real-time vulnerability mitigation in resource-constrained embedded BLE devices.

We evaluated BluePLP on three representative BLE platforms featuring Cortex-M3, Cortex-M4, and Xtensa LX7 architectures. Our experiments targeted 25 publicly disclosed packet-based vulnerabilities in open-source BLE protocol stacks and real-time operating systems, including issues related to encryption negotiation, attribute handling, and state-machine transitions. BluePLP was able to apply each patch within 2 μs of the corresponding breakpoint trigger, introducing less than 0.1% runtime overhead. Furthermore, when deployed on a commercial heart rate monitor, BluePLP reduced update latency from an average of 6.2 s required for traditional OTA updates to the microsecond level, achieving seamless patching without any service interruption.

Our main contributions are as follows.

- We design a unified patching framework that leverages hardware breakpoints to support heterogeneous BLE devices, including those based on Cortex-M3, Cortex-M4, and Xtensa LX7 architectures.
- We utilize embedded exception handlers to redirect execution from vulnerable code to RAM-resident patches, enabling live updates without requiring a system reboot.
- We show the effectiveness of BluePLP by mitigating 25 packet-based vulnerabilities across multiple real-time operating systems and BLE protocol stacks.

The remainder of this paper is organized as follows. Section 2 provides the background and outlines the threat model. Section 3 describes the system design of BluePLP in detail. Section 4 presents the evaluation results on performance and effectiveness. Section 5 discusses related work, and Sect. 6 concludes the paper.

2 Vulnerable BLE Devices

BLE devices are typically built on low-power, microcontroller-based system-on-chip (MCU) from vendors such as Arm (Cortex-M series), MIPS, and Tensilica (Xtensa). The Arm Cortex-M family alone has been integrated into tens of billions of embedded and consumer devices, owing to its balanced combination of performance, energy efficiency, and robust ecosystem support [2]. Despite this hardware diversity, device manufacturers commonly adopt a small set of real-time operating systems (RTOSes) and protocol stacks, with Zephyr (featuring its NimBLE component) and FreeRTOS among the most widely deployed. Notably, the Zephyr Project [22] now supports more than 750 distinct development boards across multiple architectures.

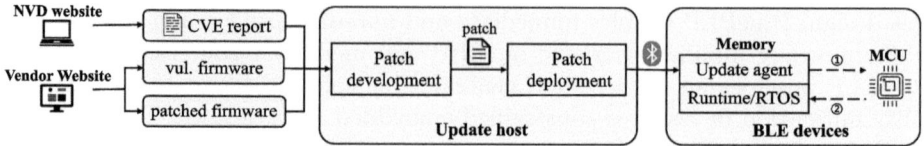

Fig. 1. Overview of BluePLP's working flow.

However, the widespread use of shared BLE protocol stacks and limited validation logic introduces significant security risks. Attackers can craft malicious LINK-LAYER or L2CAP packets to exploit missing boundary checks, state machine inconsistencies, or malformed attribute values, often leading to denial-of-service conditions or remote code execution. The practical impact of these threats has been demonstrated by Mantz et al. [9], who reported real-world instances of remote code execution in Bluetooth stacks, including the Bleeding-Bit vulnerabilities affecting Broadcom chips.

Typically, embedded MCUs are equipped with built-in exception handling mechanisms. Exception vectors can be triggered by hardware events or configured through breakpoint units, such as Arm's Flash Patch and Breakpoint (FPB) module. By registering custom exception handlers, firmware can reroute execution from the original routine to alternative code regions in RAM. This capability can be leveraged to verify patch integrity through a host-to-device debug channel and to dynamically redirect control flow at runtime, all without requiring a full system reboot. This mechanism forms the foundation for enabling immediate, non-disruptive security patching in BLE-enabled embedded devices.

3 BluePLP Design

3.1 Overview

Figure 1 depicts the overall workflow of BluePLP. The framework consists of three primary components: patch development, patch deployment, and the update agent. The patch development process comprises three stages: vulnerability detection, patch generation, and patch verification. To minimize resource usage and maintain device performance, the processes of patch generation and deployment are carried out on an external update host, while only the update agent operates on the BLE device.

The patch development process in BluePLP includes three stages: vulnerability localization, patch generation, and patch verification. This process accepts as input a CVE report, the vulnerable firmware image, and the patched firmware image, ultimately producing a binary patch. Once the patch is generated, the update host establishes a Bluetooth connection to the device and transmits the patch data. Upon receiving the patch, the update agent parses the patch metadata and configures a hardware breakpoint at the specified instruction address (①). When program execution reaches this address, the microcontroller dynamically redirects control flow to the patch code stored in RAM, allowing the vulner-

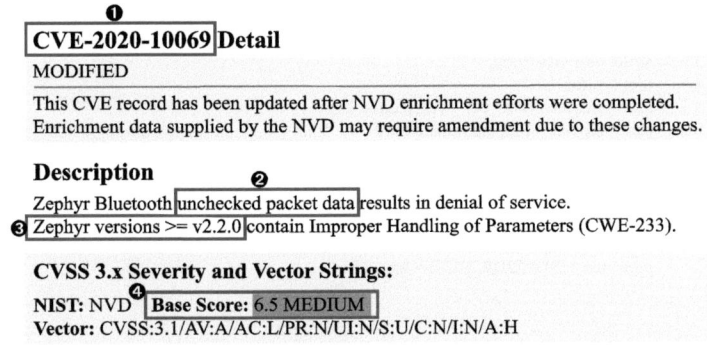

Fig. 2. Example of a CVE report for a packet-based BLE vulnerability from the NVD. The report provides key information, including the CVE ID (①), root cause of the vulnerability (②), affected RTOS or library version (③), and CVSS 3.0 score (④).

ability to be remediated without interrupting other running tasks or requiring a system reboot (②).

3.2 Vulnerability Detection

We begin by manually retrieving CVE reports through the National Vulnerability Database (NVD) API[1]. Relevant entries are filtered based on the real-time operating system and Bluetooth protocol stack. For each selected entry, we extract the CVE identifier, root cause, affected software version, and CVSS 3.0 score, and save the report as a JSON file in our project directory. Figure 2 presents an example report for a vulnerability in Zephyr OS v2.2.0, which has a CVSS score of 6.5. This vulnerability allows denial-of-service attacks by exploiting malformed packets resulting from missing boundary checks.

Then, we obtain both the vulnerable firmware image and its corresponding official patch from the vendor's archive, verifying their integrity using the published SHA-256 hashes. By performing a binary diff between the two images, we identify the precise memory address where the patch modifies the code and record the entry point of the affected function. As shown in Fig. 3, the function `ull_scan_rsp_set` is implicated in the vulnerability CVE-2021-3581. This function is configured through host-controller interface commands. It fails to validate the length parameter, which results in a potential memory overflow. We pinpoint the missing boundary check for the parameter `len` at line 5 and designate the corresponding instruction address as the vulnerability entry point.

Finally, we establish a debug connection between the update host and the device's microcontroller to facilitate communication and validation. A breakpoint verifier is deployed to collect and report contextual information back to the host. By writing the entry point address into the hardware comparator register, a breakpoint is programmed at the identified location. Each breakpoint hit is confirmed

---
[1] https://nvd.nist.gov/developers/vulnerabilities.

```
 1  uint8_t ull_scan_rsp_set (struct ll_adv_set *adv, uint8_t len, uint8_t const *const data)
 2  {   struct pdu_adv *prev;
 3      struct pdu_adv *pdu;
 4      uint8_t idx;
 5 bkpt prev = lll_adv_scan_rsp_peek(&adv->lll);
 6      pdu = lll_adv_scan_rsp_alloc(&adv->lll, &idx);
 7      pdu->type = PDU_ADV_TYPE_SCAN_RSP;
 8      pdu->rfu = 0;
 9      pdu->chan_sel = 0;
10      pdu->tx_addr = prev->tx_addr;
11      pdu->rx_addr = 0;
12      pdu->len = BDADDR_SIZE + len;
13      /*omit*/ }
```

Fig. 3. The vulnerable function for CVE-2021-3581. The vulnerability is caused by a missing length check on the parameter `len`. The "+" indicates that the official patch adds a boundary check for the function parameter `len`.

by monitoring the messages sent from the BLE device to the host during execution, ensuring accurate vulnerability localization and reliable patch deployment.

3.3 Patch Generation

Figure 4 illustrates the patch generation process in BluePLP. First, we conduct binary differencing at the assembly instruction level to identify changes between the vulnerable code and the official patch (Step ❷). Each instruction is classified into one of four categories: addition, deletion, unchanged, or changed. Additions are instructions present only in the official patch, deletions are present only in the vulnerable code, and unchanged instructions share identical opcodes and operands. Instructions differing in either opcode or operand are categorized as changed.

Next, we record the MCU state transitions between the two code versions (Step ❸). We map registers holding equivalent values in both versions to restore the correct execution context when transitioning from the vulnerable code to the patched code (Step ❹). For example, as shown in Fig. 4, register R7 in the vulnerable code and register R8 in the patched code serve the same function and contain identical values. This correspondence (R7 → R8) is annotated and used to recover the appropriate register state in the final patch.

In some cases, instruction sequences may differ syntactically yet be functionally equivalent. We identify and mark such segments as semantically equivalent, and exclude them from the final patch to reduce unnecessary changes. Only the additions, deletions, and true modifications are retained. For instance, in Fig. 4, semantic equivalence is illustrated by mapping (i) to (ii) + (iii), and in the adjusted patch, we reverse this mapping to restore the original behavior, mapping (ii) + (iii) back to (i).

Finally, to avoid potential stack corruption, we refrain from inserting a POP instruction in the adjusted patch. Instead, we use LDR PC, =loc_800333 to return control to the original instruction address. This method is more reliable, as

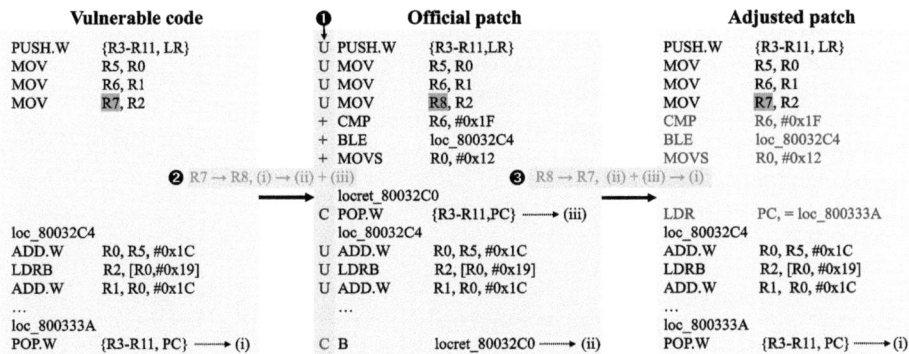

Fig. 4. Patch generation process for CVE-2021-3581. Step ❶: Compare the vulnerable code and the official patch, where U denotes unchanged instructions, C denotes changed instructions, and "+" indicates added instructions. Step ❷: Record MCU state changes between the two code versions. Step ❸: Restore the MCU state in the patched code by mapping corresponding registers and program behavior from the vulnerable version. Instructions highlighted in red indicate the final adjustments included in the patch.

the exception handler's call stack may not match that of the original vulnerable code, and direct stack manipulation could compromise program stability.

3.4 Patch Verification

After generating each patch, we conduct manual verification to ensure that it does not interfere with the original system's functionality. A patch is considered valid only if it preserves all data structures and does not modify any memory outside the stack frame of the vulnerable function. Altering data structures, such as changing macro definitions or modifying struct layouts, can shift the device's memory layout and introduce unpredictable behavior. Similarly, writing to memory regions outside the intended stack frame may cause system instability or failures. For example, in dual-stack configurations, corrupting the inactive stack may prevent the processor from properly restoring its context when switching stacks. We also verify that the patch logic does not alter the execution context of the update agent itself. Any changes to its code or data regions may trigger recursive exceptions, which can lead to device lockup [25].

To further validate safety and correctness under real-world conditions, each patch is deployed on a physical BLE device. We establish a debug link between the update host and the device under test, and pair it with a second BLE device to simulate typical interactions. On the test device, a continuously blinking LED serves as an indicator of normal peripheral activity. After applying the patch at runtime, we perform differential testing by comparing the behavior of the patched device with that of the unpatched version. If the LED stops blinking or the device loses connectivity during testing, the patch is deemed unsafe, as it disrupts standard device operations. This rigorous validation ensures that only safe and effective patches are applied in live environments.

Algorithm 1. Control Flow Redirection.

1: *breakpoint_address* ← get breakpoint address;
2: *exception_type* ← get exception type;
3: ExecuteExceptionHandler ();
4: SaveContext ();
5: SetMode (*exception_type*);
6: *patch_address* := BinarySearch (*breakpoint_address*);
7: ExecutePatch (*patch_address*);
8: ModifyLR (*return_address*);
9: RecoveryContext ();

3.5 In-Field Deployment

The update host transmits patch information over an encrypted Bluetooth channel to resource-constrained BLE devices that do not support native firmware updates. Each update packet contains four elements: the vulnerability entry address, the binary patch code, the target address for patch application, and the return address. A lightweight update agent running on the device as the lowest-priority task receives these packets without interfering with ongoing device operations.

Upon receiving a patch, the update agent establishes a mapping between the original entry address and the patch address. It then writes the patch code into a reserved region of memory. Subsequently, the agent configures the hardware breakpoint register with the specified vulnerability entry address. The microcontroller continuously monitors the program counter (PC); when its value matches the configured breakpoint address, the processor triggers an exception.

The patch operation is implemented by extending the system's built-in exception handler. On Cortex-M3 and Cortex-M4 cores, hardware breakpoints invoke the `DebugMon_Handler`. Algorithm 1 details the redirection process. When an exception occurs, the handler is selected from the exception vector table based on the exception type (line 3). The processor saves its current execution context, including all register values and operating mode, and switches to handler mode (lines 4–5). A binary search is then performed to locate the appropriate patch address in the mapping table (line 6). Control is redirected to the patch code, and upon completion, the handler writes the precomputed return address into the link register (lines 7–8). Finally, the processor restores the saved execution context and resumes normal operation by loading the link register value into the program counter (line 9).

4 Evaluation

In this section, we first introduce our experimental setup and then present a comprehensive evaluation of BluePLP. Specifically, we assess both the effectiveness and performance of the proposed framework. Effectiveness refers to BluePLP's ability to remediate vulnerabilities in BLE devices successfully. Performance is evaluated in terms of runtime latency and memory overheads introduced by BluePLP during operation.

Table 1. Hardware specifications of evaluation boards.

Device MCU	Architecture	Frequency	Flash	SRAM
STM32F103	Cortex M3	72 MHz	64 KB	20 KB
nRF52840	Cortex M4	64 MHz	1 MB	256 KB
ESP32S3	Xtensa LX7	240 MHz	384 KB	512 KB

4.1 Experimental Setup

All experiments were conducted using a Windows 10 computer as the update host and three development boards representing different microcontroller architectures: the STM32F103 (Cortex-M3), the nRF52840 (Cortex-M4), and the ESP32 (Xtensa LX7). Table 1 summarizes the clock frequencies and available hardware resources for each board.

For the heart rate monitoring scenario, we integrated the nRF52840 board with a photoplethysmogram (PPG) sensor. The sensor measures blood flow by detecting changes in light absorption from red and infrared LEDs. It produces one hundred digital samples per minute, which are streamed in real time from the microcontroller to the host computer over Bluetooth. The host then displays the waveform and computes the heart rate by counting signal peaks within a fixed time interval.

To evaluate runtime overhead, we utilized each board's on-chip timer, which operates at the same frequency as the processor. We measured the execution time of critical routines both with and without BluePLP enabled to quantify any added latency. Memory consumption was assessed by examining the firmware's symbol table before and after integrating the BluePLP patch, allowing us to determine the additional memory footprint introduced by the framework.

4.2 Effectiveness

Vulnerabilities Patching. To evaluate the effectiveness of BluePLP in defending BLE devices against packet-based attacks via runtime live patching, we conducted a comprehensive experimental study. We compiled a dataset of security vulnerabilities affecting widely used BLE real-time operating systems (RTOSes), such as Zephyr OS and Contiki-NG, as well as the NimBLE protocol stack. This dataset includes 27 real-world CVEs sourced from the National Vulnerability Database (NVD). Table 2 provides a summary of these CVEs, detailing the affected RTOS or library versions and their corresponding CVSS v3 scores. Among them, 13 CVEs are classified as High severity (CVSS scores 7.0–8.9), and two are rated as Critical (CVSS scores 9.0–10.0).

Table 2. The effectiveness of BluePLP in defending against common RTOS vulnerabilities on STM32F103, nRF52840, and ESP32S3 boards. Two CVEs cannot be fixed, which are marked as - in the last two columns.

CVE-ID	OS/Lib	Ver.	CVSS	Vulnerability description/Reason why cannot fix	Before patch	After patch
CVE-2019-9506	Zephyr OS	2.0.0	7.6	Macro definition modification are involved	-	-
CVE-2020-10061	Zephyr OS	2.2.0	8.8	Improper full-buffer handling	System crash	Safe
CVE-2020-10065	Zephyr OS	2.2.0	8.8	Missing Size Checks in Bluetooth HCI over SPI	System crash	Safe
CVE-2020-10066	Zephyr OS	2.2.0	5.7	Incorrect Error Handling in Bluetooth HCI core	System crash	Safe
CVE-2020-10068	Zephyr OS	2.2.0	6.5	Back-to-back duplicate packets cause errors	System crash	Safe
CVE-2020-10069	Zephyr OS	2.2.0	6.5	Unchecked packet data results in denial of service	System crash	Safe
CVE-2021-3433	Zephyr OS	2.5.0	4.0	Invalid channel map in CONNECT_IND	System crash	Safe
CVE-2021-3436	Zephyr OS	2.5.0	6.5	Need to add too many sanity checks	-	-
CVE-2021-3581	Zephyr OS	2.5.0	8.8	Buffer access with incorrect length value	System crash	Safe
CVE-2021-3966	Zephyr OS	3.0.0	9.6	Buffer overflow in net_buf_add_mem	System crash	Safe
CVE-2022-1041	Zephyr OS	3.0.0	8.8	Out-of-bound write during provisioning	System crash	Safe
CVE-2022-1042	Zephyr OS	3.0.0	8.8	Out-of-bound write during provisioning	System crash	Safe
CVE-2023-4264	Zephyr OS	3.4.0	9.6	Potential buffer overflow vulnerabilities	System crash	Safe
CVE-2023-4424	Zephyr OS	3.4.0	8.8	No boundary checks for BLE advertising packets	System crash	Safe
CVE-2023-5753	Zephyr OS	3.4.0	8.8	Disabling assertions may lead to a buffer overflow	System crash	Safe
CVE-2024-3332	Zephyr OS	3.6.0	6.5	Specially ordered packets can trigger DoS attack	System crash	Safe
CVE-2024-5931	Zephyr OS	3.6.0	6.5	Lack of user input validation	System crash	Safe
CVE-2024-6135	Zephyr OS	3.6.0	7.6	Lack of boundary check for buffer length	System crash	Safe
CVE-2024-6259	Zephyr OS	3.6.0	7.6	Improper discarding in adv_ext_report	System crash	Safe
CVE-2024-6444	Zephyr OS	3.6.0	6.5	No proper validation of the length of user input	System crash	Safe
CVE-2024-8798	Zephyr OS	3.7.0	7.5	No proper validation of the length of user input	System crash	Safe
CVE-2024-47248	NimBLE	1.7.0	6.3	Buffer copy without checking the size of input	System crash	Safe
CVE-2024-47249	NimBLE	1.7.0	5.0	Improper array index validation	System crash	Safe
CVE-2024-47250	NimBLE	1.7.0	5.0	Missing validation of HCI advertising report	System crash	Safe
CVE-2024-51569	NimBLE	1.7.0	7.5	Missing validation of HCI completed packets	System crash	Safe
CVE-2020-12140	Contiki-NG	4.4	8.8	Lack of boundary check for payload length	System crash	Safe
CVE-2022-41972	Contiki-NG	4.8	6.5	Missing validation for channel ID metadata	System crash	Safe

Our analysis confirmed that BluePLP can effectively mitigate over 90% (25/27) of these vulnerabilities. We manually reproduced each CVE listed in Table 2 on three BLE devices featuring different MCU architectures. For each vulnerability, we verified that the BluePLP live patch restored device functionality in a manner consistent with the official patch, without introducing any additional errors or instability. These results show that BluePLP provides robust mitigation against packet-based attacks and can be applied across heterogeneous BLE platforms.

There are two primary reasons why BluePLP cannot address the remaining two vulnerabilities. First, BluePLP does not support patches that require modifications to data structure definitions or global variables (e.g., CVE-2019-9506 in Zephyr OS). Such changes may significantly alter the memory layout of the device, causing subsequent data addresses to shift and increasing the risk of incorrect memory accesses. Second, BluePLP is not suitable for scenarios that require simultaneous patching of a large number of vulnerable functions (e.g.,

CVE-2021-3436 in Zephyr OS), due to the limited number of hardware breakpoints available on BLE devices, which is typically no more than eight [25].

Heart Rate Monitor Patching. To further assess the availability and practical effectiveness of BluePLP, we conducted a real-time patching experiment on a heart rate monitoring system—a scenario in which both security and high availability are essential. The heart rate sampling task in this system contained a representative memory corruption vulnerability (specifically, an out-of-bounds read) that could be exploited to compromise device integrity. In our experimental setup, we implemented both a heart rate sampling task and an LED blinking task, and deployed the BluePLP update agent on the device. The sampling task was configured to run at 10-millisecond intervals to emulate a real-time monitoring workload.

Using BluePLP, we generated a secure patch to remediate the memory corruption vulnerability. After deploying the patch and configuring the corresponding hardware breakpoint, the system automatically initiated the live patching process. Before patch application, the update host displayed erratic photoplethysmogram (PPG) waveform signals, which were the result of out-of-bounds memory accesses during sampling. Once the patch was applied, BluePLP inserted the necessary boundary checks, restoring normal operation of the heart rate monitoring system. The LED blinking task continued uninterrupted, indicating that device functionality and stability were maintained throughout the patching process.

During the entire experiment, BluePLP successfully remediated the vulnerability in the heart rate monitor without disrupting either the sampling or safety monitoring tasks. These results show that BluePLP can deliver effective security updates with minimal impact on system availability and runtime performance in critical real-world applications.

4.3 Performance

We evaluated the computation overhead and memory usage of BluePLP, and conducted a direct comparison with BlueSWAT to assess their relative efficiency and resource requirements.

The computational overhead of BluePLP refers to the additional runtime latency incurred during dynamic vulnerability patching. We measured this overhead using the high-precision on-chip timer available in BLE devices, which increments automatically with each microcontroller clock cycle. By recording the timer values immediately before and after the live patching operation, we calculated the total patching delay according to the formula

$$T = C/f,$$

where T is the total runtime delay, C is the number of MCU clock cycles elapsed, and f is the processor frequency.

To assess the runtime impact of BluePLP, we measured the latency introduced during the patching of 25 real-world CVEs on the nRF52840 development board. On average, BluePLP introduced only 1.64 μs of additional delay per patch. This minimal overhead is attributable to the direct use of hardware features available in BLE devices. However, this hardware-based design also imposes certain constraints, most notably the limited number of hardware breakpoints supported by embedded platforms.

The memory overhead introduced by BluePLP is also minimal. The exception handler implementation for all three supported architectures (Cortex-M3, Cortex-M4, and Xtensa LX7) requires an average of only 83 Bytes. For the 25 patches evaluated on the nRF52840 board, the average memory usage per patch is approximately 41 bytes. The update agent of BluePLP is highly compact, implemented in just a few dozen lines of C code.

For comparative analysis, we also implemented BlueSWAT's patching approach on the nRF52840 development board and conducted a side-by-side performance evaluation using the same set of 27 CVEs (Table 2) under identical RTOS configurations. In terms of computational overhead, we compared the total vulnerability patching latency between BluePLP and the Just-In-Time (JIT) mode of BlueSWAT, which is faster than BlueSWAT's interpreter mode. BluePLP achieved an average patching latency of 2.6 μs, while BlueSWAT exhibited an average overhead of 4.7 μs. Regarding memory consumption, BluePLP added only about 1KB of Flash and 0.6KB of SRAM, with each patch requiring an average of 41 bytes. By contrast, BlueSWAT required approximately 13KB of Flash and 0.8KB of SRAM, with each patch averaging 103 bytes.

Overall, BluePLP shows superior performance and lower resource consumption compared to BlueSWAT. Nonetheless, BluePLP also has some limitations. First, unlike BlueSWAT, it lacks defense capabilities against session-based attacks, such as replay attacks, in which earlier packets set preconditions for exploiting later vulnerabilities. Second, due to the limited number of hardware breakpoints, BluePLP cannot simultaneously patch a large number of vulnerabilities. We recommend that device manufacturers increase the number of hardware breakpoints in future BLE device designs. Third, similar to prior live patching frameworks for BLE devices, the patch generation process in BluePLP currently requires manual intervention. In future work, we plan to automate patch generation to enhance scalability and facilitate large-scale deployment.

5 Related Work

5.1 Bluetooth and BLE Security

Early research on Bluetooth security primarily focused on device identification and detection of malicious behavior by analyzing features at the physical and network layers. BlueID [1] introduced fingerprinting techniques based on temporal characteristics, such as packet interval and advertisement interval, to prevent attackers from forging timestamps for deceptive purposes. Subsequent approaches for detecting deception attacks [4,17] have leveraged both physical features (e.g., RSSI, carrier frequency offset) and network features (e.g., broadcast interval, broadcast status) to identify spoofed devices and dynamically detect anomalous behaviors.

Several methods for detecting man-in-the-middle attacks [21,23] have extracted physical and timing features—including RSSI, packet intervals, channel utilization, and communication distance—to model Bluetooth communication behavior and identify abnormal patterns. More recently, researchers have developed intrusion detection systems (IDS) specifically tailored for BLE environments. For example, the BrakTooth detection system [11] monitors LMP layer traffic and analyzes protocol features to identify vulnerabilities in master-controller interactions. I4Tech [13] applies machine learning by collecting data via ESP32 devices to train models that recognize common denial-of-service attacks in Bluetooth Mesh networks.

In addition, several solutions employ static or differential analysis techniques. FirmXRAy [16] has enabled Bluetooth fuzz testing, leading to the discovery of critical vulnerabilities, including remote code execution. Further efforts to enhance security in Linux and Android environments include LBM [14] and Profile Binding [19], which strengthen access control and pairing logic for peripheral devices. Collectively, these works have significantly advanced the state of Bluetooth and BLE security across multiple system layers.

5.2 Live Patching

Ksplice [3] introduced live patching frameworks capable of dynamically replacing Linux kernel modules, thereby enabling vulnerability remediation without interrupting system operation. To address the security risks posed by delayed updates in Android systems, solutions such as PatchDroid [10], KARMA [7], and InstaGuard [6] extend live patching techniques to user-space services and applications, enhancing patch responsiveness and enabling timely security fixes for mobile devices. However, these kernel-level and user-space live patching tools are not suitable for resource-constrained BLE devices, which lack the processing power and memory required to support such frameworks.

Within the domain of BLE device security, PatchScope [24] extracts program memory objects and their associated input fields to provide patch context information for security analysts. However, this approach exhibits a false positive rate of approximately 15%, which is unacceptably high for safety-critical control systems. VULMET [20] generates patches by calculating the semantic equivalence of official patches and determining the weakest preconditions for patch correctness. Nonetheless, VULMET is not applicable to BLE devices that use static linking, as these platforms are highly sensitive to address mapping and cannot tolerate address offsets during the patching process, thereby limiting their practical deployment.

6 Conclusion

In this paper, we present BluePLP, a lightweight and dynamic vulnerability patching framework tailored for BLE devices. BluePLP leverages the built-in exception handling mechanisms of embedded systems to redirect execution flow from vulnerable code to live patches at runtime. By utilizing hardware breakpoints available in BLE devices, BluePLP ensures compatibility and portability across different processor architectures. We implemented BluePLP on three distinct platforms, each based on a different microcontroller architecture, to validate its adaptability and effectiveness. In our experiments, BluePLP successfully mitigated 25 real-world vulnerabilities in BLE protocol stacks and was deployed in a commercial heart rate monitor, demonstrating its practical applicability. The framework introduces negligible runtime overhead and minimal memory consumption, making it especially suitable for deployment on resource-constrained embedded devices.

Acknowledgments. We would like to thank the anonymous reviewers for their insightful and detailed feedback. This work was supported in part by the National Natural Science Foundation of China (NSFC) under Grant No. 62402225.

Disclosure of Interests. The authors have no competing interests to declare that are relevant to the content of this article.

References

1. Albazrqaoe, W., Huang, J., Xing, G.: Practical bluetooth traffic sniffing: systems and privacy implications. In: Proceedings of the 14th Annual International Conference on Mobile Systems, Applications, and Services, MobiSys 2016, pp. 333–345. Association for Computing Machinery, New York (2016). https://doi.org/10.1145/2906388.2906403
2. ARM: ARM Cortex-M. Wikipedia. https://en.wikipedia.org/wiki/ARM_Cortex-M (2025)
3. Arnold, J., Kaashoek, M.F.: Ksplice: automatic rebootless kernel updates. In: Proceedings of the 4th ACM European Conference on Computer Systems, EuroSys 2009, pp. 187–198. Association for Computing Machinery, New York (2009). https://doi.org/10.1145/1519065.1519085
4. Cai, H.: Securing billion bluetooth devices leveraging learning-based techniques. In: Proceedings of the Thirty-Eighth AAAI Conference on Artificial Intelligence and Thirty-Sixth Conference on Innovative Applications of Artificial Intelligence and Fourteenth Symposium on Educational Advances in Artificial Intelligence. AAAI 2024/IAAI 2024/EAAI 2024. AAAI Press (2024). https://doi.org/10.1609/aaai.v38i21.30544
5. Che, X., He, Y., Feng, X., Sun, K., Xu, K., Li, Q.: Blueswat: a lightweight state-aware security framework for bluetooth low energy. In: Proceedings of the 2024 on ACM SIGSAC Conference on Computer and Communications Security, CCS 2024, pp. 2087–2101. Association for Computing Machinery, New York (2024). https://doi.org/10.1145/3658644.3670397
6. Chen, Y., et al.: InstaGuard: instantly deployable hot-patches for vulnerable system programs on android. In: 25th Annual Network and Distributed System Security Symposium, NDSS 2018, San Diego, California, USA, 18–21 February 2018. The Internet Society (2018). https://www.ndss-symposium.org/wp-content/uploads/2018/03/ndss2018_08-2_Chen_paper.pdf

7. Chen, Y., Zhang, Y., Wang, Z., Xia, L., Bao, C., Wei, T.: Adaptive android kernel live patching. In: Proceedings of the 26th USENIX Conference on Security Symposium, SEC 2017, pp. 1253–1270. USENIX Association, USA (2017)
8. He, Y., et al.: RapidPatch: firmware hotpatching for real-time embedded devices. In: 31st USENIX Security Symposium (USENIX Security 22), pp. 2225–2242. USENIX Association, Boston (2022). https://www.usenix.org/conference/usenixsecurity22/presentation/he-yi
9. Mantz, D., Classen, J., Schulz, M., Hollick, M.: Internalblue - bluetooth binary patching and experimentation framework. In: Proceedings of the 17th Annual International Conference on Mobile Systems, Applications, and Services, MobiSys 2019, pp. 79–90. Association for Computing Machinery, New York (2019). https://doi.org/10.1145/3307334.3326089
10. Mulliner, C., Oberheide, J., Robertson, W., Kirda, E.: PatchDroid: scalable third-party security patches for android devices. In: Proceedings of the 29th Annual Computer Security Applications Conference, ACSAC 2013, pp. 259–268. Association for Computing Machinery, New York (2013). https://doi.org/10.1145/2523649.2523679
11. Nandikotkur, A., Traore, I., Mamun, M.: Detecting braktooth attacks. In: SECRYPT, pp. 787–792 (2023)
12. Niesler, C., Surminski, S., Davi, L.: Hera: hotpatching of embedded real-time applications. In: NDSS. The Internet Society, Virtual (2021)
13. Sivanandam, N., Ananthan, T.: Intrusion detection system for bluetooth mesh networks using machine learning. In: 2022 International Conference on Industry 4.0 Technology (I4Tech), pp. 1–6 (2022). https://doi.org/10.1109/I4Tech55392.2022.9952758
14. Tian, D.J., Hernandez, G., Choi, J.I., Frost, V., Johnson, P.C., Butler, K.R.B.: LBM: a security framework for peripherals within the Linux kernel. In: 2019 IEEE Symposium on Security and Privacy (SP), pp. 967–984 (2019). https://doi.org/10.1109/SP.2019.00041
15. Wang, F., et al.: ProFactory: improving IoT security via formalized protocol customization. In: 31st USENIX Security Symposium (USENIX Security 22), pp. 3879–3896. USENIX Association, Boston (2022). https://www.usenix.org/conference/usenixsecurity22/presentation/wang-fei
16. Wen, H., Lin, Z., Zhang, Y.: FirmXRay: detecting bluetooth link layer vulnerabilities from bare-metal firmware. In: Proceedings of the 2020 ACM SIGSAC Conference on Computer and Communications Security, CCS 2020, pp. 167–180. Association for Computing Machinery, New York (2020). https://doi.org/10.1145/3372297.3423344
17. Wu, J., Nan, Y., Kumar, V., Payer, M., Xu, D.: BlueShield: detecting spoofing attacks in bluetooth low energy networks. In: 23rd International Symposium on Research in Attacks, Intrusions and Defenses (RAID 2020), pp. 397–411. USENIX Association, San Sebastian (2020). https://www.usenix.org/conference/raid2020/presentation/wu
18. Wu, J., Wu, R., Xu, D., Tian, D.J., Bianchi, A.: SoK: the long journey of exploiting and defending the legacy of king harald bluetooth. In: 2024 IEEE Symposium on Security and Privacy (SP), pp. 2847–228066 (2024). https://doi.org/10.1109/SP54263.2024.00023
19. Xu, F., Diao, W., Li, Z., Chen, J., Zhang, K.: Badbluetooth: breaking android security mechanisms via malicious bluetooth peripherals. In: 26th Annual Network and Distributed System Security Symposium, NDSS

2019, San Diego, California, USA, 24–27 February 2019. The Internet Society (2019). https://www.ndss-symposium.org/ndss-paper/badbluetooth-breaking-android-security-mechanisms-via-malicious-bluetooth-peripherals/
20. Xu, Z., et al.: Automatic hot patch generation for android kernels. In: 29th USENIX Security Symposium (USENIX Security 2020), pp. 2397–2414. USENIX Association (2020). https://www.usenix.org/conference/usenixsecurity20/presentation/xu
21. Yurdagul, M.A., Sencar, H.T.: BLEKeeper: response time behavior based man-in-the-middle attack detection. In: 2021 IEEE Security and Privacy Workshops (SPW), pp. 214–220 (2021). https://doi.org/10.1109/SPW53761.2021.00035
22. Zephyr: Zephyr Project (2025). https://zephyrproject.org/
23. Zhang, B.: Real-time bluetooth man-in-the-middle attack detection system based on graph neural network. Authorea Preprints (2025)
24. Zhao, L., Zhu, Y., Ming, J., Zhang, Y., Zhang, H., Yin, H.: PatchScope: memory object centric patch diffing. In: Proceedings of the 2020 ACM SIGSAC Conference on Computer and Communications Security, CCS 2020, pp. 149–165. Association for Computing Machinery, New York (2020). https://doi.org/10.1145/3372297.3423342
25. Zhou, M., Wang, H., Li, K., Zhu, H., Sun, L.: Save the bruised striver: a reliable live patching framework for protecting real-world PLCs. In: Proceedings of the Nineteenth European Conference on Computer Systems, EuroSys 2024, pp. 1192–1207. Association for Computing Machinery, New York (2024). https://doi.org/10.1145/3627703.3650068

Author Index

A
Ameen, Muhammad 267

B
Bu, Yanling 1, 17

C
Chan, Qizhong 31
Chang, Zhaoxin 160
Chen, Guo 458
Chen, Han 114
Chen, Xing 282
Chen, Yin 190
Cheng, Jianhong 348, 364

D
Dai, Donghui 48
Ding, Dawei 532, 545
Duan, Wei 238

F
Fang, Siyuan 282
Fu, Congle 411
Fu, Rao 442

G
Gao, Jiahe 224
Gao, Xiang 82
Gong, Yu 209
Gu, Chen 458
Guan, Lishen 282
Guo, Bin 174
Guo, Yi 333
Guo, Yifan 174

H
Han, Saibing 1
Hao, Zhanjun 147, 477
He, Bing 411
He, Li 238

He, Xinyuan 1
Hou, Qing 348, 364
Hou, Yibo 224
Hu, Ruihan 238
Hu, Wenjing 99
Hu, Xupu 557
Huang, Fangwan 424, 458
Huang, Honglan 411
Huang, Wenhao 190
Huang, Yongbin 424

I
Ito, Takashi 190

J
Jin, Zhongfeng 557
Jouaber, Badii 160

K
Khyzer, Hira 298

L
Le, Yi 82
Li, Chongrong 224
Li, Chun 298
Li, Dongbo 224
Li, Fenfang 477
Li, Juan 65
Li, Runze 505
Li, Shijie 493
Li, Shuangping 318
Li, Xinhai 31, 333
Li, Yiwei 477
Liu, Kaixiang 493
Liu, Keli 348
Liu, Lixin 17
Liu, Lu 65
Liu, Wangcan 532
Liu, Wenping 48
Liu, Xuanyun 458

Long, Tingting 397
Lu, Xiaozhen 17
Lu, Yunan 333
Lu, Yuru 147
Lv, Shichao 493

M
Ma, Jinfeng 114
Ma, Xujun 160
Mansoor, Atif 129
Meng, Chenxu 31, 48
Miao, Yumeng 129

N
Nakazawa, Jin 190

P
Pan, Yuzhu 267
Pan, Zhiwen 493

Q
Qian, Chaoma 532, 545
Qin, Liangwei 224

R
Ruan, Xiaoli 348, 364

S
Shen, Yiran 442
Song, Zishuo 267
Sun, Hao 82
Sun, Limin 493
Sun, Nan 411
Sun, Zhuo 174

T
Tang, Xianghong 348, 364
Tang, Yin 380
Tian, Meng 99
Tian, Peng 99

W
Wang, Chonghua 557
Wang, Guowei 147, 477
Wang, Haitian 129
Wang, Jiangtao 318
Wang, Ling 209
Wang, Pei 160
Wang, Peng 520

Wang, Pengfei 267
Wang, Ruifeng 318
Wang, Xiangyu 147
Wang, Xinjie 493
Wang, Xinyu 129
Wang, Yiren 129
Wang, Yuanyuan 532, 545
Wang, Yuejiao 147, 477
Wang, Yunfeng 99
Wang, Zhenwei 267
Wang, Zhicheng 31
Wang, Zhu 174
Wei, Tongjie 557
Wu, Qihui 17
Wu, Yangbo 255
Wu, Yuezhong 424, 458

X
Xiang, Sha 505
Xie, Huojin 255
Xu, Ke 318
Xu, Zhelin 411
Xue, Xinyu 160

Y
Yan, Haoran 48
Yan, Yu 298
Yang, Banglie 505
Yang, Huan 380
Yang, Jing 348, 364
Yang, Lei 31, 48, 333
Yang, Zongli 532, 545
Ye, Wei 380
Yin, Bo 224
Yin, Hao 520
Yu, Kang 99
Yu, Zhiwen 174
Yu, Zhiyong 424, 458
Yuan, Yao 48

Z
Zang, Tianzi 65
Zeng, Lingcheng 333
Zeng, Qingzhu 333
Zhang, Bolei 114
Zhang, Changkai 114
Zhang, Daqing 160
Zhang, Deyu 380, 397

Zhang, Fusang 160
Zhang, Haoyu 174
Zhang, Jiang 477
Zhang, Jinrui 380
Zhang, Jun 99
Zhang, Miao 282
Zhang, Peng 364, 557
Zhang, Xianyang 411
Zhang, Yu 129, 298
Zhang, Yuliang 129

Zhao, Guangrong 442
Zhao, Jingwen 532, 545
Zhao, Mingshu 267
Zhao, Shihan 31
Zhao, Tianqi 238
Zheng, Liyang 282
Zhou, Ming 557
Zhou, Xingshe 298
Zhu, Shuo 505
Zhu, Yuan 545

MIX
Papier aus verantwortungsvollen Quellen
Paper from responsible sources
FSC® C105338

If you have any concerns about our products,
you can contact us on
ProductSafety@springernature.com

In case Publisher is established outside the EU,
the EU authorized representative is:
**Springer Nature Customer Service Center GmbH
Europaplatz 3, 69115 Heidelberg, Germany**

Printed by Libri Plureos GmbH
in Hamburg, Germany